Wetlands of Tropical and Subtropical Asia and Africa

Wetlands: Biodiversity, Livelihoods and Conservation

Series Editor : Thammineni Pullaiah
Sri Krishnadevaraya University
Anantapur, India

1) **Wetlands of Mountainous Regions: Biodiversity, Livelihoods and Conservation**
   By Thammineni Pullaiah
   ISBN: 9781394235209
2) **Wetlands of Tropical and Subtropical Asia and Africa: Biodiversity, Livelihoods and Conservation**
   By Thammineni Pullaiah
   ISBN: 9781394235247
3) **Wetlands of Tropical and Subtropical South and Central America: Biodiversity, Livelihoods and Conservation**
   By Thammineni Pullaiah
   ISBN: 9781394307135

Wetlands of Tropical and Subtropical Asia and Africa

Biodiversity, Livelihoods and Conservation

Edited by

Thammineni Pullaiah
Department of Botany
Sri Krishnadevaraya University
Anantapur
Andhra Pradesh
India

WILEY

This edition first published 2025

Registered Offices
John Wiley & Sons, Inc., 111 River Street, Hoboken, NJ 07030, USA
John Wiley & Sons Ltd, New Era House, 8 Oldlands Way, Bognor Regis, West Sussex, PO22 9NQ, UK

For details of our global editorial offices, customer services, and more information about Wiley products visit us at www.wiley.com.

The manufacturer's authorized representative according to the EU General Product Safety Regulation is Wiley-VCH GmbH, Boschstr. 12, 69469 Weinheim, Germany, e-mail: Product_Safety@wiley.com.

Wiley also publishes its books in a variety of electronic formats and by print-on-demand. Some content that appears in standard print versions of this book may not be available in other formats.

Library of Congress Cataloging-in-Publication Data Applied for:

Hardback ISBN: 9781394235247

Cover Design: Wiley
Cover Image: Courtesy of Krishna Prosad Mondal

Set in 9.5/12.5pt STIXTwoText by Straive, Pondicherry, India
Printed and bound by CPI Group (UK) Ltd, Croydon, CR0 4YY

C9781394235247_150325

Contents

About the Editor

Dr. T. Pullaiah is a former professor at the Department of Botany at Sri Krishnadevaraya University in Andhra Pradesh, India, where he has taught for more than 35 years. He has held several positions at the university, including Dean, Faculty of Biosciences, Head of the Department of Botany, Head of the Department of Biotechnology, and Member of Academic Senate. He was the president of Indian Botanical Society (2014), president of the Indian Association for Angiosperm Taxonomy (2013), and fellow of Andhra Pradesh Akademi of Sciences. Under his guidance 54 students obtained their doctoral degrees. He has authored 70 books, edited 40 books, and published over 340 research papers. His books include *Redsanders: Silviculture and Conservation* (Springer), *Genetically Modified Crops* (Springer), *Sandalwood: Silviculture, Conservation and Applications* (Springer), *Advances in Cell and Molecular Diagnostics* (Elsevier), *Camptothecin and Camptothecin Producing Plants* (Elsevier), *Paclitaxel* (Elsevier), *Monograph on Brachystelma and Ceropegia in India* (CRC Press), *Ethnobotany of India* (5 volumes, Apple Academic Press), *Global Biodiversity* (4 volumes, Apple Academic Press) and *Invasive Alien Species* (4 volumes, Wiley Blackwell). He was also a member of Species Survival Commission of the International Union for Conservation of Nature (IUCN). Professor Pullaiah received his PhD from Andhra University, India, attended Moscow State University, Russia, and worked as post-doctoral fellow during 1976–1978.

List of Contributors

Ouéda Adama
Laboratoire de Biologie et Ecologie animales (LBEA), UFR/SVT
Université Joseph KI-ZERBO (Burkina Faso)
Ouagadougou, Burkina Faso

Adeyemi Ojutalayo Adeeyo
Institute of Nanotechnology and Water Sustainability
Florida, South Africa

Adebola Rashidat Adewale
Department of Fisheries
Faculty of Science, Lagos State University
Lagos, Nigeria

Gordon N. Ajonina
Cameroon Wildlife Conservation Society
CWCS Coastal Forests
Mangrove & Marine Programme
Yaoundé, Cameroon

Institute of Fisheries and Aquatic Sciences
University of Douala (Yabassi)
Douala, Cameroon

Shehu Akintola
Department of Fisheries
Faculty of Science Lagos State University
Lagos, Nigeria

Mercy Adewumi Alabi
Department of Microbiology
School of Life Sciences
University of KwaZulu-Natal
Durban, South Africa

Al-Jubaer
Institute of Remote Sensing and GIS
Jahangirnagar University
Dhaka, Bangladesh

Thiri Dae We Aung
Biodiversity and Nature Conservation Association
Yangon, Myanmar

Taïbou Ba
Department of Education, Science, Technology & Innovation African Union Commission
Addis Ababa, Ethiopia

Nitin Bassi
Council on Energy, Environment and Water (CEEW)
New Delhi, India

Andleeb Batool
Department of Zoology
Government College University
Lahore, Pakistan

Nard Bennas
Faculty of Sciences
Abdelmalek Essaâdi University
Tetouan, Morocco

Gnansounou S. Constant
Laboratoire de Biomathématiques et d'Estimations Forestières
Faculté des Sciences Agronomiques
Université d'Abomey-Calavi
Cotonou, République du Bénin

Christopher J. Curtis
Department of Geography, Environmental Management and Energy Studies
University of Johannesburg
Johannesburg, South Africa

Akodékou A. David
Laboratoire de Biomathématiques et d'Estimations Forestières
Faculté des Sciences Agronomiques
Université d'Abomey-Calavi Cotonou
République du Bénin

Abdoul Aziz Diouf
Ecological Monitoring Center
Research & Development Program
Dakar, Senegal

Thomas E. Efole
Institute of Fisheries and Aquatic Sciences
University of Douala (Yabassi)
Douala, Cameroon

Mohamed El Haissoufi
Polydisciplinary Faculty of Taza
Sidi Mohamed Ben Abdellah University
Fez, Morocco

Isa Olalekan Elegbede
Department of Fisheries
Faculty of Science Lagos State University
Lagos, Nigeria

Department of Environmental Science
National Open University of Nigeria
Abuja, Nigeria

Department of Environmental Planning
Brandenburg University of Technology
Cottbus, Germany

Abdeslam Ennabili
Superior School of Technology
Sidi Mohamed Ben Abdellah University
Fez, Morocco

Mathieu Gueye
Department of Botany and Geology Laboratory of Botany, IRL 3189 Environnement Santé et Société, IFAN Ch. A. Diop
Dakar, Senegal

Saiba Gupta
Council on Energy, Environment and Water (CEEW)
New Delhi, India

Ibrahim Mohammed Hashim
Sudanese Wildlife Society
Khartoum, Sudan

Dassou G. Hospice
Département de Biologie Végétale
Laboratoire des Sciences du Végétal et Pharmacopée, Ecole Doctorale des Sciences de la Vie et de la Terre
Université d'Abomey-Calavi, Cotonou
République du Bénin

Amira M. Hotaiba
Department of Environmental Sciences
Faculty of Science
Alexandria University, Alexandria, Egypt

Mohamed Elmekki Ali Elbadawi Hussien
Department of Wildlife
University of Sinnar, Suki, Sudan

Kaboré Idrissa
Laboratoire de Biologie et Ecologie animales (LBEA), UFR/SVT
Université Joseph KI-ZERBO (Burkina Faso)
Ouagadougou, Burkina Faso

Emily Osa Iduseri
Department of Environmental Science
National Open University of Nigeria
Abuja, Nigeria

Sheikh Tawhidul Islam
Institute of Remote Sensing and GIS
Jahangirnagar University
Dhaka, Bangladesh

Ayushi Kashyap
Council on Energy
Environment and Water (CEEW)
New Delhi, India

Abdelmajid Khabbach
Faculty of Sciences, Sidi Mohamed Ben Abdellah University
Fez, Morocco

Jasper Knight
School of Geography, Archaeology and Environmental Studies
University of the Witwatersrand
Johannesburg, South Africa

Bramley Jemain Lemine
Department of Chemical Engineering
Cape Peninsula University of Technology
Bellville, South Africa

Department of Water Science
University of Western Cape
Bellville, South Africa

Mohamed Libiad
Faculty of Sciences, Abdelmalek Essaâdi University
Tetouan, Morocco

Mauro Lourenco
National Geographic Okavango Wilderness Project, Wild Bird Trust
Johannesburg, South Africa

A.H. Magdon Jayasuriya
EML Consultants PLC
Pitakotte, Sri Lanka

Rodgers Makwinja
Department of Geography, Environmental Management and Energy Studies
University of Johannesburg
Johannesburg, South Africa

Department of Fisheries, Ministry of Natural Resources & Climate Change
Lilongwe, Malawi

Thomas Marambanyika
Department of Geography Environmental Sustainability and Resilience Building
Midlands State University
Gweru, Zimbabwe

Sawadogo Yabyouré Marc-Florent
Laboratoire de Biologie et Ecologie animales (LBEA)
UFR/SVT, Université Joseph KI-ZERBO (Burkina Faso)
Ouagadougou, Burkina Faso

Paterne Arnaud Bernard Mingou
Department of Botany and Geology, Laboratory of Botany
IFAN Ch. A. Diop Dakar, Senegal

Krishna Prosad Mondal
Institute of Remote Sensing and GIS
Jahangirnagar University
Dhaka, Bangladesh

Titus A.M. Msagati
Institute of Nanotechnology and Water Sustainability
Florida, South Africa

Maryam Mukhtar
Department of Zoology
University of the Punjab
Lahore, Pakistan

Oshneck Mupepi
Department of Geography Environmental Sustainability and Resilience Building
Midlands State University
Gweru, Zimbabwe

Tatenda Musasa
Institute of Water Studies, Faculty of Natural Sciences
University of the Western Cape
Cape Town, South Africa

Department of Geography Environmental Sustainability and Resilience Building
Midlands State University
Gweru, Zimbabwe

Naseeba Mustafavi
Department of Zoology
Government College University
Lahore, Pakistan

Thet Zaw Naing
Myanmar Bird and Nature Society
Yangon, Myanmar

Mishal Nawaz
Department of Zoology
Government College University
Lahore, Pakistan

Barthelemy Ndongo
Department of Rural Engineering
Faculty of Agronomy & Agricultural Sciences,
National Focal Point of Ramsar Convention
University of Dschang
Dschang, Cameroon

Tinyiko Rivers Nkuna
Department of Earth Science
Faculty of Science Engineering and Agriculture
University of Venda
Limpopo, South Africa

Yaa Ntiamoa-Baidu
Centre for Biodiversity Conservation Research
University of Ghana
Accra, Ghana

Department of Animal Biology and
Conservation Science
University of Ghana
Accra, Ghana

Abdul Rahamon Olodo
Department of Environmental Planning
Brandenburg University of Technology
Cottbus, Germany

Kehinde Moyosola Ositimehin
Department of Environmental Studies
Ohio University, Athens
OH, USA

Asia Parveen
Department of Zoology
Government College University
Lahore, Pakistan

Kelvin S.-H. Peh
School of Biological Sciences
University of Southampton
Southampton, UK

Syed Hafizur Rahman
Department of Environmental Sciences
Jahangirnagar University
Dhaka, Bangladesh

Dania Razzaq
Department of Zoology
Government College University
Lahore, Pakistan

Konaté Sidiki Roland
Laboratoire de Biologie et Ecologie animales
(LBEA), UFR/SVT
Université Joseph KI-ZERBO (Burkina Faso)
Ouagadougou, Burkina Faso

Glèlè Kakaï Romain
Laboratoire de Biomathématiques et
d'Estimations Forestières
Faculté des Sciences Agronomiques
Université d'Abomey-Calavi Cotonou
République du Bénin

Abdulwakil Olawale Saba
Department of Fisheries
Faculty of Science Lagos State University
Lagos, Nigeria

Soumic Samad
Research Consultant, Remote Sensing
Division, Center for Environmental and
Geographic Information Services (CEGIS)
Dhaka, Bangladesh

Monishankar Sarkar
Institute of Disaster Management and
Vulnerability Studies
University of Dhaka
Dhaka, Bangladesh

Zanvo M. G. Serge
Laboratoire de Biomathématiques et d'Estimations Forestières
Faculté des Sciences Agronomiques
Université d'Abomey-Calavi Cotonou
République du Bénin

Byomkesh Talukder
Department of Global Health
Florida International University
Miami, FL, USA

Emmanuel Nii Attram Taye
Centre for Biodiversity Conservation Research
University of Ghana, Accra, Ghana

Department of Animal Biology and Conservation Science
University of Ghana, Accra, Ghana

Solomon G. Tesfamichael
Department of Geography, Environmental Management and Energy Studies
University of Johannesburg, Johannesburg
South Africa

Minette Eyango Tomedi-Tabi
Institute of Fisheries and Aquatic Sciences
University of Douala (Yabassi)
Douala, Cameroon

Avi Uzan
Israel Nature and Parks Authority
Jerusalem, Israel

Salako K. Valère
Laboratoire de Biomathématiques et d'Estimations Forestières
Faculté des Sciences Agronomiques,
Université d'Abomey-Calavi
Cotonou, République du Bénin

Bancé Victor
Laboratoire de Biologie et Ecologie animales (LBEA)
UFR/SVT, Université Joseph KI-ZERBO (Burkina Faso)
Ouagadougou Burkina Faso
Centre Universitaire de Manga
Université Norbert ZONGO
Koudougou, Burkina Faso

Ding Li Yong
BirdLife International (Asia)
Tanglin International Centre
Singapore, Singapore

Preface

Wetlands are globally recognized as important habitats for wildlife and human productivity. Wetlands provide numerous ecosystem services including carbon sequestration, water filtration, nutrient retention, and flood mitigation. In addition, wetlands are important migratory stops for birds and mammals and breeding habitats for amphibians, birds, and some reptiles. Wetlands are threatened due to anthropogenic activities such as wetland draining or filling, hydrological alterations, chronic degradation due to non-point source pollution, and invasion of exotic species. There are increasing concerns regarding the effects of climate change, such as sea-level rise, salination, and desertification. If these wetland resources are not conserved with heart and soul with the help of Governments and International Agencies, they may be lost forever. Wetlands have experienced an estimated 50% loss over the last century; their conservation importance is particularly recognized in several multi-lateral environmental agreements, such as the Water Framework Directive and the Ramsar Convention. Information on wetlands from around the world is available but scattered in grey literature, local languages, and on various aspects on a piecemeal basis, namely plant and animal biodiversity, ecotourism, livelihoods, and peoples' dependence. This book attempts to formulate the information together in a concise and systematic manner which will benefit the readers. This will also benefit academics, researchers, and a wide range of interested people. This book gives an account of the wetland ecosystem and its ecology, the resources and potentials of wetlands, conservation efforts, wetland ecosystem services, and threats to conservation.

I wish to express my grateful thanks to all the contributors for their cooperation and erudition. Since it is a voluminous subject, we might have missed some of the aspects. Readers are requested to give their suggestions for improvement in the coming edition. We could not cover all the countries either due to the shortage of expertise or due to the busy schedule of the experts. I thank Dr. Frank Weinreich of Wiley Publishers for his thought-provoking inputs during the book formulation.

Thammineni Pullaiah

1

Recommendations for Sustainable Management of Wetlands in Indian Tropics

Nitin Bassi, Ayushi Kashyap, and Saiba Gupta

Council on Energy, Environment and Water (CEEW), New Delhi, India

1.1 Introduction

India has a highly variable climate, with substantial differences in both spatial and temporal extent of rainfall and temperature. The north and northeastern regions receive high rainfall and experience low temperatures in comparison to the western, central, and peninsular India. As a result, India is unique in terms of water availability, with perennial rivers in the north and north-eastern regions and seasonal rivers in the rest of India. Also, there is high year-on-year variability in rainfall and temperature in different parts of the country.

The natural water systems in India include rivers, lakes, estuaries, lagoons, mangrove swamps, and many other water bodies. Most of them qualify as wetlands as per the Ramsar Convention on Wetlands, an international treaty signed in 1971 for national action and international cooperation for the conservation and wise use of wetlands and their resources. As per Article 1.1 of the Ramsar Convention, wetlands are "areas of marsh, fen, peatland or water, whether natural or artificial, permanent or temporary, with water that is static or flowing, fresh, brackish or salt, including areas of marine water the depth of which at low tide does not exceed six metres." As of 22 March 2024, there are 172 contracting parties to the Ramsar Convention, and 2513 wetlands with an area of about 257 million hectares (m ha) have been designated as Ramsar sites. Out of these, 1133 sites are in Europe: 431 sites in Asia; 425 sites in Africa; 216 sites in South America; 222 sites in North America; and 86 sites in Oceania (Ramsar Secretariat 2024). However, the area designated as Ramsar sites is only a small proportion in comparison to estimates of global wetland area, which is about 1.5–1.6 billion hectares (Davidson and Finlayson 2019).

As of 2020, India has about 5% of its area under wetlands (Space Application Centre 2011) and 80% Ramsar sites (Ramsar Secretariat 2024). The total wetland area under Ramsar sites in India is about 1.3 m ha (Ramsar Secretariat 2024). Initial attempts to prepare the wetland inventory of India were made between the 1980s and early 1990s. As per the Country Report of Directory of Asian Wetlands (Woistencroft et al. 1989) and the Directory of Indian Wetlands 1993 (WWF India and Asian Wetland Bureau 1993), the areal spread of wetlands in India was around 58.3 m ha. Paddy fields accounted for nearly 71% of this wetland area. Subsequently, the Ministry of Environment and Forests (1990) estimated the wetlands area to be about 4.1 m ha, but excluded the mangrove swamps.

Wetlands of Tropical and Subtropical Asia and Africa: Biodiversity, Livelihoods and Conservation,
First Edition. Edited by Thammineni Pullaiah.

The first scientific mapping of wetlands of the country was carried out using satellite data of 1992–1993 by the Space Applications Centre (SAC), Ahmedabad, which classified wetlands based on the Ramsar Convention definition. As per this inventory, the area of wetlands was about 7.6 m ha (Garg et al. 1998). These estimates did not include paddy fields, rivers, canals, and irrigation channels. Subsequently, the National Wetland Atlas 2011 was prepared by SAC and the entire country was considered for assessment. A total of 201,503 wetlands were identified and mapped on a 1 : 50,000 scale (SAC 2011). In addition, 555,557 wetlands (area <2.25 ha, which is smaller than the minimum measurable unit) were identified as point features. As per these estimates, a total wetland area of 15.3 m ha was estimated. The latest inventory was prepared for 2017–2018 and released in 2021 (SAC 2011). As per these estimates, the total wetland area in India is about 15.98 m ha.

Wetlands offer several benefits and services, which are categorized into provisioning (for instance, freshwater and food), regulating (for instance, climate resilience), supporting (such as nutrient cycling), and cultural (like recreation and tourism) functions. However, these unique ecosystems are under threat due to anthropogenic pressures and climate change. The main anthropogenic reasons include diversion for agricultural and nonagricultural purposes and pollution.

To improve their conservation, many policies (schemes and programs) were adopted in India. The most recent scheme launched in the budget of 2023–2024 called the *Amrit Dharohar* targets to maintain a healthy and well-managed network of identified Ramsar sites by promoting their conservation values over the course of three years (Ministry of Environment, Forest and Climate Change 2023). The activities within the scheme focus on enhancing multiple ecosystem services such as carbon sequestration, pollution abatement, flood control, biodiversity support, and cultural significance generated by the Ramsar sites.

In 2022, another policy called the *Amrit Sarovar* Mission was launched to increase water conservation by rejuvenating 75 ponds in each district of the country (Government of India 2023a). The Mission was completed on 15 August 2023, and nearly 50,000 ponds across the country were rejuvenated. On the back of the success of this policy, the target was doubled. As of April 2024, work on 66% of the identified sites has been completed under the Amrit Sarovar Mission (Mission Amrit Sarovar 2024). Both *Amrit Sarovar* and *Amrit Dharohar* place importance on increasing people's participation in conservation activities and local livelihood generation.

Further, specific to coastal wetlands, the Mangrove Initiative for Shoreline Habitats and Tangible Incomes (MISHTI) was launched in 2023 that aims to develop a 540 km^2 stretch of mangroves through the sharing of best practices between mangrove hosting Indian states and enable resource mobilization through Public Private Partnership (Government of India 2023b).

An equally important centrally sponsored scheme for the conservation and restoration of wetlands is the National Plan for Conservation of Aquatic Ecosystems (NPCA), which was launched in 2015. The scheme follows a four-pronged approach of baseline information collection, rapid assessment of wetland health, stakeholder engagement, and development of an Integrated Management Plan for their conservation (Ministry of Environment, Forest and Climate Change 2020). NPCA is being upscaled from 33 wetlands in the first cycle to 1000 in the next phase to enhance water quality and improve biodiversity within ecosystems (Khanduja et al. 2023).

Similarly, the National Mission for Clean Ganga Programme specifically focuses on wetland conservation planning within the Ganga basin. There are other policies as well, such as the Atal Mission for Rejuvenation and Urban Transformation 2.0, that seek to improve the quality of wetlands and rejuvenate deteriorated water bodies by investing in the proper disposal of sewage, restoring polluted drains and planning the better management of landscapes linked to wetlands (Ministry of Housing and Urban Affairs 2021).

Implementation of new schemes has been instrumental in increasing the focus on wetland management within the overall water agenda. However, there is a need for more comprehensive management strategies (Bassi et al. 2014) to achieve target 6.6 of Sustainable Development Goal

(SDG) 6, which deals with protecting and restoring water-related ecosystems such as wetlands. For this to happen, the established and planned institutional mechanisms must operate in coherence to counter the anthropogenic and climate-related threats to wetlands and their associated landscapes.

Following the introduction, the second section provides the methodology. The third section provides a detailed account of wetland status in India, it compares two time periods 2006–2007 and 2017–2018 for which the wetland inventories were analyzed. The fifth section provides a detailed account of the challenges faced by the wetland ecosystems in India. The last section provides recommendations to strengthen actions on wetland management in India.

1.2 Methodology

An eclectic approach was followed for developing the chapter. For analyzing the status of wetlands in terms of their type, number, and area, the national inventories for 2006–2007 and 2017–2018 were compared.

As discussed in the introduction (Section 1.1), India has so far undertaken two rounds of National Wetland Inventory and Assessment. It is a geo-databased approach to track the evolving status of wetland distribution in the country. A national-level database is maintained by classifying wetlands based on the Ramsar definition (Table 1.1) and tracking the spatial and temporal transition of wetlands at different regional scales. The first inventory was prepared for 2006–2007.

Table 1.1 Wetlands classification as per the National Wetland Inventory of India.

Level I	Level II	Level III
Inland wetlands	Natural: High-altitude Himalayan lakes, wetlands lying in the flood plains of major rivers, saline and temporary wetlands lying in the arid to semiarid regions	Lakes
		Ox-bow lakes/cutoff meanders
		High-altitude wetlands
		Riverine Wetlands
		Waterlogged (natural)
		River/Stream
	Man-made: Wetlands constructed for storing water, fish cultivation, or for recreation	Reservoirs/Barrages
		Tanks/Ponds
		Waterlogged (man-made)
		Salt Pans (inland)
		Aquaculture ponds (inland)
Coastal Wetlands	Natural	Lagoons/Backwaters
		Creek
		Sand/Beach
		Intertidal mud flats
		Salt marsh
		Mangroves
		Coral Reefs
	Man-made	Salt Pans (coastal)
		Aquaculture ponds (coastal)

Source: India Wetland Portal and the Ministry of Statistics and Programme Implementation.

The second round of the National Wetland Inventory and Assessment was undertaken by the Space Applications Centre, ISRO Ahmedabad using data available for the years 2017–2018. Subsequently, the National Wetland Decadal Change Atlas, 2017 (Gupta et al. 2021) was published to compare the change in wetland statistics since the first assessment undertaken for 2006–2007. As of 2017–2018, a total of 231,195 wetlands have been mapped in the country having an area of 15.98 m ha or 4.86% of the total geographical area of India. Rivers and streams, reservoirs, intertidal mud flats, and tanks and ponds comprise the majority of wetland types in the country (Figure 1.1).

For assessing the challenges faced by the natural wetland ecosystem in India, a thorough review of the existing published studies was undertaken. The selection of review papers was based on the purpose and use case, thus following a more traditional review approach rather than a systematic one.

Based on the findings from the assessment of wetlands status and a review of the challenges faced by them, recommendations to ensure their management on a sustainable basis were suggested. To a large extent, the recommendations were based on the authors' learnings and understanding of the governance, financial, data and information, and knowledge-sharing landscape in India.

1.3 Status of Wetlands in India

The decadal analysis, between 2006–2007 and 2017–2018, shows that there has been an increase in the overall number of wetlands in India by around 9%. In the same period, the wetland area has increased by about 4%. This has been due to the expansion of existing wetlands as well as an increase in the formation of new wetland areas. Further, a small proportion of the previously existing wetland area (about 25 thousand ha) has disappeared.

In terms of state-wise distribution, there has been a shift in the number (Figure 1.2) and area (Figure 1.3) of wetlands, with the majority of states recording an increase in the total wetland cover. As per the second assessment (2017–2018), the highest number of wetlands is recorded in Tamil Nadu (11.6%), Maharashtra (11.2%), Andhra Pradesh (10.4%), Uttar Pradesh (8%), and Gujarat (7.6%). In terms of area coverage, Gujarat (21.9%) takes the lead, followed by Maharashtra (7.2%), Andhra Pradesh (7.1%), West Bengal (7.1%), and Uttar Pradesh (6.9%). The highest change in wetland numbers between the two assessment periods is observed in Maharashtra.

Further, the distribution of wetland types is not uniform throughout the country. For instance, 42% of all lakes in India lie in the state of Tamil Nadu. Also, the distribution of wetland areas within each state is subject to geomorphological and prevailing socioeconomic conditions. For example, the majority (about 86%) of the area under the high-altitude wetlands is within the state of Ladakh. Similarly, 95% of the area under man-made inland salt pans is in Rajasthan due to an expanding salt production industry. The states of Gujarat, West Bengal, Andhra Pradesh, and Tamil Nadu host several diverse wetlands, such as creeks, salt pans, salt marshes, intertidal mudflats, mangroves, and aquaculture ponds, in comparison to other states.

1.4 Inland and Coastal Wetlands

According to the 2017–2018 assessment, inland and coastal wetlands make up about 74 and 26% of the total wetland area, respectively. The decadal net change in the number and area of inland wetlands has seen a positive increase of about 8 and 5%, respectively, whereas for coastal wetlands it

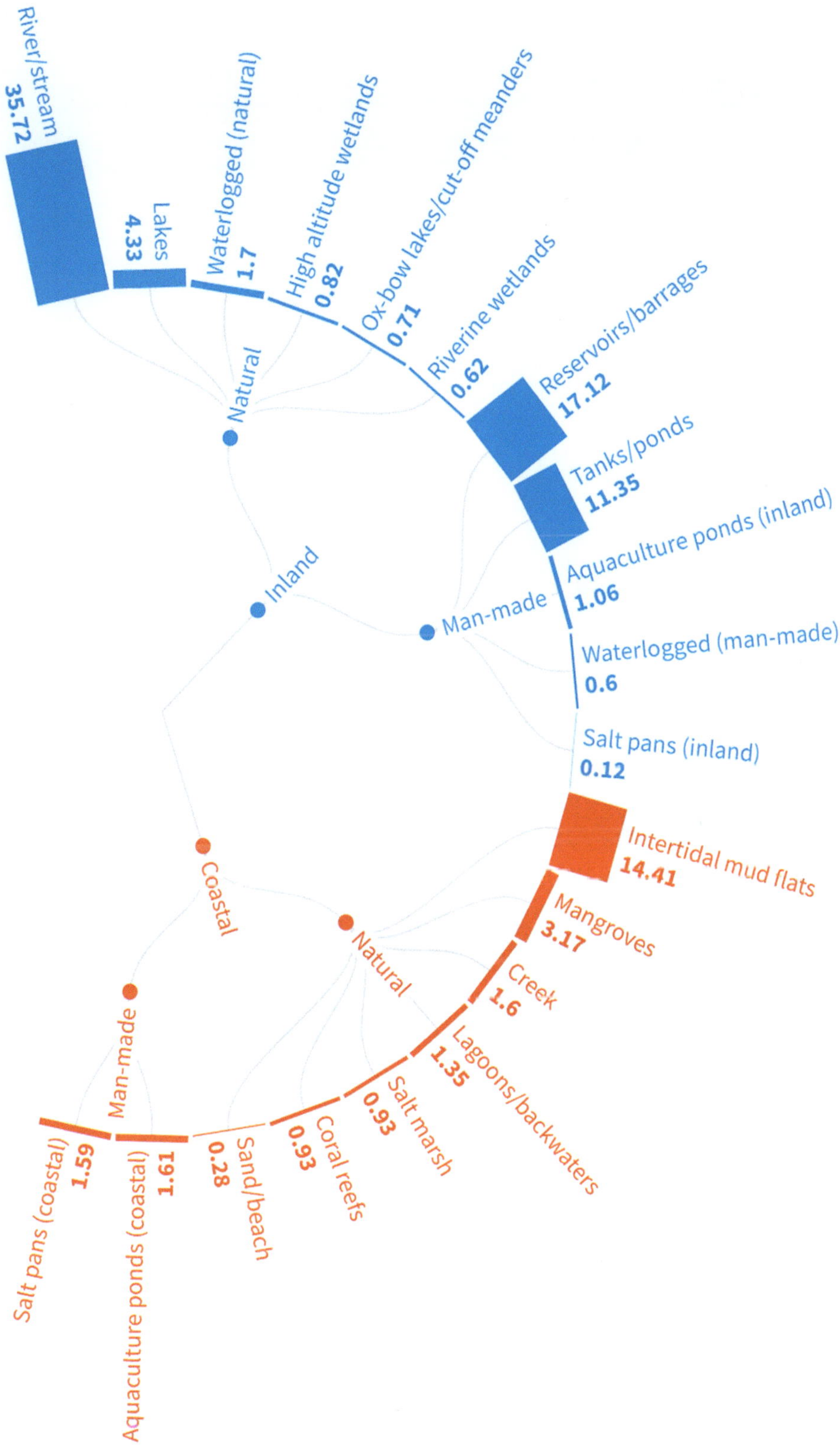

Figure 1.1 Percentage distribution of wetland types according to National Wetland Inventory for 2017–2018. *Source:* Adapted from Gupta et al. (2021).

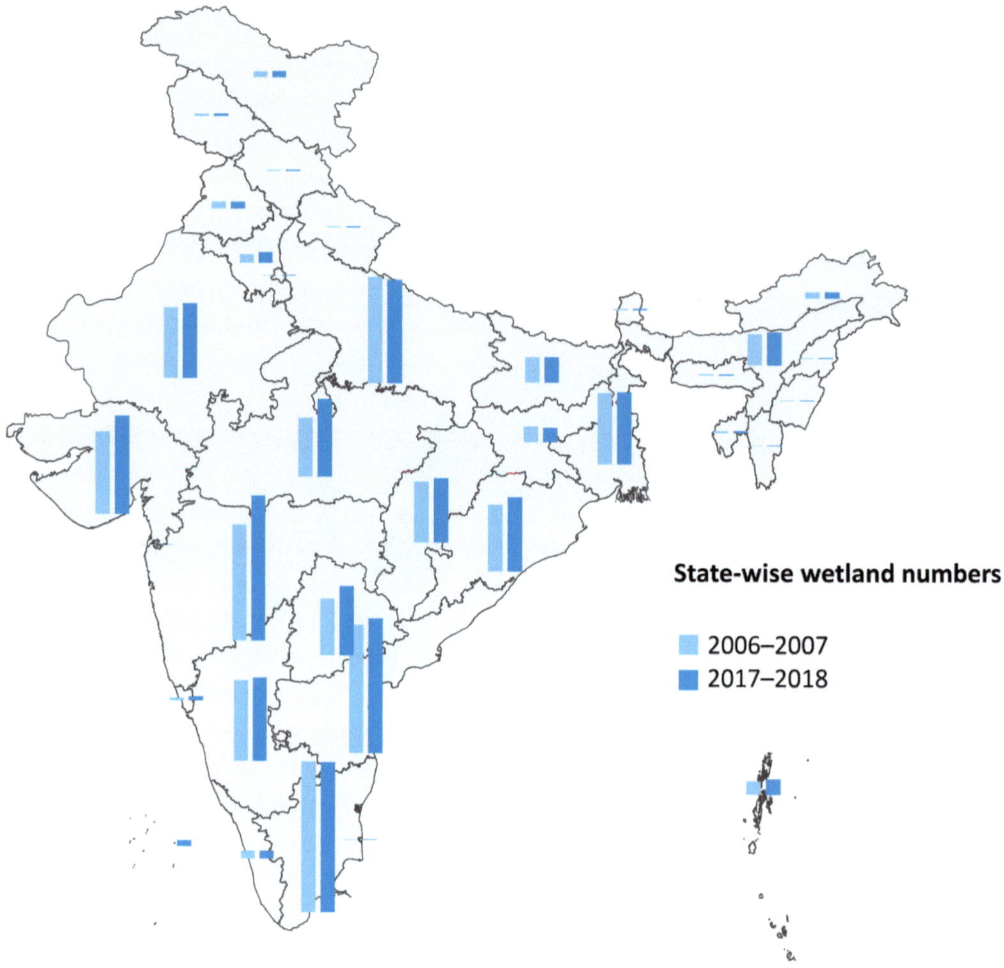

Figure 1.2 State-wise change in the number of wetlands between 2006–2007 and 2017–2018. Note: The values for the union territories of Daman and Diu and Dadar and Nagar Haveli are spatially represented together. *Source:* Adapted from Gupta et al. (2021).

has been about 20 and 1%, respectively (Figure 1.4). It is to be noted that an increase in the number of wetlands may not always translate to larger area coverage, as demonstrated in the case of coastal wetlands. Percentage increase under different categories of wetlands is presented in Figure 1.4.

In 2017–2018, the highest number of inland wetlands was reported in Uttar Pradesh and Maharashtra (Figure 1.5). Madhya Pradesh jumped from the seventh position in 2006–2007 to the third position by recording a 13.5% increase in the area under inland wetlands (Figure 1.6). Gujarat recorded the highest number of coastal wetlands under both assessment periods. In 2017–2018, almost 23% of the total number of such wetlands were recorded in Gujarat.

1.5 Natural and Man-Made Wetlands

The increase in the wetland area between 2006–2007 and 2017–2018 can be mainly attributed to an increase in man-made wetlands such as tanks, reservoirs, and aquaculture ponds in both inland and coastal regions. The existing area (in 2017–2018) under man-made wetlands is about 33%, an

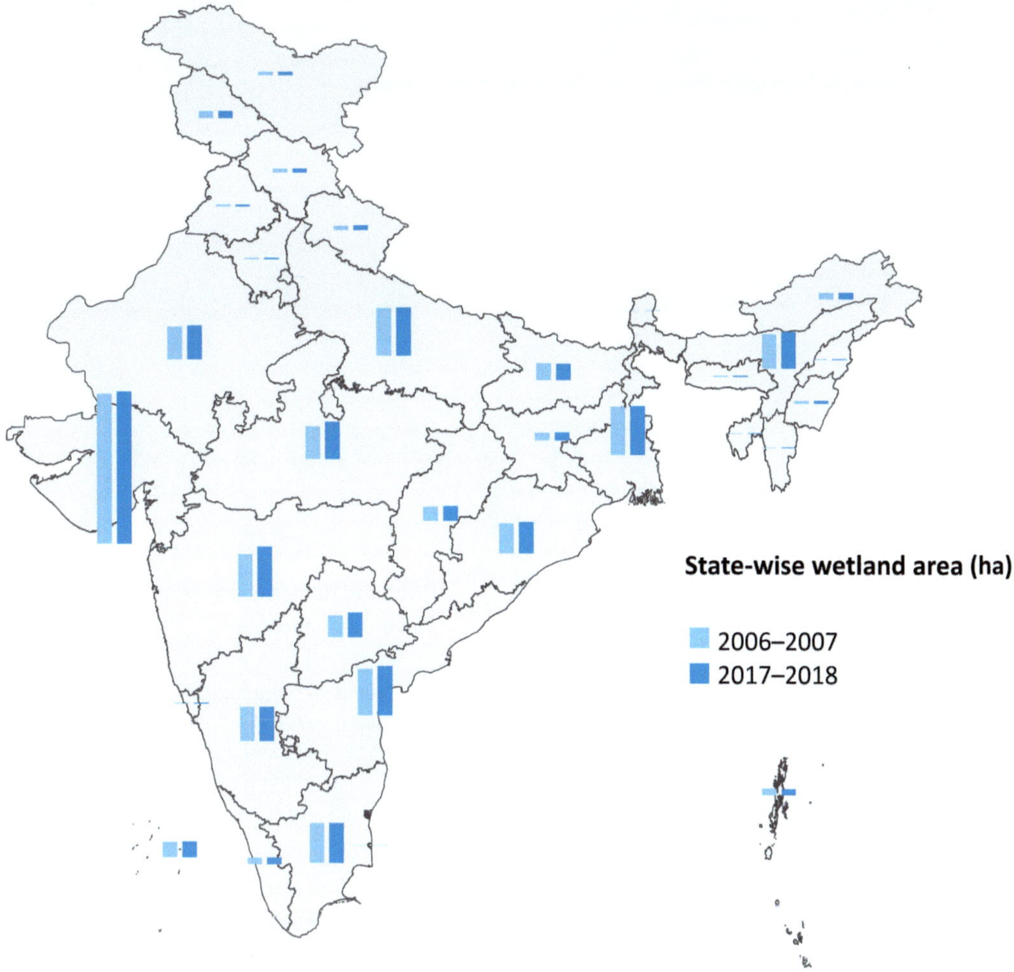

Figure 1.3 State-wise change in area under wetlands between 2006–2007 and 2017–2018. Note: The values for the union territories of Daman and Diu and Dadar and Nagar Haveli are spatially represented together. *Source:* Adapted from Gupta et al. (2021).

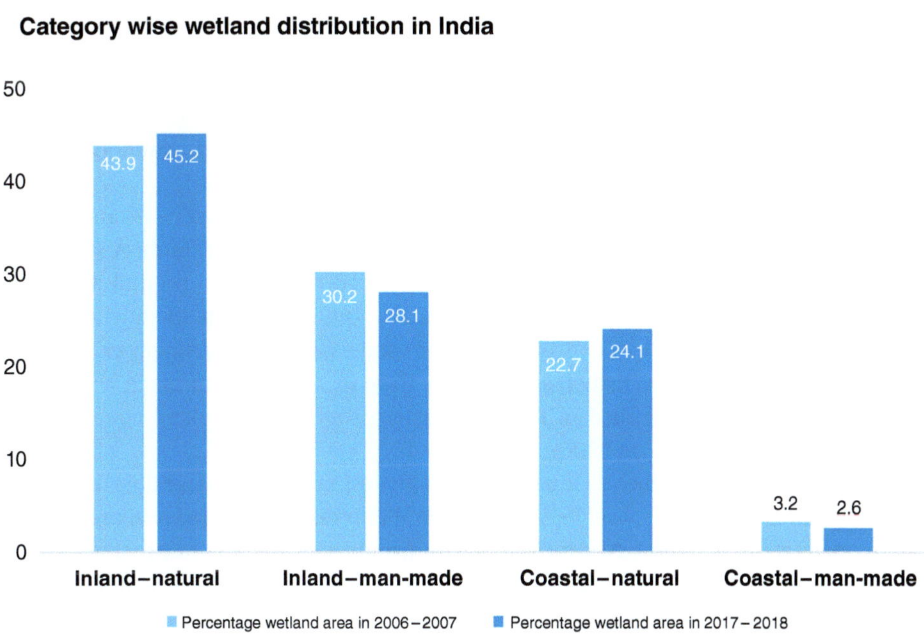

Figure 1.4 Percentage change in wetland distribution between assessment years. *Source:* Adapted from Gupta et al. (2021).

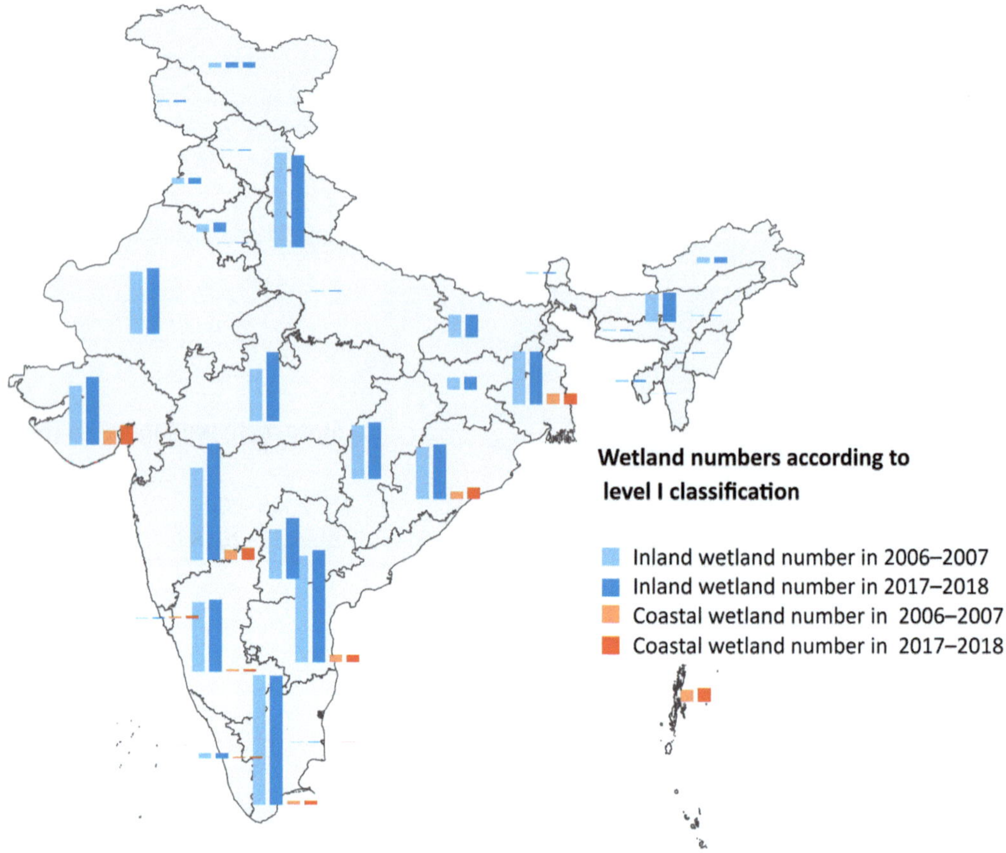

Figure 1.5 State-wise change in the number of wetlands according to Level I classification. Note: The values for the union territories of Daman and Diu and Dadar and Nagar Haveli are spatially represented together. *Source:* Adapted from Gupta et al. (2021).

increase of about 13% from 2006 to 2007. The number of man-made wetlands has increased as well by about 11% between 2006–2007 and 2017–2018. Nevertheless, the natural wetlands comprise about 67% of the total area under wetlands with no substantial change in the distribution during the two assessment periods.

Further, the analysis shows that there is a substantial increase in area under reservoirs and tanks across all states leading to an increase in the percentage share of man-made wetlands. A substantial increase in the construction of new reservoirs was recorded in Maharashtra and Madhya Pradesh, salt pans in Gujarat, and inland aquaculture ponds in Andhra Pradesh. In Gujarat, large areas of natural marshlands have been converted to man-made salt pans, causing a change in distribution between wetland types. Further, a high percentage increase in man-made wetlands, in both numbers and area, was recorded in the northeastern states where inland aquaculture activities have increased (Figures 1.7 and 1.8).

In line with the 2006–2007 assessment, Gujarat had the highest number of natural wetlands in 2017–2018 as well. This is followed by West Bengal and Uttar Pradesh. The area under inland rivers has increased in many states including Assam, Maharashtra, and Uttar Pradesh. The major reason for this is due to river braiding or an increase in stream width. The area under coastal mangroves

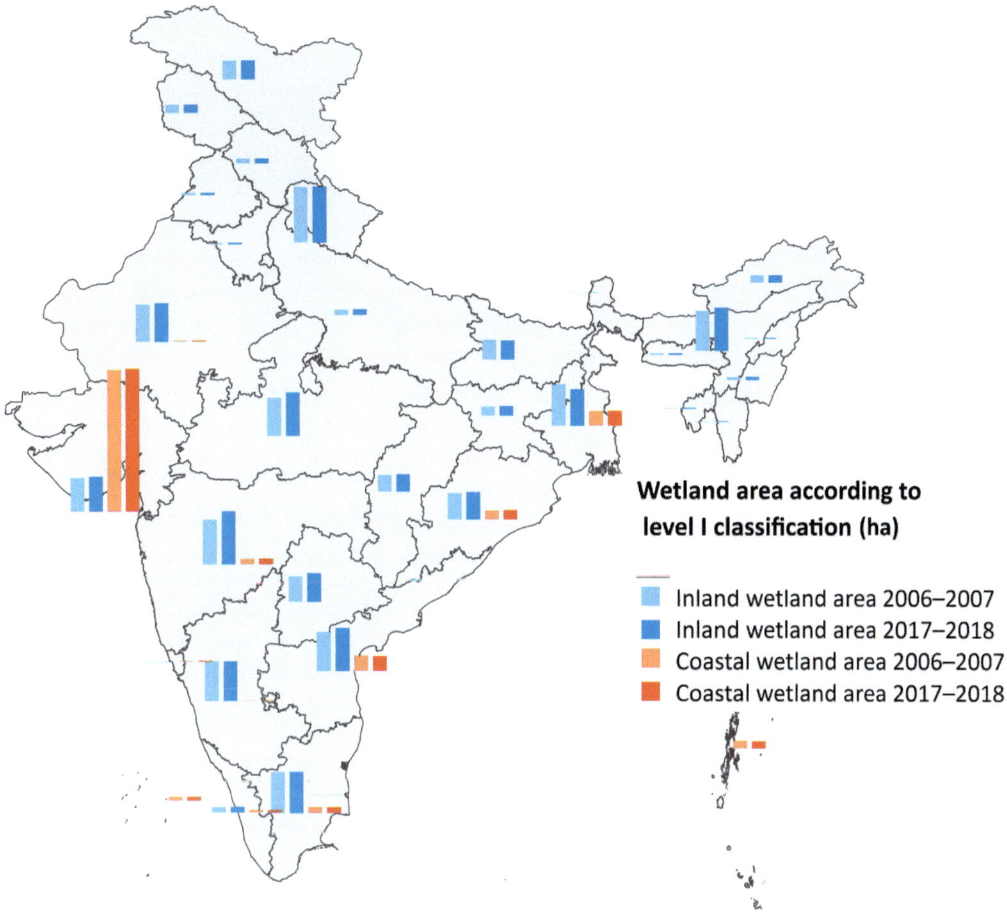

Figure 1.6 State-wise change in area under wetlands according to Level I classification. Note: The values for the union territories of Daman and Diu and Dadar and Nagar Haveli are spatially represented together. *Source:* Adapted from Gupta et al. (2021).

increased by about 4% between the two assessments. The major increase in the area under coastal mangroves was recorded in Gujarat, West Bengal, and the union territory of Andaman and Nicobar.

1.6 Overall Findings

The indicator used to track the progress toward achieving SDG target 6.6 is a measure of the change within the aquatic ecosystem over time. The exercise of periodic mapping and assessment of wetland status through a systematic methodology is therefore important for establishing baseline data, recording underlying trends in the wetland ecosystem, and developing scientifically informed management strategies for wetland protection. The findings of the National Wetland Inventory and Assessment reveal that most of the wetland types have shown positive change in area as well as in numbers. The number of Ramsar sites in India has also steadily increased. The country now has 80 diverse Ramsar sites with a total area of about 1.3 m ha, which equals 9% of the total wetland area of India (Ramsar 2024).

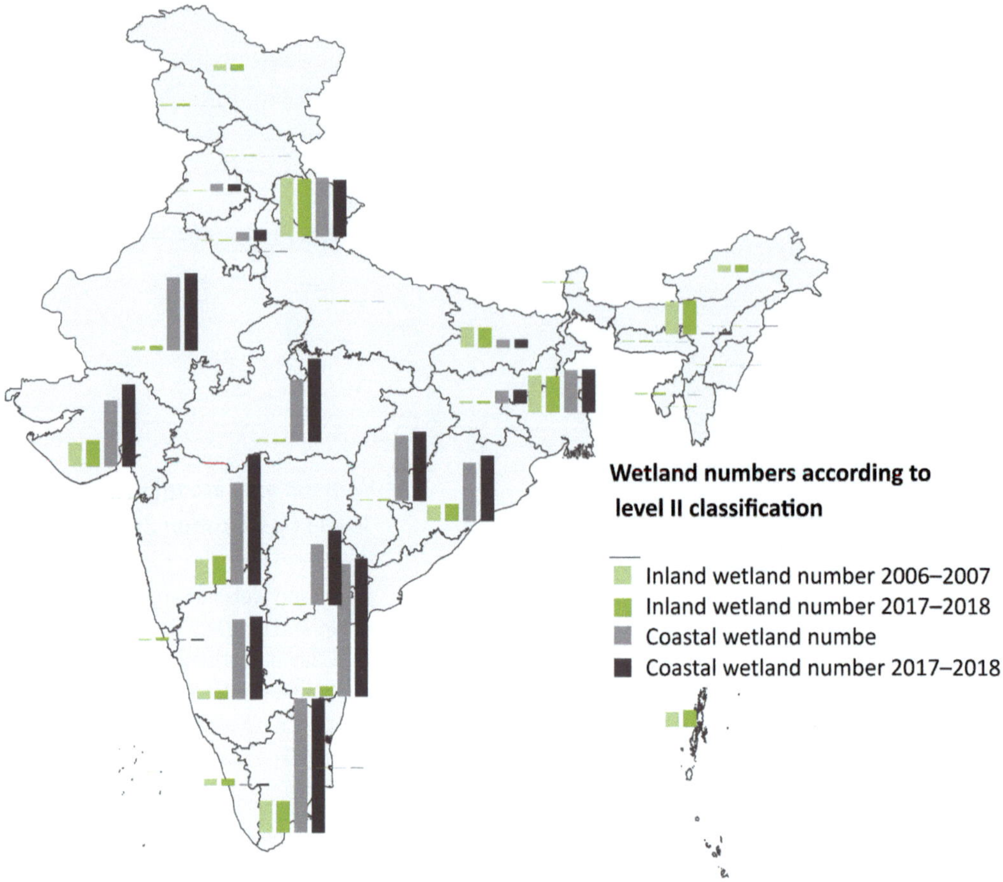

Figure 1.7 State-wise change in the number of wetlands according to Level II classification. Note: The values for the union territories of Daman and Diu and Dadar and Nagar Haveli are spatially represented together. *Source:* Adapted from Gupta et al. (2021).

The positive change in wetlands is majorly led by the growth of areas under man-made wetlands such as tanks, reservoirs, and aquaculture ponds in India. One of the likely reasons is the need for having such storage to fulfill the growing needs of irrigation, industrial and domestic water demands, fisheries, and flood control. There have been instances of natural wetlands being converted to man-made wetlands, mainly in the states on the western coast. While man-made wetlands do harbor biodiversity and provide provisional, regulatory, and cultural services, a healthy network of natural wetlands is important to maintain ecosystem productivity. Nevertheless, there is a need to corroborate national-level findings with regional and local on-ground assessments for concrete wetland management planning.

1.7 Challenges Faced by Natural Wetland Ecosystem in India

Wetlands in India and across the globe are facing increased anthropogenic pressures. This is mainly due to unplanned urbanization, land use changes, infrastructure development, and pollution from industrial effluent and agricultural runoff (Bassi et al. 2014). Further, climate change

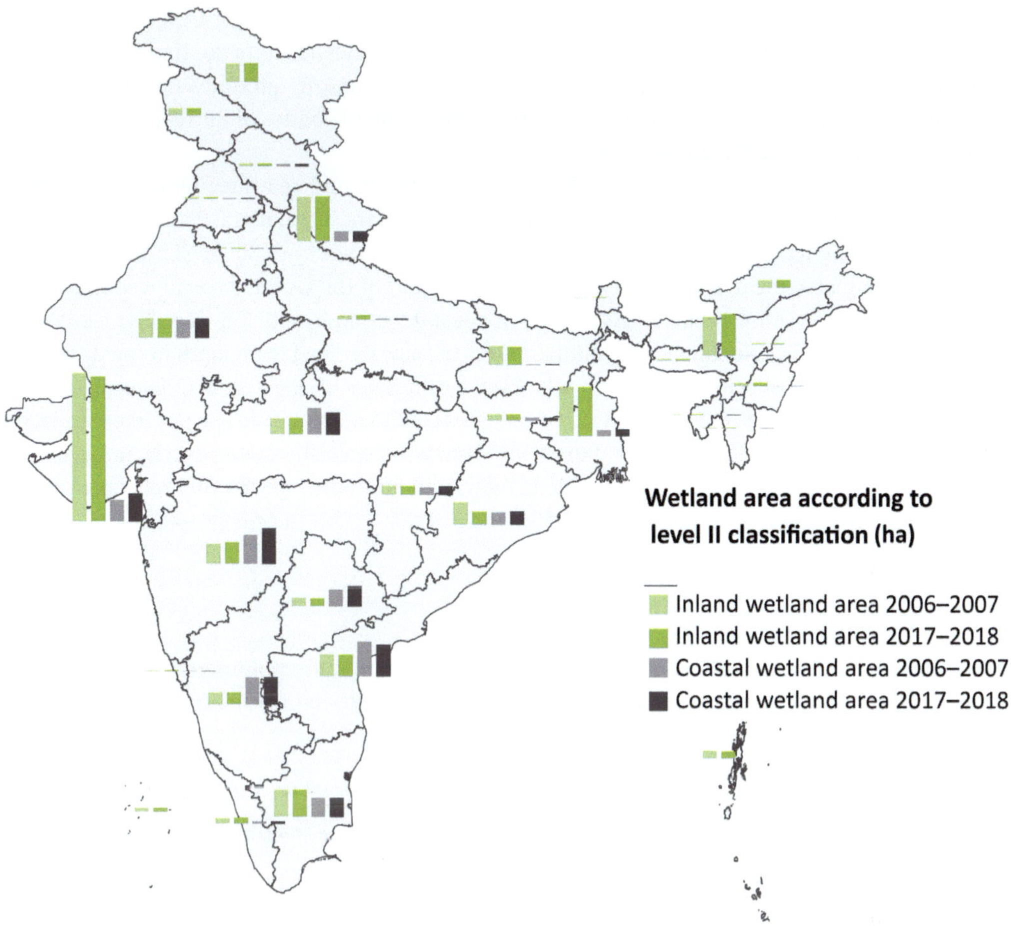

Figure 1.8 State-wise change in area under wetlands according to Level II classification. Note: The values for the union territories of Daman and Diu and Dadar and Nagar Haveli are spatially represented together. *Source:* Adapted from Gupta et al. (2021).

and variability have been identified as significant threats to wetlands, with rising temperatures and altered hydrological patterns (Erwin 2009). Some of the main challenges faced by wetland ecosystems in India are discussed below.

1.7.1 Climate Change Related

Climate-induced change is causing frequent extreme hydrological events, such as floods and droughts, impacting freshwater ecosystems globally (IPCC 2022). As per the latest progress on SDG 6 reported in 2023, one in five river basins in the world (over 20%) experienced rapid changes in the area covered by surface waters over a period of five years (2015–2020), compared to a 20-year reference period (UN-Water 2023). Some river basins are experiencing a rapid increase in their surface water area due to flooding, and others are experiencing a rapid decline due to the drying up of water bodies. Further, about 2.4 billion people in the world live in countries with water stress (UN-Water 2023). Estimates suggest that in India too, 11 out of the 15 major river basins will be water stressed by 2025 (Bassi et al. 2023).

Water quality is also impacted in addition to the quantity of water entering freshwater ecosystems. The occurrence of climate-induced extreme hydrological events alters the timing and magnitude of water received by water bodies (from surface run-off, groundwater flows, and precipitation), as well as the degree of dilution or concentration of solutes, thus altering the water quality of freshwater systems (Jones 2013).

Among other climate-induced threats, the rise in sea levels and seawater temperatures has received increased attention as of late. According to the Intergovernmental Panel on Climate Change (IPCC), the global sea level is projected to rise by 8–25 cm by 2050 (IPCC 2019). As per estimates, a 1 m global mean sea level rise could threaten half of the world's coastal wetlands. For India, this will lead to an estimated loss of 84% of coastal wetlands and 13% of saline wetlands (Blankespoor et al. 2016). Further, rising temperatures of seawater lead to changes in the distribution patterns of aquatic species and biodiversity. Even a difference of one degree Celsius in seawater temperature can impact the distribution of fish stock, with some tropic species likely to face regional extinction (Rao and Vivekanandan 2008). This will especially have adverse impacts on aquatic species that cannot relocate to suitable habitats and migratory species that rely on various wetland types throughout their life cycle (Bassi et al. 2014).

1.7.2 Anthropogenic

Anthropogenic factors such as population growth, urbanization, water pollution, gray infrastructure development, and reduced blue cover, combined with climate-induced risks, impact wetland resources. Some of these factors are discussed below.

1.7.2.1 Urbanization and Land Use Changes

India has steadily urbanized over the past decades. The percentage of the total population living in urban areas has doubled from 17% in 1960 to over 35% as of 2022 (World Bank 2022). As a result, the area under agricultural and nonagricultural use (residential and commercial) has significantly increased between 1950–1951 and 2008–2009 to meet the rising water, food, and housing demands of the growing population (Bassi et al. 2014). This has been at the cost of changes in land use of floodplain areas, primary forests, grasslands, and associated freshwater ecosystems in most of the major river basins (Zhao et al. 2006). For instance, almost 50% of the wetland area in the National Capital Region of Delhi has decreased over a period of three decades (1989–2019). This is on account of the increase in area under cropland and built-up due to rapid urban expansion (Venkatesh 2020). Another instance is that of the Greater Bengaluru metropolitan area, where due to massive concretization over the years, there has been a loss of almost 40% of the area under water bodies between 1973 and 2020. Over just six years between 2011 and 2017, the storage capacity of the city's water bodies has reduced by 76% due to siltation (IISC 2017).

1.7.2.2 Agricultural, Municipal, and Industrial Pollution

Untreated runoff from agricultural land (consisting of pesticides and fertilizers) and untreated wastewater from urban areas adversely impact water quality due to siltation, excess nutrient loadings, and changes to water flows.

The intensification of agriculture in India over the last four decades has increased fertilizer consumption from 6 million tonnes in 1981–1982 to almost 30 million tonnes in 2021–2022 (Ministry of Agriculture & Farmers Welfare 2022). These nutrients are discharged into surface water bodies through agricultural runoff and lead to eutrophication and harmful algal bloom. Algae consume the oxygen in water on decomposition, causing a reduction or elimination of fish stock and other organisms (EPA 2024).

In addition to agricultural runoff, the discharge of untreated urban wastewater is a significant reason for high levels of pollution in many of the major Indian rivers, primarily concentrated in the stretches passing through urban areas (Gupta et al. 2023). For instance, the stretch of the Yamuna from Wazirabad to Okhla in Delhi, which is less than 2% of the total river length, accounts for 80% of its total pollution load (NGT 2012). The situation is no different for other rivers passing through urban areas, with the Central Pollution Control Board identifying 311 polluted stretches across India with Biological Oxygen Demand (an indicator of fecal contamination and organic pollution) above 3 mg/l (CPCB 2022).

As of 2021, over 72,000 million liters per day of domestic wastewater was generated from India's urban areas alone. The installed treatment capacity was 44%, but only 28% of the wastewater generated was actually treated (CPCB 2021). Many sewage treatment plants (STPs) do not function at maximum capacity or do not meet prescribed effluent water quality standards. In some cases, STPs are underutilized since the sewerage networks do not cover unauthorized colonies and suburban areas. Further, many urban utilities are unable to scale up wastewater treatment capacity as the capital and operating costs of building infrastructure are high. As a result, untreated and partially treated wastewater from urban areas is discharged into freshwater bodies leading to high levels of pollution (Gupta et al. 2023).

1.8 Recommendations for Managing Wetlands Sustainably

1.8.1 Improve Governance

Improvement in governance is perhaps the single most important factor for sustainable management of wetlands. As discussed in Section 1.1, India has several wetland management policies (programs and schemes). However, there is no specific regulation governing their use (Bassi et al. 2014). It is indirectly influenced by several other legal instruments. Important ones include Water (Prevention and Control of Pollution) Act, 1974; Forest (Conservation) Act, 1980; Environmental (Protection) Act, 1986; Wildlife (Protection) Amendment Act, 1991; Biodiversity Act, 2002; and Scheduled Tribes and Other Traditional Forest Dwellers (Recognition of Forest Rights) Act, 2006 (Ministry of Environment and Forests 2007; Prasad et al. 2002; Raju 2012; Bassi 2016).

Provisions under these Acts range from the protection of water quality and notification of ecologically sensitive areas to contributing toward conserving, maintaining, and augmenting the floral, faunal, and avifaunal biodiversity of the country's aquatic bodies. However, the term wetland was not used specifically in any of these legal instruments.

Based on the directives of the National Environment Policy, 2006, and recommendations made by the National Forest Commission, the Central Government of India notified the Wetlands (Conservation and Management) Rules, 2010. As per them, the Central Wetlands Regulatory Authority (CWRA) was constituted. An Expert Group on Wetlands was constituted to examine management action plans for the newly identified wetlands (Ministry of Environment and Forests 2012). The rules have restricted activities such as reclamation, setting up industries in the vicinity, solid waste dumping, manufacture or storage of hazardous substances, discharge of untreated effluents, and any permanent construction within the wetlands. It also regulates activities (which will not be permitted without the consent of the state government) such as hydraulic alterations, unsustainable grazing, harvesting of resources, releasing treated effluents, aquaculture, agriculture, and dredging.

However, only some selected wetlands based on the significance of the functions performed by them for the overall well-being of the people were being regulated under these rules (Bassi 2016). These belong to the following six categories: wetlands selected under Ramsar Convention; wetlands in ecologically sensitive and important areas; wetlands recognized as UNESCO World Heritage sites; high-altitude wetlands (at or above an elevation of 2500 m with an area equal to or greater than 5 ha); wetland complexes below an elevation of 2500 m with an area equal to or greater than 500 ha; and, any other wetland identified by the Authority (Government of India 2010).

In 2017, a revision to the existing Wetlands (Conservation and Management) Rules was notified. The 2017 revision did away with the six categories and instead included all the Ramsar sites and those notified by the Central Government, State Government, and Union Territory (UT) Administration for conservation and management, except for those under forest and coastal regulation zones. Further, the revised rules led to the constitution of the Wetlands Authority in each State and the UT with the major responsibility of preparing a list of wetlands in their jurisdiction within three months and a list of wetlands to be notified for conservation and management within six months from the date of publication of the 2017 rules. Also, a National Wetlands Committee was constituted with the primary responsibility of monitoring the implementation of the rules by the respective wetland authority (Government of India 2017).

To date, only a handful of states and UTs have notified wetlands for conservation. They include Uttar Pradesh, Chandigarh, Goa, and Rajasthan. Even in these states and UTs, only some important wetlands have been notified. Though the states and UTs are making progress on updating their wetlands inventory, they need to fasten the wetland notification process as well. This step is important as it provides legal status to the water body and leads to the demarcation of its boundaries, and identification of its zone of influence where anthropogenic activities need to be regulated. Thus, creating an enabling environment for sustainable management and wise use of wetlands.

Moreover, considering that wetlands are considered nature-based solutions that can provide numerous services of great social, economic, and environmental value to humankind (Thorslund et al. 2017), it is important to protect them from pollution due to anthropogenic activities. One of the main sources of pollution is untreated wastewater from the domestic and industrial sectors that is discharged into such systems. The urban wastewater management policy must take cognizance of the linkage between urban wastewater and its impact on wetland health. In urban areas having no demand for treated wastewater, wetlands can be recipients of such water which will help them rejuvenate especially during summers when they receive no freshwater flow. This will mainstream a circular economy approach to wastewater management and contribute to SDG target 6.3 on substantially increasing recycling and safe reuse of treated wastewater.

1.8.2 Enable Finance

The global economic value of ecosystem services provided by wetlands annually stands at over USD 47 trillion (Canning et al. 2021). Yet there is a pervasive under-valuing of water resources and water-based ecosystems and their benefits. An ecosystem-based approach (EbA) can enable the protection and rejuvenation of water-based ecosystems, including mountains, forests, wetlands, rivers, aquifers, and lakes. EbA was highlighted as an essential strategy for reducing disaster risk by the Disaster Risk Reduction Working Group under the Indian G20 Presidency in 2023. The current Brazilian Presidency prioritizes payment for ecosystem services (PES) mechanisms by recognizing the benefits provided by nature, with a focus on strengthening developing countries' capacities to develop and adopt PES mechanisms. Such a market mechanism can provide a viable avenue for funding wetland restoration via direct benefits to those who preserve them (individuals, organizations, companies, etc.).

Along similar lines, the Government of India incentivizes the voluntary adoption of environment-friendly practices by individuals, industries, and local bodies by awarding tradable green credits under the Green Credit Programme. These credits can be earned through various activities identified under the program, including water management (water conservation, water harvesting, and water use efficiency, including treatment and reuse of wastewater) and mangrove restoration and conservation. Water management practices such as water harvesting and treatment and reuse of wastewater promote wetland health, and hence such schemes can be leveraged for wetland restoration projects.

Further, ecosystem-level solutions require partnerships between different pools of finance – private companies, communities, civil society organizations, local governments, national governments, and international financial institutions. Urban local bodies are essential players in the management of urban water bodies. Hence, cities must explore and leverage blended finance options such as municipal bonds, public equity, loans from multilateral institutions, value-captured finance, and viability gap funding, to diversify investment streams for funding wetland conservation and rejuvenation projects (Barbarwar et al. 2023).

1.8.3 Strengthen Data and Information

Many wetlands are experiencing land use changes in their catchment leading to either reduced or contaminated inflow or both into them (Bassi et al. 2014). The land use changes are mainly from forest areas to agricultural land and from agriculture to built-up areas. It is important to regulate large-scale land-use changes in the catchment area of wetlands to ensure water inflow and prevent them from getting polluted. This is important to maintain their hydrological and ecological integrity. To enable this, the development of catchment areas has to be undertaken with a proper management plan so that there are no major adverse impacts (in the form of reduced flows or pollution) in water bodies located downstream. One of the basic needs for the preparation of sound management plans is proper data on hydrology, physical and social changes, and economic processes in the catchment. This will need the setting up of proper monitoring systems to measure runoff and streamflow in the local catchments, and frequent updating of the land use and land cover spatial maps.

Moreover, water quality assessments are presently undertaken intermittently and that too only for a select number of wetlands, mainly rivers and lakes. The assessments do not cover small wetlands that play an important role in the hydrological cycle. Therefore, an effective and proper water quality-monitoring strategy needs to be developed for all wetlands with significant ecological value and should be made part of all the existing wetland conservation programs. The generation of reliable data on the present condition of wetlands in terms of their water quality status will help in devising and monitoring schemes for improvement in water quality management.

1.8.4 Drive Cooperation, Knowledge Sharing, and Stakeholder Engagement

International cooperation and knowledge sharing and stakeholder engagement can make substantial contributions to the sustainable management of wetlands. Two platforms can play an important role in this. First is the G20 Water Platform and the other is the Global River Cities Alliance (GRCA).

The G20 Water Platform is the main tool for sharing the outcomes of the G20 Water Dialogues. It can be developed as a partnership facilitation portal by providing details of the policy and legal landscape governing the wetland management of each G20 country, the existing partnerships between G20 member countries, and stakeholders who have demonstrated excellence in protecting and conserving wetlands (governments, nongovernmental organizations, multilateral organizations, funders, and the local community). Such knowledge sharing can enable the development of pilots and research on wetlands, and drive investments which in turn can boost partnerships for economic growth, jobs, innovation, knowledge generation, and dissemination.

The GRCA, launched at COP-28 by the Government of India, has members from 11 countries. It is a voluntary platform to discuss and exchange knowledge and experiences on sustainable management of river stretches passing through the urban areas. It has the potential to be an example of a global water partnership that can enable South–South collaboration for managing wetlands, especially in those in urban areas.

1.9 Conclusion

To conclude, for sustainable management of wetlands in India, four components are crucial. First, "innovative governance reforms" including regulations, institutional setup, and policies leading to their notifications can make a difference to the business-as-usual way of managing wetlands. These together can shift focus to considering wetlands as a living entity and take cognizance of the co-benefits they offer in terms of maintaining aquatic life and providing water security for domestic and productive purposes. Second, "viable financial models" can help strengthen private investments in catchment protection, wastewater treatment, and reuse infrastructure leading to water conservation, reduction in demand for freshwater, and pollution abatement through checks on contamination of freshwater including wetlands. Third, "reliable data and information" on the inflow and quality status of wetlands in urban areas and the changes that are shaping the land use and land cover. Fourth, "cooperation and knowledge sharing including community engagement" for creating awareness on wetland sensitive development, especially in urban areas, and behavioral change on water use leading to healthy and living wetlands.

References

Barbarwar, S., Gupta, S., and Parmer, A. (2023). Evaluating nature-based solutions (NbS) as a tool for urban resilience in the global south. In: *The Empathic City* (ed. N. Biloria, G. Sebag, and H. Robertson). Cham: S.M.A.R.T. Environments, Springer 10.1007/978-3-031-32840-4_10.

Bassi, N. (2016). Implications of institutional vacuum in wetland conservation for water management. *IIM Kozhikode Society & Management Review* 5 (1): 41–50.

Bassi, N., Kumar, M., Sharma, A., and Pardha-Saradhi, P. (2014). Status of wetlands in India: a review of extent, ecosystem benefits, threats and management strategies. *Journal of Hydrology: Regional Studies* 2: 1–19.

Bassi, N., Gupta, S., and Chaturvedi, K. (2023). *Reuse of Treated Wastewater in India: Market Potential and Pointers for Strengthening Governance*. New Delhi: Council on Energy, Environment and Water.

Blankespoor, B., Dasgupta, S., and Laplante, B. (2016). Sea-level rise and coastal wetlands: impacts and costs. World Bank Policy research working paper 6277.

Canning, A., Jarvis, D., Costanza, R. et al. (2021). Financial incentives for large-scale wetland restoration: beyond markets to common asset trusts. *One Earth* 4 (7): 937–950.

Central Pollution Control Board (2021). *National Inventory of Sewage Treatment Plants in India*. New Delhi: Ministry of Environment, Forest and Climate Change, Government of India.

Central Pollution Control Board (2022). *Polluted River Stretches for Restoration of Water Quality – 2022*. New Delhi: Ministry of Environment, Forest and Climate Change, Government of India.

Davidson, N.C. and Finlayson, C.M. (2019). Updating global coastal wetland areas presented in Davidson and Finlayson (2018). *Marine and Freshwater Research* 70 (8): 1195–1200.

EPA (2024). The effects: dead zones and harmful algal blooms. https://www.epa.gov/nutrientpollution/effects-dead-zones-and-harmful-algal-blooms#:~:text=Excess%20nitrogen%20and%20phosphorus%20cause,in%20the%20water%20is%20consumed (accessed 15 April 2024).

Erwin, K.L. (2009). Wetlands and global climate change: the role of wetland restoration in a changing world. *Wetlands Ecology and Management* 17: 71–84.

Garg, J.K., Singh, T.S., and Murthy, T.V.R. (1998). *Wetlands of India*. Ahmedabad: SAC, Indian Space Research Organisation.

Government of India (2010). *Wetlands (Conservation and Management) Rules 2010*. New Delhi: Ministry of Environment and Forests, Government of India.

Government of India (2017). *Wetlands (Conservation and Management) Rules 2017*. New Delhi: Ministry of Environment, Forest and Climate Change, Government of India.

Government of India (2023a). Amrit Sarovar Scheme. Press release (5 December 2023).

Government of India (2023b). MISHTI scheme promotes development of 540 Sq. Kms mangroves across 11 states and 2 union territories. Press release (6 April 2023).

Gupta, P.K., Patel, J.G., Singh, R.P. et al. (2021). *Space Based Observation of Indian Wetlands*. Ahmedabad: Space Applications Centre, ISRO.

Gupta, S., Chaturvedi, K. and Bassi, N. (2023). The value of treated wastewater. Urban update (11 July 2023).

IISC (Indian Institute of Science) (2017). Rejuvenation Blueprint for Lakes in Vrishabhavathi Valley. ENVIS Technical Report 122. Environmental Information System, CES, Indian Institute of Science, Bangalore 560012.

IPCC (2019). Sea level rise and implications for low-lying islands, coasts and communities. In: *IPCC Special Report on the Ocean and Cryosphere in a Changing Climate* (ed. H.-O. Pörtner, D.C. Roberts, V. Masson-Delmotte, et al.), 321–445. Cambridge and New York: Cambridge University Press.

IPCC (2022). *Water*. In: *Climate Change 2022: Impacts, Adaptation and Vulnerability. Contribution of Working Group II to the Sixth Assessment Report of the Intergovernmental Panel on Climate Change* (ed. H.-O. Pörtner, D.C. Roberts, M. Tignor, et al.), 551–712. Cambridge and New York: Cambridge University Press.

Jones, I. (ed.) (2013). *The Impact of Extreme Events on Freshwater Ecosystems*. British Ecological Society: London.

Khanduja, E., Negi, U., Chaturvedi, K. et al. (2023). *Ripples: India's Sustainable Water Management Story*. New Delhi: Council on Energy, Environment and Water.

Ministry of Agriculture and Farmers Welfare (2022). *Agricultural Statistics at a Glance 2022*. New Delhi: Ministry of Agriculture and Farmers Welfare, Government of India.

Ministry of Environment and Forests (1990). *Wetlands of India – A Directory*. New Delhi: Ministry of Environment and Forests, Government of India.

Ministry of Environment and Forests (2007). *Conservation of Wetlands in India: A Profile (Approach and Guidelines)*. New Delhi: Ministry of Environment and Forests, Government of India.

Ministry of Environment and Forests (2012). *Annual Report 2011–12*. New Delhi: Ministry of Environment and Forests, Government of India.

Ministry of Environment, Forest and Climate Change (2020). National Plan for Conservation of Aquatic Ecosystems (NPCA). https://indianwetlands.in/our-work/national-plan-for-conservation-of-aquatic-ecosystems/ (accessed 12 April 2024).

Ministry of Environment, Forest and Climate Change (2023). *Amrit Dharohar: Implementation Strategy*. New Delhi: Ministry of Environment, Forest and Climate Change, Government of India.

Ministry of Housing and Urban Affairs (2021). *Atal Mission for Rejuvenation and Urban Transformation 2.0: Operational Guidelines*. New Delhi: Ministry of Housing and Urban Affairs, Government of India.

Mission Amrit Sarovar. (2024). Total number of sites identified and completed under Amrit Sarovar. https://amritsarovar.gov.in/Masterreport (accessed 10 April 2024).

NGT (National Green Tribunal) (2012). Restoration and Conservation of River Yamuna. NGT Final Report.

Prasad, S.N., Ramachandra, T.V., Ahalya, N. et al. (2002). Conservation of wetlands of India – a review. *Tropical Ecology* 43 (1): 173–186.

Raju, K.D. (2012). *The Wetlands Jurisprudence in India: A Case Study of the West Bengal Conservation and Management of Wetlands and Water Bodies Policy 2012*. Kharagpur: Rajiv Gandhi School of Intellectual Property Law.

Ramsar (2024). Ramsar Sites Information Service. https://rsis.ramsar.org/ris-search/?language=en&f[0]=regionCountry_en_ss per cent3AIndia&pagetab=2 (accessed 12 April 2024).

Rao, P. and Vivekanandan, E. (2008). Impact of climate change on Indian marine fisheries. Bay of Bengal News (June 2008).

Secretariat, R. (2024). *The List of Wetlands of International Importance*. Gland: The Secretariat of the Convention on Wetlands.

Space Applications Centre (2011). *National Wetland Atlas*. Ahmedabad: Indian Space Research Organisation.

Thorslund, J., Jarsjo, J., Jaramillo, F. et al. (2017). Wetlands as large-scale nature-based solutions: status and challenges for research, engineering and management. *Ecological Engineering 108*: 489–497.

UN-Water. (2023). *Blueprint for Acceleration: Sustainable Development Goal 6 Synthesis Report on Water and Sanitation 2023*. United Nations.

Venkatesh, S. (2020). Understanding the impact of urbanization on the wetlands of Yamuna floodplains in Delhi-NCR. Wetlands International (18 July).

Woistencroft, J.A., Hussain, S.A., and Varshney, C.K. (1989). India: introduction. In: *A Directory of Asian Wetlands* (ed. D.A. Scott). International Union for Conservation of Nature: Switzerland.

World Bank. (2022). Urban population % of total population. https://data.worldbank.org/indicator/SP.URB.TOTL.IN.ZS (accessed 15 April 2024).

WWF India and Asian Wetland Bureau (1993). *Directory of Indian Wetlands, 1993*. New Delhi: WWF India.

Zhao, S., Peng, C., Jiang, H. et al. (2006). Land use change in Asia and the ecological consequences. *Ecological Research* https://doi.org/10.1007/s11284-006-0048-2.

2

Wetlands of Bangladesh: Biodiversity, Livelihoods, and Conservation

Krishna Prosad Mondal[1], Al-Jubaer[1], Soumic Samad[2], Monishankar Sarkar[3], Syed Hafizur Rahman[4], Sheikh Tawhidul Islam[1], and Byomkesh Talukder[5]

[1] *Institute of Remote Sensing and GIS, Jahangirnagar University, Dhaka, Bangladesh*
[2] *Research Consultant, Remote Sensing Division, Center for Environmental and Geographic Information Services (CEGIS), Dhaka, Bangladesh*
[3] *Institute of Disaster Management and Vulnerability Studies, University of Dhaka, Dhaka, Bangladesh*
[4] *Department of Environmental Sciences, Jahangirnagar University, Dhaka, Bangladesh*
[5] *Department of Global Health, Florida International University, Miami, FL, USA*

2.1 Introduction

Bangladesh is predominantly wetland, with only 20% of its terrain comprising tablelands and highlands. Wetlands are integral to Bangladesh's biodiversity, environment, and livelihoods (Islam 2010; PFC 2016). Bangladesh's significant coastal and interior wetlands are the vast delta systems and floodplains of the Ganges, Brahmaputra, and Meghna rivers (Ratner et al. 2013; Al-Amin et al. 2021). Wetlands are widespread in Bangladesh, covering about thirteen to eight million hectares or roughly half of the country's total land area (PFC 2016; Islam 2021). However, the increasing population in Bangladesh is putting pressure on its wetlands due to its dense population of 1258 people per square kilometer (BBS 2022). While projects like irrigation, drainage, and flood control have already destroyed over 2.1 million hectares of wetlands within the Ganges-Brahmaputra floodplain (Islam 2010, 2021; PFC 2016), shrimp farming in coastal regions contributes to the degradation of coastal wetlands, with intense erosion in catchment areas linked to shrimp cultivation (Alam et al. 2021; Al-Amin et al. 2021; Sharmin et al. 2021).

Bangladesh boasts a diverse array of wetland types, each with its unique biodiversity and livelihoods. Understanding these types of wetlands is crucial for effective wetland conservation efforts and ensuring the sustainability of Bangladesh's wetlands (Islam 2010).

2.2 Types of Wetlands in Bangladesh

In Bangladesh, wetlands encompass various water bodies such as rivers, haors, beels, jheels, ponds, dighis, floodplains, and lowland areas (PFC 2016; Islam 2021). Following the definition set forth by the Ramsar Convention (Ramsar Convention Secretariat 2013), the Department of Bangladesh Haor and Wetlands Development defines wetlands as "areas of marsh, fen, peatland, or water,

Wetlands of Tropical and Subtropical Asia and Africa: Biodiversity, Livelihoods and Conservation,
First Edition. Edited by Thammineni Pullaiah.

whether natural or artificial, permanent or temporary, with water that is static or flowing, fresh, brackish or salt, including areas of marine water the depth of which at low tide does not exceed six meters."

Bangladesh hosts a plethora of wetlands (Figure 2.1), including freshwater marshes, swamps, floodplains, peatlands, riverine and coastal ecosystems, and the distinct northeastern haor regions (Islam 2010, 2021; PFC 2016). Each type of wetland showcases unique characteristics, ecological functions, and biodiversity. However, the wetlands of Bangladesh can be categorized into four broad categories, which are presented below.

2.2.1 Riverine Wetlands

The riverine wetland in Bangladesh constitutes the landward boundary of the Riverine System, encompassing the channel bank and both natural and artificial levees (PFC 2016). These areas undergo fluctuating water levels across seasons, providing habitats for various plant and animal species. Riverine wetlands in Bangladesh are predominantly found along the Ganges, Brahmaputra, and Meghna floodplains, critical ecological zones supporting high biodiversity and serving as sanctuaries for numerous species (Islam 2010). During the monsoon season, when rivers swell, these wetlands absorb excess water, mitigating the impact of floods, and serve as vital water sources for crops and local communities during dry seasons (PFC 2016). According to Bangladesh Water Development Board (BWDB), Bangladesh has 405 rivers (Bangladesh Water Development Board, 2020), including tributaries, with a combined length of approximately 24,000 km (World Bank, 2023). On the other hand, The National River Conservation Commission (NRCC) claimed that Bangladesh has 1008 rivers (National River Conservation Commission Bangladesh, 2023). These rivers are distributed in eight hydrological units of Bangladesh that is depicted in Figure 2.2. It is important to note that the boundaries of the hydrological divisions are marked by the alignment of major rivers of the country. Except south-eastern hilly region, the rivers of Bangladesh flow through low-gradient floodplains. The presence of trees, bushes, persistent emergent mosses, emergent lichens, or upland vegetation characterizes riverine wetlands in Bangladesh.

2.2.2 Floodplain Wetlands

Floodplain wetlands in Bangladesh are expansive areas primarily located along the Ganges, Brahmaputra, and Meghna River deltas and their associated tributaries (Rana et al. 2009; Rahman and Hickey 2019). These wetlands undergo seasonal inundation and form vast biological zones integral to the nation's landscape. During the wet monsoon months, such as the Haor (Figure 2.3), floodplain wetlands transform into extensive inland seas, providing essential breeding grounds and habitats for a diverse array of aquatic and semiaquatic species, including fish and migratory birds (Rahman and Hickey 2019; Tima et al. 2021; Nahiduzzaman et al. 2023; Sohel et al. 2023).

In the dry season, after the floods recede, floodplains like those along the Padma River become highly fertile and suitable for cultivating rice, jute, and other staple crops, benefiting from the nutrient-rich silt deposited (Rana et al. 2009; Islam 2010; Al-Amin et al. 2021). Floodplain wetlands are characterized by high biodiversity, crucial in filtering water and reducing pollution.

2.2.3 Coastal Wetlands

Coastal wetlands in Bangladesh encompass a vast stretch along the southern coastline, extending from the renowned Sundarbans in the southwest to the Teknaf Peninsula in the southeast, bordering Myanmar (Menéndez et al. 2020; Rahman 2022). The Sundarbans, recognized as the largest

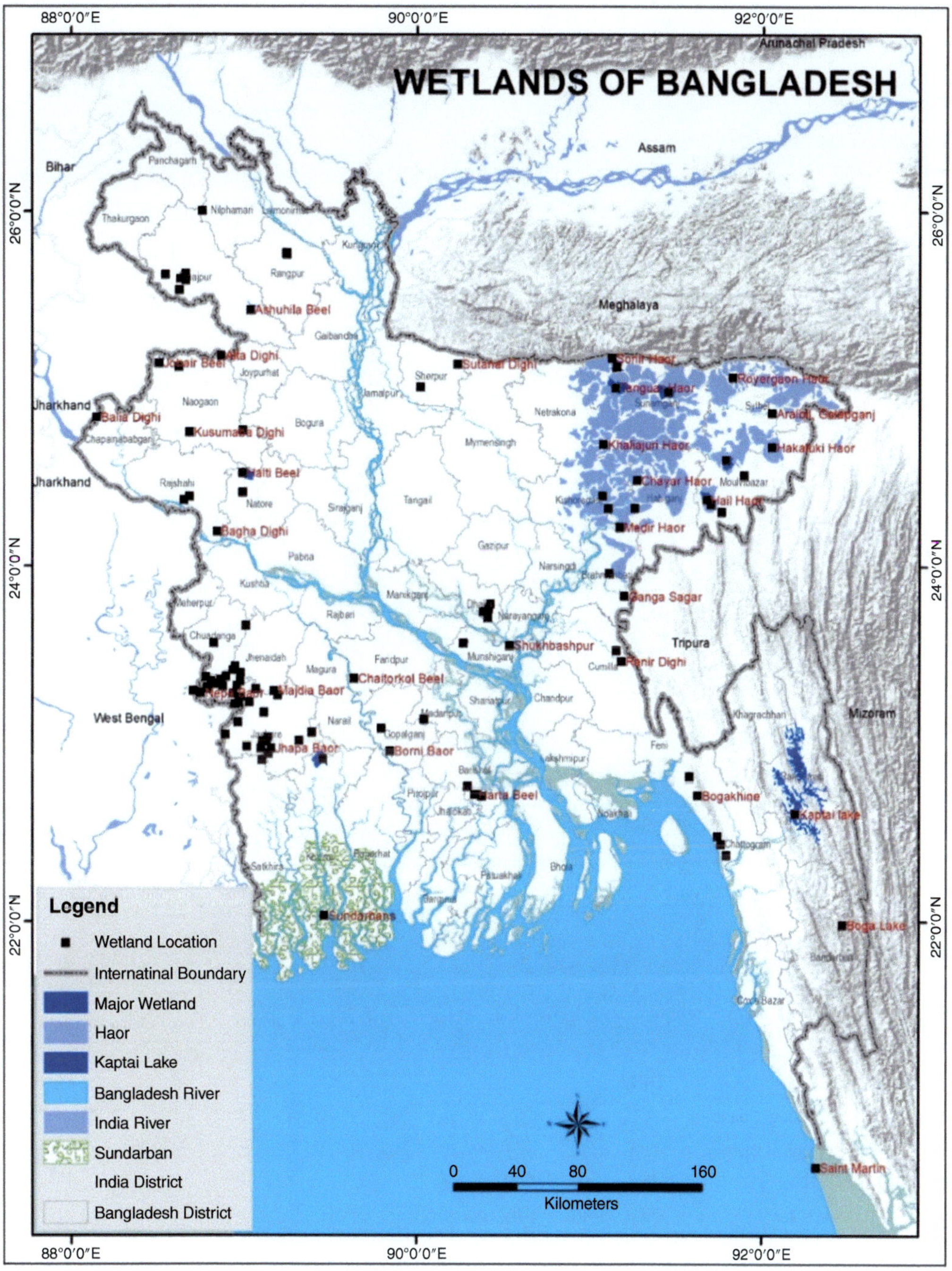

Figure 2.1 Wetlands distribution in Bangladesh.

mangrove forest globally and a UNESCO World Heritage site, along with the Cox's Bazar coastal strip boasting the lengthiest natural sandy coastline, are key features of these coastal wetlands. Comprising sandy beaches, mudflats, salt marshes, and mangrove forests, these wetlands harbor diverse ecosystems supporting a rich array of flora and wildlife.

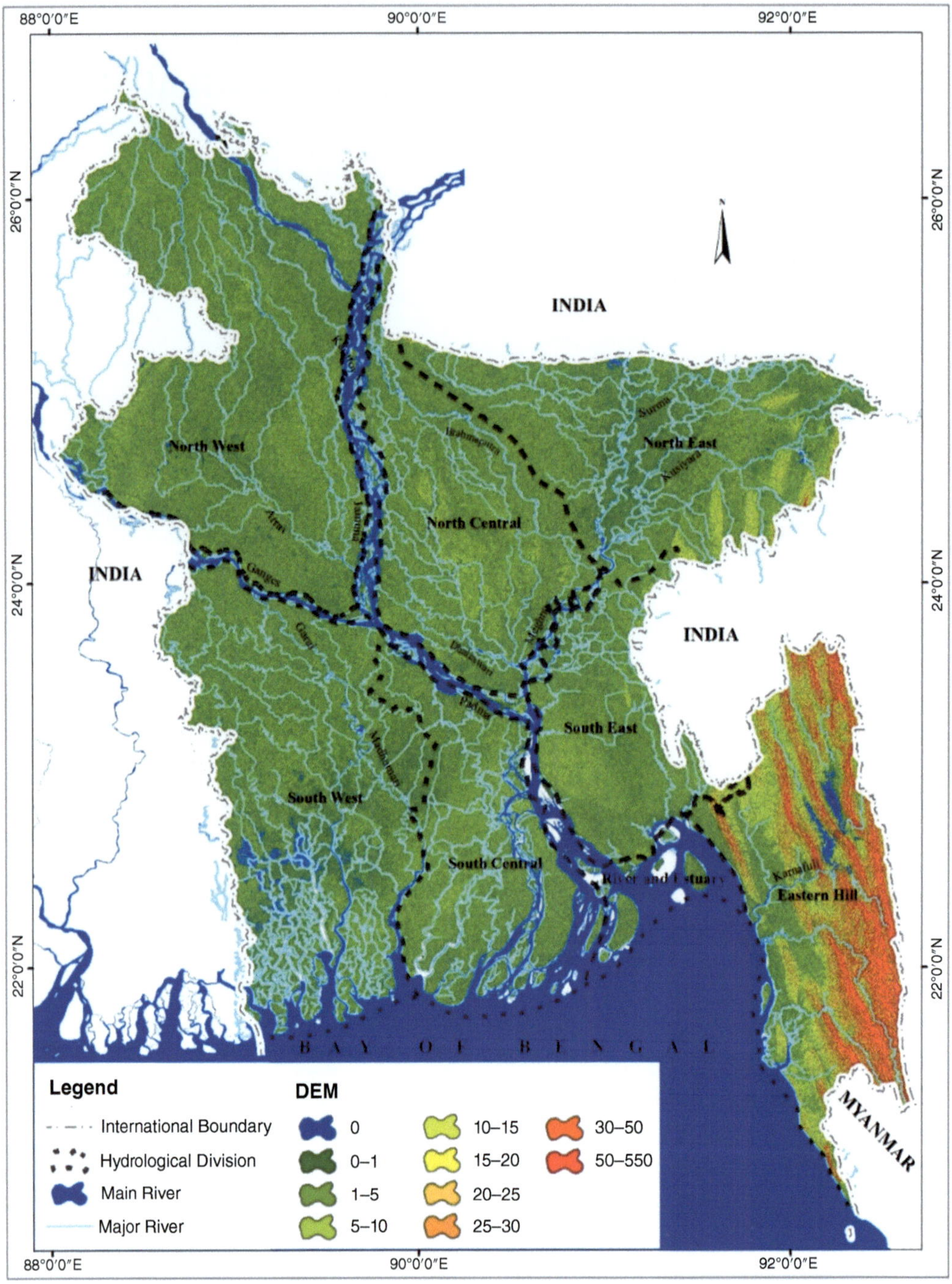

Figure 2.2 Eight hydrological divisions in Bangladesh with digital elevation model.

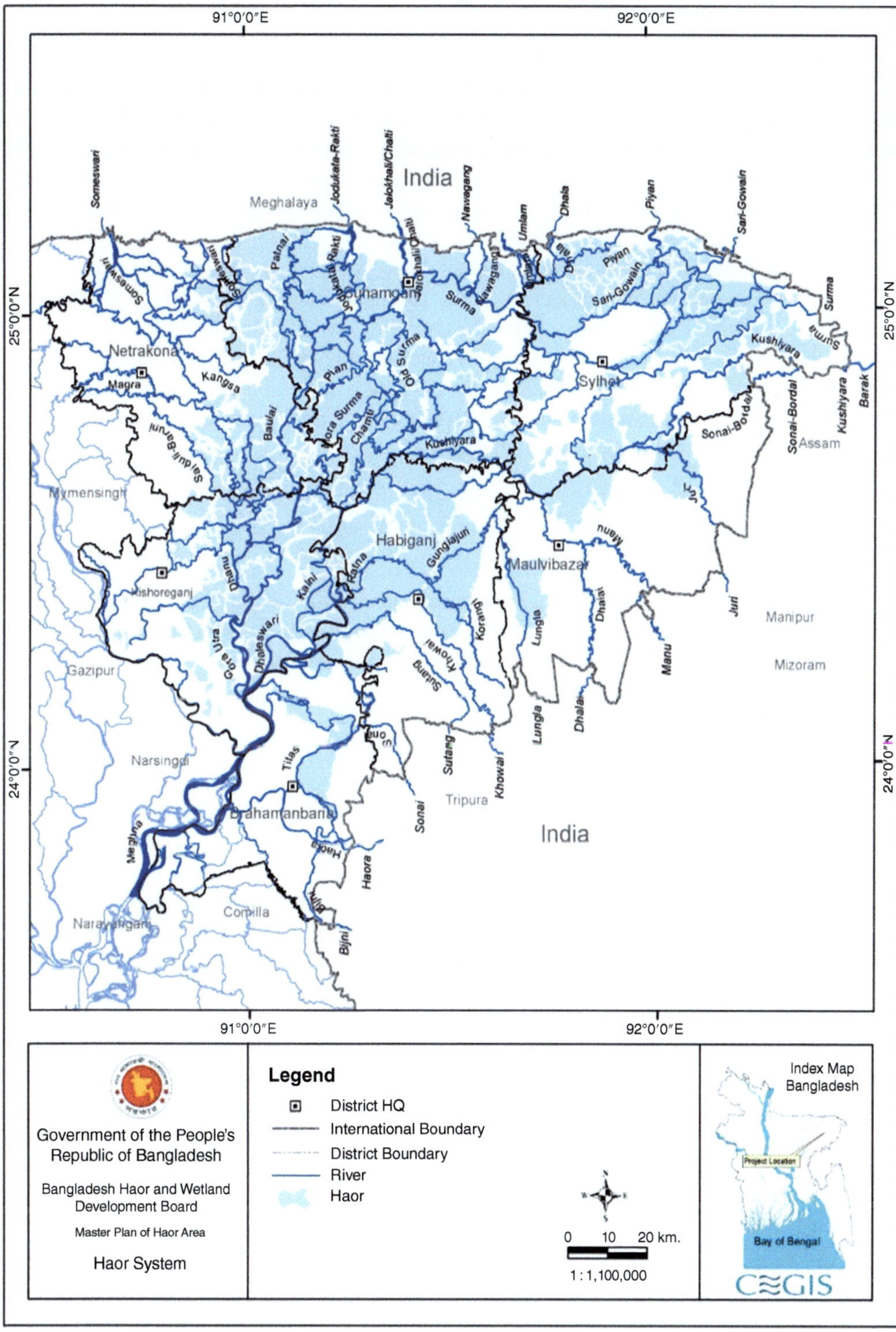

Figure 2.3 Haor distribution is in the northeastern part of Bangladesh. *Source:* CEGIS (2012)/Center for Environmental and Geographic Information Services.

2.2.4 Marshes and Swamps, Oxbow Lakes, and Beels

Marshes and swamps, typical wetland types in Bangladesh, are predominantly located in low-lying areas prone to seasonal floods. Notable examples include Tanguar Haor and Hakaluki Haor in the Sylhet basin and the Barind Tract in Rajshahi Division (CEGIS 2012). Marshes are characterized by grassy environments subject to seasonal or continuous flooding, while swamps are wooded areas often containing extensive freshwater swamp forests. These wetlands remain waterlogged for most of the year, supporting a diverse range of semiaquatic species (Rabbani et al. 2013).

Oxbow lakes and beels are prevalent in Bangladesh's low-lying floodplains, formed by meandering rivers in the central and northeastern regions. Examples include Arial Beel in Dhaka and Chalan Beel in Rajshahi Division, representing historic river meanders cutoff. Shallow water bodies within these wetlands host diverse ecosystems teeming with aquatic plants and animals. Beels (Figure 2.4), larger than oxbow lakes, such as Hakaluki Haor and Hail Haor, undergo significant size variations between the monsoon and dry seasons. Fluctuations in water levels impact local ecology and human activities (CEGIS 2012; PFC 2016).

2.3 Biodiversity

Bangladesh boasts a rich diversity of wetland flora and fauna, with the composition of species and ecosystems varying based on the type of wetland. Table 2.1 summarizes the typical flora and fauna found in different wetlands. The following section provides an overview of the biodiversity across various types of wetlands in Bangladesh.

2.3.1 Riverine Wetland Biodiversity

The flora of riverine wetlands in Bangladesh contributes significantly to the biodiversity of these ecosystems. These wetlands feature a rich assortment of aquatic and semiaquatic plants. Lotus (*Nelumbo nucifera*) and water lilies (*Nymphaea* spp.) (Figure 2.5) are prevalent, adding aesthetic appeal and critical habitats for various aquatic species. The riverbanks are lined with dense grasses like *Saccharum* spp. (Figure 2.5) and reeds like *Phragmites* spp., providing ideal bird nesting grounds and contributing to water purification by trapping pollutants (Rahman et al. 2018). Additionally, unique trees such as the shola (*Salmalia malabarica*) and indigenous species like *Barringtonia acutangula* contribute to the diversity, offering valuable resources for local communities (Khan 2007; Gopal and Chauhan 2006).

The fauna in Bangladesh's riverine wetlands is diverse and abundant. The Ganges River dolphin (*Platanista gangetica*) (Figure 2.6), an endangered species, navigates these waters using echolocation, indicating the ecosystem's health (IUCN 2015). These wetlands are also home to various fish species, including prawns (*Penaeus* spp.), catfish (*Pangasius* spp.), and the iconic Hilsa (*Tenualosa ilisha*) (Figure 2.6), all vital to local fisheries. Otters (*Lutrogale perspicillata*) thrive in the shallows, while reptiles like monitor lizards (*Varanus bengalensis*) and turtles (*Batagur baska*) play significant roles in the food chain (Gopal and Chauhan 2006; Rana et al. 2009). Additionally, these wetlands serve as crucial stopovers for migratory birds, with species arriving from Central and Northern Asia during their seasonal migrations (Islam 2008; Rahman et al. 2015). The diverse fauna in these riverine wetlands underscores the ecological significance of these habitats and highlights the need for ongoing conservation efforts.

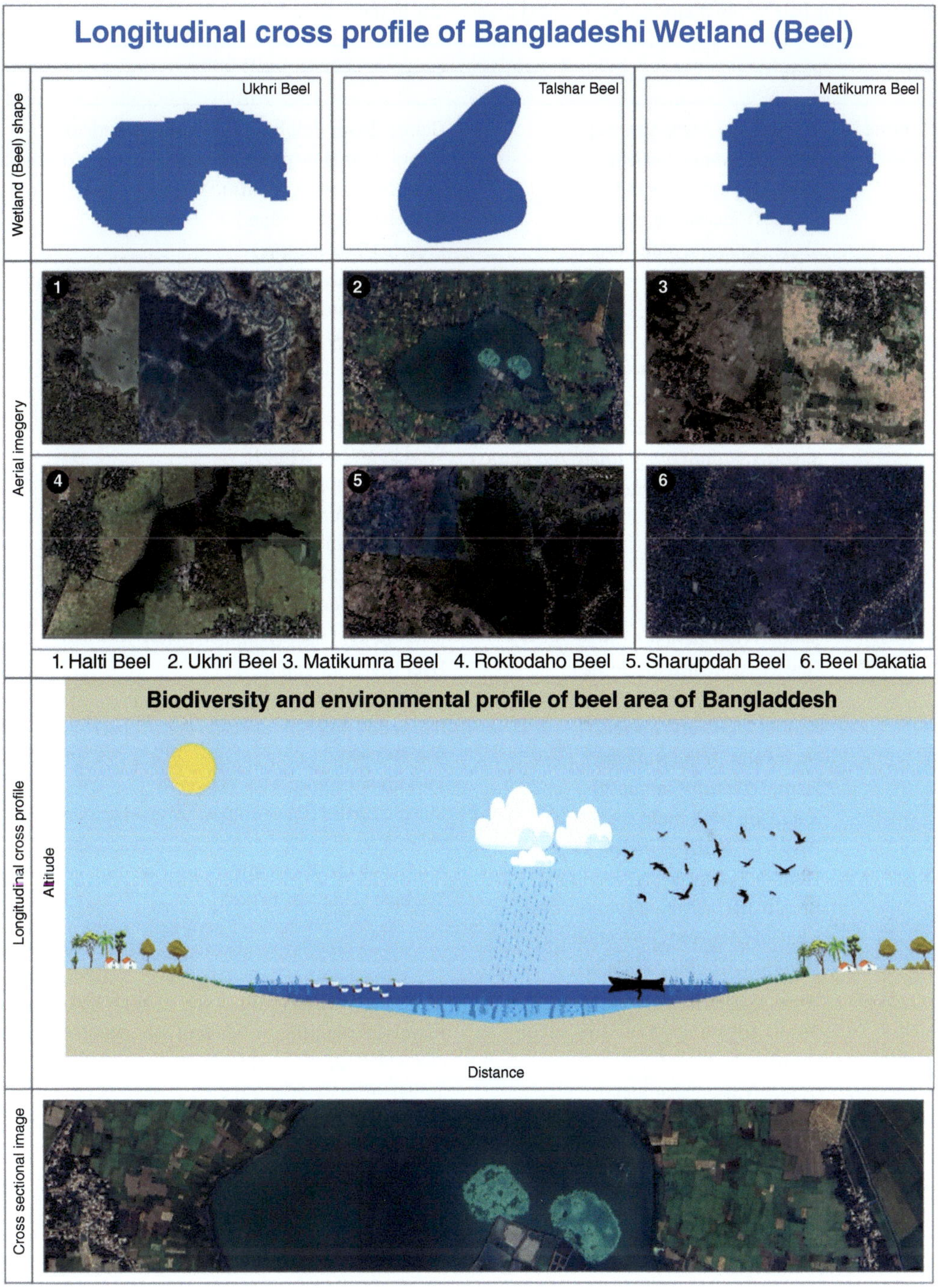

Figure 2.4 This figure provides a comprehensive overview of special wetland types in Bangladesh, specifically Beels. The figure outlines six Beels: Halti, Ukhri, Matikumra, Roktodaho, Sharupadah, and Beel Dakatia, accompanied by corresponding satellite images. Additionally, it includes a longitudinal cross-profile diagram illustrating the ecological dynamics of these wetland areas, including habitat, animal life, and human activity. The image captures the real landscape, emphasizing the rich biodiversity and environmental importance of these wetland regions.

Table 2.1 The lists of the major flora and fauna biodiversity in the wetlands.

Wetlands	Flora	Fauna
Riverine Wetland	Water lilies (*Nymphaea* spp.) Lotus (*Nelumbo nucifera*) Reeds (*Phragmites* spp.) Grasses (*Saccharum* spp.) Shola (*Salmalia malabarica*)	Ganges River dolphin (*Platanista gangetica*) Otters (*Lutrogale perspicillata*) Hilsa fish (*Tenualosa ilisha*) Catfish (*Pangasius* spp.) Prawns (*Penaeus* spp.) Turtles (*Batagur baska*) Monitor lizards (*Varanus bengalensis*) Migratory birds (Central and Northern Asia)
Floodplain Wetland	Sal trees (*Shorea robusta*) Grasses (Barti – *Panicum paludosum*) Reeds (Hogla – *Typha elephantina*) Shrubs (Satmuli – *Asparagus racemosus*) Herbs (False daisy – *Eclipta prostrata*)	Greater one-horned rhinoceros (*Rhinoceros unicornis*) Monkeys (*Cercopithecidae* spp.) Wild boar (*Sus scrofa*) Snakes (*Python molurus*) Lizards (*Varanus* spp.) Storks (*Ciconiidae* spp.) Egrets (*Ardeidae* spp.) Hornbills (*Bucerotidae* spp.)
Coastal Wetlands	Sundari (*Heritiera fomes*) Gewa (*Excoecaria agallocha*) Goran (*Ceriops decandra*) Keora (*Sonneratia apetala*) Nipa Palm (*Nypa fruticans*) Hantal (*Phoenix paludosa*) Kala Bain (*Avicennia* spp.) Passur (*Xylocarpus mekongensis*) Dhundul (*Xylocarpus granatum*) Bruguiera (*Bruguiera gymnorrhiza*) Khalshi (*Aegiceras corniculatum*) Hargoja (*Acanthus illcifolius*)	Royal Bengal tiger (*Panthera tigris*) Deer (*Cervidae* spp.) Fishing cat (*Prionailurus viverrinus*) Irrawaddy dolphin (*Orcaella brevirostris*) Various fish species Various shrimp species Marine turtles (Olive Ridley turtle – *Lepidochelys olivacea*) Mud Crab (*Scylla serrata*) Sea Bass (*Lates calcarifer*)
Marshes and Swamps	Water hyacinth (*Eichhornia crassipes*) Reeds(Nol-Khag – *Typha angustata*) Grasses Sedges (Cyperaceae spp.) Phoenix palm (*Phoenix paludosa*)	Amphibians (frogs – *Rana* spp., toads – *Bufo* spp.) Insects (Dragonflies – *Odonata* spp., Mosquitoes (*Culicidae* spp.) Waterfowl (*Anatidae* spp.) Herons (*Ardeidae* spp.)
Oxbow Lakes and Beels	Water lilies (*Nymphaea* spp.) Lotus (*Nelumbo nucifera*) Reeds (Nol-khag – Elephant grass) Grasses (Kusha grass – *Saccharum spontaneum*)	Various fish species Amphibians (frogs, toads) Water snakes (*Natrix* spp.) Ducks (*Anas* spp.) Geese (*Anser* spp.) Kingfishers (*Alcedinidae* spp.)

Figure 2.5 Water lilies and *Saccharum* sp. of the Riverine Wetland.

Figure 2.6 Ganges River dolphin and Hilsa of the Riverine Wetland. *Source:* Photos by M. A. Aziz and S. C. Das.

2.3.2 Floodplain Wetland Biodiversity

Floodplain wetlands in Bangladesh are known for their multilayered flora, featuring a diverse mix of plant life. The landscape blends aquatic and terrestrial environments, where towering Sal (*Shorea robusta*) trees are a prominent feature. These trees provide valuable timber and offer shade and cover for various bird species (Khan 2007). Beneath the canopy, a dense carpet of grasses and reeds creates grazing fields for herbivores and nesting sites for ground-dwelling birds. Shrubs and herbs contribute further to the biodiversity of the floodplains, providing food and habitat for insects, reptiles, and small mammals (Khan et al. 1994). This complex flora structure in the floodplain wetlands is essential for supporting the ecosystem's balance and promoting agricultural activities (Figure 2.7).

The fauna in Bangladesh's floodplain wetlands is equally rich and diverse particularly freshwater fish, birds, reptiles and amphibians. More than 260 fish species, such as hilsa (*Tenualosa ilisha*), rui (*Labeo rohita*) and catla (*Catla catla*), flourish in these regions, utilizing the seasonal floodwaters for reproduction and nourishment. Reptiles, including monitor lizards (*Varanus* spp.) and snakes (*Python molurus*), play critical roles as scavengers and predators, contributing to the ecological balance of the wetlands. These wetlands are crucial for both migratory and resident

Figure 2.7 Land diversity in Haor region in dry and rainy seasons.

avifauna, including lesser whistling ducks (*Dendrocygna javanica*), egrets (*Ardeidae* spp.), hornbills (*Bucerotidae* spp.), storks (*Ciconiidae* spp.) and herons (*Ardeidae*), which utilize them for foraging and nesting, particularly in winter (Islam 2008). These fauna elements, along with the rich biodiversity of the floodplains, underline the ecological significance of these wetlands and the importance of maintaining their natural habitats.

2.3.3 Coastal Wetland Biodiversity

Coastal wetlands in Bangladesh, extending from the Sundarbans in the southwest to the Teknaf Peninsula in the southeast, boast a diverse range of plant species (UNESCO 2017). The most prominent feature of these coastal regions is the mangrove forests, with iconic species like the Sundari tree (*Heritiera fomes*), the Gewa tree (*Excoecaria agallocha*), the Keora tree (*Sonneratia apetala*) (Figure 2.8). These mangroves form dense thickets that protect the coastline from erosion and storm surges and create critical habitats for aquatic life (Gopal and Chauhan 2006). The mangroves' salt tolerance allows them to thrive in coastal conditions. Beyond the mangroves, coastal wetlands include sandy beaches, mudflats, and salt marshes, all contributing to the area's unique biodiversity (Al-Amin et al. 2021). The coastal vegetation also plays a vital role in carbon sequestration, mitigating the impacts of climate change, and providing resources for local communities involved in fishing, salt production, and other coastal livelihoods (Rana et al. 2009).

The fauna of Bangladesh's coastal wetlands is equally rich and diverse. These ecosystems are home to the Royal Bengal tiger (*Panthera tigris tigris*), which has adapted to the challenging conditions of mangrove forests (Figure 2.9) (Vasudev et al. 2017). Coastal wetlands also support various aquatic species, including crabs, fish, and mollusks, essential for the local fishing industry (Figure 2.9).

Additionally, these wetlands serve as crucial stopovers for migratory birds, with species taking refuge in the mudflats and salt marshes during their seasonal journeys (Al-Amin et al. 2021). Other significant fauna in these coastal regions include the saltwater crocodile (*Crocodylus porosus*), spotted deer (*Axis axis*), and various species of monkeys (Figure 2.9), each contributing to the unique biodiversity of coastal wetlands (Gopal and Chauhan 2006). This rich fauna underscores the ecological importance of coastal wetlands and highlights the need for conservation efforts to protect these complex ecosystems.

Figure 2.8 (a) The picture represents the dense vegetation in Sundarbans. (b) Golpata tree. (c) Keora. (d) Hental. (e) Hargoja. (f) Bain.

Figure 2.9 (a) Royal Bengal tiger. *Source:* (a) Photo by Bangladesh Forest Department/Public domain. (b) Spotted deer. (c) Monkey. (d) Crocodile. (e) Mud Crab. (f) Sea Bass (Vetki).

2.3.4 Marshes and Swamps Biodiversity

Freshwater marshes and swamps in Bangladesh are rich in diverse plant life, often dominated by water hyacinth (*Eichhornia crassipes*). These wetlands feature a dense understory of reeds, grasses, and sedges, creating a unique pattern within the marshland. The occasional presence of palm trees, such as *Phoenix paludosa*, adds to the visual diversity of these ecosystems (Khan 2007). This lush plant life not only enhances the beauty of the marshes but also plays a crucial role in supporting the fauna and providing critical resources for local communities. The abundant vegetation in these areas also contributes to the overall ecological balance, supporting the food chain and providing nesting and breeding grounds for various species.

The fauna of Bangladesh's freshwater marshes and swamps is equally diverse. Amphibians such as frogs (*Rana* spp.) and toads (*Bufo* spp.) use these ecosystems as breeding grounds, with their melodic calls filling the air during the monsoon season. The marshes are a haven for insects, including mosquitoes (*Culicidae* spp.) and dragonflies (*Odonata* spp.), the latter playing a key role in controlling mosquito populations and supporting the food chain (Islam 2008). These wetlands are also frequented by various birds, such as waterfowl (*Anatidae* spp.) and herons (*Ardeidae* spp.), which feed on the abundant amphibian and insect life (Gopal and Chauhan 2006). In addition, freshwater marshes and swamps provide essential habitats for various other species, including fish and other small animals, supporting local ecosystems and livelihoods (Hossain et al. 2015; Nahiduzzaman et al. 2023).

2.3.5 Oxbow Lakes and Beels Biodiversity

In Bangladesh, oxbow lakes, which are created when a river's meanders are cut off from the main channel, and beels, which are shallow depressions within the floodplain, are essential freshwater ecosystems. Numerous fish species find a peaceful home in these placid bodies of water embellished with lotuses, reeds, grasses, and water lilies (Gopal and Chauhan 2006). Reptiles such as water snakes (*Natrix* spp.) slither among the vegetation, while amphibians such as frogs and toads seek shelter in the shallows. Amidst the gracefully gliding ducks and geese (*Anas* spp.) and kingfishers (*Alcedinidae* spp.) perched on overhanging branches, the latter plunge headfirst into the water to capture their prey. The biodiversity of Bangladesh's wetlands is greatly enhanced by these freshwater havens, despite their apparent small size. Table 2.1 lists the major biodiversity in the wetlands below.

2.4 Red Listed Species By IUCN

Out of numerous species assessment systems to estimate the vulnerability level of different species, the system proposed by IUCN is the most widely accepted one which has been carried out for more than 50 years and is categorized into 9–11 levels varying between global and national/regional assessments, but of them only three are marked as threatened categories: Critically Endangered (CR), Endangered (EN), and Vulnerable (VU) (IUCN 2015). IUCN has set a good stand working in Bangladesh since the late 1990s by categorizing the threatened species into five volumes: The Red List, as well as specific lists for amphibians and reptiles, birds, fish, and mammals (Table 2.2) and translating them into Bengali for growing mass awareness among the local people (IUCN 2015).

The rich variety of flora and fauna in Bangladesh's wetlands is protected by the IUCN's Red List, although some of these species are under threat. This is a summary of some important red-listed species that can be found in different types of wetland ecosystems:

2.5 Livelihoods of People Living in Wetland Areas

In the lower Bengal Delta region, water environments are central to the livelihoods of rural communities in Bangladesh. These riverine wetlands comprise a complex network of geomorphological, hydrological, and ecological factors that support various human activities (Al-Amin et al. 2021).

Table 2.2 Red listed species in Wetland of Bangladesh.

Wetlands	Red listed species by IUCN
Riverine Wetland	Ganges River dolphin (*Platanista gangetica*) (endangered)
Floodplain Wetland	Royal Bengal tiger (*Panthera tigris*) (endangered)
	Greater one-horned rhinoceros (*Rhinoceros unicornis*) (conservation dependent)
	Wild Water Buffalo (*Bubalis arnee*) (endangered)
	Pink-headed Duck (*Rhodonessa caryophyllacea*) (critically endangered)
	Bengal Florican (*Houbaropsis bengalensis*) (critically endangered)
Coastal Wetlands	Irrawaddy dolphin (*Orcaella brevirostris*) (endangered)
	Marine turtles (Olive Ridley turtle – *Lepidochelys olivacea*, etc.) (vulnerable to endangered depending on species)

Source: Adapted from IUCN (2015).

Fishing is a primary source of income and sustenance in riverine wetlands. Communities living along the banks of large rivers and river islands rely on open fisheries to meet their dietary needs and generate income through market sales. Women play a crucial role by repairing fishing gear and nets and sometimes joining fishing trips with their families. In some cases, women even fish alone, particularly in the southwestern mangrove forests, despite risks from crocodiles and tigers (Gopal and Chauhan 2006). Agriculture is another key livelihood in these areas. Farmers use local water sources for irrigation, leveraging the nutrient-rich sediment deposited during floods from the previous year. Due to this sediment, river islands often yield higher crop productivity than mainland fields (Ahmed and Haque 2023). Common crops include paddy, jute, vegetables, and fruits like melons. Women contribute significantly through postharvest crop processing, homestead gardening, and household-level livestock and poultry rearing (Sultana and Thompson 2007). Waterways in riverine wetlands serve as crucial transportation routes. They facilitate the movement of crops and goods, supporting local trade and broader economic activities (Hossain et al. 2015). This transportation network connects rural communities to markets and services, underpinning the local economy.

Floodplain wetlands in Bangladesh are vital to the livelihoods of many people, providing resources for various activities. Communities in these areas adopt a blend of water-based and non-water-based strategies to sustain themselves (Sarkar and Nasreen, 2023). Floodplain wetlands offer fertile land for cultivating various crops due to nutrient-rich sediment deposits from seasonal floods. Crops like paddy, jute, and various vegetables are commonly grown. Inhabitants use water from local sources such as rivers, oxbow lakes (Baors and Beels), man-made ponds, and dighis for irrigation (Byomkesh et al. 2009; Al-Amin et al. 2021). Besides crop cultivation, communities in the floodplains engage in livestock rearing and poultry farming. These activities rely heavily on water availability for feeding and maintenance, using resources from the surrounding wetlands. Floodplain wetlands support various fishing activities, with some communities depending on fishing as their primary source of income (Dey et al. 2021; Islam 2021). In addition to subsistence fishing, commercial fishing operations contribute to the local economy. Women often play a role in crafting and repairing fishing gear and nets (Gopal and Chauhan 2006). Waterways in floodplains are essential for transportation. They facilitate the movement of crops, goods, and people, supporting local and broader economic activities. This mode of transportation helps communities access markets and other services, contributing to their livelihoods. Floating cultivation has been a long-standing practice in some low-lying peat swamp areas. This technique allows communities to grow crops on floating beds, adapting to the changing water levels and providing a unique way to sustain themselves (Figure 2.10). Alongside traditional agriculture, homestead gardening is a common practice in floodplain wetlands, where small plots near homes are cultivated for personal use, enhancing nutritional support and food security for poor household (Sarkar and Kulkarni, 2022).

Coastal wetlands in Bangladesh are crucial for the livelihoods of many communities, providing a range of resources and opportunities for sustaining both subsistence and commercial activities (Figure 2.10). Coastal wetlands, including the coastal and offshore island areas of the Bay of Bengal, are central to fishing activities. Communities use small fishing gear to venture into nearby waters for subsistence-level fishing. Women play a significant role in these activities, helping craft and repair fishing gear and nets (Gopal and Chauhan 2006; Sarkar and Nasreen, 2023). Fishing in these areas can also take the form of organized commercial operations. These ventures often involve deep-sea fishing trips, with groups of fishermen going on extended expeditions. While the organizers generally reap the primary benefits, ordinary fishermen typically receive wages based on informal negotiations and market dynamics (Gopal and Chauhan 2006). Although primarily recognized for fishing, some coastal wetlands also support agriculture. The fertile soil and access to water resources allow for crop cultivation, particularly in areas where the land is relatively stable (Figure 2.10). Waterway transportation is another key livelihood in coastal wetlands, enabling the movement of goods and people. This mode of transportation supports local trade and connects coastal communities to broader economic networks (Hossain et al. 2015). Women in coastal wetlands contribute to household income and support through various activities. They assist with fishing-related tasks, such as repairing gear, and sometimes join fishing trips. Women may even fish alone in areas like the southwestern mangrove forests despite the risks involved (Gopal and Chauhan 2006). Additionally, women sell their catch in local markets to generate income for their families.

Communities in Bangladesh living in marshes and swamps, oxbow lakes, and beels wetlands areas have developed various livelihoods closely tied to their water environments (Figure 2.10). Inhabitants of low-lying peat swamp areas in southern Bangladesh often engage in floating cultivation, a long-standing practice used as a coping mechanism and means of livelihood. This technique involves growing crops on floating beds, allowing for agriculture even when water levels fluctuate (Al-Amin et al. 2021). Floating cultivation provides a sustainable method for crop production in these wetland environments, accommodating the seasonal changes in water levels. Oxbow lakes, known locally as Baors, and other types of beels (wetlands formed from abandoned river channels) are crucial sites for fishing activities. These bodies of water are rich in fish, providing communities with a source of protein and income (Byomkesh et al. 2009). Fishing in these areas varies from subsistence to commercial levels, with local markets benefiting from the catch. In these wetland regions, agriculture plays a significant role in local livelihoods. The nutrient-rich sediment deposited in the marshes, swamps, and oxbow lakes allows for successful crop cultivation. Farmers grow paddy, jute, vegetables, and fruits, often relying on water from local sources for irrigation. Livestock rearing and poultry farming are also common, with water resources supporting these activities (Al-Amin et al. 2021). Waterway transportation is a key component of livelihoods in these wetland areas. It provides a cost-effective means to transport crops and goods, facilitating local trade and supporting broader economic activities (Hossain et al. 2015). Women in these wetland regions play a crucial role in supporting household livelihoods. They are involved in postharvest crop processing, livestock and poultry rearing, and homestead gardening. Women also contribute to fishing-related tasks by repairing gear and crafting nets, demonstrating their versatility and resourcefulness in sustaining household economies (Sultana and Thompson 2007) (Figure 2.10).

Figure 2.10 The figure represents the livelihood of five types of wetland: Baor (a, b, c), Beel (d, e, f), Haor (g, h, i), Lake (j, k, l), Pond, and Dighi (m, n, o). The most common livelihoods in this area are fishing, cultivation of agricultural crops, and waterway transportation.

(m) (n) (o)

Figure 2.10 (Continued)

2.6 Conservation

Bangladesh's wetlands are degrading, and their total area is being shrinked due to human activity and natural factors. The impacts of climate change in Bangladesh have recently added significant stress to wetland ecosystems. However, these habitats face threats such as overfishing, pollution, and climate change-driven sea-level rise, necessitating conservation efforts to sustain their diverse biodiversity and ecological equilibrium for future generations (Chakraborty 2018).

Despite these challenges, Bangladesh's efforts in wetland conservation have not gone unnoticed globally. There is a growing international focus on responsible wetland management, leading to increased agreements for wetland conservation. Bangladesh, a proactive participant in these agreements, is crucial in preserving its wetlands (Byomkesh et al. 2009; Islam 2010; Al-Amin et al. 2021). The country's commitment to upholding these international agreements is evident through its National Environment Policy, which includes wetland protection measures and ratifying the Ramsar Convention. These actions are not just symbolic; they are tangible steps toward safeguarding the ecological value, diverse plant and animal species, and livelihoods dependent on wetlands (CEGIS 2012; PFC 2016).

Conservation efforts for wetlands in Bangladesh are crucial for preserving their diverse ecosystems, which play essential roles in biodiversity maintenance, flood regulation, groundwater replenishment, and providing economic opportunities for local communities. However, these habitats face significant threats from industrial pollution, inappropriate land use, and climate change, necessitating robust legal frameworks and international cooperation for their preservation and sustainable management. The Bangladesh government has enacted laws and regulations to protect wetland environments, such as the Bangladesh Environment Conservation Act (1995) and Rules (1997), which designate environmentally significant areas as "Ecologically Critical Areas" (ECAs), prohibiting harmful industrial activities and promoting sustainability. Additionally, Bangladesh is a signatory to the Ramsar Convention on Wetlands (1971), designating numerous wetland sites as Ramsar Sites and committing to their conservation through local and national strategies.

Legislation like the Wildlife (Preservation and Security) Act 2012 safeguards vital mangrove ecosystems, such as the Sundarbans Reserved Forest, against illegal logging and poaching. The National Forestry Policy of 1994 and the Forest Act of 1927 support the preservation of wooded wetlands, emphasizing participatory forest management and biodiversity conservation. Initiatives like the Community-Based Resource Management Project (CBRMP) engage local communities in sustainable wetland and fisheries management, implementing conservation strategies aligned with domestic and international legal standards. Awareness and education campaigns further reinforce these laws, educating citizens about legal protections and the importance of wetland ecosystems to promote environmental stewardship.

Wetlands in Bangladesh are safeguarded through citizen education, community engagement, international agreements, and national legislation. Despite growing environmental concerns, implementing integrated conservation strategies is imperative to ensure the preservation of these landscapes for future generations and uphold their ecological, economic, and social benefits.

2.7 Conclusion

In conclusion, wetlands are vital to Bangladesh's biodiversity, environment, and livelihood. Significant coastal and interior regions cover about half of the country's land area, essential in maintaining ecological balance and supporting local communities. However, escalating population growth and development projects have exerted immense pressure on these ecosystems, leading to the nationwide degradation and loss of wetlands. Projects like irrigation and shrimp farming have already destroyed millions of hectares of wetlands, threatening the sustainability of these vital habitats.

Nevertheless, Bangladesh has taken proactive steps toward wetland conservation, both nationally and internationally. The country's commitment to international agreements like the Ramsar Convention and the implementation of domestic laws and regulations demonstrate its dedication to preserving these invaluable ecosystems. Efforts such as the Wildlife Act and CBRMP engage local communities in conservation activities, fostering a sense of stewardship and responsibility toward wetland protection.

Despite the challenges, Bangladesh's wetlands continue to provide essential services to local communities and the broader economy. From flood regulation to supporting agriculture and fisheries, these ecosystems play multifaceted roles in sustaining livelihoods and ensuring environmental sustainability. Understanding Bangladesh's diverse types of wetlands and their unique characteristics is crucial for effective conservation efforts.

Moving forward, it is imperative to prioritize integrated conservation strategies that address Bangladesh's wetlands' complex challenges. Citizen education, community engagement, and international collaboration are essential pillars of this endeavor. By safeguarding these ecosystems, Bangladesh can secure a sustainable future for its people while preserving its wetlands' ecological integrity and resilience for generations to come.

References

Ahmed, S. and Haque, C.E. (2023). Wetland entitlement and sustainable livelihood of local users in Bangladesh: a case study of Hakaluki Haor Communities. *Environmental Management* 71 (2): 334–349. https://doi.org/10.1007/s00267-022-01741-5.

Al-Amin, A.K.M.A., Alam, K., Shrestha, U.B. et al. (2021). Ecosystems, livelihood assets and willingness to pay for wetland conservation in Bangladesh. *Environment, Development and Sustainability* 23 (12): 17503–17534. https://doi.org/10.1007/s10668-021-01398-w.

Alam, M.Z., Baki Bhuiyan, M.A., Abdullah, H.M. et al. (2021). Changes of land use and land cover with the diversity of fishes, aquatic plants, and bird's species at wetland ecosystem. *The Scientific World Journal* 2021: 7533119. https://doi.org/10.1155/2021/7533119.

Bangladesh Environment Conservation Act. (1995). Act No. 1 of 1995. Dhaka, Bangladesh: Government of Bangladesh. Retrieved from Dhaka, Bangladesh: http://bdlaws.minlaw.gov.bd/act-791.html.

Bangladesh Environment Conservation Rules (1997). Statutory Rules and Orders No. 1 of 1997. Dhaka, Bangladesh: Government of Bangladesh. Retrieved from Dhaka, Bangladesh: https://faolex.fao.org/docs/pdf/bgd19918.pdf (accessed 9 April 2024).

Bangladesh Water Development Board. (2020). Monitoring the performance of completed coastal embankment projects through remote sensing and GIS techniques [PDF]. Bangladesh Water Development Board. http://www.hydrology.bwdb.gov.bd/img_upload/ongoing_project/757.pdf (accessed 9 April 2024).

BBS (2022). Population & Housing Census 2022 Preliminary Report. Retrieved from Dhaka, Bangladesh: https://sid.portal.gov.bd/sites/default/files/files/sid.portal.gov.bd/publications/01ad1ffe_cfef_4811_af97_594b6c64d7c3/PHC_Preliminary_Report_(English)_August_2022.pdf (accessed 9 April 2024).

Byomkesh, T., Nakagoshi, N., and Shahedur, M.R. (2009). State and management of wetlands in Bangladesh. *Landscape and Ecological Engineering* 5 (1): 81–90. https://doi.org/10.1007/s11355-008-0052-5.

CEGIS (2012). Master Plan of Haor Area. Retrieved from Dhaka, Bangladesh: https://dbhwd.portal.gov.bd/sites/default/files/files/dbhwd.portal.gov.bd/publications/baf5341d_f248_4e19_8e6d_e7ab44f7ab65/Haor%20Master%20Plan%20Volume%201.pdf (accessed 10 April 2024).

Chakraborty, T. R. (2018, May 11–13). Threat of climate change to nature conservation in Haor Basin of Bangladesh. Paper presented at the International Lake Conference, Pokhara, Nepal.

Dey, N. C., Parvez, M., & Islam, M. R. (2021). A study on the impact of the 2017 early monsoon flash flood: Potential measures to safeguard livelihoods from extreme climate events in the haor area of Bangladesh. *International Journal of Disaster Risk Reduction*, 59. doi:https://doi.org/10.1016/j.ijdrr.2021.102247.

Gopal, B. and Chauhan, M. (2006). Biodiversity and its conservation in the Sundarban Mangrove Ecosystem. *Aquatic Sciences* 68 (3): 338–354. https://doi.org/10.1007/s00027-006-0868-8.

Hossain, M.S., Hein, L., Rip, F.I., and Dearing, J.A. (2015). Integrating ecosystem services and climate change responses in coastal wetlands development plans for Bangladesh. *Mitigation and Adaptation Strategies for Global Change* 20 (2): 241–261. https://doi.org/10.1007/s11027-013-9489-4.

Islam, N. (2021). State of Wetlands in Bangladesh, Opinion. The Daily Sun. Retrieved from https://www.daily-sun.com/printversion/details/530098.

Islam, S.N. (2010). Threatened wetlands and ecologically sensitive ecosystems management in Bangladesh. *Frontiers of Earth Science in China* 4 (4): 438–448. https://doi.org/10.1007/s11707-010-0127-0. Accessed 9 April 2024.

Islam, S. (2008). Biodiversity of Bangladesh: 現状と課題 (Genzai to kadai). *Journal of Asian Ecology* 21 (1): 1–8. (in Japanese with English summary).

IUCN (2015). Red List of Bangladesh Volume 1: Summary. Retrieved from Dhaka, Bangladesh: https://portals.iucn.org/library/sites/library/files/documents/RL-549.3-003-v.1.pdf (accessed 10 April 2024).

Khan, M.S. (2007). *Flora of Bangladesh*, 51. Asiatic Society of Bangladesh.

Khan, M.S., Huq, E., Rahman, A.A. et al. (1994). *Wetlands of Bangladesh*. Dhaka, Bangladesh: Bangladesh Centre for Advanced Studies (BCAS).

Menéndez, P., Losada, I.J., Torres-Ortega, S. et al. (2020). The global flood protection benefits of mangroves. *Scientific Reports* 10 (1): 4404. https://doi.org/10.1038/s41598-020-61136-6.

Nahiduzzaman, M., Karim, E., Hossen, M. et al. (2023). Diversity and seasonal variation of fish assemblages of Dingapota Haor an eutrophic wetland of northeastern Bangladesh. *Jordan Journal of Biological Sciences* 16: 403–412. https://doi.org/10.54319/jjbs/160303.

National River Conservation Commission Bangladesh. (2023). *Definition and number of Bangladeshi rivers*. National River Conservation Commission Bangladesh. https://nrccb.portal.gov.bd/site/page/df835c45-cbd8-4397-8173-b015908d1d0a (accessed 09 April 2024).

PFC (2016). Classification of Wetlands of Bangladesh, Dhaka, Bangladesh:

Rabbani, G., Rahman, S.H., and Faulkner, L. (2013). Impacts of climatic hazards on the small wetland ecosystems (ponds): evidence from some selected areas of coastal Bangladesh. *Sustainability* 5 (4): 1510–1521. Retrieved from https://www.mdpi.com/2071-1050/5/4/1510.

Rahman, H.M.T. and Hickey, G.M. (2019). Assessing institutional responses to climate change impacts in the north-eastern floodplains of Bangladesh. *Environmental Management* 63 (5): 596–614. https://doi.org/10.1007/s00267-019-01155-w.

Rahman, M.M., Jiang, Y., and Irvine, K. (2018). Assessing wetland services for improved development decision-making: a case study of mangroves in coastal Bangladesh. *Wetlands Ecology and Management* 26 (4): 563–580. https://doi.org/10.1007/s11273-018-9592-0.

Rahman, H.M.T., Hickey, G.M., and Sarker, S.K. (2015). Examining the role of social capital in community collective action for sustainable wetland fisheries in Bangladesh. *Wetlands* 35 (3): 487–499. https://doi.org/10.1007/s13157-015-0635-5.

Rahman, M.M. (2022). Is co-management a double-edged sword in the protected areas of Sundarbans mangrove? *Biology & Philosophy* 37 (1): 4. https://doi.org/10.1007/s10539-022-09836-3.

Ramsar Convention Secretariat (2013). *The Ramsar Convention manual: A guide to the Convention on Wetlands (Ramsar, Iran, 1971)*, 6e. Gland, Switzerland: Ramsar Convention Secretariat https://www.ramsar.org/sites/default/files/documents/library/manual6-2013-e.pdf. (accessed 10 April 2024).

Ramsar Convention on Wetlands (1971). *Convention on Wetlands of International Importance especially as Waterfowl Habitat. Ramsar, Iran*. Ramsar Convention Secretariat https://www.ramsar.org/sites/default/files/documents/library/scan_certified_e.pdf. (accessed 10 April 2024).

Rana, M., Chowdhury, M., Sohel, M. et al. (2009). Status and socio-economic significance of wetland in the tropics: a study from Bangladesh. [Status and socio-economic significance of wetland in the tropics: a study from Bangladesh]. *iForest – Biogeosciences and Forestry* 2 (5): 172–177. https://doi.org/10.3832/ifor0512-002.

Ratner, B.D., Cohen, P., Barman, B. et al. (2013). Governance of aquatic agricultural systems analyzing representation, power, and accountability. *Ecology and Society* 18 (4): Retrieved from http://www.jstor.org/stable/26269448.

Sarkar, M. and Nasreen, M. (2023). Water insecurity and poverty: changing livelihood in coastal Bangladesh. In: *Coastal Disaster Risk Management in Bangladesh*, 75–104. Routledge https://doi.org/10.4324/9781003253495-7.

Sarkar, M. and Kulkarni, K.G. (2022). Pandemic and global food insecurity: experience from developing countries. *International Journal of Management* 11 (1): 56–88. https://doi.org/10.15410/aijm/2022/v11i1/167674.

Sharmin, A., Hossain, M., and Mollick, A.S. (2021). Farmers' perceptions and attitudes toward aquasilviculture in the periphery of the Sundarbans forest of Bangladesh. *Small-Scale Forestry* 20 (3): 391–405. https://doi.org/10.1007/s11842-021-09473-w.

Sohel, M.S.I., Islam, H.M.N., Ullah, M.A. et al. (2023). Ecological and economic significance of swamp vegetation nursery for successful reforestation program: an insight from Bangladesh. *Geology, Ecology, and Landscapes* 1–17. https://doi.org/10.1080/24749508.2023.2256546.

Sultana, P. and Thompson, P.M. (2007). Community based fisheries management and fisher livelihoods: Bangladesh case studies. *Human Ecology* 35 (5): 527–546. https://doi.org/10.1007/s10745-006-9092-3.

Tima, T.A., Schneider, P., Chanda, S.K. et al. (2021). Analyses implementation realities of legal frameworks for sustainable management of Tanguar Haor fisheries resources in Bangladesh. *Sustainability* 13 (16): https://doi.org/10.3390/su13168784.

UNESCO (2017). The Sundarbans. https://whc.unesco.org/en/soc/3563/ (accessed 10 April 2024).

Vasudev, D., Nichols, J.D., Ramakrishnan, U. et al. (2017). Assessing landscape connectivity for tigers and prey species: concepts and practice. In: *Methods for Monitoring Tiger and Prey Populations* (ed. K.U. Karanth and J.D. Nichols), 255–288. Singapore: Springer Singapore.

World Bank (2023). *Bangladesh - First Bangladesh Regional Waterway Transport Project: Improving domestic and regional connectivity through inland waterways (Project Profile).* World Bank Group https://documents.worldbank.org/curated/en/099554101032435615/IDU1d759f0841172c142191b2bb1189ca1d56d5d. (accessed 09 April 2024).

3

Wetlands of Myanmar: Biodiversity, Livelihoods, and Conservation

Thiri Dae We Aung[1], Thet Zaw Naing[2], Ding Li Yong[3], and Kelvin S.-H. Peh[4]

[1] *Biodiversity and Nature Conservation Association, Yangon, Myanmar*
[2] *Myanmar Bird and Nature Society, Yangon, Myanmar*
[3] *BirdLife International (Asia), Tanglin International Centre, Singapore, Singapore*
[4] *School of Biological Sciences, University of Southampton, Southampton, UK*

3.1 Introduction

Wetlands are a crucial part of our natural world, covering about 6% of the Earth's land surface. These ecosystems are dynamic, continually evolving due to both natural forces and human influences. They play a pivotal role in supporting a diverse range of important biodiversity, maintaining key ecological roles while providing benefits to human communities. Their conservation is vital for preserving biodiversity and sustaining productivity. This need calls for wise use of their resources, which is based on an ecosystem approach to wetland management (Ramsar Convention Secretariat 2010, 2012; Bishnu et al. 2014). Historically and today, numerous policies, regulations, laws, and initiatives have been developed with the objective of preventing the loss of wetlands and fulfilling the commitments to their sustainable use. It is vital to understand the current situation, status and trends regarding the extent and health of these wetlands. Understanding the factors driving changes in wetlands and assessing whether responses to these changes are appropriate are essential to guide long-term management and conservation activities. This comprehensive understanding is necessary to ensure future generations can also experience the benefits that wetlands provide (Gardner and Davidson 2011).

3.2 Wetlands in Myanmar

Wetland ecosystems in Myanmar encompass a wide and diverse range of environments, often covered in water for significant parts of the year (Scott 1989; Davies et al. 2004; Nature Wildlife Conservation Division 2019). These diverse and vital ecosystems can be found in various forms (Davies et al. 2004), from expansive tropical mangrove forests across Myanmar's coastline to seasonal ponds and marshes, floodplain lakes (including oxbows), permanent riparian swamps (riverbanks), and shallow freshwater lakes on the floodplains of Myanmar's major rivers, not to mention the marshes of large reservoirs. In hypersaline environments, there are wetlands such as salt lakes, brackish lagoons, estuaries, coastal salt marshes, and even extensive seagrass beds along coastlines

Wetlands of Tropical and Subtropical Asia and Africa: Biodiversity, Livelihoods and Conservation,
First Edition. Edited by Thammineni Pullaiah.

and coral reefs (Davies et al. 2004). Each of these habitats is unique, characterized by distinct water sources and flows of water (hydrology), water quality, soil types, and a rich array of plant and animal biota. According to the Ramsar classification under the Convention on Wetlands of International Importance (thereafter as the 'Ramsar Convention') (Ramsar Convention Secretariat 2019), the wetlands in Myanmar can be classified into three main types: inland, marine/coastal, and human-made wetlands. Each type plays a distinct role in the country's ecology and supports various forms of life.

3.2.1 Inland Wetlands

The inland freshwater wetlands of Myanmar are predominantly associated with two major river systems: the Ayeyarwaddy and Chindwin Rivers and the Sittaung River. These rivers, among the largest in mainland Southeast Asia (Tordoff et al. 2007), and their tributaries hold a rich diversity of wetland types, including rivers and streams, floodplain lakes, non-forested and forested peatlands, marshes, swamps, forested wetlands, and groundwater-dependent wetlands.

3.2.2 Marine/Coastal Wetlands

Marine and coastal wetlands are widely distributed across Myanmar's long coastline, stretching about 2300 km, and comprising three distinct geographical regions: the Rakhine Coastal Region, the Ayeyarwaddy Delta and Gulf of Mottama, and the Tanintharyi Coastal Region, the latter extending to the southern border with Thailand along the Pakchan (or Kraburi) River. These coastal landscapes host various wetland types such as intertidal flats or mudflats, some of the most extensive at the Gulf of Mottama spanning Yangon Region and Mon state, saltmarshes, mangroves, seagrass beds, coral reefs, lagoons, shallow subtidal systems, sand dunes and beaches, rocky shores, and shellfish reefs. These ecosystems are rich in biodiversity and offer critical habitats for marine species and many migratory waterbirds (Tordoff et al. 2007; Zöckler et al. 2014).

3.2.3 Human-Made Wetlands

Human-made wetlands in Myanmar include various human-modified environments such as ponds, reservoirs, rice paddies, constructed or wastewater wetlands, salt pans, and aquaculture ponds. These artificial wetlands are essential for agricultural activities, the provision of water-related ecosystem services (e.g. water for drinking, domestic use, and irrigation; flood mitigation), and the conservation of biodiversity. Some of the most extensive human-made wetlands in Myanmar are rice fields, which cover large areas of the Ayeyarwady Delta, and along the floodplains of the Ayeyarwady. Possibly among the best-known examples of a man-made wetland is the Moeyungyi Wetland Wildlife Sanctuary – a reservoir – not far from Yangon.

3.3 Important Wetland Sites in Myanmar

An inventory of wetlands for Myanmar produced by Davies et al. (2004) gives a comprehensive list of about 100 wetland sites found in the country. To date, only eight of them are recognized as the East Asian-Australasian Flyway Partnership (EAAFP) Flyway network sites. These same wetland sites may also designated as Wetlands of International Importance (or Ramsar sites), except Paleik Lake, under the Convention on Wetlands. Four of these wetland sites are also gazetted as protected areas (PAs).

Here, we present 13 important wetland sites – some of which are PAs, Ramsar sites, and/or EAAFP Flyway network sites – that are managed through a multistakeholder approach, which includes Myanmar's central government, local authorities, local communities, universities, and conservation organizations (Figure 3.1). These wetlands are Key Biodiversity Areas (KBAs) and Important Bird and Biodiversity Areas (IBAs), recognized for their significance to globally important congregations of migratory waterbirds under internationally recognized criteria as well as for their representation of wetland ecosystems under wise use (Ramsar Convention Secretariat 2010). Active participation and traditional knowledge of indigenous people and local communities are pivotal for effective conservation and sustainable use of these wetland resources (Chatterjee et al. 2008).

Since 2018, there has been an increased emphasis on involving and empowering local communities in wetland conservation efforts in government policy. Designations such as Wetlands of International Importance and the EAAFP Flyway network sites have been instrumental in highlighting areas of ecological significance and promoting their conservation. Active participation of local communities has led to the designation of various protected areas, Ramsar sites and EAAFP network sites, reflecting a growing interest and commitment to preserving Myanmar's wetland ecosystems. This collaborative approach between various stakeholders is key to ensuring the effective conservation and sustainable use of wetland resources.

Wetlands are not only of ecological significance but also are crucial for the livelihood of local communities. Apart from the EAAFP Flyway network sites and Ramsar sites, which are internationally recognized for their importance to biodiversity, there are several wetland areas directly managed by local people. These areas highlight the intricate relationship between Myanmar's diverse wetlands and the communities that depend on them.

3.3.1 Ayeyarwaddy River – Bagan Section

The Bagan section of the Ayeyarwaddy River, located in the Mandalay Region, stretches approximately 45 km from Yae Thu Gyi Kyun to Pyaw Bwe Village. This segment, covering an area of 34,200 ha in Nyaung-U, is marked by a diverse array of geological and hydrological features that contribute to its ecological richness. Characterized by its braided river channels and extensive banks composed of sand, silt, and clay, this section of the Ayeyarwaddy River exhibits dynamic water flow patterns, particularly evident during the dry season. Several channels may cut off entirely from the main Ayeyarwaddy main channel or shrink, affecting the connectivity of various parts of the river system. The flat banks and the adjacent western ridge, which runs parallel downstream of Nyaung-U before veering west, contribute to the area's unique topography. The Bagan Plain, located to the east, features elevated bluffs that overlook the river, enhancing the scenic and ecological value of the region.

The Ayeyarwaddy River Bagan section is a critical habitat for various species. At least 110 species of birds have been recorded (Naing 2009; Zöckler 2019; Zöckler et al. 2020). The area is particularly significant for its support of globally threatened species, such as the Critically Endangered Yellow-breasted Bunting (*Emberiza aureola*, Figure 3.2a) and the Endangered Black-bellied Tern (*Sterna acuticauda*, Figure 3.2b). Vulnerable species such as the Indian Skimmer (*Rynchops albicollis*) and the Greater Spotted Eagle (*Clanga clanga*) are also known to occur here, alongside Near Threatened species including the Laggar Falcon (*Falco jugger*), Eurasian Curlew (*Numenius arquata*), possibly important populations of River Lapwing (*Vanellus duvaucelii*), Great Thick-knee (*Esacus recurvirostris*), and River Tern (*S. aurantia*). Additionally, the Endangered Irrawaddy Dolphin (*Orcaella brevirostris*, Figure 3.3) is a notable mammal that inhabits this section of the river, highlighting the area's importance as a biodiversity hotspot.

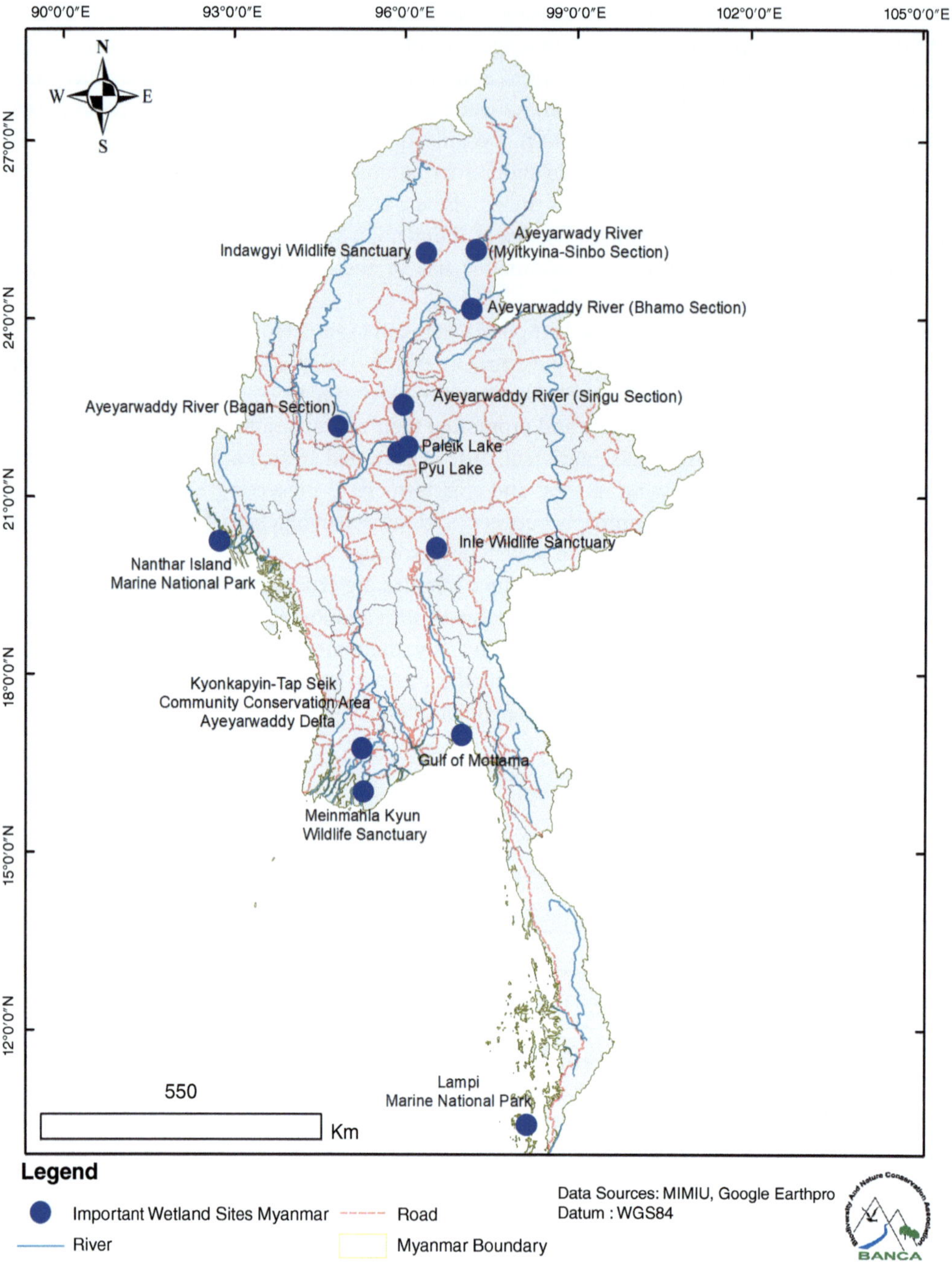

Figure 3.1 Location of the 13 important wetlands in Myanmar. *Source:* MIMU, Google Earthpro.

Despite being recognized as a KBA for its ecological importance, the Bagan section of the Ayeyarwaddy River is not currently recognized under major international conservation frameworks such as the Ramsar Convention nor as an EAAFP Flyway Network site. However, the presence of the critically endangered and vulnerable species emphasizes the urgent need for protective

Figure 3.2 Examples of globally threatened birds that occur in the wetlands of Myanmar: (a) Yellow-breasted Bunting. Critically endangered. *Source:* Photo by Thet Zaw Naing (chapter author). (b) Black-bellied Tern. Endangered. *Source:* Photo by Thet Zaw Naing (chapter author). (c) Baer's Pochard. Critically endangered. *Source:* Photo by Min Thiha Zaw. (d) White-bellied Heron. Critically endangered. *Source:* Photo by Thet Zaw Naing (chapter author). (e) Lesser adjutant. Vulnerable. *Source:* Photo by Thet Zaw Naing (chapter author). (f) Sarus Crane. Vulnerable. *Source:* Photo by Thet Zaw Naing (chapter author).

measures and enhanced conservation status to safeguard the diverse ecosystems and species that depend on this river section.

The river supports the livelihoods of communities in three towns – Nyaung-U, Bagan, and Myit Chay – and 41 villages along its banks. Most residents are engaged in agriculture, utilizing the drying sandbanks to grow crops like beans, maize, and onions. Fishing and other river-based activities

Figure 3.3 The endangered Irrawaddy Dolphin, *Orcaella brevirostris*. *Source:* Photo by Thiri Dae We Aung (chapter author).

also play a significant role in the local economy, tying the community closely to the health and stability of the river ecosystem.

This section faces several ecological threats, including habitat disruption from electric fishing, conversion of riverbanks and islands to agricultural land, and bird poaching. These activities not only threaten the local wildlife but also disrupt the ecological balance of the river system, posing long-term risks to both biodiversity and community livelihoods. Addressing the ecological and environmental challenges in the Ayeyarwaddy River Bagan Section requires a coordinated approach involving local communities, government agencies, and international conservation organizations. By promoting sustainable practices, enforcing environmental regulations, and potentially seeking international conservation designations, stakeholders can work together to ensure the preservation of this ecologically vital river section for future generations.

3.3.2 Ayeyarwaddy River – Bhamo Section

The Bhamo section of the Ayeyarwaddy River, situated in Kachin State, Myanmar, spans approximately 51 km from Thapanbin village to Sin Khan. This section, enveloping an area of 10,400 ha, is strategically positioned downstream of the first Ayeyarwaddy gorge, creating a unique ecological landscape between the first and second gorges. This stretch of the river, characterized by a major bend that shifts from south to west, is noted for its expansive width, reaching up to 3 km at certain points. The river's dynamics slow down here, forming an outwash plain where sediments from the faster-moving gorge stages are deposited. This deposition process results in extensive sandbanks, particularly around Bhamo, and creates a complex network of channels, including seasonally inundated sandbanks and permanent oxbow lakes. Additionally, there is a large marshy area in the river's inner bend, which is indicative of the typical floodplain stage of a river with significant ecological importance.

The Bhamo section is a KBA, supporting a rich biodiversity, with at least 100 species of birds (Tordoff et al. 2007; Naing 2012, 2013, 2015; Zöckler 2019). Notably, this area provides habitat for four globally threatened bird species, including three Critically Endangered species – the Baer's

Pochard (*Aythya baeri*, Figure 3.2c), White-bellied Heron (*Ardea insignis*, Figure 3.2d), and White-rumped Vulture (*Gyps bengalensis*) – and the Endangered Black-bellied Tern. Other Near Threatened species like the Ferruginous Pochard (*A. nyroca*), Asian Woolly-necked Stork (*Ciconia episcopus*), Great Thick-knee, River Lapwing, and River Tern also frequent in this area. Moreover, like the Bagan section, the endangered Irrawaddy Dolphin inhabits these waters.

The town of Bhamo and at least 14 villages are situated on the river's banks. Most of the local population is engaged in agriculture, with some relying on fishing and other river-based activities for their livelihoods. The communities' dependence on the river highlights the critical need to maintain the ecological health of the river system not only for biodiversity but also for the sustenance of local human populations.

Significant environmental threats to this section include electric fishing, human settlement on the islands and riverbanks, conversion of land to agriculture, gold mining, bird poaching, and erosion of the river's outer bends. These activities compromise the ecological integrity of the river and threaten the survival of its diverse species. While designations such as the Ramsar or EAAFP Flyway Network sites are not yet applied to the Bhamo section, the presence of Critically Endangered and Vulnerable species underlines the urgent need for formal protection measures. To safeguard the unique ecological characteristics and biodiversity of the Ayeyarwaddy River's Bhamo section, it is imperative to implement comprehensive conservation initiatives that focus on habitat preservation, sustainable resource management, and enforcement of environmental regulations to mitigate the adverse effects of human activities. Also, collaborative efforts that involve local communities, government agencies, and international conservation bodies are essential to ensure the long-term sustainability of this vital river ecosystem, promoting a balance between human use and ecological preservation.

3.3.3 Ayeyarwaddy River – Myitkyina–Sinbo Section

The Myitkyina to Sinbo section of the Ayeyarwaddy River, located in Kachin State, Myanmar, stretches approximately 119 km between the townships of Myitkyina and Sinbo. This section of the river, encompassing an area of 57,800 ha, serves as a crucial ecological zone. Starting from Pammati village near Myitkyina and ending at Sinbo, where the river's first gorge begins, this segment is characterized by its dynamic river channel and associated floodplain wetlands. The Ayeyarwaddy River along this stretch is notably wide, ranging from 700 to 800 m, with extensive banks composed of sand, silt, and gravel. Some rocky outcrops punctuate the western shores, adding to the region's diverse topography. During the dry season, the river's depth can reach about 12 m. Major tributaries like the Mogaung Chaung augment the river's flow, along with several smaller streams emanating from the eastern mountain ridges on the international border with China.

This section is rich in biodiversity, with over 130 species of birds recorded. The river and its floodplain support four globally threatened bird species (Naing 2012, 2013, Naing et al. 2015a, b; Zöckler 2019): the Critically Endangered White-bellied Heron and White-rumped Vulture and the Endangered Pallas's Fish-eagle (*Haliaeetus leucoryphus*). Additionally, the area is a habitat for other Near Threatened species such as the Lesser Adjutant (Figure 3.2e), Ferruginous Pochard, Spot-billed Pelican (*Pelecanus philippensis*), Black-necked Stork (*Ephippiorhynchus asiaticus*), Asian Woolly-necked Stork, Eurasian Curlew, Great Thick-knee, River Lapwing, and River Tern. The dynamic nature of the river, with its erosion-prone bends and oxbow lakes, creates diverse habitats that are critical for these species. While the Ayeyarwaddy River section between Myitkyina and Sinbo is recognized as a KBA, it is not yet designated under international conservation frameworks like Ramsar nor as an EAAFP Flyway Network site. The area's natural attributes and undeniable ecological significance emphasize the need for protection status to preserve its rich biodiversity and ecological functions.

The river's vicinity is home to one town, Sin Bo, and 12 villages, with a predominant reliance on agriculture. Locals engage in growing rice, bananas, coconuts, corn, and mustard. Additionally, fishing remains a vital part of their livelihood, alongside recession agriculture practiced on the silt banks exposed by lowering river levels. The unique lifestyle and practices of these communities are closely tied to the health of the river ecosystem.

This section of the Ayeyarwaddy faces significant environmental threats, including electric fishing, the construction of fishing structures that block tributaries, gold mining, bird poaching, and severe erosion at river bends. These activities not only threaten the local wildlife but also undermine the ecological stability of the river system. Conservation efforts must prioritize sustainable practices and the enforcement of environmental regulations to mitigate these threats. Collaboration with local communities, government agencies, and international conservation organizations will be key to developing strategies that balance ecological preservation with the socioeconomic needs of the local population. By fostering a comanaged approach, the Ayeyarwaddy River's Myitkyina to Sinbo section can be safeguarded for future generations, ensuring its continued role as a vital ecological corridor in Myanmar.

3.3.4 Ayeyarwaddy River – Singu Section

The Singu section of the Ayeyarwaddy River, situated in Mandalay Region's Singu Township, covers an area of 7500 ha and stretches approximately 74 km from Kyauk Myaung to Mingun village. This section, also designated as KBA, shares similar hydrological and geological characteristics with the Bagan and Bhamo sections of the Ayeyarwaddy River, such as the extensive braided channels and alluvial plains that dominate the landscape and numerous extensive sand, silt, and clay banks. The landscape is predominantly a flat alluvial plain to the east, with some low hills to the west rising to 200 m above sea level. Like the Bhamo section, this area is marked by numerous small tributaries that contribute to the river's flow and dynamic changes in channel connectivity during the dry season.

As a recognized Irrawaddy Dolphin Protected Area, the Singu section holds a significant conservation value, particularly for the Endangered Irrawaddy Dolphin. This area supports a diverse array of avian life, with at least 100 species of birds recorded, like the Bagan and Bhamo sections of the Ayeyarwaddy River (Naing 2012, 2013, 2015). The riverbanks of the Singu section host the town of Singu and 28 villages where the local economy primarily revolves around agriculture, with a significant portion of the community also engaged in fishing and other river-based activities. Echoing the socioeconomic importance of the Bagan and Bham sections, the Singu section of the river is integral to the livelihoods of these communities, emphasizing the need for sustainable resource management to ensure the continued vitality of the river ecosystem.

This section faces environmental threats similar to those observed in other parts of the river, including electric fishing, habitat conversion on the islands and riverbanks for human settlement and agriculture, gold mining, bird poaching, and erosion of riverbanks. While the section is not designated as Ramsar nor EAAFP Flyway Network site, its status as an Irrawaddy Dolphin Protected Area highlights targeted conservation efforts aimed at preserving this critical habitat for the Endangered dolphin species and other biodiversity. Ongoing and future conservation efforts in the Singu section should focus on strengthening the protection of critical habitats, enhancing community engagement in sustainable practices, and promoting conservation awareness among local and regional stakeholders. Conservation strategies must address the multifaceted challenges posed by human activities and natural river dynamics to maintain and enhance the ecological integrity of this important section.

3.3.5 Gulf of Mottama

The Gulf of Mottama (previously known as the Gulf of Martaban) is a vast estuarine wetland located at the confluence of the Sittaung, Yangon, and Thanlwin (Salween) rivers spanning Yangon, Bago, and Mon State regions along the Bay of Bengal. This large, funnel-shaped estuarine system spans an area of about 160,000 ha, and is notable for its extensive intertidal mudflats, among the most extensive in Southeast Asia and a highly dynamic ecosystem shaped by a pronounced tidal cycle. This cycle results in a powerful tidal bore phenomena, making the Gulf one of the most dynamic estuaries globally, with constant sediment redistribution, channel shifts, and substantial erosion and accretion. These mudflats experience significant ecological shifts due to the area's high tidal amplitude and speed, which contribute to its status as one of the last coastal wilderness areas in Asia.

The biodiversity of the Gulf of Mottama is exceptionally rich and diverse, driven by the high productivity of its wetland ecosystem. It is home to around 700 species of invertebrates and 39 fish species, many of which are of commercial importance. The Inner Gulf serves as an important nursery and breeding ground for the Giant Freshwater Prawn (*Macrobrachium rosenbergii*), while small crabs (*Dotilla intermedia*), occurring in densities of up to 5000 individuals/m^2, form a critical part of the diet for numerous shorebirds. One of the Gulf's most critical conservation values is its importance as one of the world's key wintering areas for the Critically Endangered Spoon-billed Sandpiper (*Calidris pygmaea*), hosting more than half of the remaining global population during the nonbreeding season (Aung et al. 2017). This has led to its designation as a Ramsar site (Site no. 2299), one of the largest in Asia, and as an IBA and an East Asian-Australasian Flyway Network Site (EAAF 117).

The local populations of the villages along the Gulf fluctuate significantly due to sedimentation and coastal erosion, with migration patterns heavily influenced by these environmental changes (Wongthong and True 2015). Communities in the region primarily depend on fishing, agriculture (including rice, vegetables, and animal husbandry), and bird hunting, which has notably contributed to the decline in bird populations (Wongthong and True 2015).

Threats to the Gulf of Mottama include habitat destruction, particularly of mangrove forests between Thaton and Paung Townships, illegal fishing practices, and the use of pesticides in nearby agricultural fields, which degrade both land and water quality (Wongthong and True 2015). Bird populations are threatened by poaching, with the use of poisons and mist nets during migration periods contributing to their decline (Zöckler et al. 2010) although conservation work led by the Biodiversity and Nature Conservation Association (BANCA) has alleviated the problem to some extent. These ongoing threats necessitate robust conservation measures. Recognizing the area as a World Heritage Site could be a pivotal step in enhancing its global conservation status and ensuring long-term ecological protection. A significant conservation initiative funded by the Swiss Agency for Development and Cooperation (SDC), in cooperation with various international and local organizations, focuses on improving the livelihoods of local communities while promoting sustainable wetland management. Capacity development for local conservation groups (LCGs) includes training on Communication, Education, Participation, and Awareness (CEPA), joint monitoring of waterbirds, and tackling illegal activities through patrolling during the wintering season.

3.3.6 Indawgyi Lake Wildlife Sanctuary

Indawgyi Lake Wildlife Sanctuary is located in Mohnyin Township, Kachin State in northern Myanmar and spans a significant area of 81,499 ha. As Myanmar's largest freshwater lake, Indawgyi Lake measures approximately 22.5 km in length and 11.3 km in width, totaling an area of 26,000 ha when fully flooded. The sanctuary, established in 2004, serves as a vital biosphere reserve and has been designated as both an IBA and an ASEAN Heritage Park due to its high ecological value.

The sanctuary's landscape is marked by the vast open waters of the lake, surrounded by natural, seasonally inundated grasslands and extensive herbaceous marshes dominated by various grasses and plants. These marshes are particularly crucial as they support the growth and sustainability of a rich tapestry of biodiversity. Eight streams originating from the surrounding mountains feed into the lake, enriching it with nutrients and supporting a diverse aquatic ecosystem. The lake and its surrounding wetlands are an indispensable winter feeding habitat for migratory and resident waterbird species. Notably, it supports species such as the Critically Endangered Baer's Pochard, along with the Vulnerable Sarus Crane (*Grus antigone*), Common Pochard (*A. ferina*), Graylag Geese (*Answer anser*), Common Coot (*Fulica atra*), Oriental Darter, Gray-headed Swamphen (*Porphyrio indicus*), cormorants, gulls, and terns during the winter season.

The biodiversity of this exceptional wetland system extends beyond avian species; the sanctuary is home to 93 recorded fish species, at least 13 species of amphibians, and 23 species of reptiles, including globally threatened turtles. The seasonally flooded grasslands around the lake are a critical breeding ground for fish that migrate from the Ayeyarwady River during the monsoon season and provide essential habitat for resident water birds and the Endangered Hog Deer (*Axis porcinus*; Lwin et al. 2016). The freshwater swamp forests adjacent to these wetlands are crucial for supporting local human settlements, paddy fields, and grazing areas. The sanctuary's significance is highlighted by its recognition as a Ramsar site (Site no. 2256) and an EAAFP Flyway Network site (EAAF118), highlighting its global ecological importance. These designations are based on the sanctuary's role in supporting endangered species, sustaining ecological processes, and maintaining water quality.

Local communities living around Indawgyi Lake, including 11 village tracts (an administrative subdivision of Myanmar's rural townships) and 36 villages with 13 directly on the lake's banks, are intricately linked to the health of this ecosystem. Their traditional livelihoods, primarily fishing and farming, depend heavily on the intact nature of their environment. Community-based ecotourism, such as tours on motor canoes, offers a sustainable way to appreciate and preserve the lake's natural beauty and biodiversity.

However, the sanctuary faces significant threats. Gold mining activities in the southwest and overfishing pose ongoing risks to the lake's ecological balance. To address these challenges, the Nature and Wildlife Conservation Division (NWCD) has implemented a comprehensive five-year management plan, with a new seven-year strategy underway focusing on sustainable fisheries, community forestry, and environmental education among other initiatives. Collaborations with NGOs and local community organizations, like Fauna and Flora International and the Indawgyi Natural Farming Association, play a critical role in the sanctuary's conservation efforts. These partnerships help facilitate research, monitoring, and community engagement projects, essential for fostering sustainable practices and raising awareness about conservation needs. Indawgyi Lake Wildlife Sanctuary stands as a beacon of biodiversity and a testament to the ongoing efforts required to protect such irreplaceable natural treasures.

3.3.7 Inle Lake Wildlife Sanctuary

Nestled within Shan State, the Inle Lake Wildlife Sanctuary spans an area of 53,373 ha, with elevations ranging from 830 to 1270 m above sea level. Established in 1985, this sanctuary was primarily designed to safeguard migratory bird habitats. It achieved further recognition as an ASEAN Heritage Park in 2003 and was declared Myanmar's first Biosphere Reserve in 2015. Inle Lake, fed by 29 streams with four major subcatchments, showcases an impressive array of habitats including wetlands and diverse forest types such as pine forests, dry forests (Indaing), and hill deciduous forests. These ecosystems support an extensive range of flora and fauna, underlining the lake's biodiversity. Notably, the lake serves as a crucial node on both the Central Asian and East Asian Australasian Flyways, providing vital feeding and resting stops for migratory waterbirds.

The sanctuary is a haven for waterbirds, including the globally Vulnerable Sarus Crane (*Grus antigone*, Figure 3.2f), and supports at least 267 bird species, of which 82 are wetland species. The lake's waters harbor 43 fish species, including 16 endemics like the culturally significant Inle Carp (*Cyprinus intha*), locally known as nga-phein. The surrounding areas are also rich in biodiversity, harboring 13 reptiles (including 3 turtle species), 9 mammals, 75 butterflies, and 20 species of snails. Additionally, the region is abundant in medicinal plants, with 527 species recorded, alongside 184 species of orchids and numerous angiosperms.

The surrounding region of Nyaung Shwe Township, comprising 36 village tracts and 446 villages, is home to approximately 172,469 residents. The Intha community, known for its distinctive leg-rowing fishing technique, is integral to the local culture. Other prevalent livelihoods include blacksmithing, goldsmithing, weaving, and tourism. The traditional weaving of lotus fiber – known locally as "kyar thingan" – is a unique product of the area. Additionally, the floating island agriculture practiced by Intha farmers, particularly for tomatoes and other vegetables, is renowned worldwide.

Despite its ecological richness, Inle Lake faces significant environmental threats. Over the years, the lake's surface area has drastically reduced, mainly due to sedimentation from deforestation and erosion, and the invasive growth of floating gardens. The excessive use of chemical fertilizers and pesticides in agriculture has led to severe water pollution, posing health risks to local communities, and threatening the lake's native species. The region also grapples with the impacts of climate change, which has altered rainfall patterns and intensified seasonal variability.

The sanctuary's management involves various stakeholders including the Ministry of Natural Resources and Environmental Conservation (MONREC) and international partners like the United Nations Development Programme (UNDP). From 2012 to 2016, significant efforts were made under the Inle Lake Conservation and Rehabilitation Project to restore environmental stability and improve local quality of life through sustainable community development initiatives. An Environmental Education Centre (EEC) was established to foster community participation and enhance conservation awareness. EEC also served as a platform for a better coordination between government departments and communities, facilitated by the Intha Literature, Cultural and Regional Development Association (Hatfield Consultants 2015). The Nature Wildlife Conservation Division (NWCD) continues to spearhead conservation efforts through comprehensive management planning, which includes biodiversity monitoring, community education, and sustainable tourism management. These initiatives aim to balance ecological preservation with the socioeconomic development of local communities, ensuring the long-term sustainability of Inle Lake's unique ecosystem.

3.3.8 Kyon Ka Pyin-Tap Seik Community Conservation Area, Ayeyarwaddy Delta

The Kyon Ka Pyin-Tap Seik Community Conservation Area (KTCCA), nestled in the Ayeyarwady Delta within Wakema Township, Ayeyarwady Region, spans approximately 17,460 ha. This area features a diverse landscape that includes riverine forests, shrublands, grasslands, croplands, settlements, bare lands, rivers, water bodies, and fish ponds. KTCCA's diverse land cover includes riverine forest (1593 ha), shrubland (9 ha), grassland (486 ha), cropland (14,125 ha), settlement (13 ha), bare land (92 ha), river and water body (929 ha), and fish pond (213 ha). This variety supports a dynamic ecosystem where water and land intertwine, creating habitats conducive to both aquatic and terrestrial biodiversity.

The conservation area is notable for its significant avian population, with at least 170 species of birds recorded. KTCCA provides critical habitat for several globally threatened species. It hosts a substantial population of the critically endangered Yellow-breasted Bunting, with an estimated 38,000 birds observed in the 2023–2024 period. Additionally, the Vulnerable species such as the Sarus Crane, Jerdon's Babbler (*Chrysomma altirostre*), and Greater Spotted Eagle are also present. Near Threatened species including the Black-headed Ibis (*Threskiornis melanocephalus*), Painted Stork (*Mycteria leucocephala*), and Asian Golden Weaver (*Ploceus hypoxanthus*) further highlight the area's ecological value. The Endangered Fishing Cat (*Prionailurus viverrinus*) and three globally threatened turtle species – the Critically Endangered Burmese Peacock Softshell Turtle (*Nilssonia formosa*), the Endangered Burmese Eyed Turtle (*Morenia ocellata*), and the Vulnerable Asiatic Softshell Turtle (*Amyda cartilaginea*) – add to the area's biodiversity significance.

KTCCA is home to five villages with more than 2000 residents whose livelihoods mainly rely on agriculture, with fishing and harvesting of other wild goods also playing significant roles. Despite its rich biodiversity, KTCCA faces several environmental threats. These include electric fishing, conversion of riverbanks and sandbanks to agricultural land, bird poaching, and habitat destruction. Grassland areas, which are privately owned, are particularly vulnerable as they have been increasingly converted to agricultural land, with grassland fires and reed collection posing additional threats.

The designation of the KTCCA as a community conservation area is a testament to the local commitment to preserving this unique ecological region. Efforts to enhance conservation include addressing the direct threats to biodiversity through community engagement and sustainable practices. The integration of community life with conservation efforts highlights the potential for sustainable resource management driven by local involvement.

Looking forward, expanding educational initiatives and strengthening local governance can further empower the communities to manage their natural resources effectively.

3.3.9 Lampi Marine National Park

Lampi Marine National Park, established in 1996, is a critical conservation area located in the Myeik Archipelago within Boke Pyin Township in the Taninthayi Region of southern Myanmar. Encompassing 20,500 ha, this park includes Lampi Island – the largest and central island of the park – along with several smaller islands and the surrounding marine environments. Lampi Island stretches 48 km in length and features hilly terrain ranging from 150 to 270 m above sea level. The island and its surroundings are characterized by a rich mosaic of ecosystems, including evergreen forests, mangroves, beach and dune forests, coral reefs, and seagrass beds, particularly prominent on the eastern side of the island. These diverse habitats contribute to the ecological complexity and high productivity of the region.

Lampi Marine National Park is a biodiversity hotspot with significant conservation value. The area is home to 232 recorded species of birds. It supports globally threatened bird species such as the Vulnerable Plain-pouched Hornbill (*Rhyticeros subruficollis*), which can occur here in large flocks – on their migrations, Wallace's Hawk-eagle (*Nisaetus nanus*) and Great Slaty Woodpecker (*Mulleripicus pulverulentus*), alongside Near Threatened species like the Black-tailed Godwit (*Limosa limosa*), Eurasian Curlew, and Nicobar Pigeon (*Caloenas nicobarica*). The park is an important sanctuary for sea turtles, harboring species such as the Endangered Green Sea Turtle (*Chelonia mydas*) and the Vulnerable Loggerhead Sea Turtle (*Caretta caretta*) and Olive Ridley Sea Turtle (*Lepidochelys olivacea*) (Naing et al. 2015a, b; Zöckler et al. 2018). Additionally, it also provides habitat for the threatened mammal species, including the Critically Endangered Sunda Pangolin (*Manis javanica*) and the Vulnerable Dugong (*Dugong dugong*); and threatened mangrove species such as the endangered Yae-kanazo mangrove (*Heritiera fomes*) (Zöckler and Aung 2019).

Recognized as an ASEAN Heritage Park on 18 December 2003, Lampi Marine National Park's designation emphasizes its importance in regional conservation efforts (Oikos and BANCA 2012). The NWCD spearheads the park's management, focusing on sustainable fisheries, research and monitoring, patrolling, community-based ecotourism, environmental education, waste management, and livelihood development. Collaborative projects with national and international NGOs like Istituto Oikos and several community-based organizations are crucial to the sustainable management of the park.

The park's vicinity hosts four villages within its boundaries and one outside, with a combined population exceeding 3000 residents. The local community predominantly engages in fishing and harvesting aquatic resources. These activities, while central to their livelihood, pose significant challenges to conservation efforts, particularly when unsustainable practices such as the use of gill nets and hunting terrestrial animals are employed. Significant threats to Lampi Marine National Park include overfishing, logging, hunting, and the improper disposal of garbage and solid waste. These activities jeopardize the delicate balance of the marine and terrestrial ecosystems and threaten the survival of the park's diverse species.

Efforts to protect and preserve Lampi Marine National Park must continue to evolve, incorporating scientific research, enhancing enforcement of conservation laws, and increasing community engagement. Education and empowerment of local communities to adopt sustainable practices are essential for the long-term preservation of this ecological treasure. By fostering a deeper connection between the community and its natural environment, Lampi can continue to thrive as a model of conservation excellence in Southeast Asia.

3.3.10 Meinmahla Kyun Wildlife Sanctuary

The Meinmahla Kyun Wildlife Sanctuary, established in 1993, is a vital wetland reserve located in Bogale Township, Ayeyarwady Region of Myanmar and sits on a vast estuarine region on the Ayeyarwady Delta. Covering an area of 13,670 ha, this sanctuary includes an island approximately 25.7 km long and 9.7 km wide, predominantly covered in mangrove forests, some of which have been degraded due to natural disasters like Cyclone Nargis and ongoing illegal human activities such as logging and fishing. The sanctuary is a mangrove ecosystem reserve, featuring a rich diversity of mangrove species. With 29 recorded species of mangrove trees, the area includes the Critically Endangered species, such as *Bruguiera hainesii* (Figure 3.4) and *Sonneratia griffithii* (Zöckler and Aung 2019). The ecological diversity of the sanctuary extends to its wildlife, being one of the few places with regular sightings of Fishing Cats and saltwater crocodiles, along with significant numbers of coastal Irrawaddy Dolphins (Vulnerable), with about 20–30 individuals observed.

Figure 3.4 The Critically Endangered mangrove, *B. hainesii*. *Source:* Photo by BANCA.

Meinmahla Kyun is recognized for its importance to migratory species, acting as a key staging and wintering site for many species in the EAAF. The sanctuary hosts numerous migratory waterbirds during the winter months from November to February. At least 117 bird species have been documented, including the Critically Endangered Spoon-billed Sandpiper, small numbers of the Endangered Spotted Greenshank (*Tringa guttifer*) annually, and Great Knot (*Calidris tenuirostris*) (Zöckler 2016). The Near Threatened Lesser Adjutant, Eurasian Curlew, Black-tailed Godwit, and Black-headed Ibis also occur. In terms of mammalian biodiversity, the sanctuary records include 18 species such as the Endangered Asian Wild Dog (*Cuon alpinus*), Vulnerable Fishing Cat (*Prionailurus viverrinus*) and Smooth-coated Otter (*Lutrogale perspicillata*). The area is also significant for the breeding of several sea turtle species, including Olive Ridley, Leatherback, Loggerhead, and Green Sea Turtles, highlighting its importance in marine conservation.

Designated as an ASEAN Heritage Park in 2003, a Ramsar site in 2017 (Site no. 2280), and an EAAFP Flyway Network site in 2018 (EAAF140), Meinmahla Kyun's protected status emphasizes its importance for global biodiversity conservation, especially for its intertidal mudflats which support a large number of migratory waterbirds. The NWCD has implemented a five-year management plan in collaboration with the Myanmar Environmental Rehabilitation-concentrated Network (MERN), focusing on sustainable fisheries, environmental education, and community-based ecotourism.

The sanctuary is surrounded by 15 villages, home to a population where 34% engage in fishing, 22% in agriculture, and 38% in various other jobs. The local communities are integral to the ongoing conservation efforts, although challenges such as overfishing and the capture of juvenile and low-value commercial species indicate a near-collapse state of the local fisheries (Macintosh 2016).

Initiatives such as the FAO's "Sustainable small-scale fisheries and aquaculture livelihoods in coastal mangrove ecosystems" have introduced co-management systems involving 20 villages along the Bogale Riverbank to address these challenges.

Significant threats to the sanctuary include illegal logging, fishing, and the harvesting of other natural resources, which have contributed to the degradation of its mangrove forests and other ecosystems. Additionally, environmental challenges such as cyclones and human-induced habitat changes pose ongoing risks to its biodiversity. Continued efforts in community engagement, stricter enforcement of conservation laws, and enhanced monitoring and research are essential to sustain and build upon the conservation successes at Meinmahla Kyun Wildlife Sanctuary. Strengthening the cooperation between local communities, government bodies, and international organizations remains critical to the effective management and preservation of this unique mangrove ecosystem.

3.3.11 Nanthar Island Marine National Park

Nanthar Island Marine National Park, established in 2022, is situated 20 km northwest of Sittwe and south of Yathaedaung Townships, isolated from the Bay of Bengal. This park encompasses a diverse array of coastal and marine ecosystems including mangrove forests, low-lying sandbanks, intertidal mudflats, and man-made fish ponds, crucial for the conservation of marine and avian biodiversity (Zöckler et al. 2014). Nanthar Island is bordered by the Mayu, Kaladan (Kispanadi), and Lemyo rivers, with their many tributaries forming a natural boundary between India, Bangladesh, and western Myanmar. The Kaladan River, Rakhine's largest river, is vital for the livelihoods of thousands in the region, emphasizing the island's role in regional ecological and economic sustainability (Zöckler et al. 2012).

The marine national park is internationally recognized for its biodiversity, and especially migratory species. It supports about 5% of the global population of the Critically Endangered Spoon-billed Sandpiper, along with the Endangered Spotted Greenshank and Great Knot. Additionally, the Vulnerable Indian Skimmer also winters here annually. The intertidal mudflats around Nanthar Island are key habitats for these migratory shorebirds, underscoring its global ecological importance. Marine turtles find a safe haven for nesting on Nanthar Island, with more than 120 nests of three sea turtle species – Loggerhead Sea Turtle (*Caretta caretta*), Green Sea Turtle (*Chelonia mydas*), and Olive Ridley Sea Turtle (*Lepidochelys olivacea*), all globally threatened – recorded each year. The presence of Irrawaddy Dolphins further highlights the island's rich marine biodiversity. Nanthar Island has been designated as an IBA, a site within the East Asian-Australasian Flyway Network of Sites (EAAF 139), and the sixth Ramsar site in Myanmar (Site no. 2421). These designations reflect its importance as a critical stopover and breeding site for various migratory species and its role in global biodiversity conservation, despite its relatively small area.

The surrounding area includes five villages with a total population exceeding 5300 people across about 970 households. The local economy is predominantly based on fisheries, with the communities heavily dependent on this resource for most of the year. Villagers engage in aquaculture, fishing, and the collection of shellfish and crabs. Additionally, during the rainy season from June to October, residents cultivate rice, chilies, watermelons, and betel. Some local practices, such as the active capture of sea turtles, pose conservation challenges.

Significant threats to Nanthar Island's ecosystems include overfishing, habitat destruction, and the illegal capture of marine turtles. Addressing these challenges requires robust management strategies that involve local communities, fostering sustainable practices that align with conservation goals. Ongoing efforts to protect Nanthar Island Marine National Park should focus on enhancing community engagement, implementing sustainable resource management practices, and continuing to monitor and protect the critical habitats found within the park. Education and empowerment of local communities are essential to ensure the sustainability of their livelihoods and the conservation of the park's unique biodiversity.

3.3.12 Paleik Lake

Paleik Lake, locally known as Paleik Inn ("Inn" meaning "Lake" in the local language), is a natural freshwater wetland and an oxbow lake situated within Kyaukse Township, Mandalay Region. Covering an area of 480 ha, this compact floodplain lake is linked to the Ayeyarwaddy River through the Myit Nge River, playing a crucial role in supporting agriculture through river irrigation. Paleik Lake is a permanent open water body surrounded by marginal vegetation and tall grasses. The lake's bottom is primarily composed of clayey soil. The wetland is an essential component of the riverine ecosystem, providing significant ecosystem services including water purification, flood mitigation, and habitat for wildlife.

Recognized as an important bird and biodiversity area (IBA) and KBA, Paleik Lake supports a rich diversity of avian life. It is particularly vital for globally threatened species such as the Critically Endangered Baer's Pochard, the Vulnerable Common Pochard, and other widespread waterbird species including the Greylag Goose, Glossy Ibis (*Plegadis falcinellus*), and Black-headed Ibis. The lake also serves as a breeding ground for various waterbird species, with recent counts recording between 2450 and 4430 waterbirds of up to 49 species.

Paleik Lake is pivotal for the local economy, supporting agriculture and fish farming activities for four surrounding villages with a total population of 2560. Local farmers rely on the lake for irrigation during both the rainy and dry seasons, cultivating rice and other seasonal crops. Additionally, aquaculture is practiced in inundated pools and small streams around the lake, providing a subsistence income for the community.

Despite its ecological importance, Paleik Lake faces several threats. The most significant is the extensive use of electric fishing gear, which has surged in recent years. Other challenges include ineffective law enforcement against illegal fishing practices, the potential threat of invasive species like the Vermiculated-sailfin Catfish (*Pterygoplichthys disjunctivus*), and local practices such as using harmful substances to hunt migratory waterbirds. Furthermore, local agriculture heavily relies on chemical fertilizers and pesticides due to the high cost and low availability of natural fertilizers.

The current management of Paleik Lake involves multiple government entities including the Department of Fisheries, the Department of Agriculture, and the General Administration Department, under the oversight of the MONREC. Recent conservation initiatives have focused on education and awareness campaigns about wetland and biodiversity value, enforcement of wildlife protection laws, and community engagement in waste management practices. In November 2023, Paleik Lake was designated as a site within the EAAF Flyway Network (EAAF154), reflecting its growing importance in international conservation efforts. This designation is a step toward enhancing the global recognition of Paleik Lake's biodiversity and ecological importance, aiming to bolster future conservation and sustainable management efforts in collaboration with the EAAFP. Continued efforts in community engagement, strengthened enforcement of environmental laws, and international cooperation are vital to ensure the preservation and enhancement of this critical wetland sanctuary.

3.3.13 Pyu Lake

Pyu Lake, located in Tada-U township of the Mandalay Region, is a significant artificial wetland ecosystem within Myanmar's Central Dry Zone. This 1900-ha lake, primarily fed by the Kinda Dam on the Pang Laung River, plays a crucial role in both local agriculture and biodiversity conservation. Pyu Lake is a permanent water body that has been transformed into a vital resource through its connection with Kinda Dam, making it a key component of the region's irrigation system. The lake supports extensive agricultural activities by distributing water to the surrounding farmland through a network of irrigation canals. The landscape around Pyu Lake includes extensive areas of aquatic vegetation dominated by water lilies (*Nymphaea*) and *Ottelia cordata*, with marginal vegetation like *Typha angustifolia* and various algae, which contribute to the ecosystem by providing food and shelter for aquatic fauna and nesting grounds for waterbirds.

Annually, Pyu Lake provides refuge for the Critically Endangered Baer's Pochard and supports an impressive diversity of up to 40 waterbird species, totaling at least 5200 individuals in recent counts. The lake functions as a crucial stopover and breeding site within the EAAF (Figure 3.5), emphasizing its importance in global avian biodiversity conservation networks. Additionally, the lake harbors 28 species of fish, underscoring its ecological richness. In recognition of its ecological importance, Pyu Lake was designated as the seventh Ramsar site in Myanmar in 2023 (Site no. 2533) and as an EAAF Network site (EAAF155). These designations highlight the lake's critical role in supporting migratory and resident waterbird populations and its contribution to global biodiversity conservation efforts.

Around Pyu Lake, seven villages with a population of 5600 rely predominantly on agriculture, animal husbandry, and local fishing practices that are uniquely adapted to the region's traditions and ecological conditions. The community's engagement in these practices reflects a deeply rooted dependence on the lake's resources for their economic and subsistence needs. However, Pyu Lake

Figure 3.5 Migratory waterbirds in Pyu Lake. *Source:* Photo by Thiri Dae We Aung (Chapter author).

faces several conservation challenges. These include the use of harmful fishing practices, such as the use of poison baits to catch migratory waterbirds during the bird migration season from January to April. Additionally, the widespread use of chemical fertilizers and pesticides in agriculture poses a threat to the lake's water quality and biodiversity. Local knowledge gaps concerning sustainable agricultural practices further exacerbate these issues.

Efforts to protect Pyu Lake's biodiversity include initiatives led by LCGs such as the Shwe Kan Thayar Nature Conservation Association, in partnership with the Biodiversity and Nature Conservation Association (BANCA). These initiatives focus on monitoring waterbird populations, educating the local community about conservation, and enforcing wildlife protection laws. However, political instability has recently hindered the effectiveness of these conservation activities. In the long term, it is crucial to revitalize and expand conservation efforts at Pyu Lake. Strengthening community involvement in sustainable practices, enhancing the enforcement of environmental regulations, and continuing international cooperation under frameworks like the Ramsar Convention and the EAAF are essential to ensure the long-term preservation of this vital wetland ecosystem.

3.4 Wetland Management Policy and Practices in Myanmar

3.4.1 Evolution of Policy: From the Past to the Present

Myanmar's approach to PA management, including wetlands, has evolved significantly over time, influenced by its unique history and biodiversity (Aung 2007). The country's PA system started nearly 150 years ago under royal patronage, with policies and practices developed through 19 pieces of legislation during and after the colonial period (Aung 2007). This history has shaped the current framework for conserving Myanmar's diverse ecosystems.

The initial stages of Myanmar's PA system date back to the introduction of Buddhism in the 11th century AD, with Myanmar's kings designating forest areas ("Bemetaw") where animals were protected from harm (Lwin et al. 1990). The British colonial period also contributed to the establishment of PAs, focusing on forest conservation for timber production and wildlife sanctuaries (Bryant 1996). However, these sanctuaries often conflicted with the needs of local communities and faced challenges in effective management and enforcement (Bryant 1996). The postindependence period saw slow progress in PA management due to economic and political difficulties (Aung 2007). The game warden's role, although resurrected in 1972, remained limited in effectiveness, especially in wildlife management within parks. Significant strides were made with the Nature Conservation and National Parks Project (NCNPP) from 1981 to 1985, which developed conservation institutions and expanded the PA network (Aung 2007).

In recent years, Myanmar has made significant progress in conserving its natural habitats, paying particular attention to its coastal wetlands. At the present time, the country has protected 6% of its land as PAs, and there are plans to increase this protection to 17% (Zöckler et al. 2018). This expansion of PAs reflects a growing recognition of the ecological and socioeconomic importance of these habitats (Zöckler et al. 2018). A notable advancement in Myanmar's conservation efforts occurred in 2017 with the addition of two internationally important coastal wetlands to the Ramsar list of Wetlands of International Importance. This move highlights the country's commitment to safeguarding its diverse and valuable wetland ecosystems. Myanmar has also been proactive in recognizing the importance of migratory bird flyways. By 2018, the country had designated three coastal sites as Flyway network sites under the EAAFP, further highlighting its role in global biodiversity conservation.

Government efforts extend beyond wetlands to include marine areas, with a significant push to expand the Marine Protected Area (MPA) network. This expansion is critical, given the previously limited coverage of marine/coastal areas within Myanmar's conservation framework (<3%; Zöckler et al. 2018). Both national and international nongovernmental organizations (NGOs) have been instrumental in this process, proposing several sites for inclusion in the MPA network (e.g. see Zöckler et al. 2018). Their involvement demonstrates the collaborative nature of conservation efforts in Myanmar, which blends governmental initiative with nongovernmental expertise and support.

3.4.2 Wetland Management for Biodiversity

Myanmar's geographical position along both the Central Asian Flyway (CAF) and the EAAF makes its wetlands important staging and end destinations for more than 150 migratory species. Annually, these birds, which migrate large distances across continental Asia and over the Tibetan Plateau and eastern China, find refuge in Myanmar's diverse inland and coastal wetlands during the winter months. Short-distance migrants such as storks, Masked Finfoot *Heliopais personatus*, Glossy Ibis and several heron species also undertake localised movements across mainland Southeast Asia in response to the monsoons. Although Myanmar is not yet a party to the Convention on Migratory Species, significant steps have been taken to protect these migratory species. This includes the establishment of PAs, Flyway network sites, and Ramsar sites at several wetlands known to be of importance to migratory waterbirds.

Since 1994, Myanmar has been actively involved in the Asian Waterfowl Census, contributing to the collation of valuable data on waterfowl populations. This initiative supports the conservation of several species of migratory waterbirds, such as the Spoon-billed Sandpiper, Baer's Pochard, Spotted Greenshank, Great Knot, and Yellow-breasted Bunting. These species are regularly observed during their wintering period at key wetland sites in Myanmar.

The Forest Department of Myanmar plays a pivotal role in collaborating with relevant departments for the conservation of these wetlands. This includes efforts to protect diverse ecosystems such as mangrove forests, freshwater, and coastal areas. These habitats are not only crucial for migratory birds but also support a wide range of fauna, such as the Irrawaddy Dolphin, sea turtles, and seagrasses.

In 2019, the government took a significant step in supporting habitat restoration for various protected areas, including essential wetland sites such as Lampi Marine National Park, Indawgyi Lake, and Inle Lake. These efforts aim to enhance and preserve the ecological integrity of these vital habitats. In recent years, a shift toward community-led management of biodiversity in wetlands has been initiated in Myanmar. This approach involves empowering local communities to assume responsibility for wetland areas and actively participate in their sustainable management. This community-centric model is integral to ensuring long-term conservation and responsible use of wetland resources, aligning ecological preservation with local livelihoods and cultural practices (Saluja et al. 2023).

3.4.3 Wetland Management for Livelihoods

In Myanmar, the management of wetlands has significant implications for the livelihoods of local communities and the broader economy (Saluja et al. 2023). For example, the Moeyungyi Wetland Wildlife Sanctuary (hereafter, the "Moeyungyi wetland"), Myanmar's first Ramsar site, encompasses a wide range of ecosystem services that greatly benefit the surrounding communities and have an estimated annual economic value of approximately USD 22 million (Aung et al. 2021b). This value stresses the importance of the Moeyungyi wetland, not only for biodiversity conservation but also for supporting the livelihoods of approximately 12,000 households.

The Moeyungyi wetland supports various livelihood activities, including fishing, agriculture (mainly rice cultivation), and grazing of buffalo (Aung et al. 2021b). A significant portion of the local population depends on these activities to make their income. The resources of the wetland, such as fish, lotus stalks, and water for domestic use and irrigation, play a crucial role in sustaining these livelihoods (Aung et al. 2021b). For example, fish harvested from the wetland provides a substantial income for many families, with an estimated annual economic benefit of approximately USD 15.4 million. Similarly, lotus harvesting and buffalo grazing within the wetland contribute significantly to the local economy (Aung et al. 2021b).

Agriculture, particularly rice cultivation, is one of the wetland ecosystem services of the wetland (Aung et al. 2021b). Water storage capacity supports irrigation for rice paddies in the area, which is crucial for local food security and the economy (Aung et al. 2021b). The potential expansion of rice cultivation, which is anticipated due to Myanmar's plans to increase rice production, highlights the role of freshwater wetlands in supporting agricultural practices. However, any changes in wetland water management for irrigation need careful consideration, as they could affect other ecosystem services and the overall biodiversity of the area (Conallin et al. 2019).

Effective wetland management requires balancing the conservation of its biodiversity with the sustainable use of its resources for livelihoods. This management approach aligns with Myanmar's National Wetland Policy and its emphasis on the "wise use of wetlands." The assessment of the Moeyungyi wetland, including its contribution to local livelihoods, serves as a model for integrating ecosystem services into wetland management and policy development in Myanmar (Aung et al. 2021b). This approach can guide the management of other wetlands in the country, ensuring their conservation, while also supporting the livelihoods of local communities.

The Moeyungyi wetland illustrates the intricate link between wetland ecosystems and human livelihoods (Aung et al. 2021b). Its management and conservation are essential not only for protecting biodiversity but also for sustaining the economic well-being of local communities. Integrating ecosystem services into wetland management policies ensures that these critical habitats continue to provide valuable resources to local livelihoods while maintaining their ecological functions.

The restoration of biodiversity in degraded wetlands also goes hand in hand with the livelihoods of local communities in Myanmar. Initiatives are being taken to create jobs and provide sustainable livelihoods that do not compromise the environment and the health of the ecosystem. These efforts aim to strike a balance between the use of natural resources and the conservation of wetlands. National and international NGOs play a critical role in supporting the livelihoods of local villagers who rely on wetlands. They have been actively involved in educating these communities on sustainable practices of wetland use. The focus is on encouraging local participation in conservation activities, ensuring that their dependence on these ecosystems does not lead to degradation.

One of the key strategies in this regard involves the identification and pilot implementation of community-managed fisheries areas, particularly within the Ayeyarwady proposed Ramsar site. Conservation organizations are facilitating local fishing communities to develop and manage these areas, which include designated fish conservation zones. This process involves developing management plans for these zones and seeking legal designation from the Department of Fisheries. Such initiatives empower communities to manage their resources responsibly, aligning their livelihoods with conservation goals. Furthermore, wetland managers are advocating sustainable agricultural practices among farmers in Ramsar and the proposed Ramsar sites. The adoption of sustainable rice production (SRP) standards is being encouraged and efforts are underway to form SRP farmer groups at the site level. Organizational development training is provided to these groups, along with technical training in sustainable agriculture based on SRP standards. This approach aims to pilot sustainable agricultural systems that facilitate the wise use of wetlands.

Recently, there has been a focus on the adoption of internal control systems and the certification of SRP practices. Efforts are being made to develop a value chain for SRP-certified rice, which includes linking these practices to broader conservation programs such as the Wildlife Conservation Society's SRP and facilitating access to farmers' markets in Mandalay, Myanmar. This comprehensive approach to sustainable agriculture within wetland areas not only preserves biodiversity but also ensures the economic well-being of the local communities that depend on these vital ecosystems.

3.4.4 Wetland Management for Conservation

In Myanmar, the conservation of biodiversity in wetlands, particularly within protected areas, emphasizes better management and the wise use of these valuable ecosystems. The country currently hosts seven Ramsar sites, covering a total area of 278,913 ha, which constitutes about 4.4% of Myanmar's total wetland area. Two of these Ramsar sites, the Gulf of Mottama and Pyu Lake, lie outside the existing network of protected areas. In these regions, conservation efforts are primarily driven by community-led initiatives.

The management of Ramsar sites that are designated as protected areas falls under the purview of the Forest Department. This department oversees these protected areas under the guidance of the National Wetland Committee reformed in 2022. This committee adheres to the guidelines established in the National Wetland Policy and Strategic Action Plan of Myanmar, 2019 (Ramsar Convention Secretariat 2010, 2015).

Beyond the bounds of protected wetland areas, the central dry zone of Myanmar is one of the most impoverished and climate-vulnerable regions in Southeast Asia. The lifeline of this region is the Ayeyarwady River, Asia's last large, undammed river, vital for agriculture, drinking water, and groundwater replenishment. The river, especially the stretch between Mandalay and Bagan, along with its associated wetlands, such as Pyu and Paleik lakes, is crucial for the congregation of migratory water birds. These areas support significant populations of various bird species, including those that account for more than 1% of flyway populations such as the Ruddy Shelduck, Northern Pintail, Indian Spot-billed Duck, Black-headed Ibis, Glossy Ibis, and Greylag Goose. They are also habitats for globally threatened species such as the Baer's Pochard (Critically Endangered), Yellow-breasted Bunting (Critically Endangered), Black-bellied tern (Endangered), Common Pochard (vulnerable), Burmese narrow-headed turtle (Critically Endangered), and Burmese peacock softshell turtle (Critically Endangered) (Aung et al. 2016, 2019, 2021a). The wetlands of the Ayeyarwady are identified as priority sites for biodiversity conservation in Myanmar's National Biodiversity Strategy and Action Plan (NBSAP).

Myanmar's National Wetland Policy highlights the global significance of the Central Ayeyarwaddy wetlands, advocating for their designation as proposed Ramsar sites (National Wetland Policy 1995). This policy emphasizes the "wise use of wetlands" approach and prioritizes conservation initiatives, including sustainable and climate-smart rice farming as part of Myanmar's National Adaptation Programme of Action (NAPA) (Nature Wildlife Conservation Division 2019).

Communication, Education, Participation and Awareness (CEPA) strategies are essential in wetland management, with the aim of building awareness among various stakeholder groups.

The government of Myanmar, with its unique and diverse forest ecosystems, seeks cooperation with the Ramsar Convention and related organizations to further the conservation, wise use, and management of its wetlands. The assistance of these international entities is sought to promote conservation, manage wetlands, and build the capacity of Forest Department staff.

The experience in managing wetlands within protected areas in Myanmar has emphasized the vital role of local communities in sustainable conservation efforts. Community-based wetland conservation actively involves local communities in and around wetland areas, fostering capacity building, education, and awareness. This approach includes empowering locals to participate in the monitoring and evaluating biodiversity within these wetlands, ensuring a more holistic and inclusive approach to conservation.

3.5 Challenges in Wetland Management in Myanmar

In 2003, Myanmar's median PA size was less than $50\,km^2$, and the total PA size was only approximately 2.1% of the country area (Aung 2007). Today, the country has reached a level of 6% of its land area under protection (Zöckler et al. 2018), a laudable improvement. However, protected areas remain mostly terrestrial, with a limited representation of inland and coastal wetlands, especially mangroves, and marine habitats. This reflects an uneven representation of Myanmar's ecosystems, with certain biounits such as the Burmese coast being under-represented under the current PA system (Aung 2007).

Myanmar's wetlands, world-renowned and important ecosystems, face multifaceted challenges in their management and conservation. The recent COVID-19 pandemic and the political changes in 2021 have significantly affected both local communities and management frameworks, particularly with respect to government institutions. However, despite these adversities, there has been a sustained commitment from local stakeholders, supported by conservation initiatives, to community-based management of these wetlands.

One of the primary threats to Myanmar's wetlands is illegal and unsustainable fishing practices. Local fishermen in coastal wetlands often employ micro-small nets, while electric shock fishing is prevalent in ponds and rivers. These practices not only disrupt ecological balance but also endanger the diverse aquatic life that these wetlands support. Agricultural intensification also poses another significant threat to freshwater wetlands, with long-term consequences. The increasing use of pesticides and chemical fertilizers has increased the levels of pollution in these delicate ecosystems. This is compounded by the expansion of agricultural land, which often encroaches on wetland areas. Additionally, activities such as mining and the introduction of invasive species further stress these environments. Unregulated tourism, climate change, and tree cutting for fuel wood in watershed areas also contribute to the degradation of these vital habitats.

These challenges are exacerbated by inadequate law enforcement, a lack of cohesive collaboration between various stakeholders, limited budgets, understaffing, and inadequate training for PA staff (Aung 2007; Saluja et al. 2023). Fifty-eight percent of park wardens are educated as foresters, with an average tenure of four years in a park, which is less than ideal for management continuity (Aung 2007). The synergy between government bodies, conservation organizations, private sectors, and local communities is crucial to effective conservation of wetlands. However, currently this coordination is insufficient, leading to ineffective responses to the threats these ecosystems face (Aung 2007; Saluja et al. 2023). Addressing these challenges requires a multifaceted approach that includes strengthening legal frameworks, improving collaborative efforts among all stakeholders, and adopting sustainable practices in both fishing and agriculture. Community participation and education are also key components, as they can lead to better stewardship of these vital natural resources. The preservation of Myanmar's wetlands demands a concerted effort that combines regulatory enforcement with active participation from all sectors involved.

The conservation of wetlands should also involve dealing with the challenges of law enforcement and fostering positive community relationships. Illegal activities such as hunting, logging, and agricultural encroachment are common threats to protected wetlands. There is a need for systematic antipoaching patrols and law enforcement, as well as community involvement, to mitigate conflicts and encourage sustainable resource use (Yong et al. 2023). Budgetary support for PAs, although increased by 11% since 1996, remains insufficient to address the needs of parks, especially in remote and understaffed areas (Aung 2007). Innovative and sustainable means of support, including partnerships with nongovernmental organizations and local organisations, are essential for the future of Myanmar's wetlands.

3.6 Conclusion

Effective wetland management requires a deep and comprehensive understanding of various aspects. It starts with recognizing the different living environments and the species that they support. Understanding how these habitats and species interact to form functional ecosystems is crucial. Identifying the natural processes that sustain these ecosystems is also essential. Equally important is being aware of potential threats, including the impact of past and present human activities on these natural processes. However, sustainable wetland management involves more than simply understanding these elements. It requires devising a full range of measures to sustain these environments. This includes integrating the wetlands into the broader community context and being prepared to respond to potential threats from developments in surrounding areas. The goal is to achieve a balance between the needs of wetlands and those of the people who depend on them, striving for a sustainable coexistence that benefits both nature and human communities.

For Myanmar's wetlands to persist, it is crucial to reformulate policies, increase the representation of diverse wetland habitats in the PA system, improve management planning, and standardize law enforcement protocols. Leadership training for wardens, community engagement, and innovative funding strategies will play a vital role in the effective conservation of these critical ecosystems.

References

Aung, U.M. (2007). Policy and practice in Myanmar's protected area system. *Journal of Environmental Management* 84 (2): 188–203.

Aung, T.D., Naing, T.Z., Moses, S. et al. (2016). *An Assessment of the Wintering Population of Baer's Pochard in Central Myanmar*. Biodiversity and Nature Conservation Association.

Aung, P.P., Tun, T.A., Tun, A.M. et al. (2017). Monitoring Programme for the Critically Endangered Spoon-Billed Sandpiper on its Most Important Wintering Grounds in Myanmar. Report for BANCA.

Aung, T.D., Zaw, T.S., Win, L. et al. (2019). *The Study of Baer' Pochard in Central Myanmar*. Biodiversity and Nature Conservation Association.

Aung, T.D., Zaw, T.S., Aung, E.T. et al. (2021a). *Climate Change Vulnerable Assessment of Pyu Lake and Paleik Lake, Mandalay Region*. Biodiversity and Nature Conservation Association.

Aung, T.D.W., Kyi, S.W., Suzue, K., et al. (2021b). Rapid ecosystem service assessment of a protected wetland in Myanmar and implications for policy development and management. *Ecosystem Services* 50: 101336. https://doi.org/10.1016/j.ecoser.2021.101336

Bishnu, B.B., Nakamura, R., Tolentino, A.S. Jr., and Suzuki, S. (2014). *Catalyzing the Wise Use of Wetlands in Myanmar: Efforts and Ways Forward*. Ramsar Center Japan.

Bryant, R.L. (1996). *The Political Ecology of Forestry in Burma*. Honolulu: University of Hawaii Press.

Chatterjee, A., Phillips, B., and Stroud, D.A. (2008). *Wetland Management Planning. A Guide for Site Managers*. WWF, Wetlands International, IUCN & Ramsar Convention.

Conallin, J.C., Baumgartner, L.J., Lunn, Z. et al. (2019). Migratory fishes in Myanmar rivers and wetlands: challenges for sustainable development between irrigation water control infrastructure and sustainable inland capture fisheries. *Marine and Freshwater Research* 70 (9): 1241–1253.

Davies, J., Sebastian, A.C., and Chan, S. (2004). *A Wetland Inventory for Myanmar*. Ministry of Environment: Tokyo.

Gardner, R.C. and Davidson, N.C. (2011). The Ramsar Convention. In: *Wetlands: Integrating Multidisciplinary Concepts* (ed. B.A. LePage), 189–203. Dordrecht: Springer.

Hatfield Consultants (2015). *Inle Lake Conservation and Rehabilitation Project – End of Project Evaluation. Final Report*. Yangon: UNDP Myanmar.

Lwin, T., Uga, and Khaing, S.T. (1990). *Wildlife Conservation in Myanmar*. Yangon: Forest Department, Ministry of Agriculture and Forests.

Lwin, N., Linkie, M., Harihar, A. et al. (2016). Initiating conservation of a newly discovered population of the endangered hog deer *Axis porcinus* in Myanmar. *Oryx* 52 (1): 126–133.

Macintosh, D.J. (2016). *Assessment of Fisheries Management Needs an Sustainable Livelihood Opportunities in the Villages surrounding Meinmahla Kyun Wildlife Sanctuary. TCP Report*. Fauna & Flora International.

Naing, T.Z. (2009). *Bird observations near Bagan in 2008. Field Feathers* 2009.

Naing, T.Z. (2012). Bird Survey Along the Ayeyarwaddy River Between Mandalay and Singu in Central Myanmar, April & May 2012. Report for Wildlife Conservation Society.

Naing, T.Z. (2013). Bird Survey Along the Ayeyarwaddy River Between Mandalay and Bhamo in Central and North Myanmar, February 2013. Report for Wildlife Conservation Society.

Naing, T.Z. (2015). Bird Survey Along the Ayeyarwaddy River Between Mandalay and Bhamo in Central and North Myanmar 2013 & 2015. Report for Wildlife Conservation Society.

Naing, T.Z., Aung, P.P., and Htoo, S.L. (2015a). *Bird Surveys in Taninthayi Region, February to December 2014*. Report for Wildlife Conservation Society.

Naing, T.Z., Aung, P.P., and Htoo, S.L. (2015b). *Bird surveys in Taninthayi Region, January to June 2015*. Report for Wildlife Conservation Society.

National Wetland Policy (1995). The Republic of the Union of Myanmar.

Nature Wildlife Conservation Division (2019). *Republic of the Union of Myanmar: National Wetland Inventory. Technical Data*, vol. 1. NWCD: Nay Pyi Taw.

Oikos, I. and BANCA (2012). *Myanmar Protected Areas Context, Current Status and Challenges*. BANCA: Yangon.

Ramsar Convention Secretariat. (2010). *National Wetland Policies: Developing and implementing National Wetland Policies. Ramsar Handbooks for the Wise Use of Wetlands*, 4th edition, Ramsar Convention Secretariat vol. *2*.

Ramsar Convention Secretariat (2012). Resolution XI.12. Wetlands and health: taking an ecosystem approach. *11th Meeting of the Conference of the Parties to the Convention on Wetlands (Ramsar, Iran, 1971). Wetlands: Home and Destination* (6–13 July 2012). Romania: Bucharest.

Ramsar Convention Secretariat (2015). *National Biodiversity Strategy and Action Plan (2015–2020)*. Switzerland: Gland.

Ramsar Convention Secretariat (2019). Guidance on information on national wetland extent, to be provided in Target 8. "National Wetlands Inventory" of the Ramsar National Report for COP14. Switzerland: Gland.

Saluja, R., Prasad, S., Lwin, T.H., et al. (2023). Assessment of community dependence and perceptions of wetlands in the Upper Chindwin Basin, Myanmar, *Resources* 12: 112. https://doi.org/10.3390/resources12100112

Scott, D.A. (1989). *A Directory of Asian Wetlands. IUCN*. Switzerland: Gland.

Tordoff, A.W., Appleton, T., Eames, J.C. et al. (2007). Avifauna surveys in the lowlands of Kachin. *Natural History Bulletin of the Siam Society* 55 (2): 235–306.

Wongthong, P. and True, J. (2015). Updated situation analysis of the Gulf of Mottama. Community-Led Coastal Management in the Gulf of Mottama Project.

Yong, D.L., Jain, A., Jones, V.R. et al. (2023). *Bird Hunting in Mainland Southeast Asia. Situation Analysis and Recommendations for Conservation Action*. Singapore: BirdLife International (Asia). 52 pp.

Zöckler, C. (2016). *Bird Survey and Training Report Ayeyarwady Delta*. Myanmar: Fauna & Flora International.

Zöckler, C. (2019). *Winter Bird Monitoring at the Ayeyarwady River, Myitkyina – Bagan, Myanmar 2017, 2018 and 2019*. Manfred-Hermsen Stiftung and Fauna Flora International.

Zöckler, C. and Aung, C. (2019). The mangrove of Myanmar. In: *Sabkha Ecosystems. Tasks for Vegetation Science*, vol. 49 (ed. B. Gul, B. Böer, M. Khan, et al.), 253–268. Cham: Springer.

Zöckler, C., Hla, T.H., Clark, N. et al. (2010). Hunting in Myanmar is probably the main cause of the decline of the Spoon-billed Sandpiper *Calidris pygmeus*. *Wader Study Group Bulletin* 117 (1): 1–8.

Zöckler, C., Hla, T.H., and Bräunlich, A. (2012). Status of Bar-Headed Geese (*Anser indicus*) wintering in western Myanmar. *Kasarka* 15 (1): 63–65.

Zöckler, C., Naing, T.Z., Moses, S. et al. (2014). The importance of the Myanmar Coast for Water Birds. *Stilt* 66: 37–51.

Zöckler, C., Aung, P.P., Grindley, M. et al. (2018). *Coastal Wetlands in Myanmar – A Directory of Important Sites for Biodiversity*. Cambridge: ArcCona Ecological Consultants.

Zöckler, C., Lwin, N., Tun, T.Z. et al. (2020). Surveys of riverine birds along the Ayeyarwady River in 2017–2019 and conservation implications. *Forktail* 36: 1–15.

4

Wetlands of Plains of Pakistan

Andleeb Batool[1], Asia Parveen[1], Mishal Nawaz[1], Dania Razzaq[1], Maryam Mukhtar[2], and Naseeba Mustafavi[1]

[1] *Department of Zoology, Government College University, Lahore, Pakistan*
[2] *Department of Zoology, University of the Punjab, Lahore, Pakistan*

4.1 Introduction

A plain is a large area of terrain that is mostly level or slightly slopes in elevation. Plains occupy over one-third of the planet's land part. There are numerous varieties of plains, and each has certain qualities of its own. Among the most typical varieties are floodplains that are formed by the regular flooding of rivers. They are typically fertile and well-suited for agriculture. Coastal plains are located along coastlines and are often formed by the accumulation of sediment from rivers and erosion of the land. Piedmont plains are located at the foot of mountains and are often formed by the erosion of the mountains. Intermountain plains are located between mountains and are often formed by the filling of valleys with sediment (Adnan and Khan 2009). Pakistan has a highly diverse geographical area, which includes majestic mountain ranges of the Himalayas, Karakorum, and Hindu Kush, the Indus plains, deserts, and coastal areas. Altitudinal variations range from the world's second-highest mountain (K2) in the north to the sea level in the south.

4.2 Wetlands in Plains of Pakistan

A region of land that is saturated or completely submerged in water is called a wetland. The distinct plants known as hydrophytes characterize wetland habitats. Wetlands are defined as "areas of marsh, fen, peat land or water, whether natural or artificial, permanent or temporary, with water that is static or flowing, fresh, brackish or salt, including areas of marine water the depth of which at low tide does not exceed six meters" (Chaudhry 2010). Pakistan boasts significant and distinctive wetlands that sustain a broad spectrum of globally significant biological diversity, providing habitat for both native and endangered animals, and functioning as vital staging and breeding grounds for migratory bird populations, including those of globally threatened species, that travel through the Central Asian Flyways (Khan et al. 2012). According to reports, Pakistan is home to 225 wetland and lake locations, having a combined surface area of around 9.7%.

Wetlands of Tropical and Subtropical Asia and Africa: Biodiversity, Livelihoods and Conservation, First Edition. Edited by Thammineni Pullaiah.

The Convention on the Conservation of Waterfowl and Wetlands is also known as the Ramsar Convention (Chaudhry 2010).

The vital core of Pakistan is the 650,000 km^2 Indus Plain, which stretches from the Himalayas to the Arabian Sea. Ancient civilizations, agriculture, and a thriving populace all call it home. The Indus Valley Civilization, which lived between 3300 and 1300 BCE, is well known for its sophisticated drainage systems, urban planning, and creative expression. Cities with illustrious histories include Mohenjo-daro and Harappa. Its lifeline, the Indus River, offers natural irrigation, turning the terrain into a patchwork of fields for sugarcane, wheat, rice, and cotton (Abakode 2024). The Indus Plain has various topographies as well, with dry plains in the south and verdant valleys and rocky hills in the north (Watto and Mugera 2016). The Indus Plain's wetlands are incredibly diverse, with each having a distinct personality and ecological importance. The Upper Indus Basin has two (2) sub-basins divided by the Indus River, the Kohat sub-basin to the west and the Potwar sub-basin to the east. Similarly, the Lower Indus Basin encompasses the central and southern sub-basins (Nawaz et al. 2018).

4.3 The Haleji Wetland

Haleji Lake, one of Asia's largest bird sanctuaries and a significant wintering habitat for migratory birds, is located close to Karachi. It is Pakistan's second-largest freshwater lake and an important migratory bird habitat. The Haleji Wetland is one of the Ramsar sites and is situated around 15 km northwest of Thatta City and 85 km east of Karachi. Its coordinates are 24°48′55″N latitude and 67°45′37″E longitude. Both birdwatchers and waterfowl sanctuaries can be found here (Whitehead et al. 2009). There is a 5 km route from the National Highway to the Lake, and there is a 10 km wetland drive around it. At an elevation of 60 m, its catchment area is around 1704 ha. Pakistan's Sindh province is home to the sizable freshwater lake known as Haleji Lake. When water from seasonal sources was collected in a depression, it was once a saltwater lake. The soldiers stationed in Karachi during World War II needed more water. The lake was filled in with fresh water from an irrigation canal system after the salt water was removed and an embankment was built around it. As a consequence, Haleji developed a vital water source for Karachi's growing population and a sensitive haven for ducks. Siberian migratory birds use this water reservoir as a habitat from November to February. In the winter, it is a haven for bird lovers. To the great joy of bird enthusiasts, migrating birds visit this lake in considerable numbers throughout the winter. Haleji Lake is home to a wide variety of wildlife, including numerous endangered species. It is also one of Sindh's most significant locations for waterfowl breeding, staging, and wintering, with an annual population of 50,000–100,000 birds. In 1971, Haleji Lake was established as the first protected area as a game refuge. In 1997, it was designated as a wildlife sanctuary. Ramsar Advisory paid a visit to the location in 1990. The expedition concluded that the ecosystem of the Haleji Wetland is seriously threatened (Pratolongo 2013). The main factors that led to the deterioration of the Ramsar site were inadequate freshwater intake, poor management techniques, and lax monitoring and protection efforts. It is advised that the contaminated lake water be drained and that freshwater from the canal be released in order to preserve its quantity and quality, rebuild the destroyed ecosystems, and increase the number of wetland's endangered species (Mahessar et al. 2015) (Figure 4.1).

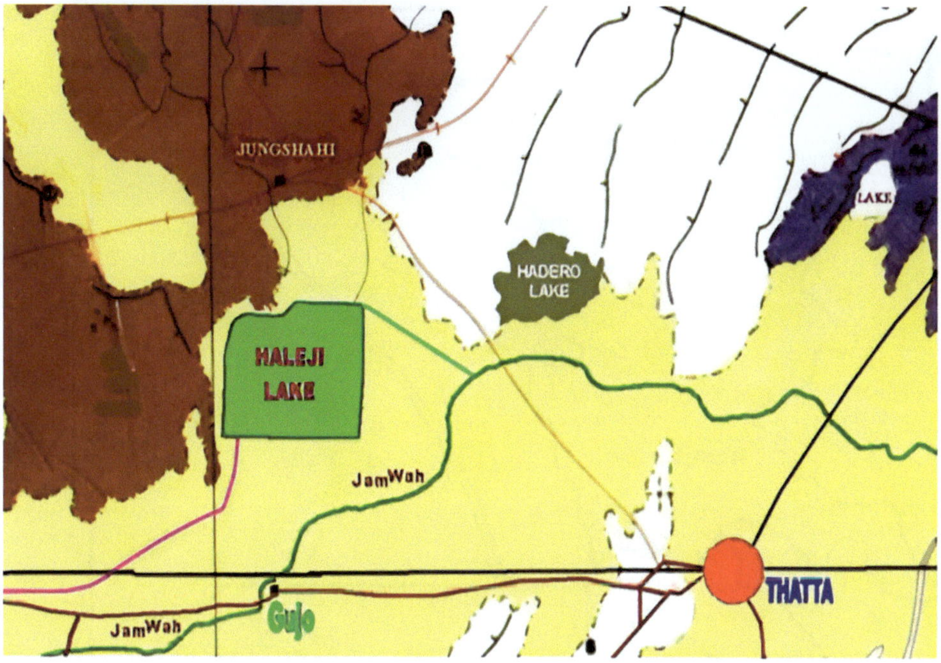

Figure 4.1 Location of Haleji Lake. *Source:* Adapted from Khan et al. (2012).

4.4 The Keenjhar Lakes

Keenjhar Lake was created by linking two natural lakes, Keenjhar and Sunheri ("Green" and "Golden") by dynamiting the separating hills in 1958 to create a single lake that was dubbed Kalri Lake at first. This man-made tropical lake was renamed Keenjhar Lake in 1972. The lake is located at 24°47′N and 68°2′E, around 120 miles from Hyderabad (Korai et al. 2008). It spans an area of 80 km^2 and is 27.2 km long. It is supplied with water by the River Indus by the Kalri Baghar Feeder, a canal that begins at the Ghulam Mohammad Barrage. The lake was formerly close to 25 m deep, but ongoing siltation and the buildup of decomposing vegetation have reduced the depth to 6–8 m (Figure 4.2).

The fish species found in Keenjhar Lake are as follows: omnivore fish accounted for 13.74%, carnivorous fish for 40.05%, and planktivores for 39.21% of the total. The eutrophic Keenjhar Lake produces a large amount of phytoplankton and a comparatively large amount of zooplanktons due to its high productivity and richness in plant nutrients. Fish that get this diet have good development rates and are highly nutritious. Thirty-nine species of aquatic plants were identified from Keenjhar Lake in the Chotiari Reservoir district, Sanghar. *Hydrilla verticellata, Najas indica, Myriophyllum spicatum, Potamageton* spp. (three species), *Vallisneria spiralis, Ceratophyllum demersum, Eichhornia speciosa, Chara* spp., *Typha angustata, Scirpus* spp., and *Cyperus difformis* have been previously reported from Keenjhar Lake. Prawns of families Palaemonidae and Atyidae have been previously reported by Keenjhar Lake (Siddiqui et al. 1973; Leghari et al. 1999; Jahangir et al. 2000). Eight species of family Penaeidae and 33 fish species

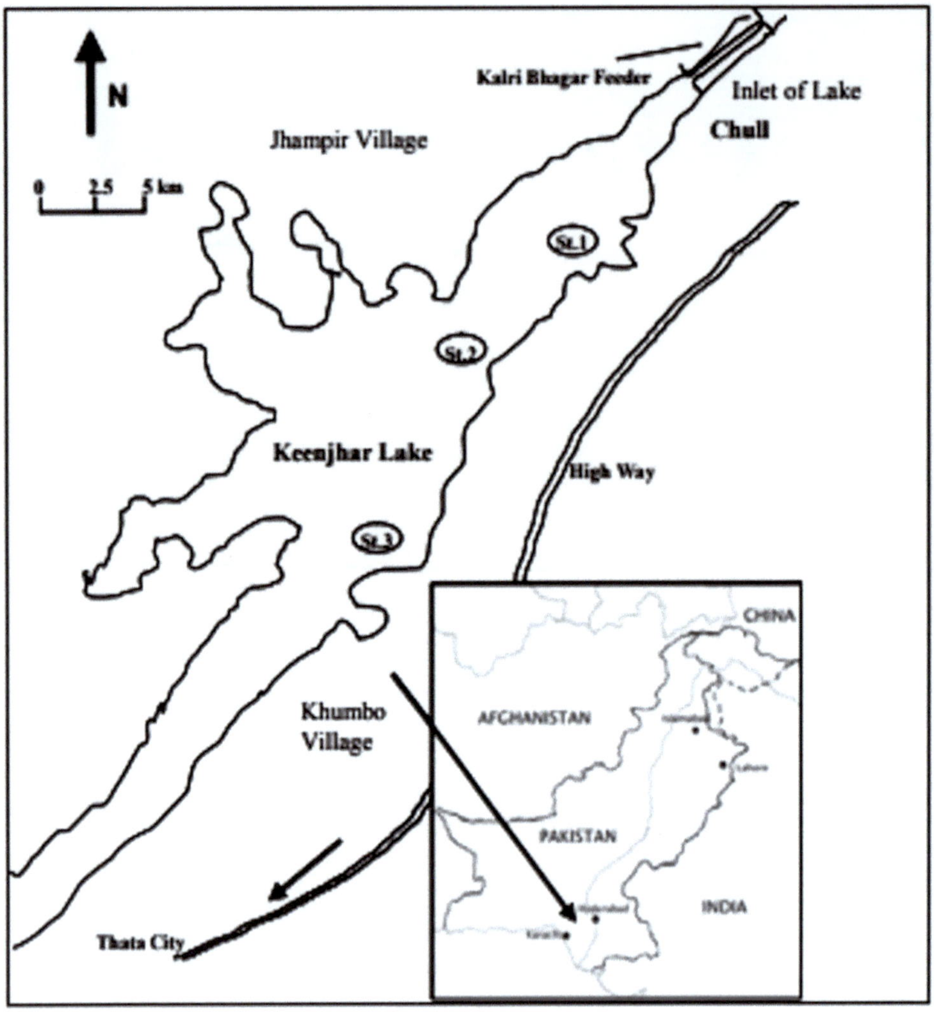

Figure 4.2 The Keenjhar Lake. *Source:* Korai et al. (2008)/Central Fisheries Research Institute.

representing 24 genera and 14 families (Sufi 1957; Leghari et al. 2000) were recorded. Overall, 51 fish species were identified from the Keenjhar Lake (Korai et al. 2008).

4.5 Chashma Barrage Wildlife Sanctuary

This sanctuary was created in 1991 and is situated in Punjab, Pakistan, next to the Chashma Barrage on the Indus River. Many permanent and migratory species, such as ducks, geese, pelicans, cormorants, and shorebirds, can find home there. Chashma Barrage is situated 32°25′N, 71°22′E; 25 km southwest of Mianwali, off the Mianwali – Dera Ismail Khan road, Punjab Province. Also includes areas located in the NWFP-Dera Ismail Khan District. The area of 34,099 ha and an altitude of 225 m were found in the Indus Ganges Monsoon Forest. A large-sized embankment or flood bund system on the Indus River that, during low tide, splits the reservoir into five small lakes, each covering up to 250 ha. The building project was finished in 1971. Springtime brings with it the most floods. The extent of the main river channel ranges in depth from 4.6 to 8.8 m, while the five lakes range in depth from 0.2 m during the dry season to 5.0 m at its peak during the flood season. The range of pH readings is 6.5–7.2. The arid subtropical environment with chilly winters and scorching summers. The range of yearly rainfall is 300–500 mm, while the range of relative humidity is 22–85%. In June, the average maximum temperature is 41 °C, while in January, the average lowest temperature is 4.5 °C.

The aquatic vegetation's species are *Typha angustata, Typha elephantina, Phragmites australis, Saccharum spontaneum, Nymphaea lotus, Myriophyllum* sp., *Nymphoides cristata, Potamogeton crispus,* and *Zannichellia palustris.* Among the native vegetation types of the area are species like *Olea ferruginea, Acacia modesta, A. nilotica, Adhatoda vasica, Dodonaea viscosa, Gymnosporea royleana, Prosopis cineraria, Reptonia buxifolia, Salvadora oleoides, Tamarix aphylla, T. dioica, Zizyphus mauritania, Z. nummularia, Chrysopogon aucheri, Lasiurus hirsutus, Heteropogon contortus,* and *Panicum antidotale. Prosopis glandulosa* has been introduced to the area. Most of the natural thorn forest on the plains east of the Indus has been destroyed for agricultural purposes and irrigated plantations of *Dalbergia sissoo* and other plants.

In 1974, the marsh was first designated as a 33,084 ha Wildlife Sanctuary. Since the Sanctuary was notified twice more, in January 1989 and July 1984, there has been a significant improvement in the degree of protection. The sanctuary did not cover several significant regions on the reservoir and the right bank of the River Indus. After a comprehensive examination of the region, it has been suggested that the borders of the sanctuary should cover a 3 km stretch that runs the whole length of the reservoir, including on both its sides and downstream of the Barrage constructions.

The water storage from projects like the Diamer-Bhasha Dam in Pakistan serves multiple purposes, including fishing, energy production, and agriculture. WAPDA grants leases for fishing operations. For usage in regional cottage industries, kana (*Saccharum* spp.) and reeds (*Phragmites australis* and *Typha elephantina*) are gathered. At the reservoir, a fish hatchery was created and put into service in 1987. The surrounding lands are utilized for forestry, agriculture, and cattle grazing.

A large storage dam is intended to be built in Kala Bagh, upstream of Chasma. The building of this dam may restrict the amount of water that can be stored there and change the Chasma Barrage's water regime. This construction could impact fishing activities in the marsh and cattle grazing on the shoreline, leading to significant disturbances in the area.

Significant disturbance occurs along the coastline, and the harvesting of reeds and notable variations in water level negatively impact aquatic flora. The development of fisheries, energy generation, flood management, and water storage for agriculture are the main priorities. In 1992, almost a thousand metric tons of fish were collected. In recent years, there has been a significant rise in the yield. The nature of the marshes sustains the local weavers. Additionally, the barrage offers fantastic chances for scientific study and environmental education.

A significant wintering and staging place for a diverse range of birds. In mid-winter, the marsh sustains about 50,000 Anatidae and coots, and in certain years, a significantly higher number. There were more than 114,000 birds in January 1975, around 100,000 in January 1987 and January 1988, and over 200,000 in January 1989, 1990, and 1991. The species that are most prevalent include *Fulica atra, Aythya ferina, A. penelope, A. crecca*, and *A. acuta*. The highest recorded numbers of *Fulica atra* were 165,400 in January 1989, 85,600 in 1990, and 82,400 in 1991. *Anser indicus* (recent peaks of 277 in 1982 and 65 in 1985) and a tiny wintering flock of *Anser anser* are infrequently seen during passage and winter. Peak tallies have consisted of 710 *Phalacrocorax niger*, 890 *Egretta garzetta*, 220 *Ardea cinerea*, 500 *Tadorna ferruginea*, 7436 *A. strepera*, 3900 *A. platyrhynhos*, 29,840

A. clypeata, 25,000 *Aythya ferina*, 165,418 *Fulica atra*, 560 *Larus ridibundus*, 800 *Nycticorax nycticorax*, 1800 *E. alba*, 220 *Platalea leucorodia*, 13,000 *Anas penelope*, 62,104 *A. crecca*, 10,100 *A. acuta*, 3156 *Netta rufina*, 3100 *A. fuligula*, 130 *Philomachus pugnax*, and 160 *Sterna aurantia*.

In 1992 and 1993, just 11,000 and 29,000 birds were counted on the Reservoir, which had an impact on mid-winter waterfowl counts in January. This is most likely the outcome of the water management initiatives that the Punjab Department of Power and Irrigation and WAPDA undertook. In January, barrage gates close and the water level rises above average, burying the aquatic vegetation. As a result, diving ducks have less food available, which eventually has an impact on the number of waterfowl. The changes in the population are being attentively observed.

In lower numbers, *Podiceps cristatus, P. nigricollis, Phalcrocorax carbo*, and other shorebird species, gulls, and terns are among the other frequent winter visitors. Cranes (*Grus grus* and *Anthropoides virgo*) use the barrage as a major staging location in the spring and fall. There have been reports of other ducks as well, including *Mycteria leucocephala, Ciconia nigra, Himantopus himantopus, Recurvirostra avosetta,* and *Glareola lactea*.

In the summer, the sanctuary serves as a refuge for birds that breed. In the summer, over 50 species are typically documented in a single day. This area has records of the internationally endangered long-tailed Warbler, *Prinia burnessi*. Though not recently spotted, the Indus Dolphin *Plantanista minor* has been documented in Chashma and is found in some sections of the Indus River upstream of the Barrage. *Sus scrofa cristatus, Axis porcinus, Canis aureus, Felis libyca,* and *Lutra perspicillata* are among the other animals that may be found in the region. *Gadusa chapra, Notopterus chitala, Catla catla, Cirrhinus mrigala, C. rego, Labeo rohita, L. microphthalamus, Puntius ticto, P. stigma, Barilius vagra, Wallago attu, Rita rita, Bagarius bagarius, M. seenghala, Heteropneustes fossilis, Eutropiichthys vacha, Nandus, Mastacembelus armatus, M. pancalus, Ambassis nama, A. ranga,* and *Channa punctatus* are among the abundant fish species. Among the other water animals were *Lissemys punctata, Hirudinaria* sp., *Rana tigrina, Kachuga smithi, Trionyx gangeticus, P. lamarrei,* and *Palaeomon dayanus* (https://rsis.ramsar.org/RISapp/files/RISrep/PK816RIS.pdf).

4.6 Taunsa Barrage Wildlife Sanctuary

This 3568-ha sanctuary was created in 1972 and is located in Punjab, Pakistan, next to the Taunsa Barrage on the Indus River. With over 250 kinds of birds identified, the marsh is home to a diverse array of wildlife, including songbirds, raptors, shorebirds, and waterfowl (Haider et al. 2022).

4.7 Astola Island

Astola, sometimes referred to as Haft Talar (meaning seven hills) locally, is an island in the Arabian Sea with a latitude of 25°7′21.51″N and a longitude of 63°50′51.53″E. Situated around 39 km off Pasni, Balochistan, it is the largest offshore island in Pakistan (Iqbal et al. 2018) (Figure 4.3).

It is home to colonies of approximately 22 coral types, and its sandy beach serves as breeding sites for various bird species, including plovers, coursers, and gulls, in addition to the endangered green sea turtle (Figure 4.4). The saw-scaled viper is native to the island. The waters around the island host a diverse range of cetaceans, including, whales, dolphins, and other fish species. The island is used by local fishermen to anchor their boats throughout the fishing season. During each year, Astola serves as a seasonal base for mainland fishermen to catch oysters and lobsters.

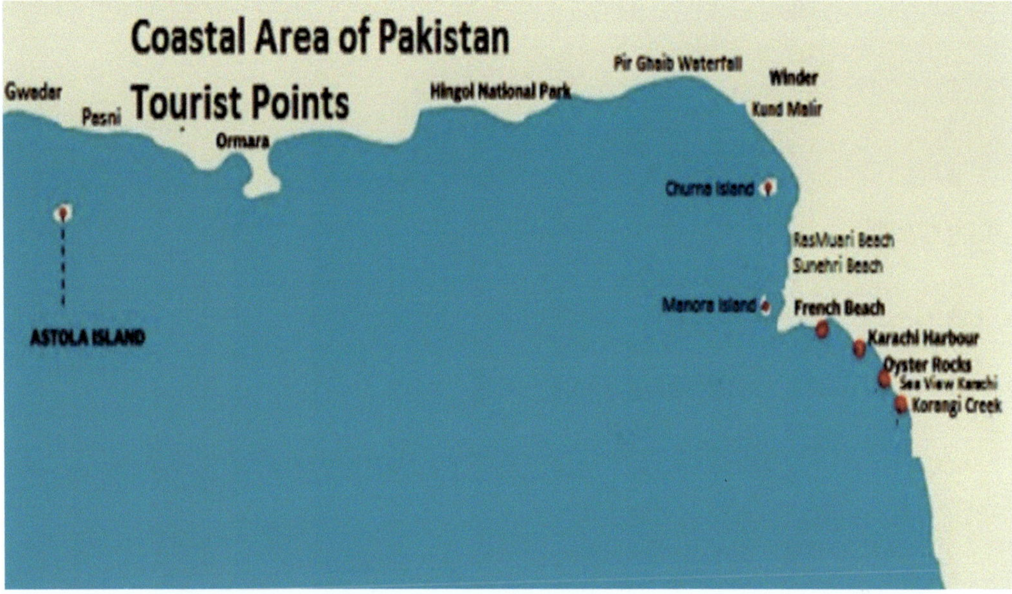

Figure 4.3 Astola Island. *Source:* Shahzad (2020)/International Journal of Multidisciplinary and Current Research.

Figure 4.4 Green Sea turtles. *Source:* Olive Ridley Project/https://oliveridleyproject.org/wp-content/uploads/2021/03/two-green-hatchlings-ghost-net_astola-island.jpg (last accessed on 19 June 2024).

4.8 Jiwani Wetland

Jiwani is located on the Makran coast of Balochistan, in the extreme southwest, close to the Pakistan–Iran border. The location is an accurate representation of a dry climate (Figure 4.5).

It features various habitat types, including desert, marine, freshwater, tropical thorny woods, mangroves, and scrub zones. It serves as one of Pakistan's nineteen existing Ramsar sites.

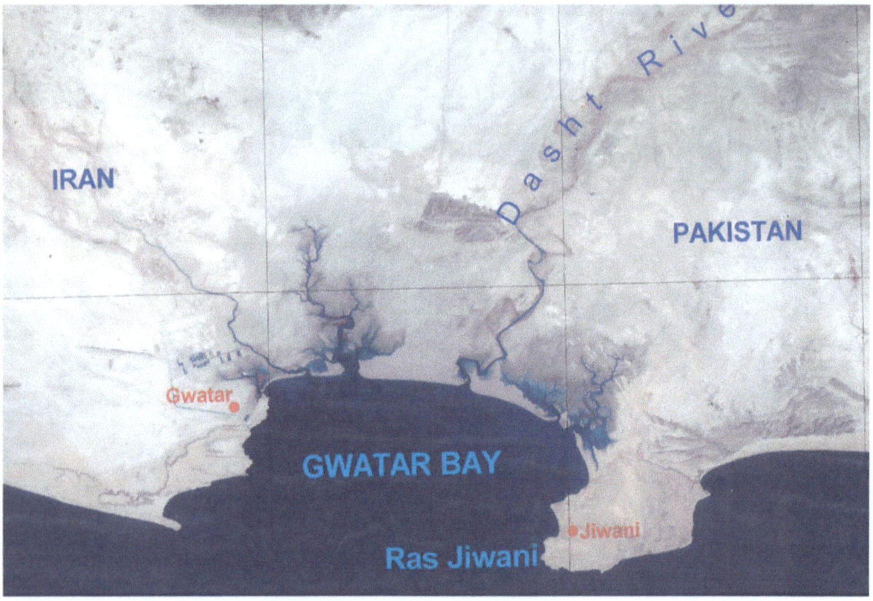

Figure 4.5 Jiwani Island. *Source:* Pandrani et al. (2005)/Zoological Society of Pakistan.

The area is of remarkable importance because it sustains endangered marine turtles, mangroves, marine animals, marsh crocodiles, and other birds (Pandrani et al. 2005). The Jiwani Coast is an important passageway and wintering habitat for migrating waterbirds. During the winter, terns, gulls, flamingos, waders, and pelicans are regular sights. Migratory ducks use the nearby water ponds and reservoirs during the winter. Grey Partridge, Sand grouses, and See-see Partridge are valuable game birds since they are widely spread and very common (Begum et al. 2016). The green turtle (*Chelonia mydas*), which breeds on the sandy coastlines of Sindh and Balochistan regions on a regular basis, is the most common of the five marine turtle species found in Pakistan's shoreline waters. The loggerhead, hawksbill, leatherback, and turtles olive ridley are among the remaining four species, but they are rarely seen. Local residents rely on fishing and seaport activities, which have led to stresses on marine biodiversity via excessive harvesting as well as pollution. This has harmed the numbers of olive ridleys and green turtles nestling along that coast (Khan 2013).

4.9 Miani Hor

Miani Hor (25°312′N, 66°202′E) is a swamp located around 90 km from Karachi at the easternmost portion of the Balochistan coastline (Figure 4.6).

The bay's overall size is 363 km^2, with 2500 ha of mangrove forest consisting of the following species: *Ceriops tagal, Rhizophora mucronata*, and *Avicennia marina*. This is the only place near Pakistan's coastline where three varieties of mangroves occur naturally (Rasool 2008). Mullets, Sardines, Sparids Somiani, and Carangids were the most numerous species. Catfishes, Crockers, white pomphret, electric rays, and ladyfishes (*Sillago sihama*) were among the more prevalent species. Miani Hor is one of the probable fish landing sites, consisting of four different fishing villages: Dam Bunder, Baloch Goth, Bhira, and Miani Hor. These are all settlements along the coast, primarily depending on fisheries (Siddiqui and Amir 2011) (Figure 4.7).

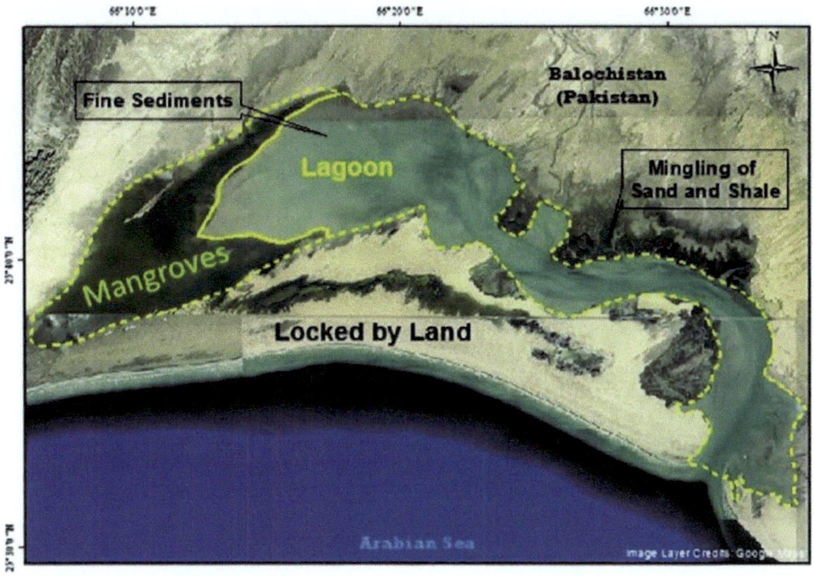

Figure 4.6 The Miani Hor Lagoon (60 km away from SW of Karachi) with mangroves and mud flats. *Source:* Solangi et al. (2019)/Elsevier.

Figure 4.7 Mangrove plantation in Miani Hor wetland. *Source:* World Wildlife Fund (WWF)/https://wwfeu.awsassets.panda.org/downloads/wwfdgispakistan2.pdf (last accessed on 19 June 2024).

4.10 Uchhali Complex

Uchalli Wetlands Complex is situated in Division Khushab, Pakistan, and consists of three lakes: Uchalli, Khabeki, and Jahlar. The Uchalli Wetlands Complex in Pakistan is one of the Ramsar sites with worldwide importance (Bhatti et al. 2018) (Figure 4.8).

Figure 4.8 Uchali complex. *Source:* WWF Pakistan/Dialogue Earth/CC BY-NC-ND 4.0/https://dialogue.earth/en/water/uchhali-pakistans-forgotten-ramsar-wetland (last accessed on 19 June 2024).

Figure 4.9 Flamingos in Uchhali complex. *Source:* WWF Pakistan/Dialogue Earth/CC BY-NC-ND 4.0/https://dialogue.earth/en/water/uchhali-pakistans-forgotten-ramsar-wetland (last accessed on 19 June 2024).

Located in the heart of the Salt Range, three distinct brackish to saline lakes with fluctuating levels, surrounded by agricultural fields, are a significant, winterizing area for the rare or vulnerable Ferruginous Duck, White-headed Duck, flamingos, and Greylag Goose. Villagers rely on the wetland for local water supply, and human activities consist of fishing, grazing livestock, recreation, and illegal hunting (Figure 4.9).

4.11 Indus River Basin Wetlands

The Indus Delta is the fifth-largest delta globally, formed by the River Indus and tidal creeks at Turshian and Khobar near Keti Bunder and the old delta near Karachi. It covers an area of 472,800 ha and comprises 17 significant creeks. The delta consists of various habitats, including freshwater lagoons, marshes, mangroves, sea bays, estuaries, creeks, mud flats, and salt flats. The ecosystem supports unique vegetation types like seaweeds, mangroves, and grasses. The delta hosts 147 fish species, with 22 being endemic. *Avicennia marina* is the principal mangrove species in the Indus (Loucks 2019).

The Indus River Delta is a vital ecosystem in the Indus Plain, providing essential services to local communities and supporting biodiversity. Some critical aspects of the national importance of the delta include its role in sustaining agriculture, providing livelihoods through fishing, and serving as a habitat for diverse flora and fauna, making it a crucial ecological and economic asset for Pakistan. The Indus Delta in the Indus Plain holds significant national importance due to its unique features and dynamic potential. The delta is a crucial ecosystem, covering agricultural land, lakes, riverine, and mangrove forests. It plays a vital role in the region's agriculture, industry, and commerce (Loucks 2019). However, the sustainability of the Indus Delta faces challenges such as erosion and environmental impacts due to dam construction and sediment rights negotiations. Natural processes and anthropogenic activities influence the morphological evolution of the Indus shelf region offshore of the delta (Khan et al. 2021). Efforts are needed to implement restoration schemes, data collection, and monitoring to ensure sustainable development in the face of climate change and economic pressures. The agricultural system in the Indus basin, closely linked to the delta, faces issues like over-extraction of groundwater, which leads to soil salinization and environmental impacts. Sustainable management practices are crucial for balancing water and food security while preserving the ecosystem health of the Indus Delta region (Loucks 2019).

The biodiversity of the Indus Delta includes a rich variety of flora and fauna, making it a significant ecological region. The Indus Delta is home to the largest arid mangrove forests globally, covering an area of around 600,000 ha. The mangrove forests have seen preservation and plantation initiatives, leading to an increase in their numbers over the years. The dominant mangrove species in the delta is *Avicennia marina*, which accounts for about 90% of the total mangroves in the region. The delta supports a diverse range of wildlife, including various bird species, fish, and mammals. A faunal biodiversity survey conducted in the area recorded a total of 38 finfish species, 21 shellfish species, 75 bird species, 10 mammal species, and 8 reptile species. Notable species found in the region include the Fishing Cat (vulnerable status), Indus River Dolphin (endangered species), Pangolin (near threatened), Asiatic Jackal (near threatened), and Bengal Fox (near threatened). Conservation initiatives in the Indus Delta focus on wetland restoration, habitat security enhancement, sustainable land use practices, and improved awareness raising and advocacy. These efforts aim to address threats to biodiversity and promote the sustainable management of this vital ecosystem. Restoration projects have been successful in replanting mangroves and engaging local communities in conservation activities (Memon 2005).

The people living in the Indus Delta region depend significantly on the river for their livelihoods and sustenance. The coastal communities in the delta rely heavily on marine resources for their livelihoods, which have been impacted by changes in water availability and environmental factors (Jamali et al. 2023). Inland sections of the Indus Delta see residents cultivating small tracts of land for agriculture, with fishing or outmigration often being the only viable option for many due to shrinking vocational opportunities (Karrar 2021). The livelihoods of around 300,000 residents in the Indus Delta are intricately tied to the river and its resources, with memories of a past abundant with water continuing to shape their vulnerabilities and challenges. The Indus River Basin supports the livelihoods of approximately 268 million people in India and Pakistan, with Pakistan particularly reliant on the river's water resources for 80% of its total water supply (Nabeel and Cheema 2021).

The depletion of water resources due to climate change impacts, rising water demand, and poor water management poses severe threats to the economic and water security of both countries, emphasizing the critical importance of efficient and fair management of the Indus River's resources for the well-being of the populations dependent on it (Nabeel and Cheema 2021). Efforts to address these challenges and ensure sustainable water management are crucial to safeguarding the livelihoods and economic security of the communities relying on the Indus Delta.

The Indus Delta faces various threats that impact its ecological balance and the livelihoods of the people dependent on it. The Indus Delta has shrunk by 92% since 1833, leading to significant environmental challenges and impacting the ecosystem (Indus River Delta 2023). Higher glacial melt is causing increased river flow, resulting in more frequent flooding and droughts, which pose significant threats to the river and the surrounding areas (Awais et al. 2022). Human activities such as dam construction, land clearance, and coal mining contribute to environmental degradation in the delta region, affecting biodiversity and ecosystem health (Jamali et al. 2023).

Poor water management practices, high water demand, and limited freshwater reaching the delta negatively affect the ecosystem and the people relying on it for their livelihoods (Raza et al. 2023). Efforts to address these threats through sustainable water management practices, conservation initiatives, and multisector transformations are crucial to safeguarding the Indus Delta's ecosystem and ensuring the well-being of the communities' dependent on its resources.

Conservation strategies for the Indus Delta are crucial to protect its ecosystem and the livelihoods of the people dependent on it. Initiatives like the "Indus River Rights Act" aim to recognize the inherent rights of nature in the Indus River, ensuring protection for its flow, essential functions within its ecosystem, and pollution-free (Raza et al. 2023).

Rejuvenating the river's natural course by dismantling dams and adopting a free-flowing river model can help revive the delta's ecosystem and maintain ecological integrity. Promoting water-efficient agricultural practices can save water for environmental flows back into the river, supporting the delta's health and biodiversity (ReliefWeb 2019). Preserving mangrove forests is essential as they form the backbone of the delta's delicate ecosystem, providing breeding grounds and food for various species (Raza et al. 2023). Implementing well-regulated water management systems, conducting research, and scientifically planning for future water needs are vital steps to ensure sustainable water use and conservation in the region (ReliefWeb 2019).

4.12 Mekran Coastal Wetlands

The Makran Coastal Wetlands Complex is present in the Indus Plain. It is located on the Balochistan coast, extending from the Basol River to Jiwani on the border with Iran (http://wwf.org.pk/pwp/mcwc). The Makran Coastal Wetlands Complex consists of five wetlands, including Astola Island (Haft Talar). It is internationally recognized as a Ramsar site and provides various ecological services and economic values to the region (Ali et al. 2008). The Mekran Coastal Wetlands Complex is approximately 708 km long and varies in width from about 9–23 km, covering an area of around 9700 km^2. This wetland area is a critical habitat for various bird species, with up to 66 different species using the region as a principal flyway for migration, including threatened species like the Sarus Crane. The Indus River within the Mekran Coastal Wetlands Complex supports over 95% of the surviving population of the endangered and endemic Indus River Dolphin, highlighting the ecological significance of this area. The socioeconomic conditions in the region surrounding the wetlands show a high level of poverty, income disparity, and dependence on agriculture and natural resource-based incomes, with farming, fishing, and livestock ownership being major sources of livelihood. Traditional fishing methods are still prevalent in the area, involving a large number of people under contract systems that often exploit poor fishermen (Ali et al. 2008) (Figure 4.10).

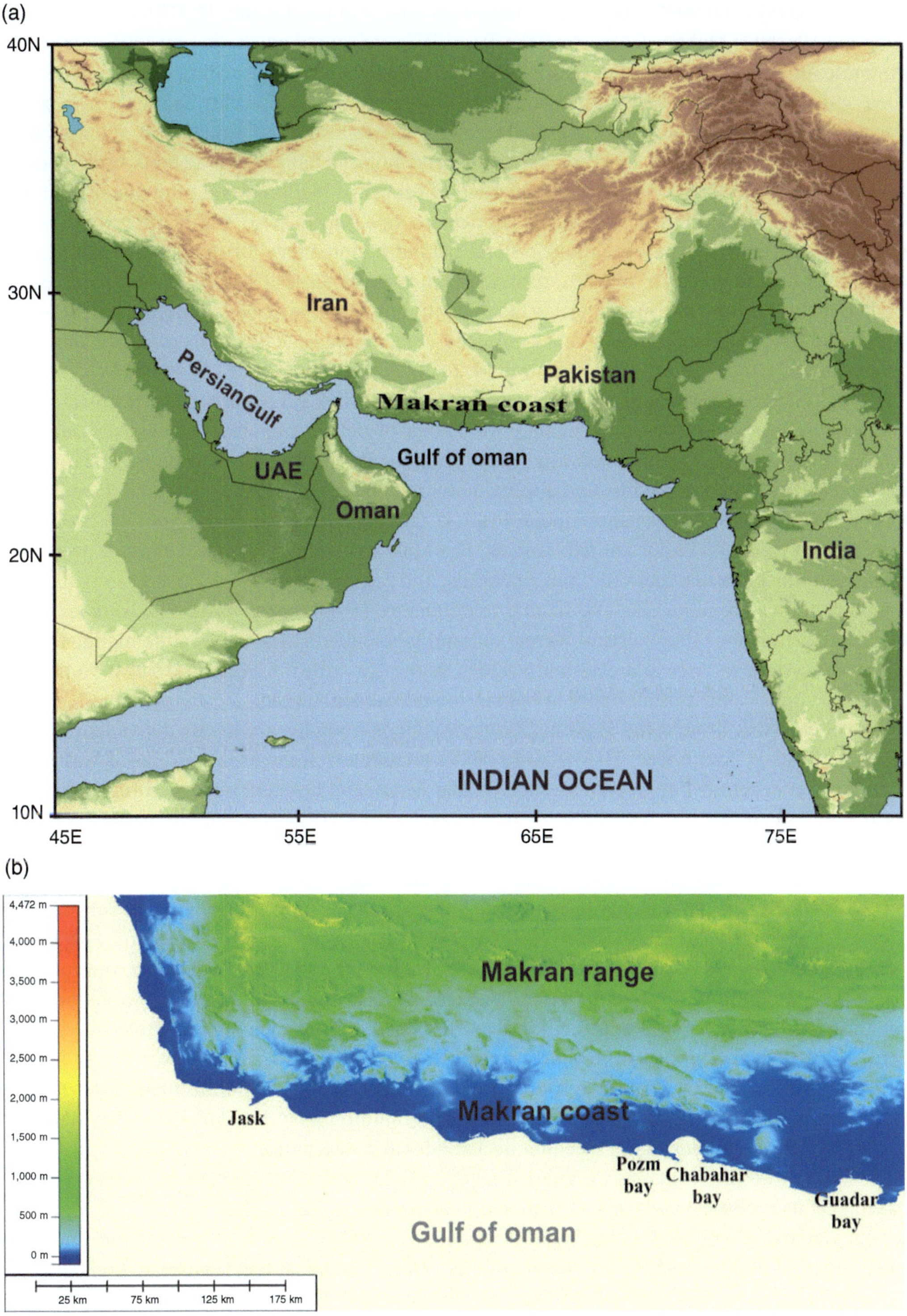

Figure 4.10 Location map of the Makran coast in the northwestern Indian Ocean (a) and SRTM-DEM of the Iranian coast of Makran (b). *Source:* Ghanavati et al. (2021) / MDPI / CC BY 4.0.

The Makran Coastal Wetlands Complex covers an area of approximately 12,774 ha and extends along Baluchistan's Makran Coast from the Basol River to Jiwani on the border with Iran. The economic functions of the Makran Coastal Wetland Complex include providing services like fuelwood and raw materials, fishing for commercial purposes, and biodiversity support for marine species like green turtles and hawksbill turtles. The wetlands also serve regulatory functions such as nature protection, maintenance of migration, and nursery habitats for various bird species (Nawaz et al. 2018). The complex offers production functions like fuelwood, raw materials, and commercial fishing, with estimated values per hectare for each function. Additionally, it provides regulatory functions like nature protection and habitat maintenance for migratory birds, along with carrier functions such as tourism, recreation, and grasslands. The Makran Coastal Wetlands Complex plays a crucial role in supporting biodiversity, providing livelihood opportunities through fishing and raw material collection, and serving as a recreational site due to its unique ecological features and historical significance (Nawaz et al. 2018).

The Makran Coastal Wetlands Complex, located in Balochistan, Pakistan, is a significant area for biodiversity conservation. The complex supports a range of globally important biodiversity, such as significant areas of coral, mangrove swamps, and seagrass. It is home to threatened marine turtles, the endangered Marsh Crocodile, and approximately 10 species of marine cetaceans. Jiwani Sea Mount and the areas of rocky shoreline, estuaries, and associated mangroves are vital breeding areas for commercially important fish species. The Makran Coastal Wetlands Complex provides production functions like fuelwood, raw materials, and commercial fishing as well as regulatory functions like nature protection and habitat maintenance for migratory birds. The complex also offers career functions like tourism, recreation, and grasslands. These data points highlight the diverse ecosystem services and economic values associated with the Makran Coastal Wetlands Complex, emphasizing its importance at both local and national levels.

The people living around the Makran Coastal Wetlands Complex are highly dependent on its resources for their livelihoods. The majority of the population in the Makran Coastal Wetlands Complex is concentrated around coastal towns like Jiwani and Gwadar, where fishing and port-based activities are the main occupations. Fishing is the principal occupation in the area, carried out by traditional fishermen using small diesel-powered boats and by deep-sea trawlers from large-scale national and international commercial fishing industries. Harmful fishing practices, such as the use of illegal nets and fishing during spawning seasons, are prevalent, leading to substantial by-catch of endangered marine species like turtles and cetaceans. Other forms of resource dependence include fuelwood collection from mangroves, hunting for sport and subsistence purposes, and coastal pollution from diesel boat fuel leakage, all contributing to negative impacts on the wetlands. The low literacy rate and cultural norms restricting women's social mobility hinder women's participation in natural resource management regimes, exacerbating poverty and perpetuating dependence on natural resource-based livelihoods (https://portals.iucn.org/). These data points highlight the significant reliance of local communities on the resources of the Makran Coastal Wetlands Complex for their economic activities and sustenance.

The Makran Coastal Wetlands Complex, located in Balochistan, Pakistan, faces several threats that impact its ecological balance and the livelihoods of the people dependent on it. The Makran Coastal Wetlands Complex is facing habitat loss due to coastal erosion, land reclamation, and the construction of harbors and ports. Coastal pollution from diesel boat fuel leakage and oil adds to the negative impact on the wetlands (https://portals.iucn.org/). Harmful fishing practices, such as the use of illegal nets and fishing during the spawning season, lead to substantial by-catch of endangered marine species like turtles and cetaceans. Climate change impacts, such as rising sea levels and more frequent extreme weather events, pose significant threats to the wetlands and the species they support. Weak social indicators for the provision of health, sanitation services, and access to drinking water exacerbate poverty and perpetuate dependence on natural resource-based livelihoods.

Conservation strategies for the Makran Coastal Wetlands Complex are essential to protect its biodiversity and ensure sustainable livelihoods for local communities. Over the past 30 years, there has been an increase in wetlands areas within the Makran Coastal Wetlands Complex, indicating positive trends in wetland conservation and restoration efforts. Developing sustainable livelihoods for communities in Ramsar sites like the Makran Coastal Wetlands Complex is crucial for promoting conservation practices and ensuring the well-being of local residents (Qamer and Hassan 2007). Recent years have seen the implementation of short-term conservation initiatives in the Makran Coastal Wetlands Complex, indicating ongoing efforts to protect and manage this critical ecosystem. The Makran Coastal Wetlands Complex has been recognized as a Ramsar site, with specific areas like Astola Island and Jiwani Coastal Wetlands receiving protected status. Conservation programs focusing on turtle and mangrove conservation have been initiated to safeguard biodiversity in the region. Efforts to conserve globally important biodiversity within the complex, including coral reefs, mangrove swamps, seagrass, threatened marine turtles, endangered species like the Marsh Crocodile, migratory birds, and marine cetaceans, are crucial for maintaining ecological balance and species diversity (Qamer and Hassan 2007).

References

Abakode (2024). Geopolitics of water in India and Pakistan: Ongoing tensions over the . . . Mondo Internazionale. https://mondointernazionale.org/focus-allegati/geopolitics-of-water-in-india-and-pakistan-ongoing-tensions-over-the-indus-river (accessed 10 March 2024).

Adnan, S. and Khan, A.H. (2009). Effective rainfall for irrigated agriculture plains of Pakistan. *Pakistan Journal of Meteorology* 6 (11): 61–72.

Ali, Z., Butler, J., Meadows, P., and Meadows, A. (2008). Developing sustainable livelihoods for communities in a Ramsar site: the Makran coastal wetlands complex, Pakistan. In: *Asian Wetland: Global Position* (ed. Y.-H. Pang, W.-T. Fang, and I.T. Savillo), 176–181. Taipei: Construction and Planning Agency.

Awais, M., Vinca, A., Parkinson, S. et al. (2022). Replenishing the Indus Delta through multi-sector transformation. *Frontiers in Environmental Science* 10: https://doi.org/10.3389/fenvs.2022.958101/full.

Begum, A., Khan, M.Z., Ghalib, S.A. et al. (2016). Distribution, status and current trends in the population of coastal birds of Balochistan. *Canadian Journal of Pure & Applied Science* 10 (2): 3853–3864.

Bhatti, S.G., Tabinda, A.B., Yasin, F. et al. (2018). Spatio-temporal variations in physico-chemical parameters and potentially harmful elements (PHEs) of Uchalli Wetlands Complex (Ramsar site), Pakistan. *Environmental Science and Pollution Research* 25: 33490–33507.

Chaudhry, A.A. (2010). Wetlands in Pakistan: what is happening to them. *World Environment Day* 5.

Ghanavati, E., Shah-Hosseini, M., and Marriner, N. (2021). Analysis of the makran coastline of Iran's vulnerability to global sea-level rise. *Journal of Marine Science and Engineering* 9 (8): 891.

Haider, M.Z., Ahmed, S., Sial, N. et al. (2022). Avian diversity and abundance of taunsa barrage ramsar site in punjab, pakistan. *Journal of Zoological Systematics and Evolutionary Research* 2022 (1): 4736195.

Indus River Basin (2024). Water security and sustainability. https://www.researchgate.net/publication/330452214_Indus_River_Basin_Water_Security_and_Sustainability (accessed 4 March 2024).

Indus River Delta (2023). Wikipedia. https://en.wikipedia.org/wiki/Indus_River_Delta (accessed 10 March 2024).

Iqbal, A., Qazi, W.A., Shahzad, N., and Nazeer, M. (2018). Identification and mapping of coral reefs using Landsat 8 OLI in Astola Island, Pakistan coastal ocean. *International Conference on Emerging Technologies (ICET)* (accessed 21–22 November 2018).

Jahangir, T.M., Khuhawar, M.Y., Leghari, S.M. et al. (2000). Some studies on water quality and biological life at Kinjhar and Haleji lakes of district Thatta, Sindh [Pakistan]. *Pakistan Journal of Biological Sciences (Pakistan)* 3 (11): 1965–1972.

Jamali, H., Jamali, M., Hayat, S., and Iqbal, J. (2023). Indus River Delta: contested narratives in the climate of change. *Asian Journal of Social Science* 51 (4): 215–226.

Karrar, H.H. (2021). The Indus delta between past and future: precarious livelihoods and neoliberal imaginaries in a parched coastal belt. *Journal of Indian Ocean World Studies* 5 (1): 47–67.

Khan, A. (2013). Pakistan wetlands programme's marine Turtle conservation efforts on Daran Beach, Jiwani, Pakistan. *Indian Ocean Turtle Newsletter* 17: 26–30.

Khan, M.Z., Abbas, D., Ghalib, S.A. et al. (2012). Effects of environmental pollution on aquatic vertebrates and inventories of Haleji and Keenjhar Lakes: Ramsar Sites. *Canadian Journal of Pure & Applied Science* 6 (1): 1759–1783.

Khan, U., Janjuhah, H.T., Kontakiotis, G. et al. (2021). Natural processes and anthropogenic activity in the Indus River Sedimentary Environment in Pakistan: a critical review. *Journal of Marine Science and Engineering* 9 (1): 1109.

Korai, A.L., Sahato, G.A., Lashari, K.H., and Arbani, S.N. (2008). Biodiversity in relation to physicochemical properties of Keenjhar Lake, Thatta district, Sindh, Pakistan. *Turkish Journal of Fisheries and Aquatic Sciences* 8 (2): 259–268.

Leghari, S.M., Jafri, S.I.H., Mahar, M.A., et al. (1999). Biodiversity of Chotiari Reservoir, District Sanghar Sindh, Pakistan. *Proceedings of the Seminar on Aquatic Biodiversity of Pakistan,* pp. 139–157 (accessed 25–27 February 2014).

Leghari, S.M., Jafri, S.I.H., Mahar, M.A. et al. (2000). Limnological study of Sonharo, Mehro Pateji and Cholari lakes of district Badin, Sindh, Pakistan. *Pakistan Journal of Biological Sciences* 3 (11): 1904–1909.

Loucks, D.P. (2019). Developed River Deltas: are they sustainable? *Environmental Research Letters* 14 (11): 113004.

Mahessar, A.A., Qureshi, A.L., Mukwana, K.C., and Jakhrani, A.Q. (2015). Study of environmental impacts and threats to the Ramsar Haleji Lake, Sindh, Pakistan. *International Journal of Applied Environmental Sciences* 10 (5): 1577–1590.

Memon, A.A. (2005). Devastation of the Indus river delta. *Impacts of Global Climate Change* https://doi.org/10.1061/40792(173)500.

Nabeel, F. and Cheema, M.J.M. (2021). Pakistan's transboundary water challenge. In: *Water Resources of Pakistan. World Water Resources*, vol. 9 (ed. M.A. Watto, M. Mitchell, and S. Bashir). Cham: Springer https://doi.org/10.1007/978-3-030-65679-9_3.

Nawaz, R., Dehlavi, A., and Bajwa, N. (2018). Indus river basin wetlands. In: *The Wetland Book* (ed. C. Finlayson, G. Milton, R. Prentice, and N. Davidson). Dordrecht: Springer https://doi.org/10.1007/978-94-007-4001-3_219.

Pandrani, A., Hasnain, S.A., Ghalib, S.A., and Ahmad, E.J.A.Z. (2005). Observations on the waterbirds of Jiwani Wetland Complex, Makran Coast (Balochistan). *Pakistan Journal of Zoology* 37 (4): 301.

Pratolongo, P.D. (2013). The ecology of coastal wetlands. *Marine Ecology in a Changing World* 140.

Qamer, F. and Hassan A. (2007) Pakistan wetlands programme. WWF.ORG.PK. http://wwf.org.pk/pwp/reports/2_Database.pdf (accessed 14 March 2024).

Rasool, F. (2008). Observation on the Avifauna of Miani Hor, Balochistan. *Pakistan Journal of Marine Sciences* 17 (2): 99–106.

Raza, S.M., Aslam, M., and Vazhayil, A.M. (2023). A tail-end tale: the story of the Indus Delta. In: *River Culture – Life as a Dance to the Rhythm of the Waters* (ed. K.M. Wantzen), 213–240. Paris: UNESCO Publishing.

ReliefWeb (2019). How to bring the Indus Delta back to life – give it water – Pakistan. https://reliefweb.int/report/pakistan/how-bring-indus-delta-back-life-give-it-water (accessed 14 February 2024).

Sciencedirect (2024). The Indus delta – catchment, river, coast, and people – sciencedirect. https://www.sciencedirect.com/science/article/abs/pii/B9780128140031000125 (accessed 9 March 2024).

Shahzad, S.M. (2020). De-constructing the problematic maritime tourism in Pakistan: opportunities and challenges. *International Journal of Multidisciplinary and Current Research* 8: 378–387.

Siddiqui, P.J., and Amir, S.A. (2011). Baseline Survey of Fish Diversity at Miani Hor, Balochistan. Report on Pakistan Wetland Programme. https://wwf.org.pk/pwp/reports/Baseline%20Survey%20of%20Fish%20Diversity%20at%20Miani%20Hor.pdf.

Siddiqui, P.A., Baqai, I.U., and Iqbal, M. (1973). Check list of fishes of Keenjhar (Kalri) lake with notes on environment conditions and fisheries potential. *Agriculture Pakisthan* 24 (2): 201–220.

Solangi, S.H., Nazeer, A., Abbasi, S., and Zulkifil, M.M. (2019). Morphological features of continental shelf margin: examples from the Pakistan Offshore. *Geodesy and Geodynamics* 10 (1): 77–91.

Sufi, S.M.K. (1957). Fish fauna of the Kinjer Lake (West Pakistan) with an account of the major fishing implements employed by the local fishermen. *Agriculture Pakistan* 8: 208–229.

Watto, M.A. and Mugera, A.W. (2016). Groundwater depletion in the Indus Plains of Pakistan: imperatives, repercussions and management issues. *International Journal of River Basin Management* 14 (4): 447–458.

Whitehead, P.G., Wilby, R.L., Battarbee, R.W. et al. (2009). A review of the potential impacts of climate change on surface water quality. *Hydrological Sciences Journal* 54 (1): 101–123.

5

Wetlands of Sri Lanka: Biodiversity, Livelihoods, and Conservation

A.H. Magdon Jayasuriya

EML Consultants PLC, Pitakotte, Sri Lanka

5.1 Introduction

Wetlands are transitional between terrestrial and aquatic systems where the water table is usually at or near the surface, or the land is covered by shallow water. They are characterized by the presence of water, either permanently or periodically. While the soils are waterlogged, they support vegetation specifically adapted to wet conditions and lack flood-intolerant vegetation. Water saturation (hydrology) largely determines how the soil develops and the types of plant and animal communities living in and on the soil. Wetlands may support both aquatic and terrestrial species. The prolonged presence of water creates conditions that favor the growth of specially adapted plants (hydrophytes) and promote the development of characteristic wetland (hydric) soils. Wetlands vary widely because of regional and local differences in soils, topography, climate, hydrology, water chemistry, vegetation, and other factors, including human disturbance. Indeed, wetlands are found from the tundra to the tropics and on every continent except Antarctica. Two general categories of wetlands are recognized: coastal or tidal wetlands and inland or nontidal wetlands (USEPA 2023). The taxonomy and nomenclature basically follow Dassanayake et al. (1980–2000) and Senaratna (2001), but are updated according to ME, NH, & DNBG (2020).

Coastal/tidal wetlands are more pronounced in temperate climates than in tropical habitats. They are closely linked to estuaries where seawater mixes with freshwater to form an environment of varying salinities. The saltwater and the fluctuating water levels (due to tidal action) combine to create a rather difficult environment for most plants. Consequently, many shallow coastal areas are unvegetated mud flats or sand flats. Some plants, however, have successfully adapted to this environment. Certain grasses and grass-like plants that adapt to the saline conditions form the tidal salt marshes that are found along the Atlantic, Gulf, and Pacific coasts. Mangrove swamps, with salt-loving shrubs or trees, are common in tropical climates. Some tidal freshwater wetlands form beyond the upper edges of tidal salt marshes where the influence of saltwater ends (USEPA 2023).

Inland/nontidal wetlands are most common on floodplains along rivers and streams (riparian wetlands), in isolated depressions surrounded by dry land (for example, playas, basins, and "potholes"), along the margins of lakes and ponds, and in other low-lying areas where the

Wetlands of Tropical and Subtropical Asia and Africa: Biodiversity, Livelihoods and Conservation, First Edition. Edited by Thammineni Pullaiah.

groundwater intercepts the soil surface or where precipitation sufficiently saturates the soil (vernal pools and bogs). Inland wetlands include marshes and wet meadows dominated by herbaceous plants, swamps dominated by shrubs, and wooded swamps dominated by trees (USEPA 2023).

5.2 Biodiversity

The classification of Wetlands of Sri Lanka follows the system presented in the recently published Diversity of Terrestrial Vegetation in Sri Lanka (Jayasuriya 2023). Therefore, they are presented in topographic order starting at the seashore (in all climatic zones) in the direction of the highest altitude in the island, i.e. from the peripheral lowlands to the central highlands (Figure 5.1). Marine and freshwater aquatic ecosystems are not included in this treatment of the wetlands.

Fourteen main wetland formations (or types) can be recognized in Sri Lanka as follows and they are abbreviated as in Jayasuriya (2023):

1) Salt Marsh (SALM)
2) Mangrove Forest (MANF)
3) Basin Mangrove Forest (BMAF)
4) Brackish Water Marsh (BRWM)
5) Brackish Water Fern Marsh (FERM)
6) Permanently Flooded Inland Marsh (PMIF)
7) Seasonally Flooded Inland Marsh (SFIM)
8) Thermal Marsh (THEM)
9) Spring Forest (SPRF)
10) Villu Grassland (VILG)

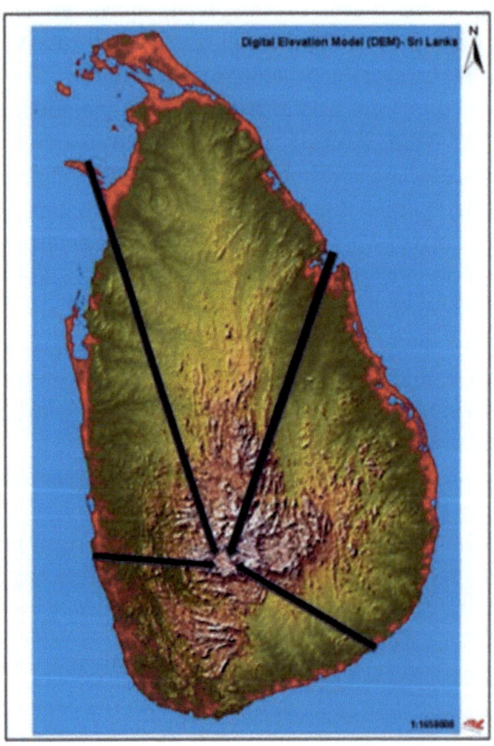

Figure 5.1 Geographical sequence of presentation of wetland formations according to rise in elevation – from peripheral lowlands to Central Highlands (Jayasuriya 2023).

11) Gilgai Wetland (GILW)
12) Dry Riverine Evergreen Forest (DREF)
13) Dry Canal-associated Evergreen Forest (DCEF)
14) Montane Wet Patana Grassland (MWPG)
    14.1) Dwarf Bamboo Assemblage (DWBA)
    14.2) Marsh–Peat Assemblage (MAPA)

In the following account, livelihoods and conservation of each wetland formation are described together with information on biodiversity as appropriate. Such applied information is elaborated only in wetland formations that are relatively large or extensive and furthermore if they are ecologically and/or economically important, e.g. salt marshes and mangrove forests.

5.2.1 Salt Marsh

This edaphic vegetation formation occurs in the northwestern (Mannar and Puttalam Districts) and southeastern (Hambantota District) semiarid zones, which normally receive between 650 and 800 mm of annual rainfall, but rarely exceeding 900 mm. They are influenced by high temperatures of 30–35°C during April–September causing extremes of drought, salt sprays from the ocean and high evaporation which brings about salt crystallization in shallow lagoons. Soils consist of Solodized-Solonetz, Solonchak, and Grumosol. Plants that thrive in this harsh environment are herbs and low shrubs that exhibit xerophytic characteristics as they constitute succulent stems and leaves (Figure 5.2a, b).

Pemadasa (1986) presented a comprehensive account on the "Saltmarsh communities" and recognized several such communities such as *Suaeda* community, *Halosarcia* (*Arthrocnemum*) community, saltmarsh pastures, and depressions within the saltmarsh formation. Some of the essential floristic elements in this formation are *Sueda maritima* ("Umiri") (T), *S. monoica, S. vermiculata, Halosarcia indica* ("Kotanai") (T), *Cynodon dactylon* ("Ruha"), *Salicornia brachiata, Evovulus alsinoides* ("Vishnu-kranti"), *Portulaca tuberosa* ("Uru-genda"), and *P. wightiana*. Saltmarsh pastures occur on somewhat elevated grounds with sand flats and three groups, viz. Wet saltmarsh pasture, dry saltmarsh pasture, and mixed pasture were described (Pemadasa 1986). The saltmarsh pastures are the major feeding grounds for livestock in Mannar, Puttalam, Hambantota, and Yala and, therefore, play a major role in the livestock industry in these regions (Pemadasa 1986).

5.2.1.1 Salt Marshes in Mannar District

Ranawana et al. (2023) estimated the total extent of salt marshes in Mannar District as 4703.07 ha that includes 4266.41 ha of undisturbed salt marsh, 148.13 hectares of encroached salt marsh, 127.51 ha of salt marsh including scattered grasslands, scrub and *Prosopis*, and 161.02 ha of salt marsh with scattered Mangroves. Ranawana et al. (2023) further recognized five saltmarsh communities in Mannar District, based on visual observations on physiognomy and species composition, namely, (A) Chenopod succulent Dominant, (B) Sedge Dominant, (C) Grass Dominant, (D) Forb Dominant, and (E) Mixed Saltmarsh Communities.

Fauna: During the rainy season, an algal mat consisting mainly of nitrogen-fixing microorganisms covers the bottom substrate of shallow water body. This provides habitat and shelter for fish species and other aquatic organisms such as the Grilled Horn Shelled Snail (*Ceritidopsilla cingulata*) and smaller fish species such as Milk Fish fry (*Chanos chanos*) and Flat head grey mullet (*Mugil cephalus*). Milk Fish fry is collected for brackish water aquaculture in the Mannar and Puttalam areas (Arulpragasam 2000).

Figure 5.2 Wetlands of Sri Lanka. (a) Salt marsh (VW). (b) Salt marsh. (c, d) Mangrove forest (VW). *Source:* Photo by Vimukthi Weeratunga. (e, f) Basin mangrove. *Source:* Photo by A.H. Magdon Jayasuriya (chapter author).

Lizards: During the long dry season (April to September) groups capable of tolerating dry, sandy, and saline conditions, especially reptiles, e.g. endangered Saw-scaled Viper (*Echis carinatus*) and the endemic Devaka's Fan-throated lizard (*Sitana devakai*) (MOE 2012).

Birds: Birds become the most abundant species in salt marshes during the wet season. They are vital wintering sites for migratory waterbirds. During the period between 2018 and 2020, 39 migratory waterbird species, 5 species having migratory and resident populations, and 13 resident waterbird species were recorded in the Vankalai Sanctuary and Videthalthivu Nature Reserve. Of them, Little Ring Plover (*Charadrius dubius*) and Kentish Plover (*Charadius alexanadrinus*) have resident and migratory populations (Ranawana et al. 2020). They are listed as endangered in the National Red List 2021 – Conservation Status of Birds of Sri Lanka (2021).

Coastal wetlands of Mannar District fall within the Palk Bay Coastal Avifaunal Zone of Sri Lanka (MoMD & E 2018) and several species inhabiting the area considered zone-restricted species. When salt marshes are filled with water a large flock of Greater Flamingo (*Phoenicopterus roseus*) could be seen in Mannar marshes. This species, an uncommon winter migrant and a summer loiterer in some areas, is mainly restricted to Palk Bay Coastal Avifaunal Zone at present. The Spot-billed Duck (*Anas poecilorhyncha*), together with Grey francolin (*Francolinus pondicerianus*), a landfowl species, are considered restricted to this special avifaunal zone of Sri Lanka (MoMD & E 2018).

Ungulates: The salt marshes sustain many grazing ungulates, especially free-grazing cattle owned by livestock herders, along with some buffaloes. In addition, there are small populations of feral donkeys inhabiting the salt marshes (Ranawana et al. 2020).

5.2.1.1.1 Livelihoods and Conservation Ranawana et al. (2023) provide information with many references that indicate major ecosystem services of salt marshes emphasizing the need for protecting them:

- **They provide a buffer against waves and storm surges**: Marsh vegetation helps in stabilizing the coast by trapping sediments from the land runoff. Over time, increased intertidal surface level and vertical leaf blades provide a form of sea defense by reducing wave energy, velocity, and duration. At times, marshes support reducing storm surge, duration, and height by providing extra water uptake.
- **Nursery areas for coastal biotas**: Salt marsh ecosystems play a major role in fisheries by providing nursery grounds. The complex and dense vegetation structure of marshes provides a safe habitat for commercially important young fishes, as this area is mostly inaccessible to large fishes. Sometimes, they act as high tide refuges bird feeding on adjacent mudflats and are used as a source of food for marsh-dependent birds.
- **Filters for nutrients and pollutants**: Marshes are called natural filters as they effectively purify water entering the estuary. As water from upstream passes through marshland, it loses velocity and suspended particles get deposited on marsh surface. This natural filtration mechanism facilitates nutrient and pollutant uptake and recycling processes.
- **Facilitate carbon sequestration**: Salt marshes are known as one of the most productive ecosystems in the world serving as an organic carbon sink sequestering millions of tons of carbon annually (Mitsch and Gosselink 1986).
- **Enhance cultural values**: The aesthetics of the salt marshes and the high diversity of migratory and passerine birds support tourism, bird watching, and related research.

5.2.2 Mangrove Forests

The mangrove forests, as described below, represent the typical characteristics of this main edaphic and hydrological vegetation formation.

Syn. *Mangals* (Macnae 1968)
Intertidal Forests (Saenger et al. 1983)
Coastal woodlands (Saenger et al. 1983
Mangrove (Mepham and Mepham 1985)
Tidal Forests (Mepham and Mepham 1985)

Being an island in the Indian Ocean with more than hundred rivers originating from the central highlands and radiating toward the sea, Sri Lanka possess many lagoons and estuaries along the coastline of 1700 km. As the coastline runs through different climatic zones, the diversity of mangroves and species therein are remarkably high, and they represent an important coastal ecosystem of the island. The number of true mangrove species reported from Sri Lanka is nearly one third of the global mangrove diversity. However, as the total annual range of tidal variation is less than 1 m, the mangrove coverage of the island is small (Jayathissa et al. 2012). Their extents in Sri Lanka have been variously estimated as 6083; 9530; 11,000; 12,189; and 15,670 ha (Pemadasa 1986; Gunatilleke et al. 2018), i.e. 0.1–0.2% of the total land area.

Mangrove forests in Sri Lanka are mainly found in the northern coast along Jaffna, Vadamarchchi, and Thondiamannar lagoons (Jaffna District); Trincomalee, Kathiraveli, Valaichenai, Batticaloa, and Pothuvil in the eastern coast; southern coasts of Weligama and Gintota and Balapitiya, and Bentota, Negombo, Chilaw, and Puttalam lagoons; and Mannar in the western and northwestern coasts. The Kala Oya estuary (Puttalam District) appears to hold the country's largest extent of mangroves, covering an area of over 1837 ha (de Silva and de Silva 1998).

The average air temperature prevailing in the MANF ranges from about 30 to 35 °C. The relative humidity is very high and ranges from 80% to >90%. The salinity in mangrove waters could vary from almost nil to more than that of seawater. The lagoons and estuaries contain alluvial deposits, and the soil consists mainly of silt and fine clay, although in some areas sandy soil is found. The soil is waterlogged and consequently poorly aerated and in some locations, there is stagnant water. Salinity depends mainly on the pattern and amount of freshwater discharge from associated rivers, the tidal amplitude, and the topography and the extent of the estuary or lagoon. Depending on the region, several species of seagrasses are found in the fore-mangrove, and various brackish water marsh or salt marsh are found in the back mangrove while strand vegetation sometimes occurs alongside the mangrove frontier. The seagrass beds are common in shallow waters in larger lagoons such as those in Puttalam, Batticaloa, and Jaffna.

The mangroves generally grow up to about 10 m in height, although in some localities, trees could grow to a height of 15–20 m. A typical mangrove formation consists of two groups of plant species, namely, the "true mangroves" (sometimes referred to as "core" or "exclusive" mangrove species) that are exclusively found in mangrove habitats and "mangrove associates" (or "back-mangrove" species) that occur also in habitats other than mangroves (Figure 5.2c–f; Table 5.1). Twenty-three species of true-mangrove trees and shrubs have been recorded in Sri Lanka (de Silva and de Silva 1998).

The true mangroves usually grow only on the intertidal shores of lagoons and estuaries, whereas mangrove associates usually occur in association with mangrove habitats, but on higher elevation away from tidal influence. However, this concept cannot be strictly applied as incidences of true mangrove species occurring in purely freshwater habitats away from the tidal influence and other instances where some mangrove associates occurring among true mangrove species have been often reported.

Various trees and shrubs grow in the back mangrove, and they are often known as "mangrove associates"; the important mangrove associates are as follows: *Acrostichum aureum*, *Acanthus ilicifolius*, *Volkameria inermis*, *Thespesia populnea*, *Hibiscus tiliaceus*, *Cerbera odollam*, *Dolichandrone spathacea*, *Ardisia elliptica*, *Tamarix indica*, and *Barringtonia racemosa etc.* In general, many true mangrove species are commonly found in all sites, but the floristic composition of the dry zone MANFs is somewhat different from that of the wet zone MANFs. Because of the seasonality of rains and the lower freshwater discharge in the dry zone, the MANFs in this zone experience higher salinity as opposed to those in the wet zone, which experience much higher and evenly distributed annual rainfall and freshwater discharge and therefore lower salinity. The palm *Nypa*

Table 5.1 Important species of mangroves of Sri Lanka (de Silva and de Silva 1998 – taxonomy updated by author according to ME, NH, & DNBG (2020).

True mangroves	Mangrove associates (fore mangroves – seagrasses)	Back mangroves
Acanthaceae (Avicenniaceae)	Cymodoceaeae	Aizoaceae
Acanthus ilicifolius	*Cymodocea serrulata*	*Seuvium portulacastrum*
Avicennia marina	*Halodule uninervis*	Amaranthaceae
A. officinalis	*Syringodium isoetifolium*	*Tecticornia indica*
Arecaceae	Hydrocharitaceae (Najadaceae)	*Salicornia brachiata*
Nypa fruticans	*Enhalus acoroides*	*Sueda maritima*
Combretaceae	*Halophila decipiens*	*S. monoica*
Lumnitzera littorea	*H. ovalis*	*S. vermiculata*
L. racemosa	*Najas graminea*	Annonaceae
Euphorbiaceae	*N. marina*	*Annona glabra*
Excoecaria agallocha	*Thalassia hemprichii*	Apocynaceae
Fabaceae	Potamogetonaceae	*Cerbera odollam*
Cyanometra iripa	*Struckenia pecinata*	*Ochrosia oppositifolia*
Lythraceae	Ruppiaceae	Arecaceae
Sonneratia alba	*Ruppia maritima*	*Cocos nucifera*
S. apetala		Asteraceae
S. calceolaris		*Sphaeranthus africanus*
Malvaceae (Sterculiaceae)		*S. amaranthoides*
Heritiera littoralis		*S. indicus*
Meliaceae		Bignoniaceae
Xylocarpus granatum		*Dolichandrone spathacea*
X. rumphii		Calophyllaceae
Primulaceae (Myrsinaceae)		*Calophyllum inophyllum*
Aegireras cornicultum		Cyperaceae
Rhizophoraceae		*Cyperus rotundus*
Bruguiera cylindrica		*Fimbristylis ferruginea*
B. gymnorhiza		*Schoenoplectus littoralis*
A. sexangula		Fabaceae
Ceriops decandra		*Aganope heptaphylla*
A. tagula		*Dalbergia candanatensis*
Rhizophora apiculata		*Derris trifoliata*
R. mucronata		*Pongamia pinnata*
Rubiaceae		Hernandiaceae
Schyphiphora hydrophyllacea		*Hernandia nymphaeifolia*
		Lamiaceae
		Premna serratifolia
		Volcameria inermis
		Lecythidaceae
		Barringtonia asiatica
		B. racemosa
		Loranthaceae
		Dendrphthoe falcata hemiparasitic on *Aegiceras corniculatum*
		Lythraceae
		Pemphis acidula
		Malvaceae
		Hibiscus tiliaceus
		Thespesia populnea
		Ardisia elliptica

(*Continued*)

Table 5.1 (Continued)

True mangroves	Mangrove associates (fore mangroves – seagrasses)	Back mangroves
		Pandanceae *Pandanus odorifer* Primulaceae (Myrsinaceae) *Ardisia elliptica* Poaceae *Zoysia matrella* Tamaricaceae *Tamarix indica* Viscaceae *Viscum orientale* on *Excoecaria agallocha* Pteridophyta (Ferns) Adiantaceae *Acrostichum aureum*

fruticans is found only in the wet zone, and its distribution extends to the southwestern coast from the estuary of Ranna Oya (Hambantota District) in the south to the estuary of Gin Oya (Puttalam District) in the north. *Rhizophora mucronata* and *Bruguiera cylindrica* are more common in the dry zone, whereas *Rhizophora apiculata* and *Bruguiera sexangula* are more common in the wet zone. The differences in the floristic compositions reflect the salinity tolerances of individual species. In the dry zone, MANFs often have patches of saltmarsh vegetation in the back mangroves (Pemadasa 1986). These edaphic mosaics are influenced by salinity, and they consist of exclusively herbaceous species dominated by species such as *Suaeda* spp., *Salicornia brachiata*, and *Tecticornia indica*. In the wet zone, however, the back mangroves are influenced by brackish conditions supporting mangrove associates such as *Acrostichum aureum,* a robust shrubby fern, several tree species, e.g. *Dolichandrone spathacea, Cerbera odollam*, and *Hibiscus tiliaceus* and shrubs such as *Volkameria inermis, Premna serratifolia,* and less often *Pandanus odorifer*. In some disturbed localities, the colonization by *Annona glabra*, an invasive alien treelet/large shrub, can be seen.

Zonation of trees can be seen in larger undisturbed MANFs. Generally, a zone of *Rhizophora, Sonneratia*, and *Bruguiera* occurs at the waterfront, with *Rhizophora* usually occupying the lowest position, followed by a mixed mangrove community of other true mangrove species such as *Aegiceras corniculatum, Ceriops tagal*, and *Lumnitzera racemosa. Avicennia marina* is found in the mixed zone, usually toward the inland margin, but may extend to the edge of water as well. The concept of steep-shore and gentle-shore mangroves was proposed by Pemadasa (1986). In smaller and narrower MANFs as well as in those that have been much affected by human interference, this zonation is disturbed and all species may occur intermixed.

de Silva and de Silva (1998) recognized two broad MANF communities in Sri Lanka, "Low saline" and "High saline." The low-saline mangrove community is usually present in river estuaries where salinity is low, often below 10 ppt, and more importantly, where this low salinity prevails throughout or during most of the year. It shows a relatively more luxuriant growth and communities of *Sonneratia caseolaris, Bruguiera sexangula*, and *Rhizophora apiculata* as well as the mangrove associates such as *Dolichandrone spathacea* and *Annona glabra* are characteristic of the low saline conditions. An estuary or lagoon-associated with a low-saline mangrove community usually contains communities of freshwater phytoplanktons, zooplanktons, macrophytes, and macrofauna. High-saline mangrove community is usually present on the shores of lagoons, islands within lagoons and estuaries in which the salinity is more than 25 ppt, at least during a part of the year. They are usually characterized by the presence of *Rhizophora mucronata* and *Avicennia marina*; but *Pemphis acidula*, *Acanthus ilicifolius*, *Aegiceras corniculatum*, and *Excoecaria agallocha* are also common. The lagoons and estuaries with associated high-saline MANFs usually contain marine planktons and seagrasses. These two mangrove communities are comparable to the classification proposed by Amarasinghe and Balasubramaniam (1992), namely, the Riverine Mangrove Forest (=Low Saline MANF) occurring in estuaries and Fringing Mangrove Forest (=High Saline MANF) occurring on the shores of lagoons and islands in lagoons. However, the inaptness of the latter classification is evident by the relationship of the species distribution with salinity (chemical) rather than geography (de Silva and de Silva 1998).

Fauna: Most of the fauna live either on or near mangrove root systems. Some invertebrates thrive in the mangrove canopy, of which the most abundant are the crabs. https://www.globalnature.org/bausteine.net/f/6462/VimukthiWeerathunga-MangrovesofSL.pdf?fd=2.

Faunal diversity includes the following:

- Snails, barnacles, bryozoans, tunicates, sponges, polychaete worms, isopods, amphipods, and jellyfish.
- **Mollusk's**: Such as Mud Whelks (Terebrala), Periwrinkles (Littoria), and shipworms.
- **Crustaceans**: Such as crabs and prawns (most dominant crustaceans)
- **Insects**: Mostly found in noninundated areas
- Fish
- **Reptiles and Amphibians**: Including snakes, lizards, crocodiles, and sea turtles.
- **Birds**: Birds are the most abundant terrestrial vertebrates found among mangroves, which turn to these habitats in search of shelter and food.
- **Mammals**: Monkeys, deer, otters, and small cats. Marine mammals such as dugongs and dolphins have also been observed in Puttalam Lagoon adjacent to mangrove habitats.

5.2.2.1 Livelihoods

Mangroves are highly productive ecosystems with an array of benefits to both animals and humans (Veetil et al. 2023).

- Mangroves play an important role in carbon sequestration by absorbing up to four times more carbon per hectare than an average tropical forest.
- They play a major role in weathering the adverse effects of climate change, as they are equipped with the ability to mitigate and reduce the impacts of coastal erosion, storms, hurricanes, and tsunamis. It is a well-known fact that mangroves played a major role as a protective barrier in some of the coastal areas to mitigate the effects of the tidal waves when Sri Lanka was affected drastically by tsunami in 2004.
- Mangroves decrease pollution of near-shore coastal waters by trapping pollutants.
- It is a place where freshwater, marine, and terrestrial habitats meet and therefore it serves to provide a home to thousands of different species of plants and animals.
- They support coastal fishing communities including shrimp farming and traditional fisheries such as brush piles and fish kraals. The highest dependence of local communities on mangrove

fisheries has been recorded from Puttalam lagoon, Mi Oya estuary, Chilaw lagoon, and Negombo lagoon, respectively. Their benefits on offshore fisheries have also been noted, as many fish species use mangroves as nursery areas.

- They provide surrounding communities with timber for construction of houses, boats, fishing gear, and for fuelwood.
- They provide some subsidiary food items to local communities. Fruits, such as *Sonneratia*, are used to make drinks and ice cream. *Nypa* stalks are used in herbal teas and their fruits are consumed in some Asian countries.
- Some mangrove species are also used in leather processing, as they are rich in tannins, a chemical commonly used in this industry.

5.2.2.2 Conservation

Between the years 1980 and 2005, approximately 25% of mangrove habitats are estimated to have been destroyed. Main causes include conversions to aquaculture, chiefly prawn farming, agriculture, and other urban uses. Approximately 3000 ha of mangroves have been recorded lost in the Puttalam District due to the expansion of aquaculture. Deforestation for timber and pollution are also causes for mangrove degradation and loss. Pollution especially from wastewater from aqua ponds further degrades these forests. Fringing mangroves are mostly what remains now in Sri Lanka around estuaries and lagoons. During dry seasons, cattle graze on lands adjacent to mangroves damaging these ecosystems (Veetil et al. 2023); https://climatefactchecks.org/mangrove-degradation-in-sri-lanka-a-story-from-the-east.

Sri Lanka Mangrove Conservation Project Sri Lanka Seacology, a nonprofit environmental conservation organization, is helping Sri Lanka become the first nation in history to preserve and replant all its mangrove forests. As a result, Sri Lanka is on its way to being a model for other countries wishing to build conservation capacity and economic prosperity in communities where healthy mangroves are most critical – small fishing villages. Seacology helps mitigate poverty in coastal communities by providing sustainable livelihoods training and microloans in exchange for leadership in community mangrove conservation cooperatives. Program participants become leaders in their communities, working on mangrove propagation, reforestation, and protection. The livelihood training and microloan program enables beneficiaries to create or expand sustainable businesses, contributing to greater financial stability, better nutrition, and health; increased economic security for families; and improved resilience against destructive storm surges. Key facts: (i) Over 14,500 ha of mangrove forest have been demarcated and educational sign boards have been installed in almost 50 lagoons and estuary systems; (ii) in 2017, the project propagated 703,800 mangrove seedlings; (iii) more than 7900 women and young people have been trained in mangrove conservation and 2893 have received microloans to create or expand sustainable businesses and foster financial stability (https://unfccc.int/sites/default/files/resource/Sri%20Lanka%20Mangrove%20Conservation%20Project.pdf).

In 2015, Sri Lanka announced that it would protect mangroves. Therefore, the Department of Wildlife Conservation, Forest Department, Central Environmental Authority, and Coast Conservation Department in Sri Lanka have taken solid initiatives to preserve the valuable mangrove vegetation of the country. The establishment of the Nature Resource Center at Pambala and Mangrove Education Center at Maduganga, establishment of mangrove nurseries, restoration of mangrove forests in lagoons and estuaries, and training community leaders while educating, developing, and improving skills and attitudes of the coastal population on the benefits of preserving the mangrove forests are some of the initial actions in this regard. The National Adaptation Plan of Sri Lanka also focuses on mangroves under Sri Lanka's climate actions.

5.2.3 Basin Mangrove Forest

Basin Mangrove Forest is classified as a vegetation association within the main vegetation formation, namely, the mangrove forests. They develop as inland depressions or basins, often behind fringe mangroves, and the surrounding soil is washed into the basin (de Silva and de Silva 1998). They are often isolated from all but the highest tides. Flooded water can remain for a long period of time due to the topography of the area and the soil therefore is waterlogged and poorly aerated with high organic content. The vegetation is like that of high-saline mangrove community with *Rhizophora mucronata* and *Avinennia marina* dominating the flora. Basin mangroves are essentially natural, and they are known to occur in the Jaffna peninsula, particularly in Vadamarachchi lagoon, "but has not been observed on the west, south or east coasts" (de Silva and de Silva 1998).

However, the occurrence of patches of Basin Mangroves was discovered in abandoned limestone quarry areas in Aruwakkalu (Puttalam District) and their vicinities on the west coast bordering the Puttalam lagoon (EML 2017). In contrast with natural and typical Basin mangroves in the Jaffna peninsula, those in Aruwakkalu were of recent origin and evidently developed anthropogenically following the excavation of quarries by the Cement Corporation which were abandoned about 50 years ago (EML 2017). Therefore, the Basin Mangroves at Aruwakkalu, although analogous with its natural precursor in the Jaffna peninsula, has been the result of the development of an edaphically modified habitat that has exposed excavated soil strata at the sea level, or below which has been influenced by relatively high salinity. This is clearly indicated by the appearance of thin layers of crystalized salt on the soil surfaces during dry seasons (Figure 5.2e, f).

Further, the recently evolved Aruwakkalu site indicates transitional vegetation communities centered around Basin Mangrove dominated by populations true mangroves, *Lumnitzera racemosa* while species such as *Acrostichum aureum*, *Excoecaria agallocha*, and *Premna serratifolia* are mangrove associates or "back-mangrove" species developed on slightly elevated fringes above the true-mangrove zone. The recent origin of the Basin Mangrove is evidenced by the absence of *Rhizophora* and *Avinennia*, which appear only in mature mangrove stands.

During wet seasons, the seepage of freshwater from surrounding elevated peripheries of the quarry that ascend to heights of 4–10 m has apparently caused the development of an edaphic and hydrological transition at the bottom areas of the quarry indicated by patches of Salt Marsh vegetation with species such as *Tamarix indica*, *Zoysia matrella* ("Mal-etora"), and *Salvadora persica*. This Salt Marsh zone occurs way above the Basin Mangrove zone, which is at the bottom and often remain inundated. Semiaquatic or periodically inundated conditions with brackish water table are indicated by the presence of species such as *Bacopa monniera* ("Lunu-wila"), *Phyla nodiflora*, *Fimbristylis ferruginea* ("Boru-pan"), and *Typha angustifolia* ("Hambu-pan").

5.2.4 Brackish Water Marsh

The marshes bordering lagoons are influenced by salinity variations in space and time. The Muthurajawela marsh in the Gampaha District, with an extent of 3068 ha, is well known and being adjacent to the Negombo Lagoon, is influenced by brackish conditions (CEA/Euroconsult 1994). The marsh habitats are permanently waterlogged and anaerobic promoting the development of anaerobic bacteria that liberate noxious gasses such as methane and hydrogen sulfide making these sites unpleasant. The reducing chemical conditions facilitate reduction of ferric and manganic ions to ferrous and manganous status, respectively, which are detrimental to mesophytes and hydrophytes but promote the colonization of species such as *Lagenandra ovata* ("Ketala"), *Colocasia esculenta* ("Gahala"), *Alocasia macrorrhiza* ("Habarala"), *Lasia spinosa* ("Kohila"), *Typha angustifolia*, and *Hanguana malayana* ("Induru") – in some sites. Marshes are often colonized by amphibious species, e.g. *Hygrophila schulli* ("Katu-ikiriya"), *Monochoria vaginalis* ("Diya-habarala"), *Lasia spinosa*, and *Ludwigia peruviana* ("Beru-diyanilla"). Considerable extents of the marsh habitats in the Muthurajawela have been already colonized by the invasive, exotic tree, *Annona glabra*. Bundala (Hambantota District) is another well-known wetland complex consisting of five lagoons and associated marshes with varying salinity levels. *Typha angustifolia* occurs in large patches in both brackish water and freshwater conditions (Figure 5.3a–d).

5.2.5 Fern Marsh

As observed by the author, FERM is a special vegetation association within the vegetation formation of BRWM. In some areas of the Muthurajawela wetland, the development of this special vegetation association, here named as Fern Marsh, dominated by the fern *Acrostichum aureum* ("Kerae or Keran-koku"), was noted. Accordingly, this vegetation is locally known as "Kerawwa," which occurs in large gregarious patches with extents of up to six hectares in some areas within the Muthurajawela wetland (Gampaha District) and possibly in other similar habitats such as Bundala. The shallow brackish water is influenced by the adjacent Negombo Lagoon, often forming backwaters of the lagoon. Within the Fern Marsh some associate species, commonly *Annona glabra* and *Hibiscus tiliaceus* and occasionally *Pandanus odoratissimus* were seen. On marginal elevated habitats, populations of *Phragmites karka* ("Nala gas" or "Bata") were observed while *Annona glabra*, a short-stature tree, is observed to be rapidly invading this habitat with a prediction of becoming the dominant species at the expense of *Acrotrichum*, sooner or later (Figure 5.3c, d).

5.2.6 Permanently Flooded Inland Marsh

This vegetation formation is essentially freshwater and permanently flooded marshes occurring in association with inland rivers. The Mahaweli Ganga flood plain system around Janaranjana wewa (Trinncomalee District) constitutes parts of this special wetland. Its essential components are riverbanks, river outflows, back slopes, backwaters, and marshes that are deeper than freshwater marshes and villus (usually up to a meter or more), often separated by patches of alluvial plains, cutoff channels, and meandering river channels that finally return to the river downstream. This special wetland habitat mainly supports floating aquatic plants such as *Eichhornia crassipes* ("Japan-jabara") in invasive scale, small populations of *Nymphaea pubescens* ("Olu"), *N. nouchali* ("Manel"), etc.

This hydrologic and edaphic vegetation formation is essentially a wetland. Such marshes occur mainly in the lowlands inward of the coastal zone and sometimes in inland habitats associated with rivers or other water bodies. During rainy seasons, the marshes are flooded to the capacity and enriched with nutrient-rich alluvium that promote luxuriant growth of marsh vegetation (Figure 5.3e).

Figure 5.3 Wetlands of Sri Lanka. (a) Brackish Water Marsh. (b) Brackish Water Marsh – with a population of *Nypa* palms. (c, d) Brackish Water Fern Marsh. (e) Permanently Flooded Inland Marsh. *Source:* Photo by A.H. Magdon Jayasuriya (chapter author).

5.2.6.1 Colombo Wetlands

According to the proposed classification, Colombo Wetlands are placed within the vegetation formation (type) No. 6 PFIM. Colombo was accredited as a Ramsar Wetland site during the 13th Conference of the Parties to the Ramsar Convention on Wetlands (COP13) in Dubai on 25 October 2018. The Ramsar convention entered into force in Sri Lanka on 15 October 1990. At present, Ramsar has accredited six sites in Sri Lanka as wetland sites, namely, Vankalai Sanctuary, Wilpattu Ramsar Wetland Cluster, Annaiwilundawa Tanks Sanctuary, Madu Ganga, Bundala, and Kumana Wetland Cluster (https://www.ramsar.org/country-profile/sri-lanka). Its latest addition Colombo Wetlands comprises the following wetlands: Beddagana wetland, Diyasauru wetland, Heen-ela wetland, Kolonnawa wetland, Kotte wetland, Maddinnagoda wetland, Mulleriyawa wetland, and Thalangama wetland.

Urban wetlands make cities livable in many important ways. They reduce flooding, replenish drinking water, filter waste, provide urban green spaces, and are a source of livelihoods. Colombo wetlands significantly benefit the urban residents. The income for households in and around the wetlands is less than 40% of the average income for the Metropolitan Colombo Region. Over 60% of these local households directly benefit from livelihoods and products derived from the wetlands and 100% will be receiving indirect benefits from flood mitigation, climate cooling, and pest regulation.

Threats on Colombo wetlands: Wetland loss and degradation continues across Colombo today. Despite all the benefits that the wetlands provide, they are threatened with progressive degradation and the current rate of wetland loss in the Colombo Metropolitan Region is approximately 1.2% per annum. If this trend continues, the area of wetlands would be reduced by one third by 2038 and by half by 2070. This will also produce a proportional decrease in flood storage capacity. If all the wetlands were infilled, flood water levels would increase by 2 m during the most severe rainfall events resulting in devastation across the city.

Invasive alien species present a significant threat to the native biodiversity of the wetlands in the city. Eleven species of alien invasive plants are currently known to be present in the wetlands including *Eichhornia crassipes*, which is widespread and chokes canals and waterways, and *Annona glabra*, which occurs in almost every wooded wetland in the city and results in significant changes in the native ecological character of the wetlands.

5.2.7 Seasonally Flooded Inland Marsh

This rare wetland, considered a distinct vegetation formation, occurs in wet lowlands in association with slow moving perennial streams with nearly flat and wide streambanks supporting rich Wet Evergreen Forest. With each rainy season, the landscape becomes flooded to various depths, often up to several meters. The terrain remains flooded for a limited period accumulating much alluvium. Much useful data was gathered, and many botanical records were made by the author during the National Conservation Review (IUCN and WCMC 1997). The canopy was 15–20 m high and the profile was that of a typical Wet Evergreen Forest. About 115 woody plants were recorded and this unique marsh forest accommodated many extremely rare species, some critically endangered and known only from this habitat. Therefore, it serves as the only known refugium for the protection of such species, e.g. *Stemonoporus moonii* ("Hora-wel") and *Mesua stylosa* ("Suwanda"). Some other very rare species recorded were *Areca concinna* ("Lena-theri-puwak"), *Vatica paludosa* ("Mandora"), *Diospyros atrata* ("Kalu-kenda"), and *Semecarpus moonii* ("Badulla"), while some less rare or common species were *Artocarpus nobilis* ("Bedi-del"), *Barringtonia racemosa* ("Midella"), *Calamus digitalis* ("Kukulu-wel"), *Calamus rotang* ("Heen-wewel"), *Dipterocarpus hispidis* ("Bu-hora"), *Humboldtia laurifolius* ("Gal-karanda"), *Mallotus fuscescens, Mesua ferrea* ("Na"), *Shorea oblongifolia* ("Pana-mora"), etc. (Figure 5.4a, b).

Figure 5.4 Wetlands of Sri Lanka. (a, b) Seasonally Flooded Inland Marsh. (c, d) Spring Forest. (e) Villu Grassland – Natural. (f) Villu Grassland – Anthropogenic. *Source:* Photo by A.H. Magdon Jayasuriya (Chapter author).

5.2.8 Thermal Marsh

Syn. Thermal spring, Geothermal spring, Hot water spring (Ekanayake et al. 2015).

Thermal marshes, a group of geologically influenced vegetation formation, are the springs that ascend from deep ground layers and discharge water at noticeably higher temperatures than that of the surrounding grounds and the atmosphere. They are a natural phenomenon linked to geothermal energy when the surface water percolates and contact with hot rocks lying at deeper levels. In Sri Lanka, thermal springs or hot springs are distributed along a narrow Eastern lowland belt running from Hambantota to Trincomalee in the dry zone. Several hot water springs have been reported in Ankihiriya, Galwewa (Nelumwewa), MahaOya, Marangala, Kiwulegama (Jayanthiwewa), Kinniya, Kapurella, Wahawa, Mahapelessa, and Muthugalwela (Ekanayake et al. 2015).

Kapurella thermal spring located in the Mahaoya flood plain (Ampara District) has been investigated in reasonable detail by Ekanayake et al. (2015). This thermal marsh has been estimated to be about 3 ha. The discharge temperature of water was 58 °C and appreciable amounts of salinity and lime have been detected. The marshy vegetation is said to be determined mainly by edaphic factors such as flat land form, waterlogged soil, and salinity and this habitat is surrounded by low grassland (10–25 cm). Abundance of flood-tolerant species, e.g. *Alternanthera sessilis* ("Mukunuwenna"), *Cyperus difformis*, *Cyperus iria* ("Wel-hiri"), *Hygrophila auriculata* ("Katu-ikiriya"), and *Ludwigia perennis* ("Piduru-wella") were noted. As the thermal marsh is year-round waterlogged and therefore an anaerobic habitat, characteristic aquatic plants, or terrestrial plants are not able to establish permanently. However, the establishment of amphibious flora adapted to thermal conditions have been recorded by Ekanayake et al. (2015). *Calamus rotang* ("We-wel"), *Acrostichum aureum* ("Keran-koku") – an indicative of high salinity, *Thelypteris confluens* – a rare fern, and *Panicum repens* occur in gregarious stands while much of the back marsh supports sedges like *Cyperus difformis* ("Tunessa"), *C. haspan* ("Halpan"), *C. pilosus*, *C. platystylis*, and *C. tenuiculmis*. Several climbers such as *Canavalia cathartica* ("Avara"), *Causonis trifolia* ("Wal-rat-diya-labu"), *Combretum ovalifolium* ("Kaduru-ketiya-wel"), *Ipomoea obscura* ("Thel-kola"), and *Mikania cordata* ("Watu-palu") were common.

5.2.9 Spring Forest

This is a unique vegetation formation occurring on lower slopes of hills in the dry zone. They essentially develop in association with perennial springs that are generated rarely within the interior of such hills at higher elevations with quartzite rocks and varying quantities of limestone that undergo slow disintegration. This process results in the formation of subterranean cavities capable of retaining and gradually releasing water, thus creating perennial springs even at the heights of the dry seasons (Personal communication, Dr. S. Somasiri, Agridev Consultants).

The spring forests are mainly hydrologically influenced, and they occur in small areas or zones, and they rapidly transform peripherally into normal dry forests (Dry Mixed Evergreen Forest) when hydrological influences diminish with elevation or distance from the spring. The plant diversity of the spring forest is rich with high canopies reaching 30 m with trees such as *Myristica ceylanica* ("Malaboda"), *Horsfieldia iriya* ("Eeriya"), *Mangifera zeylanica* ("Etamba"), and *Mesua ferrea* ("Na"). It was interesting to note the production of breathing roots by "Eeriya" trees, indicating its preference and adaptation to survive in perennial semiaquatic habitats. The subcanopy constitutes *Ficus callosa* ("Wal-gonna"), *Berrya cordifolia* ("Halmilla"), *Hydnocarpus venenatus* ("Makulu"). and *Diospyros malabarica* ("Timbiri"), while *Saraca asoca* ("Asoka"), *Mallotus resinosus*

("Ma-endaru"), *Suregada lanceolata* ("Kakkaipalai") (T), *Leea indica* ("Burulla"), *Ficus hispida* ("Kota-dimbula"), etc. form the lower layers of trees and shrubs. *Dalbergia rostrata* ("Bambara-wel"), *Uvaria macropoda* ("Balal-beti"), *Anodendron paniculatum* ("As-wel"), and *Canthium rheedii* ("Etamburu") were high lianas noted, while *Piper fallax* ("Mala-miris-wel"), *Pothos scandens* ("Pota-wel"), and the fern *Stenochlaena paustris* were frequent epiphytes on tree trunks.

The spring forest patch at the Na-maluwa spring in Kahalla (Matale District) is also known as Kahalla Ulpotha and the villagers have named this spring as "Bubula," meaning spring and therefore it is appropriately named for this unique vegetation formation. It covers about 2 ha and it also includes an interesting marsh (about 0.5 ha) occurring due to a network of active streams beyond the periphery of the forest. *Hedychium flavescens*, *Lasia spinosa* ("Kohila"), *Ludwigia perennis*, *Mikania cordata* ("Wathu-palu"), and the fern *Nephrolepis exaltata* in gregarious populations were observed here (Figures 5.4c, d).

Generally, spring forests have affinities with Moist Mixed Evergreen Forest (Intermediate Forest), as indicated by the occurrence of species such as *Mangifera zeylanica* and *Anodendron paniculatum*. However, it is interesting to find that spring forests also accommodate some wet zone plants, which are absent or rarely reported in the dry zone, e.g. *Stenochlaena palustris* and *Dalbergia rostrata*. A comprehensive account of this vegetation formation is given in DWC (1996).

5.2.10 Villu Grassland

These are wetland grassland formations that are believed to have originated from different predecessor ecosystems. One is a system of natural and shallow depressions forming lakes in which VILG have developed (Figure 5.4e, f). This edaphic and hydrological climax occurs in the Wilpattu National Park (WNP), areas south of it such as Vanathavilluwa and Eluwankulama in the Puttalam District and in areas north of WNP, i.e. Mavillu Conservation Forest in the Mannar District. Most of these villus hold freshwater while some hold saline water (EML 2005a).

Sometimes, VILGs have developed in association with shallow seasonal small tanks in the dry zone, and in this case, they are anthropogenic in origin. Further, anthropogenically originated villus have been observed in the intermediate zone in association with small-scale paddy fields, abandoned for at least 50 years.

The third origin of VILG is historically abandoned large-scale paddy-cultivated areas, and hence they are anthropogenic in origin. Large tracts of VILG occur in the Mahaweli floodplains in its lower basin and they are believed to have originated since the fall of the Kingdom of Polonnaruwa (eleventh to fourteenth century CE) in which the irrigation infrastructure and agricultural development in Sri Lanka were at their zenith (EML 2019; Jayasuriya 2019). Therefore, the latter group of VILG is an anthropogenic subclimax (Figure 5.4f). The soils of VILG are waterlogged, even during the dry seasons and therefore poorly aerated due to high water table. They are also rich in humus and therefore dark brown and mineral rich. The proportion of clay and fine sand is high, promoting waterlogged properties. The plants are also phenologically characteristic, as flowering and fruiting take place around the year. Seed germination also follows the same pattern due to the availability of perpetual soil moisture (Pemadasa 1986). Succulent grasses and sedges are common, while some trees, adapted to grow in waterlogged habitats, occur sporadically on mounds and sometime forming little islands in a matrix of VILG.

Based mainly on a study conducted in the Kala Oya Basin (EML 2005a) at Wilpattu (Nelum Vila and Thala Vila), some common species in Freshwater Villus were as follows:

Grasses: *Iseilema prostratum, Setaria flavida* ("Ha-thana"), *Brachiaria mutica* ("Diya-tana"), and *Cynodon dactylon* ("Ruha").

Sedges: *Cyperus procerus, C. pulcherimus, Schoenoplectiella articulata, Fimbristylis triflora, F. cinnamometorum*, and *F. zeylanica* (point endemic in WNP).

Other herbs: *Aponogeton natans* ("Kekatiya"), *Blyxa auberti* ("Diya-hawariya"), *Crinum defixum* ("Heen-tolabo"), *Dopatrium lobelioides, Drosera burmanii* ("Kandulessa"), *Eriocaulon quinquangulare* ("Heen-kokmota"), *Hygrophila auriculata, Limnophyton obtusifolium, Crinum viviparum* ("Heen-tolabo"), *Neptunia oleracea, Nymphoides hydrophyla* ("Kumudu"), *Nelumbo nucifers* ("Nelum"), *Nymphaea pubescens* ("Olu"), *Potamogeton nodosus, Sphaeranthus indicus* ("Muda-mahana"), *Utricularia aurea* ("Diya-pasi"), *U. polygaloides*, etc.

Shrubs and climbers: *Calamus rotang, Guilandina bonduc* ("Kumburu-wel"), *Asparagus racemosus* ("Hathawariya"), *Phyllanthus reticulatus* ("Kaila"), and *Aeschynomene aspera* ("Maha-diya-siyambala").

Trees: *Piliostgma racemosum* ("Maila"), *Terminalia arjuna* ("Kumbuk"), *Vitex leucoxylon* ("Nebada"), *Phoenix pusilla* ("Indi"), *Pongamia pinnata* ("Magul-karanda"), *Barringtonia acutangula* ("Ela-midella"), and *Nauclea orientalis* ("Bak-mi").

However, in Brackish-water Villus in the Wilpattu, e.g. Kokkare villu, species such as *Najas minor, N. graminea, Phyla nodiflora, Bacopa monnieri* ("Lunu-wila,"), *Cyperus* spp., *Fimbristylis* spp., *Tamarix indica*, and *Volkameria inermis* have been noted.

5.2.11 Gilgai Wetland

This soil/vegetation formation is a gregarious system of soil mounds (hummocks) and depressions formed over seasonally inundated lowland meadows in the dry zone. It also formed from basic alluvium flanking rivers and on peripheral watershed areas of the tanks. The soil is imperfectly to poorly drained, possessing high water-holding capacity and these wetlands develop essentially on shrink-swell and cracking clay soils (vertisols) in response to alternate drying and wetting cycles (Randall and Anderson 2007). High proportions of fine clay (smectitite), its expansive properties, and capillary action promoting upward churning of fine particles result in the genesis of this specific microrelief (Personal communication, Dr. P.B. Dharmasena, Department of Agriculture). Vertisols are sticky in the wet season but become very hard in the dry season.

Although this edaphic phenomenon was known to some soil scientists in Sri Lanka, the biological aspects of this special wetland landscapes have not attracted the attention of vegetation ecologists. The occurrence of this soil/vegetation formation was recorded during the studies conducted in the Mahaweli System B at Kayenkerni, E. of Welikanda (Polonnaruwa District) (Jayasuriya 1991), and later in Lower Mahaweli Left Bank region at Kalu-ganga and Janaranjana wewa (Trincomalee District) (EML 2019) (Figures 5.5a–c).

Hummocks are dome-shaped measuring about 80–100 cm in diameter and reaching a height of 50–60 cm and support distinctly different flora from that of depressions, apparently determined by hydrology. They also bear some shrub and tree species and further they provide a special habitat for several faunal species, such as lizards, crabs, beetles, and other insects.

Hummocks at Kayenkerni: *Commelina undulata* ("Tana-pala"), *Cyanotis axillaris* ("Bol-hinda"), *Mimosa pudica* ("Nidi-kumba"), *Alysicarpus vaginalis* ("Aswenna"), *Vigna trilobata* ("Bin-me"), *Urena lobata* ("Patta-epala"), *Melochia corchorifolia* ("Gal-kura"), *Calotropis gigantea* ("Wara"), etc.

Hummocks at Kalu-ganga: *Sphaeranthus indicus* ("Muda-mahana"), *Ocimum tenuiflorum* ("Maduru-tala"), *Desmodium* sp., *Flueggea leucopyrus* ("Katu-pila"), *Carissa spinarum* ("Heen-karamba"), *Dichrostachys cinerea* ("Andara") etc.

Figure 5.5 Wetlands of Sri Lanka. (a) Gilgai Wetland – in wet condition. (b) Gilgai Wetland – in dry condition. (c) Gilgai Wetland – vertical profile in dry condition. *Source:* Photo by A.H. Magdon Jayasuriya (chapter author). (d, e) Thermal Marsh (SE). *Source:* Photo by Sarath Ekanayakea.

Depressions at Kayenkerni: *Cyperus distans* ("Wel-hiri"), *Leptochloa neesii*, *Chamaecrista mimosoides* ("Bin-siyambala"), *Panicum repens* ("Etora"), etc.

- Gilgai is an Aboriginal word in Australia, meaning "water hole."

5.2.12 Dry Riverine Evergreen Forest

Syn. Running Water Communities (Pemadasa 1986).

The DREF is a hydrologic and perhaps edaphic vegetation formation influenced by high water table, and the mineral-rich alluvium that is seasonally deposited by river outflows in the vicinities of the rivers in the dry zone. This forest flanks the river and constitutes the interphase between the river and the typical dry zone terrestrial vegetation bordering the river, thus forming a characteristic linear zone of which the width can be as narrow as a few meters on steeper riverbanks or up to 100 m or more on flat to gently sloping river sides. In contrast with the surrounding dry zone vegetation, DREF is taller in stature, is greener with a majority of its tree species being evergreen and possessing mesophyllous leaves, and is therefore easily detectable in the field from long distances and furthermore easily located in aerial photographs and satellite images.

The canopy is close, reaching 10–25 m high with all the layers of a tall forest represented in its profile. The profile dominated by massive evergreen trees and its luxuriance makes it more akin to a rain forest. *Terminalia arjuna* is apparently the most important tree species, in terms of density, frequency, girth, and stature, thus giving character to the DREF. The trees are shallow-rooted and adapted to exploit ground water from the high water table that is characteristic of this ecosystem. The thick root systems of trees and shrubs that occur in the normal land-water interphase act as a protective armor forming a frontline defense against riverbank erosion. There are also some plants that are adapted to disperse their fruits and seeds by water. The subcanopy is clear, while the shrub layer is usually sparse, and the lianas are often seen (Figure 5.6a, b).

The main floristic constituents of the DREF, based on studies conducted on the biodiversity of the Kala Oya Basin (EML 2005a), Menik Ganga at the Ruhuna National Park (IUCN and WCMC 1997), and Kalu Ganga (Polonnaruwa-Trincomalee District boundary) (EML 2019; Jayasuriya 2019), are as follows:

Canopy: *Terminalia arjuna, Berrya cordifolia, Monoon longifolia ("Ovila"), Elaeodendron glaucum ("Neralu"), Vitex leucoxylon*, and *Diospyros malabarica.*

High subcanopy: *Diospyros ovalifolia, Aglaia elaeagnoidea* ("Puwangu"), *Lepisanthes tetraphylla, Syzygium cuminii*, and *Mischodon zeylanicus* ("Tammenna").

Low subcanopy: *Nauclea orientalis, Pongamia pinnata, Garcinia spicata ("Ela-goraka"), Ficus hispida, F. racemosa ("Attikka"), Hydnocarpus venenata, Dimorphocalyx glabellus, Ixora pavetta ("Maha-rathambala"), Pterospermum suberifolium*, and *Crataeva adansonii.*

Shrubs: *Mallotus eriocarpus* ("Bulu-petta"), *Croton aromaticus* ("Gas-keppetiya"), and *Huberantha korinti.*

Lianas and climbers: *Reissantia indica, Salacia reticulata ("Himbutu"), Salacia oblonga ("Kothala-himbutu"), Derris parviflora ("Sudu-Kalawel"), Connarus monocarpus ("Radaliya"), Derris scandens ("Kala-wel")*, and *Guilandina bonduc ("Kumburu-wel").*

5.2.13 Dry Canal-Associated Evergreen Forest

This new forest vegetation formation in Sri Lanka was recently described and characterized by Jayasuriya (2019). During the period of the Polonnaruwa Kingdom (eleventh to fourteenth Century

Figure 5.6 Wetlands of Sri Lanka. (a, b) Dry Riverine Evergreen Forest. (c, d) Dry Canal-associated Evergreen Forest. (e) Montane Wet Patana Grassland (UN). (f) Montane Wet Patana Grassland – Dwarf Bamboo Assemblage (UN). SE, Sarath Ekanayake; UN, Unknown source; VW, Vimukthi Weeratunga; all other photos by author.

AD), irrigation infrastructure and agricultural development in Sri Lanka were at their zenith. The irrigation system also consisted of canals conveying water from the Mahaweli River to distant storage reservoirs and farmlands. This DCEF belt was detected along one such canal deep within the Somawathiya National Park (Polonnaruwa District). Due to the perennial presence of water during the Polonnaruwa Period and abandonment since the fall of the Kingdom in the fourteenth century, these agricultural landscapes were reclaimed by the nature but not without distinct and permanent ecological changes such as rise of the soil water table, higher atmospheric humidity, amelioration of temperature, and acquisition of wetland features etc.

Being anthropogenic in origin and subsequently influenced by hydrologic and edaphic factors, the DCEF is believed to have been developed as a distinct forest formation. Although DCEF exhibits affinities with DREF, they differ in origin, physiognomy, and floristics, as amply discussed by Jayasuriya (2019). The canopy of the DCEF is 10–12 m in height with relatively dense lower strata (Figure 5.6c, d).

Canopy: *Vitex leucoxylon* ("Nebada"), *Terminalia arjuna, Diospyros malabarica, Albizzia chinensis* ("Kabal-mara"), *Berrya cordifolia, Madhuca longifolia* ("Mee"), *Mitragyna parvifolia*, and *Syzygium cuminii.*

Subcanopy: *Margaritaria indica* ("Karawu"), *Barringtonia acutangula, Hydnocarpus venenata, Crateva adansonii, Piliostigma racemosum, Stereospermum tetragonum* ("Dunu-madala"), *Cordia dichotoma*, and *Pterospermum suberifolium.*

Low trees and Shrubs: *Hibiscus tiliaceus, Casearia zeylanica* ("Wal-waraka"), *Tamilnadia uliginosa* ("Et-kukuruman"), *Dimorphocalyx glabellus*, and *Phyllanthus reticulatus.*

Lianas and climbers: *Salacia oblonga, Derris parviflora*, and *Ventilago maderaspatana.*

5.2.14 Saline Sparse and Open Scrub

This is nearly a monotypic growth of scrub vegetation in the dry zone inlands that have exclusively developed on saline soils. Such edaphically derived habitats were the results of excessive irrigation in the past, inadequate drainage and rising of the water table and they are presumably the principal causes of the increased salinity. Historically, the existence of saline patches indicates that irrigation was practiced during ancient times (Personal communication, Dr. P.B. Dharmasena, Department of Agriculture). While impacting agriculture, the saline soils also affect the subsequent forest growth, minimizing it to a level of scrub. Patches of saline scrub were recorded in the Polonnaruwa District, but notably within the Somawathiya National Park (EML 2019). Here, a large patch of Saline scrub, almost totally dominated by *Parkinsonia aculeata* (Jerusalem thorn or "Lunu-ehela" (S) or "Mulvakai" (T)), was found 2 km S. of Janaranjana wewa. This introduced and naturalized shrub or small tree with showy bright yellow flowers is usually found in coastal and saline habitats elsewhere in very dry regions indicating its preference to saline soil conditions.

5.2.15 Montane Wet Patana Grassland

Synonyms: *Wet Patana* (Holmes 1951)
Upper Montane Grassland (Perera 1967)
Upper Wet Patana (Mueller-Dombois and Perera 1971; Pemadasa 1986)
Lower Wet Patana (Mueller-Dombois and Perera 1971; Pemadasa 1986)
Wet Patana Grassland (Jayasuriya et al. 2006)

MWPG occurs in the Horton Plains and the vicinities of Nuwara Eliya above 2000 m. The mean temperature is 13–15 °C subject to marked diurnal and seasonal variation. The night temperatures during December–February may be depressed below 5 °C with frequent ground frost. The mean annual rainfall is >2000 mm, received through both monsoons and the intermonsoon periods, with only a short dry period during February. It is evenly distributed with a minimum of 100 mm of monthly rainfall. The weather is dominated by continuous, and at times, gale-force winds with persistent cloud cover. Mist is prevalent throughout the year, relatively more frequent at higher altitudes. Soils are relatively rich in organic components, minerals, and moisture while the pH is low (4.5–4.7). The soil A horizon is blackish, 30–60 cm thick, and spongy retaining much moisture (Pemadasa 1986).

A recent study of the habitat types in Horton Plains conducted by EML (2005b) recognized six grassland communities, namely, Tussock Grass Community, Carpet Grass Community, Grassland/Rhododendron Community, Dwarf Bamboo Community, Grass and Fern Community, and Marsh/

Peat Community. Of these, two communities are regarded as wetland assemblages, according to Jayasuriya 2023), and they are as follows:

5.2.15.1 Dwarf Bamboo Assemblage

Syn. *Dwarf Bamboo Community* (Mueller-Dombois and Perera 1971)

This occurs in valley bottoms, bordering depressions and streams and this boggy habitat is occupied by gregarious populations of a small slender bamboo species, *Kurana densifolia*, an endemic to the Horton Plains (Figure 5.6e, f). Other species: *Centella asiatica, Eriocaulon brownianum, Exacum trinervium* ("Binara"), *Hedyotis plantaginifolium*, and *Ranunculus sagittifolius* (Butter cup).

5.2.15.2 Marsh–Peat Assemblage

Syn. *Marsh/Peat Community* (Mueller-Dombois and Perera 1971)

This assemblage occurs in some valleys next to shallow and expanded streams in sheltered habitats. Several delicate herbaceous species grow in between clumps of grass, e.g. *Allium hookeri* ("Wal-loonu"), *Bupleurum ramosissimum* ("Wal'endaru"), *Centella asiatica* ("Gotu-kola"), *Senecio ludens, Carex* spp., *Cyperus brevifolius, Rhynchospora rugosa, Drosera burmannii, Eriocaulon brownianum, Juncus effusus, J. prismatocarpus, Pseudolycopodiella caroliniana, Alchemilla indica, Knoxia platycarpa, Neanotis monosperma, Lindernia rotundifolia, Valeriana moonii* ("Heen-kapuru"), and *Metathelypteris flaccida*.

Acknowledgments

I wish to thank Prof. Thammineni Pullaiah, Editor and Coordinator of this exercise and Former Professor, Sri Krishnadevaraya University, Andra Pradesh, India for inviting me to contribute this chapter and for his advice and guidance given to me until its completion. My thanks are also due to Mr. Avanthi Jayatillake, CEO, EML Consultants (Pvt) Ltd. for permitting me to write this chapter, initially based on the EML Report (authored by A.H.M. Jayasuriya et al. 2006, see References). The provision of some excellent photographs by Mr. Vimukthi Weeratunga, Wildlife Biologist (acknowledged in text) and the technical support provided in the preparation of the chapter by Mr. Rasika Jayasuriya are thankfully acknowledged.

References

Amarasinghe, M.D. and Balasubramaniam, S. (1992). Structural properties of two types of mangrove stands on the north-western coast of Sri Lanka. *Hydrobiologia* 247: 37–47.

Arulpragasam, K.D. (2000). Coastal and marine resources. In: *Natural Resources of Sri Lanka 2000*, 167–193. Sri Lanka: National Science Foundation.

Central Environmental Authority (CEA)/Euroconsult (1994). *Conservation Management Plan: Muthrajawela Marsh & Negombo Lagoon*, 129. Sri Lanka: Wetland Conservation Project.

Dassanayake, M.D., Fosberg, F.R., and Clayton, W.D. (ed.) (1980–2000). *Revised Handbook to the Flora of Ceylon*, vol. 1–14. New Delhi: Amerind.

Department of Wildlife Conservation (1996). *Resources Inventory of Protected Areas: Kahalla – Pallekele Sanctuary and Ritigala Strict Natural Reserve. A Report Submitted by the Agridev Consustants, Kandy.* Colombo: Department of Wildlife Conservation.

Ekanayake, S.P., Ranawana, K.B., Chandrajith, R. et al. (2015). Preliminary observations on ecological aspects of Kapurella thermal spring (thermal marsh) at Mahaoya. *Sri Lanka Naturalist* 7 (1–2): 4–12.

EML Consultants Pvt Ltd (2005a). *Kala Oya River Basin – Survey of Biodiversity, Wetland Issues and Options for Sustainable Management*, 406. Colombo: Mahaweli Authority of Sri Lanka.

EML Consultants Pvt Ltd (2005b). *Habitat Maps for Selected Areas in Sri Lanka: Horton Plains National Park*, 62. Ministry of Environment & Natural Resources: Battaramulla.

EML Consultants Pvt Ltd (2017). *Environment Impact Assessment of the Proposed Project on Metro Colombo Solid Waste Management*, 227. Central Environment Authority: Colombo.

EML Consultants Pvt Ltd (2019). *Environment Impact Assessment of Mahaweli Left Bank Lower Basin Development Project*, 163. Department of Irrigation: Colombo.

Gunatilleke, N., Sathurusinghe, A., and Gunatilleke, S. (2018). *Forest resources*. In: *Natural Resources of Sri Lanka* (ed. M.J.S. Wijeyaratne, A.H.M. Jayasuriya, and N.P. Wijayananda), 420. Colombo: National Science Foundation.

Holmes, C.H. (1951). The grass, fern, and savannah lands of Ceylon, their nature and ecological significance. Imperial Forestry Institute, University of Oxford, Institute Paper No. 28, 95pp.

IUCN and WCMC (1997). *Designing an Optimum Protected Areas System for Sri Lanka's Natural Forests*, vol. 1 & 2, 399. Battaramulla: Forest Department, Ministry of Forestry and Environment.

Jayasuriya, A.H.M. (1991). Some Soil-Vegetation Relationships in Mahaweli System B. A Report Submitted to MARD/MDS Projects, Pimburettawa.

Jayasuriya, A.H.M. (2019). A new forest vegetation type in Sri Lanka: dry canal-associated evergreen forest. *Ceylon Journal of Science* 48 (4): 375–281.

Jayasuriya, A.H.M. (2023). *Diversity of Terrestrial Vegetation In Sri Lanka*, 90. Pitakotte: EML Consultants PLC.

Jayasuriya, A.H.M., Kitchener, D., Biradar, C.M. (2006). Portfolio of Strategic Conservation Sites/ Protected Area Gap Analysis in Sri Lanka. Report submitted by EML Consultants (Pvt) Ltd) to the Ministry of Environment & Natural Resources, pp. 287.

Jayathissa, L.P. (2012). Present status of mangroves in Sri Lanka. In: *The National Red List 2012 of Sri Lanka; Conservation Status of the Fauna and Flora* (ed. D.K. Weerakoon and S. Wijesundara), 197–199. Colombo: Ministry of Environment.

Macnae, W. (1968). A general account of the fauna and flora of mangrove swamps and forests in the Indo-West Pacific region. *Advances in Marine Biology* 6: 73–270.

Mepham, R.H. and Mepham, J.S. (1985). The flora of tidal forests – a rationalization of the use of the term "mangrove". *South African Journal of Botany* 51 (2): 77–99.

Ministry of Environment, National Herbarium & Department of National Botanic Gardens (2020). *The National Red List 2020 – Conservation Status of the Flora of Sri Lanka*, 254.

Mitch, W.J. and Gosselink, J.G. (1986). *Wetlands*. New York: Van Nostrand Reinhold.

MOE (2012). *The National Red List 2012 of Sri Lanka: Conservation Status of the Fauna and Flora*. Colombo: Ministry of Environment 451 pp.

MoMD & E (2018). *Biodiversity Profile – Sri Lanka, Sixth National Report to the Convention on Biological Diversity*. Sri Lanka: Biodiversity Secretariat, Ministry of Mahaweli Development and Evironment 200 pp.

Mueller-Dombois, D. and Perera, M. (1971). Ecological differentiation and soil fungal distribution in the montane grasslands of Ceylon. *Ceylon Journal of Science* 9: 1–14.

Pemadasa, A. (1986). *The Green Mantle of Sri Lanka*, 1–242. Maharagama: Tharanjee Prints.

Perera, N.P. (1967). An evaluation of the human impact on the nature and distribution of wild plant communities in the Ceylon highlands. Ph.D. thesis, University of Leeds.

Ranawana, K.B., Wickramagamage, P., Ekanayake, S.P. et al. (2020). *Atlas of Mangroves, Salt Marshes and Sand Dunes of the Coastal Area from Malwathuoya to Pooneryn in the Northwestern Coastal Region, Sri Lanka*. Peradeniya: Ecological Association of Sri Lanka 90 pp.

Ranawana, K.B., Ekanayake, S.P., and Sanjeewa Jayaratne, V.A. (2023). *An Ecological Profile of Salt Marshes in Mannar District, Sri Lanka*, 60. Uda Peradeniya: Ecological Association of Sri Lanka.

Randall, J. and Anderson, S. (2007). *Soils: Genesis and Geomorphology*, 283. Cambridge: Cambridge University Press.

Saenger, P., Hegerl, E.J., and Davie, D.S. (1983). Global status of mangrove ecosystems. *The Environmentalist* 3 (3): 1–88.

Senaratna, L.K.A. (2001). *Checklist of the Flowering Plants of Sri Lanka*, 451. Colombo: National Science Foundation.

de Silva, M. and de Silva, P.K. (1998). Status, diversity and conservation of the mangrove forests of Sri Lanka. *Journal of South Asian Natural History* 3 (1): 79–102.

United States Environmental Protection Agency (USEPA) (2023). What is a wetland? https://www.epa.gov/wetlands/what-wetland (accessed 5 May 2023).

Veetil, B.K., Wickramasinghe, D., and Amarakoon, V. (2023). Mangrove forests in Sri Lanka: an updated review on distribution, diversity, current state of research and future perspectives. *Regional Studies in Marine Science* 62: 102932.

https://climatefactchecks.org/mangrove-degradation-in-sri-lanka-a-story-from-the-east/ (accessed 16 April 2024).

https://efl.lk/colombo-accredited-a-ramsar-wetland-city/ (accessed 9 April 2024).

https://www.globalnature.org/bausteine.net/f/6462/VimukthiWeerathunga-MangrovesofSL.pdf?fd=2 (accessed 19April 2024).

https://unfccc.int/sites/default/files/resource/Sri%20Lanka%20Mangrove%20Conservation%20Project.pdf (accessed 16 April 2024).

https://www.ramsar.org/country-profile/sri-lanka (accessed 16 April 2024).

6

Wetlands of Israel

Avi Uzan

Israel Nature and Parks Authority, Jerusalem, Israel

6.1 Introduction to the Climate, Topography, Geomorphology, and Hydrology of Israel

Israel has a Mediterranean climate in which there is a short but intense wet season and a long and relentless dry season. Annual rainfall stretches from 800 mm/year in the northern parts of the country to an average of 25 mm in the extreme desert in the south. Most of the precipitation falls between December and March.

One must remember that in Israel everything is small. We have small catchments, small streams, and small springs. Actually, the Jordan river is the only river in Israel and has the largest catchment, about 17 000 km^2. The largest karstic spring in Israel is the "Dan" (Gvirzman 2019). It is the major source of the Jordan river, with an average discharge of 7 cm/s. For comparison, the lowest river discharge listed in Wikipedia is three orders of magnitude larger.

Israel has two drainages: the western drainage and the eastern drainage (Figures 6.1–6.3). The western drainage ends at the Mediterranean Sea (Figure 6.2), and the eastern drainage ends at one terminal lake (the Dead sea) (Figure 6.3), several terminal salt pans, and the Red sea (an extension of the Indian ocean). On each different side of the divide, Israel's wetlands have somewhat different characteristics.

The backbone of Israel's topography is Judea and Samaria mountain ridge. To the west (hence western drainage), stretches a large area of soft hills, followed by the coastal plain. To the east (hence, eastern drainage), the mountains and streams drop sharply into the Dead Sea rift.

6.1.1 The Western Drainage Characteristics

The major morphologic character in relation to wetlands is the Kurkar (a type of calcified sandstone) ridges that run in parallel to the shoreline and create "troughs" along the landscape of the coastal plane. The western drainage streams had to carve their way through these ridges, which blocked their path to the Mediterranean Sea. Floodwater and baseflow that could not flow through the narrow paths in the ridges, created vast marshes and wet meadows. Today, few of them still exist. Another major character of the western drainage is its hydro-geological structure that created

Wetlands of Tropical and Subtropical Asia and Africa: Biodiversity, Livelihoods and Conservation,
First Edition. Edited by Thammineni Pullaiah.

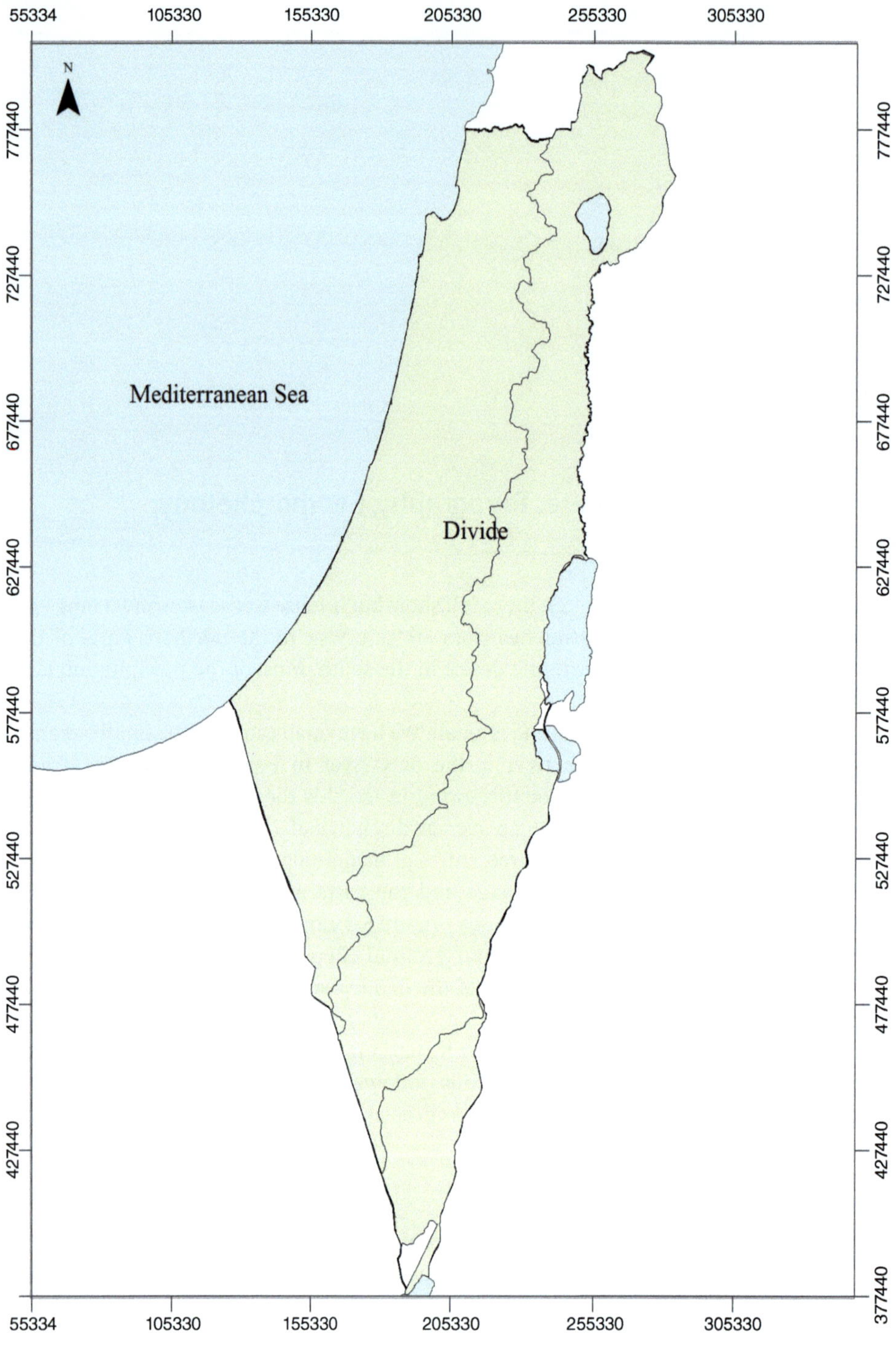

Figure 6.1 All basins of Israel.

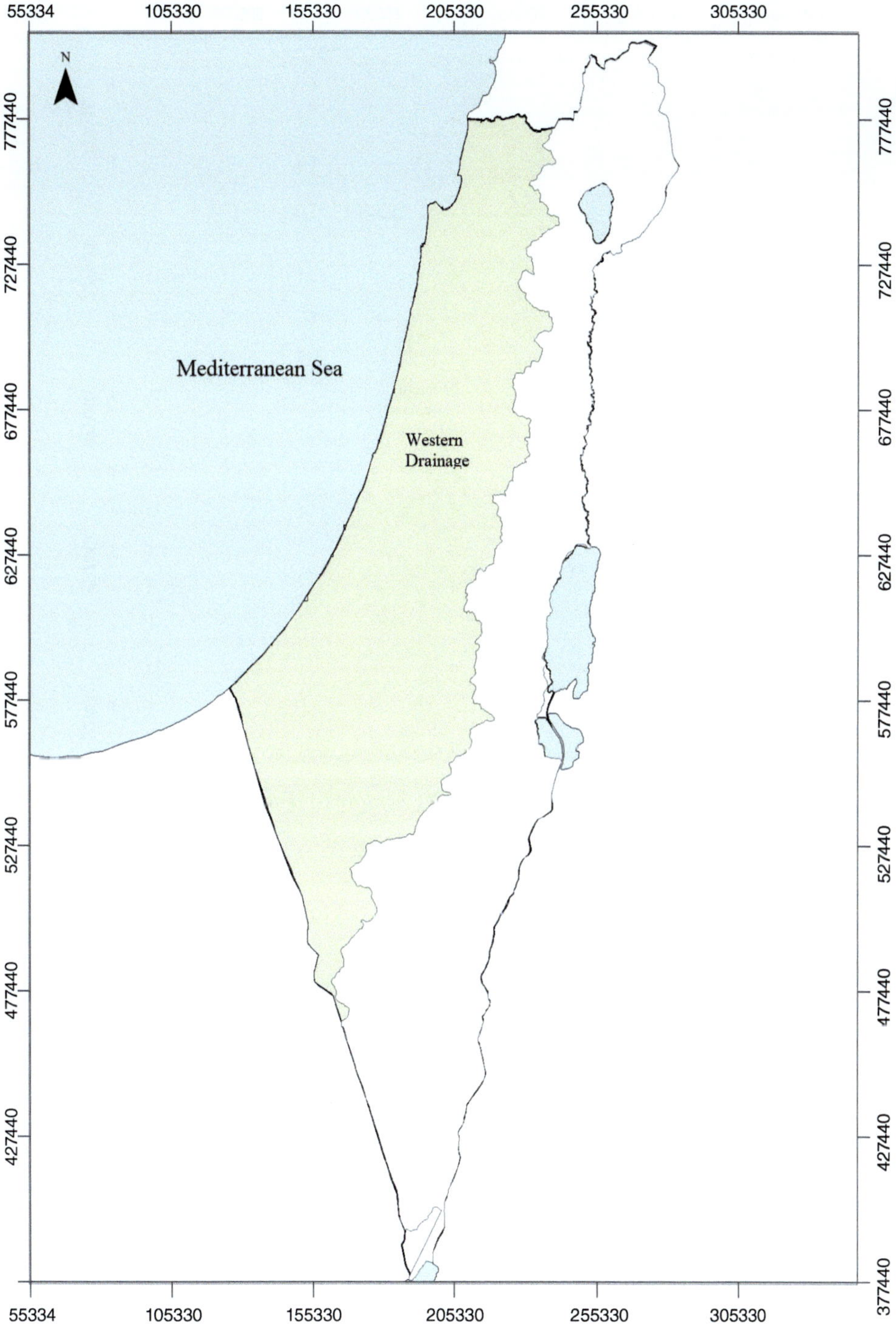

Figure 6.2 Western basin.

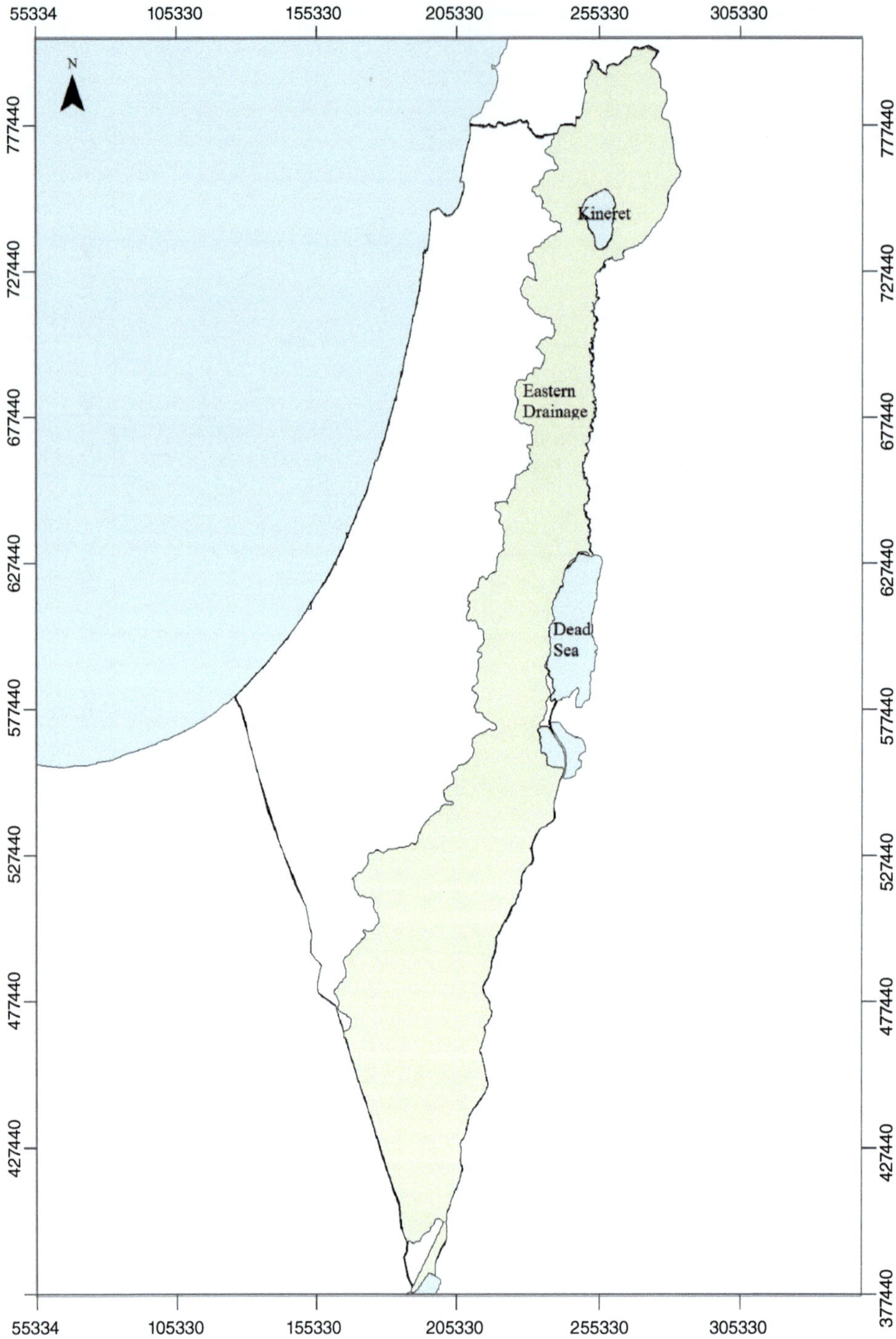

Figure 6.3 Eastern baisn.

two major regional aquifers and numerous local smaller aquifers. The two major regional aquifers are YARKON-TANINIM and CABRI-NA'AMAN (Gvirzman 2019). These Cenomanian-Turonian age limestone aquifers have natural outlets in the form of fault spring that feeds three large lowland perennial streams. The smaller aquifers mostly from the Eocene are the sources that feed minor and small/seasonal springs. They are also water sources of small mountainous streams.

In the western drainage, we also encounter a unique type of wetland, the Coastal marshes that are actually the top part of the coastal Kurkar aquifer. These aquifers have two donors. The main donor is the regional limestone aquifer that overflows (belowground) into the Kurkar, a local aquifer (Gvirzman 2019). The other donor is direct rainfall. These wetlands are actually the open roof of that local Kurkar aquifer. Its uniqueness comes from its nearshore position, which creates transit areas of fresh to brackish water. Vernal pools are a ubiquitous feature of the Mediterranean climate and one can encounter them on both sides of the divide.

6.1.2 The Eastern Drainage Characteristics

The eastern drainage has three major sections: the Jordan, the Arava, and Eilat. It has also several absolute depressions that take the form of a vast flooded saltpan once in several years. The Arava and Eilat sections and also the saltpans are flush flood system and located in arid and extreme arid parts of Israel. They do not create wetland and therefore will not be discussed in this chapter.

The Jordan River is a transboundary river stretching 223 km. It has the biggest drainage system in the eastern drainage. Its total area is 18,103 km^2. 40% of it lies in the Hashemite kingdom, 32% in Syria, 3% of it are in Lebanon, and the rest 25% are in Israel including Judea and Samaria (also known as the "West bank") (Comair et al. 2023). The Jordan drainage system is a part of the great rift valley that starts in Africa and ends in the highest peak (~2000 m ASL) of Israel, Mount Hermon. The Jordan's drainage system comprises Lowland Perennial streams, Mountainous streams, Intermittent streams, Major springs, Minor springs, Seasonal/small springs, vernal pools, marshes, wet meadows, and two lakes – the sea of Galilei and the Dead sea.

The wetlands of Israel have undergone severe degradation. Natural water sources were caped and diverted for agricultural and drinking water needs. Streams got canalized and straitened in order to minimize flood damage and to facilitate land cultivation. Marshes were drained to fight Malaria and reclaim the land. Those wetlands that were stubborn enough and kept their water got converted to fisheries.

For each habitat, there will be a description of abiotic characteristic, biodiversity, and conservation/rehabilitation measures.

6.2 Lotic Habitats

6.2.1 Lowland Perennial streams

6.2.1.1 Abiotic Characteristics

Israel's lowland streams have a slope range between 0.5 ‰ and 0.5%. But only two of them kept their sinuous nature, while all the other alluvial streams have been straitened. The first still natural meandering watercourse is the Jordan river, in its lower parts. The other one is the Yarkon stream, which used to be the biggest stream in the western drainage, with a discharge of ~220 million cubic meter per annum. Both of them have a sinuosity of ~2. It would be accurate to say that there is not even one lowland perennial stream in Israel that keeps its natural characteristics. In the western drainage, we have only five perennial streams. Three of them – Yarkon, Taninim, and Na'aman – have major springs as their sources (Gvirzman 2019).

Yarkon springs were one of the outlets of the largest lime aquifer and used to feed ~220 million cubic meter per annum, not anymore. The national water authority is managing the aquifer at low levels (below point of discharge) in order to retain the reservoir and reduce losses to the sea underground. Today, a mere 14 million cubic meters per annum is allocated as a base flow. Another 14 million cubic meters of effluents are discharged to the stream as part of the water plan to rehabilitate the drying stream.

The lower **Taninim** stream is fed by the higher outlet of the regional aquifer (Timsah springs). Its annual discharge used to be ~50 million cubic meter per annum. These waters served as the water source of the ancient town of Caesarea, more than 2000 years ago. The water was elevated by dams and directed to a series of aqueducts. During the years to come, this system was neglected, and all the area trapped between the dams, became a marsh, hence the Cabara marsh (will be discussed later). Taninim stream of today has two distinct parts. The upper freshwater part and the lower, saline water part. The upper part of the streamflow is fed by a series of minor and seasonal springs. The geology of the region, which is unique and based on layered and cracked chalk rocks, created numerous small and disconnected aquifers and semisaturated media. After a 150 mm of rainfall, the springs start to emanate and create a long-lasting seasonal flow that joins and reinforce the tenuous base flow. After leaving the soft-hilled plain, the Taninim stream enters the last part of its course to the Mediterranean, through what used to be the Cabara marsh and now is an underground drained cultivated area. In the middle of that cultivated area lies its main source, the Tim'sah springs. Its water flows through a series of drainage canals and gathers in the main, dredged, and straitened canal. At the very end of its course, the Taninim stream is flowing through the Nature reserve that keeps the remnants of the 2100-year-old water operation.

The last lowland perennial stream in the western drainage is **Na'aman** stream. Same as Yarkon and Taninim, the Na'aman stream is primarily fed by a large, limestone regional aquifer, the Cabri-Na'aman aquifer. The site of discharge is at the Ein-Afeq Nature Reserve. There is a series of springs emanating from a fault that bring the water-bearing rock formations close to the ground. Not surprisingly, this water source was also harnessed in ancient times. There is a dam, a watermill, and a fortress, from the time of the crusaders. The canal was probably maintained for navigation purposes. Later on, drainage operations were neglected, and the valley gained back its natural state as a marsh with a major flowing route in it. In the beginning of the twentieth century, the marsh was drained again, by the British mandate. The meandering Na'aman was replaced by a strait and wide canal. This operation drained most of the marsh, and its leftovers were converted to fields and fishponds.

6.2.1.2 Biotic Characteristics

Those three lowland perennial streams are close in their biotic characteristics and are described together, with some exceptions. One main characteristic of the lowland stream is the formation of vegetative belts. These belts are formed primarily by the level of soil humidity and disturbance. The drymost belt (and many other belts) is occupied largely by *Phragmites australis*. It is not a mere chance that this is the first plant to appear in this chapter. Although native to Israel, its wide physiological range and its tolerance to disturbances, both physical and chemical, made *Phragmites australis* the most dominant species in our wetlands. The thickets it creates is a blessing, and a curse, depend which expert you ask. If your field of expertise is thicket birds, then it is a blessing.

If you ask a botanist, it is a curse. The common reed supports a wide range of bird species in every season, but markedly in winter. There are as many as 40 species of birds that rely on the brown shaggy thicket. Herons, ducks, and small-body birds. When *Phragmites australis* covers a wide area, it can serve as a nesting site. Some species that are rare and in danger of extinction also need the common reed for nesting (will be discussed later). In the dry belt, we will also encounter other reeds such as *Saccharum spontaneum* and *Arundo donax*. There are not many shrubs that are related to the dry belt of the Mediterranean perennial stream. One of them is *Vitex agnus-castus*. It is blooming in the summer and its nectar-rich flowers are vital energy source when most of the other flowers have withered long ago (Zohari and Feinbrun-Dothan 1966–1986).

By entering the semiwet belt, we can find, in addition to the aforementioned, other shrubs and herbaceous plants. The shrub-like *Pluchea dioscoridis* is of tropic descent and thrives accordingly in the hotter parts of Israel. It is a dominant and landscape modulator. Because the semiwet belt is unstable in its water content and prone to disturbance, we can find many ruderal and invasive species. The most ubiquitous among the ruderal are *Malva sylvestris* and *Silybum marianum*. In the invasive section will find *Ricinus communis* and *Aster subulatus* (Zohari and Feinbrun-Dothan 1966–1986).

The wet belt is the most distinct in the plant species it harbors, the hydrophiles. Growing in a habitat that can turn from base flow to fast flashflood also requires adaptation, mostly structural. The permanent wet condition creates a root environment, which is poor in oxygen. The plants occupying this part of the stream are adapted to these conditions. The main tree that is associated with lowland perennial stream in the western drainage is *Salix acmophylla*. It is a fast-growing tree that rarely tops 8 m height. Its branches are ready to break, and when they get stuck in the mud, they rapidly turn into a new tree. Other means of adaptation to the short but intense winter in the Mediterranean stream is to bend over in the floods. Plants that use this strategy belong to the Juncaceae and Cyperaceae families. The majority of water-related species in these families are perennial herbs, which have bulbs or creeping rootstalks. This trait results many times in formation of dense growth that covers extensive areas. In the wet belt, we can find also more delicate and less "reed-like" plants such as *Epilobium hirsutum, Lycopus europaeus, Lythrum junceum,* and *Phyla nodiflora*; most of them are also perennial. These groups of plants, with their varying shapes and textures, contribute to the habitat patchiness and heterogeneity. It also creates more opportunities for rare species to cope with the fierce competition over sunlight (Zohari and Feinbrun-Dothan 1966–1986).

Last are the emergent, floating, and submerged macrophytes. This group of plants is suffering the most due to dehydration and pollution of Israel's streams. This group of plants is obligatory to perennial or at least seasonal water bodies. It is also in direct contact with the water, for the water is not only a resource but also a medium. The prominent members of that group are *Nymphaea caerulea, Nuphar lutea, Ceratophyllum submersum*, and *Potamogeton nodosus*. In the lowland perennial streams, the substrate is mostly silt. Silt provides little if any cavities, usable surface, and microhabitats. The plants that dwell inside the water create the microhabitats that create substrate for sessile organisms, a relatedly stable surface for laying eggs, and a place to hide for juveniles. It would be accurate to say that this group of plants is responsible for a significant part of the overall habitat heterogeneity of the lowland, perennial streams in Israel.

Israel does not have mega-fauna. The big (in Israeli proportions) herbivores cannot survive in the dwindling riparian zone of our lowland streams. The mammals that do survive in this habitat are wild boar *Sus scrofa*, mongoose from the species *Herpestes ichneumon*, and the rare wild cat *Felis chaus* (Figure 6.4). These species use stream-side thickets mostly as a refuge and less as a foraging habitat (BIOGIS 2024).

Figure 6.4 *Felis chaus* road cam Yarkon authority.

Figure 6.5 *Acanthobrama lissneri* Eldad Elron.

The biggest and most common fish of the lowland streams in Israel is the African catfish *Clarias gariepinus*. Its tolerance to low oxygen levels, its body size, and its wide and omnivorous feeding habits helped this species to colonize almost every water body in Israel, including artificial reservoirs. A mature *Clarias gariepinus* has no natural enemies and is a habitat modulator that increases water turbidity by means of bioturbation. The result of high turbidity is the reduction or loss of submerged and floating plants and the niche they create. The ultimate fate of a catfish-laden habitat is a reduction in biodiversity. Attempts of culling were costly, labor intensive, and did not give the desired results.

The second species worth mentioning is *Acanthobrama lissneri* (Figure 6.5). This species is endemic to the Jordan river basin and the western drainage streams of Israel. It is a small-bodied cyprinid that is associated with slow-moving and water-plant-rich streams. This is the main reason that it is also rare. In 1969, a thesis by Prof. Goren described a new species of *Acanthobrama* that got the name *Acanthobrama telavivensis* (after the city Tel-Aviv). *Acanthobrama telavivensis* is an endemic species to Israel and is designated vulnerable by the IUCN. Recovery measures were taken and a research for finding ways to keep and reproduce it in captivity was successfully completed in 2000. After securing the ability to breed this species in captivity, it was reintroduced to streams that it used to occupy in the past, but also habitats within its natural range, where there was no record of its prior existence. The basis for that description was physiological differences and partial separation of distribution areas, between *Acanthobrama lissneri* and *Acanthobrama telavivensis*. Recent genetic research has found that from a genetic point of view, it is the same species (Goren 1969).

Freshwater mollusks were the first to disappear from the polluted and drying streams of the western drainage of Israel. From the bivalves, the most abundant bivalve in the western drainage used to be *Corbicula consobrina*. Now, it is in danger of local extinction. Three more species of bivalves used to thrive in the western drainage, *Unio mancus eucirrus, Unio terminalis delicatus,* and *Potomida littoralis delesserti* all are locally extinct (Milstein et al. 2012).

The freshwater gastropods of the western drainage belong to the genus *Melanopsis*. One of them, *Melanopsis buccinoidea*, endemic to the levant and is ubiquitous in all freshwater bodies and streams in Israel that are not polluted. The other, *Melanopsis lampra,* is endemic to Israel western drainage lowland streams and not surprisingly got a vulnerable status (Milstein et al. 2012).

6.2.1.3 Conservation/Rehabilitation Measures

The lowland stream in Israel (and in other parts of the world presumably) is highly affected from anthropogenic actions and pressures. The reasons are clear and straightforward. Land is for agriculture and development and water for agriculture and a convenient way to dispose of unwanted flows. Alluvial valleys and the streams that run through them are also underpresented and Israel's protected areas. In the past decades, several state agencies and ministries decided to try and improve the structure and function of the lowland stream. In 2004, the Israeli water law was amended, to allow the allocation of water to nature. This amendment facilitated a wide range of actions: physical, regulatory, and financial. As was mentioned earlier, the lowland perennial streams get their water from regional aquifers. One of the measures taken was setting up "green" lines for water levels in the aquifers (Water Law 2004). A green line is an operational guideline for the Israeli Water Authority. Managing the aquifer according to this guideline will make sure that enough water will emanate from the springs, and the stream will flow (other actions regarding "Nature's right for water" will be discussed later) (Uzan 2010).

A natural alluvial stream carves its path in a way that reflects the flow regime, the substrate, and the biology. There are no pristine alluvial streams in Israel. The second major action taken to rehabilitate the streams is restoring at least some of their natural morphological characteristics. The most common means of intervention is converting steep banks to have gentle slopes. Steep banks hamper the establishment of vegetative belts, and by easing the slope, we allow for spontaneous or active regeneration of the typical vegetation. In many cases, we cannot rely on the spontaneous recovery of native and typical vegetation. When the degree of disturbance is high and where there is no near source for propagules, we have to engage in active vegetative rehabilitation (other means of morphological interventions will be discussed later) (Uzan 2010).

Lowland perennial streams and alluvial valleys are underrepresented in Israel (and in the world, in general). It is like that for a clear reason. These habitats are prime locations for agriculture and development. Even the small parts that are declared nature reserves that have statutory protection are narrow and include only little space for riparian zone to form. This reality drives the Israeli Nature & Parks Authority (INPA), other governmental bodies, and Non Governmental Organizations (NGO's) that want to protect and rehabilitate lowland streams, to find alternatives.

The first and one of the most important actions that was taken in the face of deteriorating streams was to articulate in a clear and comprehensive manner, a policy paper, for the conservation and rehabilitation of Israel's streams. The policy paper included a rich glossary in order to try and establish a common language between all players and stakeholders around and along the stream. The second part was a list of "Do's and don'ts principles" in the fields of flood protection, planning, infrastructures, and hands-on rehabilitation. The paper also has an elaborated appendix that includes the scientific and professional background and substantiation of the "Do's and don'ts principles."

Flood protection activities are major stressors of the lowland stream. The activities can be divided into two groups. The first one is activities that change the stream plan, profile, and cross section. The latter is activities that are aimed to maintain channel capacity. As for the first group of actions, one course of action is to try and prevent past errors. Today, drainage plans must consider ecological and eco-hydrological aspects.

Some examples are as follows:

- instead of the regular trapezoid cross section, a complex cross section that has shallow berms.
- Instead of concrete, a natural stone.
- Instead of culverts, a bridge.
- Instead of leaving exposed banks after excavations, planting and seeding local vegetation.
- Refraining as much as possible from changing the stream plan.
- Adding elements that enhance habitat heterogeneity as part of the new design of the stream (Uzan 2010).

The second group of actions is aimed to maintain channel capacity. Actions like mowing bank vegetation and excavating sediments are clogging the stream. When these actions are executed without proper timing, without the use of proper gear, and with high area coverage, it can inflict damage to the ecological system of the stream. While the first group of actions is subjected to a statutory route of approval, maintenance is carried out without any audit. In Israel, we are dealing with that lacuna in two ways. The first, to add to the drainage law, items that will regulate the maintenance of stream. Unfortunately, the last attempt has failed, because of nonrelevant reasons. The second and more successful path was to create convention maps between the INPA and the drainage authorities. The whole work needs to be done is financed by INPA and the corresponding drainage authority. The process of creating the convention maps begins with a field survey in the fields of aquatic ecology, botany, terrestrial ecology, hydrology, and geomorphology. The survey is carried out in rather uniform and well-defined sections of the stream. Findings of each section are summed, discussed, and results in a "sensitivity to maintenance" score, hence to create the "sensitivity to maintenance map." This map is used as a precursor for a discussion with the drainage authority. Each section is examined for its capacity properties and the land uses near its banks. If the land use is sensitive to flooding, then it will get a score that will counter the ecological sensitivity score. The overall score is a combination of ecological sensitivity and flood protection demands, hence creating the convention map. This map illustrates the stream network and the type, gear used, and timing of action allowed in each section of the stream. Twice a year, ecologists from the INPA and drainage authority engineers are coordinating upcoming works and actions. The convention map is completed for most of the catchments in the Mediterranean area of Israel (Uzan 2010).

In the previous paragraph, the measures taken were a reaction. In the next one, active rehabilitation measures of lowland streams will be described. One of the most desired and yet challenging actions is to bring back stream meanders that were straitened. The first thing to do is finding lost meanders. Fortunately, most of the streams were mapped before the major drainage campaigns that followed the replacement of Ottoman empire with the British mandate and the young state of Israel. The streams got their own parcel and after digitizing the old maps, we can locate lost meanders. Another way to expose and reveal meanders that were cut from the stream plan is using high-resolution DTM created in Lidar technology. These terrain models can help us uncover lost meanders, hidden under non-natural land-cover. After locating the lost meanders comes the part of actual recovery. To remeander a stream means entering farmed land. It is not easy, even when there is a legal basis. The best way is to get the farmer's consent by helping him in the direct or indirect context of the meander. For example, paving a better road, or building a culvert to ease traverse in his fields (Uzan 2010).

The next phase of the stream to deal with is the cross section. A natural alluvial stream carves its own cross section depending on the flow regime. The cross sections are not uniform and the bank slopes are correspondents of local topography, flow, and sediments regime. From an engineer's point of view, the most "efficient" cross section (maximal carrying capacity with minimal width) in earth channels is the uniform trapezoid with 1 : 3 slope of the banks. From an ecological point of view, it is one of the worst. One way to improve the structure and function of the regulated stream is to plan a complex cross section with berms, thus creating a stream within a stream. The lower stage will interact with the regular flows that occur every year or two. It will also interact with the base flow that will concentrate in a narrow thalweg, instead of spreading over to much area. It is possible and recommended to add to the designed cross section elements that will enhance the habitat heterogeneity even more. Depending on the area, we can use large boulders, gravel piles, or logs.

The last phase to rehabilitate the form of the regulated stream is the profile. Drainage engineers want to tame natural streams and tend to design a uniform, low slope profile (to slow down the velocity of the current) and to divide the drop by several reinforced hydraulic structures like chutes. If the chutes have already been built, there is very little to do in the stream profile. But, in advance, prior to installation, chutes can be substituted by a riffle-pool system or a bar-pool system in finer sediments areas. Because those formations are assembled by natural forces and are mobile, constructing them as a natural-based solution for flood protection purposes is a challenge. One way to do so is to make sure the armor that needs to keep the structure in place is buried. The other strategy, which is more intuitive and less controlled, is to place elements of disturbance on and in the stream bottom as "seeds" for in stream natural structures. These elements will collect material and over time, the overall slope of the stream will be lower and water velocity will be in a degree that erosion is minimized.

In many cases, rehabilitation of the morphology of the stream is not enough. In projects where streams have no near source for recolonization, or the extent of bare soil is high, vegetative rehabilitation is employed. The degree of intervention varies and is based on these principles:

- The prognosis for spontaneous recolonization of natural flora.
- The likelihood of invasive/aggressive (mostly *Phragmites australis*) species takeover.
- The extent of earthworks

The highest degree of intervention is "gardening" the stream banks and riparian zone. That includes conditioning the soil with compost (or other means of soil enrichment), crowded plantings of saplings of target species, from the various guilds. Planting is followed by irrigation for several years. It is also requiring hands-on maintenance of the patch to ensure the well-being of the flora, replacement of unsuccessful plantings, and the removal of nontarget species. This kind of intervention is highly expensive and should be used only in tough cases. Interventions can get less intensive and less expensive. At the end of continuous intervention, there is no vegetative rehabilitation at all, and only nonspecific maintenance like mowing once or twice a year.

6.2.2 Perennial Mountainous Streams

In the following paragraphs, Israel's perennial mountainous streams (from here will be referred to as mountainous streams) will be described and discussed with the help of particular streams that represent a defined type inside that group.

6.2.2.1 Abiotic Characteristics

In Israel, we can find mountainous streams in a range of phytogeographic regions changing over short distances. There are mountainous streams in the desert, in the Mediterranean area, and in the desert borderland. Israel mountainous stream flows mainly through limestone, less on basalt, and least on chalk. Most of the streams in this category are fed by minor springs (with one huge exception, the Dan spring, which is the largest karstic spring in the middle east). Due to their sharp profile, water velocity in these streams is high, the substrate is washed from thins, and the water is clear. A good representative of mountainous stream over limestone in the Mediterranean area is "Kziv." It is fed by a series of minor springs, with an annual average of ~9 million cubic meters. The water flows in a narrow path that opens from time to time to wider, riffle-pool areas. We hardly find any waterfalls in the western drainage, and the streams have good connectivity. Kziv stream is fed by several minor springs that are flowing in its tributaries. Their discharge creates an uninterrupted flow in winter and spring. This flow from the various and dispersed springs creates a connected network of streams and brooks. Later in the summer, the discharge is dwindling. The flow in the tributaries is receding and is feeding separated flowing stream parts and isolated pools, hence becoming seasonal streams, while the main branch keeps its water year-round. The stoney stream bed has multiple interstices that allow hyporheic flow and transit storage of water. Once in a while, the bedrock is exposed and creates little cascades and pools. Another type of mountainous stream in the Mediterranean area flows on the Golan Heights basalt. One of them, "Yehudia" stream shares the hydrological characteristics of "Kziv" but differs in its morphological characteristics. The "Yehudia" stream is divided into three distinct morphological units. The upper unit is a shallow and wide stream flowing over a landscape formed by volcanic eruptions. The lava flowed in intervals and allowed soil formation between eruptions. This "sandwich" of basalt and baked soil created numerous local, disconnected aquifers. These aquifers have many outlets that create many minor or seasonal springs. These springs are feeding the tributaries, some of them perennial and some seasonal. The substrate on which "Yehudia" stream flows is made of coarse material, mostly because it does not have a long path to get eroded. After flowing on the plane, "Yehudia" stream takes a plunge to the second morphological unit, a steep and narrow stream. At this part, the stream is cutting deep through the different lava layers and creates waterfalls (10–30 m high) and cascades. The streamflow originating from the upper reaches of the watershed is enhanced by the contribution of water from a regional aquifer. In the third part, "Yehudia" stream turns into an alluvial stream that together with neighboring streams creates a maze of channels, water bodies, and vegetate islands. Eventually, it ends in the sea of Galilee (Gvirzman 2019).

As mentioned, in Israel there are also mountainous streams in the desert. These streams are part of the eastern drainage. The eastern drainage is characterized by a much sharper slope from the divide to the base. The "Qelt" stream runs over and cuts through limestone. The geological formations vary in strength. This variance in resistance to erosion is responsible for the formation of cascades. The springs are the outlets of a regional aquifer, but their discharge is modest and gets to 4–5 million cubic meters, per annum.

6.2.2.2 Biotic Characteristics

The rapid flow of clear water supports instream fauna that have high oxygen demand. Freshwater snails like *Theodoxus jordani, Theodoxus michonii, Theodoxus karasuna*, and seven species of the genus *Melanopsis*. The biggest group of the mountainous stream is the insects. Their nymphs dwell on the bottom of the stream, in the cavities, and under the rocks. We will find members of the orders Coleoptera, Odonata, Ephemeroptera, and Trichoptera. The most prominent functional feeding groups are the predators, grazers, filter feeders, and shredders. Israel has small catchments and small streams. This is why they support only small-bodied fishes. The fishes come from the genus *Phoxinellus, Gara, Nemacheilus*, and *Varicorhinus*. These fish are associated with the stony substrate and feed mostly on biofilm and benthonic algae. Israel is the world's southern extent of range of the rare amphibian *Salamandra infraimmaculata*. Although it breeds in montane streams, it needs the more quiet and slow-flowing parts. In those bedrock or side pools, it spawns living tadpoles. The tadpoles are feeding on small crustaceans and insects' larvae. They also feed on their own kind, but show lowered aggressiveness toward their kin (Goren 1969; Milstein et al. 2012; BIOGIS 2024)

Montane streams are generally less disturbed by human activity. This is one of the reasons that trees can grow on and near its banks. The most common vegetative formation is riparian forest. The main tree is *Salix acmophylla* and it is companied by *Ficus carica*. In the northern parts of the country, we will find trees from the rare species *Platanus orientalis*. Below the trees flourishes *Nerium oleander* and *Rubus sanguineus*. Because of the shade, herbaceous plants are limited to clearings and close to the current, where trees often fall. In the riparian forest, there are also terrestrial trees like oaks, *Pistacia palaestina*, and *Rhamnus alaternus*. Due to the abundance of water, they take a more lush form.

6.3 The Jordan River

The Jordan is the only watercourse in Israel entitled to be called a river. It has the largest drainage basin, the longest path, and the biggest and most active flood plains. Although it is a transboundary river, only Israeli wetlands will be described. The Jordan river is divided into two distinct sections: the upper Jordan and the lower Jordan.

6.3.1 The Upper Jordan

6.3.1.1 Abiotic Characteristics

The upper Jordan Basin is part of the "the Sea of Galilee basin," the single freshwater lake. Most of its area is in Israel, and little parts are in Lebanon. The "upper Jordan" is the most pristine and conserved watershed in Israel and gets its own section in the chapter. It starts in "Mount Hermon" and ends in the "Kineret." The Hebrew name for the "Sea of Galilee." On its way down from fountain head to the deltas entering the "Kineret," it takes and changes forms and characteristics. The major eco-hydrological units of the upper Jordan are Sources, Brooks, Alluvial channels and the "Hula Valley," montane river, and delta.

The water gushes from the major sources and flows through the brooks that join to create the upper Jordan. From there after a short section of a united river, it enters the "Hula Valley" and separates into three canalized watercourses. The three watercourses converge again at the exit of the Hula Valley. From there, the upper Jordan takes the form of a mountain river that flows down and meets the "kinerent" with a delta, called "Ha'btieha."

6.3.1.1.1 The Upper Jordan Sources Rainfall and snowmelt percolate Mount Hermon's lime and dolomite Jurassic age outcrops and feed the Jordan river main source, the Dan springs. The Dan spring is the largest karstic spring in the middle east. Its annual discharge is approx. 250 million cubic/meter (the total spring water discharge of Israel is approx. 550 million cm/year). The Dan source has two distinct parts. The huge fountain head, where most of the water emanates one little pool, and upper step springs that are smaller numerous outflows that creates an area of sheet flows. Other two major sources are Banias, and Snir (partly in Lebanon). The Banias (after the Greek god Pan) is the second largest spring in Israel with an annual volume of 80 million cubic meters. Like the "Dan" springs, it also gets most of its water through rainfall and snowmelt of "Mount Hermon" but has also another component from the "Golan Heights" basalt outcrops. Unlike the "Dan" that discharges in flat amplitude, the "Banias" is much more annual rainfall and has a shorter "hydrological memory." The last major sources of the upper Jordan are the "Hazbani" and "Wazani" springs, both are in Lebanon. The "Hazbani" springs are fully capped and diverted out of the stream, the second, the "Wazani" is also pumped out from the stream by Lebanon, but not entirely. An annual average of 45 million cubic meters flows crosses the border and enters the upper Jordan (Gvirzman 2019).

The upper Jordan gets additional contribution of spring water along its way to the Kineret. Most of it comes from the regional basalt aquifer of the "Golan Heights." The springs are scattered along the face of the "Golan Heights" that drop toward the "Hula Valley." Most of the upper Jordan sources are utilized in various methods for irrigation and fishery. In years that rainfall is average or above, the usage of water for irrigation does not harm the natural systems, because there is plenty for everyone. But in years of drought, competition between irrigation and environmental flow requirements is fierce (will be discussed in the rehabilitation measures paragraph).

6.3.1.1.2 Brooks The water from the major sources of the Jordan river flows in a maze of interconnected streams and brooks. The network is likely originated from man-made works in early times, for irrigation. Today, this network is the skeleton of a rich and diverse ecosystem. The stream differs from one another depending mostly on length, plan, and topography. The "Hazbani" stream originates in Lebanon and has a big watershed above the sources, so it is affected by floods. It is sinuous and its bottom is made of gravel. The "Banias" stream has a very small watershed, but starts high above the other major sources. Its rapid flow carves its way through coarse material and bedrock, including a waterfall. The "Dan" stream flows only a short distance as a single trunk watercourse. The brooks that diverge from it flow in quasiperpendicular and connect to the "Hazbani" and "Banias" and create the unique maze of the upper Jordan sources and brooks.

6.3.1.1.3 Alluvial Channels and the "Hula Valley" Eighty years ago, the maze of brooks ended its way by entering a vast marsh and a lake, the "Hula." The "Hula" was the biggest (6200 ha) and most important wetland of Israel. It was created by a basalt plug at the exit of the valley that kept high water levels (up to 2 m) in the mountains surrounded valley. In the early days of the newly declared state of Israel, it was decided to drain the marsh and the lake and convert them to arable land. This decision is still considered the most devastating ecological disaster. It also triggered the first environmental struggle, the establishment of the most prominent nature conservation NGO the ISPNI. It also facilitated the declaration of the first nature reserve in Israel that was managed (and still is) by the national nature reserve authority (today, the INPA). To drain the marsh and the lake, the two main branches that entered the marsh were widened and deepened. The basalt plug was removed and perpendicular canals were excavated to drain the soggy bottom. From an agriculture point of view, the project was only partially successful. Vast areas of peatland went through underground combustion and sunk down. The marsh that used to act as a sink for nutrients turned its nature and became the source of nitrogen and phosphorous to the Jordan river and the "Kineret."

Today, the maze of brooks and streams that was described earlier converge for a short distance to a main river. But, after only 2.5 km, it splits again into two branches, the eastern canal and the western canal. Those canals were excavated as part of the drainage of the Hula marsh and lake. The two canals were designed as straight lines, with a regular trapezoid cross section. Depending on the exact location, the substrate of banks and bottom changes from mineral soils to organic soil, and even lake sediments like marl, where the canal flow through what used to be a lake. The springs from the Golan Heights that once entered the marsh and lake are collected today in yet another canal that leads them down and out of the valley. Their water joins the flow of the Jordan river only down in the area where the basalt plug used to be, and the beginning of the mountainous part of the Jordan river.

6.3.2 Montane River

The Jordan river in its mountainous part is the largest in terms of length, discharge, and wetted perimeter. It completes a plunge from 75-m ASL to −210 BSL in 20 km, an average slope of 13%. The current is swift and flows through long runs, interrupted by small riffle sections. There are only a few pools along its way from the valley, to the "Kineret." It is the only watercourse where kayakers can experience "white water" kayaking. The substrate is a mixture of eroded gravel and coarse boulders conveyed downstream by bank erosion and landslides. The river is interrupted by a weir that diverts part of the flow to a hydroelectric power station.

6.3.3 Delta

The mountainous part of the Jordan river end in a delta in the northern bank of the "Kineret." Its single thread channel splits into many braids flowing over an alluvial fan composed mostly of gravel, cobble, and spotted by boulders. There are some main stable channels in the alluvial fan accompanied by more transient, unstable ones. The Jordan delta joins the deltas of streams that originate in the "Golan Heights" to form a large unique area with a mosaic of habitats like lagoons, streams, islands, and thickets.

6.3.3.1 Biotic Characteristics

Salamandra infraimmaculata in Israel usually lives in high altitudes. The Dan springs nature reserve is an exception. Although it is situated in low altitude and warm weather, it is the largest single population of Salamandra. The dense vegetation and ample of water create a microclimate that enables this sensitive species to thrive. In the multiple flows around the hill are the habitat for the small-bodied *Oxynoemacheilus panthera,* a large variety of sensitive macro-invertebrates and a nursery for gastropods from the Neritidae family. Down from there, the main spring gushes to a small pool and then enters the mainstream. The mainstream, with its rapid flow and deep water, is the habitat of large-bodied fishes from the temperate zone like *Barbus longiceps* and *Carasobarbus canis*. It is also the habitat of the elusive *Lutra*. This species is designated as vulnerable by the IUCN. It used to inhabit all perennial and fish-bearing streams in Israel. But due to pollution and dehydration of most of the western drainage, it retreated to less-disturbed areas, such as the eastern drainage, and especially, the Jordan sources and brooks. In Israel, it is a protected species and in a status of critically endangered. After the draining of the Hula marsh and lake, fishpond became an important source of fish for the territorial *Lutra*. Economic and market changes have led to the Abandonment of most fisheries, leaving it to rely on natural habitats only. This caused yet another decline in numbers (Goren 1969; BIOGIS 2024).

The fast-flowing natural brooks are continued by artificial ditches and canals that were excavated when the Hula marsh and lake were drained. Although the watercourses of the valley are uniform and regular in shape, they still provide a habitat for rich fauna and flora. The reason is the good water quality in high discharge. Here we can also find fishes of African origin like *Sarotherodon galilaeus, Coptodon zillii*, and *Oreochromis niloticus*. The fish society in the valley also comprises the everlasting *Clarias gariepinus* and invasive species like Gambusia and domesticated carps. The abundance of water and warm temperatures encouraged the intense growth of reeds and other riparian vegetation that encroached on the canals and ditches and also helped the *Lutra*-established territories in the valley. Another predator associated with wetlands and thickets that populates the valley is wild cat *Felis chaus* (Goren 1969; BIOGIS 2024).

The Hula valley marshes and canals are the last major habitat that supports a stable population of *Cyperus papyrus*. This African sedge used to flourish also in the western drainage, but like other water-dependent (quality and quantity) species, it disappeared. The Hula Valley is today the northernmost area it occupies. The narrow riparian areas of the canals are occupied mostly by *Salix acmophylla*, *Phragmites australis*, and *Rubus sanguineus*. The soft and muddy banks and bottom is home for *Typha domingensis* and *Lythrum salicaria*. It is common to find inside the water the submerged plant *Ceratophyllum submersum* and *Potamogeton nodosus* floating over it (BIOGIS 2024).

The channels and ditches of the Hula valley harbor various macro-invertebrates. The biggest and rarest bivalve of Israel *Unio terminalis terminalis* can be found here, together with a smaller and yet endangered species *Potamida semirugata littoralis*. Both of these bivalves are endemic to the Jordan river and Orontes (Lebanon, Syria, and Turkey) that ends in the Mediterranean Sea. The fact that these species are endemic to two river systems that are disconnected and flow to different basins (Dead sea for the Jordan and the Mediterranean Sea for the Orontes) suggests that these systems were once part of one bigger system that was separated by the creation of the Dead sea rift (Milstein et al. 2012).

Before starting the drop down to the "Kineret," the Hula valley canals and ditches rejoin to a single trunk, deep and fast-flowing run. At its descent, the mountainous Jordan river appears in its full might. A developed riparian forest composed mainly of *Salix acmophylla* accompanied by *Arundo donax* and *Phragmites australis*. The line that experiences medium floods creates a shrubbery composed mainly of *Lythrum salicaria, Pluchea dioscoridis, Epilobium hirsutum,* and *Rubus sanguineus*. In the bank areas that get exposed on a yearly basis by flood currents, the herbaceous formation is ruling. Herbs like three species of the genus *Mentha*, *Lythrum junceum*, *Persicaria senegalensis*, and *Persicaria acuminata*.

The macro-invertebrate society of montane upper Jordan is diverse and rich and includes some rare and endemic species. The last survey that was conducted showed 43 taxa. 63% were insects of six orders. Coleoptera, with three different families (Dytisicidae, Hydrophylidae, and Haliplidae). In the Diptera, there were two families (Chironomidae and Ceratopogonidae). In the order Ephemeroptera, at least four families were collected (Prosopistomatidae, Heptageniidae, Baetidae, and Oligoneuriidae). The order Heteroptera yielded also four families (Vellidae, Corixidae, Gerridae, and Pleidae). The sensitive order of Trichoptera has two families representing it

(Leptoceridae and Hydropsychidae) to sum up the insect's portion, there were the Odonata with species belonging to the suborders Anisoptera and Zygoptera. The endangered species *Calopteryx syriaca* is among them.

The montane Jordan river, besides being a wetland, is also an ecological corridor. Israel is very small, and open areas are shrinking rapidly. The remaining areas suffer from segmentation caused by infrastructures. The large riparian zone enables stable populations of the endangered species *Lutra lutra* and *Felis chaus*. It is also a refuge, a territory, and a path for more common mammals like *Sus scrofa*, *Herpestes ichneumon*, *Canis aureus*, and *Hystrix indica* (BIOGIS 2024).

The delta of the upper Jordan, when it encounters the "Kineret," is the only one that survived in Israel. The delta's name "B'teiha" and it harbors most of the mammals and riparian vegetations of the montane section. But, it has also unique biotic characteristics. One major ecological function is the most important reproduction site for the lake's fishes, especially the Cichlidae. The bottom of the delta is a mixture of muddy flats where the Cichlidae builds their nests, with spots of gravels and cobbles that are suitable for fishes that need stony substrate to lay their eggs, like *Acanthobrama terraesanctae*. The hatchlings can find refuge in the matrix of emergent and floating and submerged macrophytes like *Typha angustifolia*, *Cyperus distachyos*, *Ludwigia stolonifera*, *Potamogeton nodosus*, and *Ceratophyllum submersum*. Part of the delta is a nature reserve and the nursery is safe. But, it is so important for fish reproduction, that there is a ban on fishing in the spawning season, also outside the reserve (Zohari and Feinbrun-Dothan 1966–1986).

The "Kineret" is a natural lake, but it is also one of the major state's sources for drinking water and irrigation. Its level is managed by a dam in the south of the lake and a huge pumping station in the northwestern part of the lake. The sharp and irregular changes in the lake water level have a negative impact on the delta and all of lake shores. In the past decade, Israel experienced consecutive dry years. The result was a continuous decrease in water level and exposure of the bottom of shallow parts of the lake. In the delta, large muddy flats were exposed, and lagoons got dried. The watercourses became narrow and incised their way in the alluvium, following the receding shoreline. The ecological response in the delta was an outburst of two tree species, the local *Tamarix jordanis* and the invasive *Parkinsonia aculeata*. These two species took over almost completely over the dryland that got exposed in the drought. The *Tamarix jordanis* is a habitat modulator. It secretes salt over its surroundings, making it hard for other water plants to thrive. It can also endure several consecutive years of submerged root flare, and end short reflooding is not a limiting factor.

6.3.3.2 Conservation/Rehabilitation Measures

The upper Jordan system is relatively conserved (leaving out the devastation of the drainage of the Hula marsh and lake in the 50′). There are many parts of it that were declared as nature reserves over the passing decades. The main issue in these parts of the country is how to manage water sources. The upper Jordan is an agricultural basin that relies on self-water supply. The meaning of that is every cubic meter used to irrigate the crops is taken directly from the sources or from the flowing river. The way to answer this challenge was articulated in 2003 in a policy paper named "Nature's right for water." The paper explained the rationale and the need, described negatively affected ecological systems, and principles for determining the natural needs. The policy paper served as a catalyst and an enabler for an amendment in the water law of Israel. A prominent principle that found its way to the legislature was "Headwater VS downstream." A farmer that caped a spring at the fountainhead paid a higher rate for the water he used to irrigate. On the other hand, if the farmer is willing to pump the water downstream, he will get a lower rate. Additionally, the farmer's extra expenses, electricity for pumping and filtration because of less clear water, will be reimbursed. This incentive, together with government funding, plays a role in many projects that shifted production points from the sensitive and more diverse headwaters to downstream. Other means of intervention in the Hula valley canals and ditches are morphological rehabilitation and remeandering, a complex cross section, and run-pools segments, as described above in the lowland streams section (Uzan 2010).

6.4 Lentic Habitats

Lentic habitats of Israel can be divided into three major types: Runoff-fed marshes, groundwater-fed marshes, and vernal pools. The Israeli climate and ecology do not support the creation of swamps. The main vegetated wetland type in Israel is marshes. Marshes are distinguished by their water sources and the mixture between them, the water quality (especially salinity), and their extent.

In the first decade after the establishment of the state of Israel, the alluvial valleys and the marshes that ruled their landscape undergone a massive campaign of drainage and converting wetlands to arable lands and fisheries. The campaign included canalization of streams, widening and deepening them, and cutting of their meanders. The springs capped underground, and the high-water table lowered by canalization and pumping of water from underground. All these actions have led to the loss of more than 90% of Israel's marshes. Some of the remaining few places that were not lost completely are now protected in nature reserves. But, some of them, are still in danger.

The biggest marsh in Israel was the Hula marsh ~5500 ha. After most of it was lost in the reclamation campaign that ended in 1957, a small fraction of the area was conserved as a nature reserve ~470 ha. The Hula marsh fed by runoff and stream flow. Unlike natural marshes, the Hula marsh nature reserve is higher than its surroundings. The main reason for that lies in the fact that the area conserved is situated at the outskirts of the former and now dried lake. The second reason is the sinking of the land around it because of internal combustion of the pit land exposed and oxidized in the reclamation campaign. The marsh gets its water from pumps extracting water from the western canal and holding it by dykes and levees. The water flows through the various habitats of the reserve, the wet meadow, the main lake, and the shallow marsh, and pours back to the western canal. Until 2000, the Hula nature reserve was the sole functioning remnant of the Hula marsh. In 2000, a reflooding of reclaimed land took place.

Apparently, some of the reclaimed marshlands failed as arable land. It became more and more difficult to work the land for agro-mechanic reasons. Another ecological problem that emerged from the conversion of the marsh to a field was the release of phosphorous from the pit lands. The huge amounts of phosphorous used to get trapped in the marsh soil became loose and transferable. The phosphorous found its way to the "Kineret" and apparently triggered a major and negative shift in its algal community and functionality. The excess phosphorous gave an advantage to nitrogen-fixing cyanobacteria. Some species of these cyanobacteria secrete toxins into the water. The Kineret is a major water source for drinking and agriculture. It is also a main site for recreation. To algal bloom of toxins, secreting cyanobacteria became a problem to be solved. One of the solutions for that problem was the reflooding of reclaimed pit land and the creation of a new marsh, "Agmon Hula." Unlike the Hula nature reserve, the "Agmon" is situated on the pit land of former marsh, and the water is flowing into it, by gravitation. The water flowing into the marsh is held by underground screens installed on its borders. The result is a shallow water body with a vast sedge area around it. The water overflows to a series of collecting canals that eventually leads to a pumping station. The nutrient-rich water is then pumped out of the valley. The water abstraction serves two goals. The first is limiting the phosphorous flux into the "Kineret" and the second is water for irrigation outside the valley.

6.4.1 Biotic Characteristics

The icon of the devastation that followed the draining of the Hula marsh and lake was the extinction of the frog from the Alytidae family. Before the drainage operation in 1940, a local researcher found specimens of spotted frog and identified it as *Discoglossus nigriventer*. Attempts over the years to recover more specimens of that species have failed. The species was officially declared extinct in 1996. But, in 2011, a worker at the Hula nature reserve spotted a black-spotted frog on the walking trail in the reserve. He collected it and brought it in for identification. It was a specimen of the long-considered extinct species. But things got even more interesting. Following the discovery, a small population of frogs was found. A few specimens were captured for scientific research. The researchers have found that the frog was not at all from the genus *Discoglossus*. The painted frog that was found belongs to the genus *Latonia* was known from the Paleogene and Neogene and considered extinct. The taxonomy of the frog has changed and it is now *Latonia nigriventer* (Figure 6.6), a living fossil that is also endemic to Israel (BIOGIS 2024).

The Hula marsh (both the nature reserve and the "Agmon") is a major and important site for resident, wintering, and migrating birds. Huge flocks of cranes from the species *Grus grus* spend there the winter. They use the vast shallow and open water as roosting area and feed on the surrounding fields. The vast marsh and lake that once existed in the Hula were an important stop-over site for migrating birds. For the palearctic populations of pelicans from the species *Pelecanus onocrotalus*, Israel is one of the last feeding opportunities before the long, hot, and exhausting flight over the Sahara Desert. The fisheries that replaced the marsh and lake landscape began to serve as a feeding station for the hungry pelicans. The farmers fought back and tried to deter them by nets, gas cannons, and shooting. To minimize the conflict, the INPA together with the Ministry of Agriculture are stocking reservoirs with fish bycatch to allow the pelicans the fuel they need to move on to Africa. The papyrus thickets in the Hula marsh and lake used to harbor the nesting colonies of most of the egret species that nest in Israel; *Ardea cinerea*, *Ardea purpurea*, *Ardeola ralloides*, *Ixobrychus minutus*, and *Ixobrychus minutus*. The unexplained collapse of the papyrus

Figure 6.6 *Latonia nigriventer* Uri Roll.

ended most of the nesting activity. Gradually, the papyrus was replaced by *Phragmites australis* and some of the species adapted to it. Some of them, like *Ardea cinerea,* are no longer nesting in Israel (BIOGIS 2024).

As for ducks, Israel is a wintering site for many palearctic populations of ducks. The marshes and natural water bodies allowed them to feed and grow before returning to their nesting grounds in the north. But, there are also ducks that are nesting in Israel. Two species of ducks, *Aythya nyroca* and *Marmaronetta angustirostris*, are used to nest in large numbers in the Hula marsh prior to its destruction. These populations have collapsed as a result of habitat destruction. Today, there is a slow recovery of these species as nesting species. The Hula marsh supports the biggest population of the cat *Felis chaus*. The thick and vast vegetation thickets where it can hide in the day and rear its young, the abundance and various food sources, waterfowl, rodents, fishes, and amphibians, all of that enable this species to flourish.

The western drainage has a special geomorphological feature. The coastal plain of Israel has a series of calcified sandstone ridges that run in parallel to sea shore (south–north) and perpendicular to the stream system (east–west). The streams pushed their way from the mountain top to the shore, but met with the barriers of the calcified sandstone ridges, after most of their energy had dissipated in the plain. The area between the ridges became a settling area and much of the sediment carried with floods and accumulated there over the years. The resulting landscape was flooded troughs. Some of the streams have managed to breach openings in the ridges, but some openings have been excavated as early as when the Roman empire ruled the territory. Most of the Roman drainage and water operations were neglected in the late Ottoman empire days. The combination of floods, sediment load, and narrow paths created extensive marshes in the western drainage. There were also springs and areas of with high water table that created marshes in the alluvial valleys. Most of them were capped and their water was diverted to reservoirs. This served two purposes: the first, reclaiming the marshland and converting it to arable land and the second, water for irrigation. As a matter of fact, fighting the marshes, which were considered a trope for neglection and desolation, was a source of pride and encouragement of the settlers of the newly established Israel. A good example of a runoff-fed marsh in the western drainage is the "Poleg" Marsh.

6.4.2 Abiotic Characteristics

The Poleg marsh was created by runoff water accumulated in a trough between two calcified sandstone ridges. The name Poleg means "divided." In the Bronze age, the Canaanites built a fortress in one of the ridges. In an effort to protect it, they dug a moat around the fortress, splitting it, and from there, the area got its name. Later, the Romans expanded the moat in an effort to drain the marsh. The drainage operation kept on functioning until the early Arab age. The Ottoman empire neglected the drainage operation, and the marshes had returned and prevailed for several centuries. In this period, the marsh deposited layers of vegetation that flourished in it. In the early twentieth century, the British mandate renewed the drainage operation in order to ease the malaria problem and reclaim the land for agriculture. The British operation was continued and expanded by the settlers of the young state of Israel. The operation was successful for several decades. The area became cultivated, with seasonal, and irrigated crops. But, the ancient organic soil, the legacy of the vegetation that prospered and died in the marsh, caught fire. The organic soil, exposed and oxidized after the dehydration of the marsh, started to burn in internal combustion. The fire consumed the organic matter and left the mineral fraction. The result was the creation of sinks over the reclaimed land. These sinks dropped down to the level they became even lower from the bottom of the main drainage canal. These sinks got filled with runoff and direct rainfall, but there was no way to carry out the water. The local groundwater table has risen due to a reduction in water production (high nitrate levels) and added its contribution. The 300 ha marsh started taking over the landscape again. The recreated marsh has different areas with different hydroperiods. The biggest and deepest sink became a permanent water body. The bottom of the sink is the roof of the local groundwater table. The other parts of the mosaic are seasonal marshes that support water bodies that eventually dry out. Every water body has its own time.

6.4.3 Biotic Characteristics

The main and most important (in terms of habitat structure and function) plant in the Poleg marsh is the reed *Phragmites australis*. The dense canebrake is the refuge, foraging, nesting, and roosting habitat for many bird species. The Poleg marsh is a nesting area for the ducks and egrets mentioned in the paragraph about the Hula marsh, to a smaller extent. The marsh is also the habitat for small-bodied birds. One genus of marsh-related birds, *Acrocephalus*, has five different species that live in the Poleg; *Acrocephalus melanopogon, Acrocephalus schoenobaenus, Acrocephalus scirpaceus, Acrocephalus arundinaceus,* and *Acrocephalus stentoreus*. Many species of the Genus *Acrocephalus* are known for their nesting site fidelity. The Poleg marsh is a predictable and stable habitat for returning specimens. The Cryptic *Rostratula benghalensis* is a rather common bird in southeast Asia and Africa. But the appearance of a fortiori nesting of this species in Israel is a big business for bird watchers. The Poleg marsh is the single site in Israel, where a nesting attempt (failed because of nest predation) has been recorded.

The vegetation of the Poleg marsh is diverse and changes with the type of habitat. Habitats with long hydroperiods are dominated by *Phragmites australis*. Habitats with a shorter flooding time tend to allow a more diverse plant community. Species like the rare *Persicaria lanigera* (Figure 6.7) and *Damasonium alisma* grow here. The more common habitat modulators are *Typha domingensis* and *Scirpus maritimus* (BIOGIS 2024).

Figure 6.7 *Persicaria lanigera* NPA Ranger.

6.5 Dalia Marsh

Another type of marsh is the groundwater-fed marsh, the "Dalia" marsh.

6.5.1 Abiotic Characteristics

In the mid-north of Israel, on the west side of the divide, lies the coastal area of Mount Carmel. The basic morphology of the coastal area of the western drainage is also present here. Not far from the beach, between two sandstone ridges, a calcified sandstone local aquifer is exposed. This aquifer has limited outcrops, but has underground connections with the bigger, regional limestone aquifer. The water that emerged became trapped in the trough between the two sandstone ridges, or a ridge and sand dunes, creating a groundwater-fed marsh.

This aquifer has outlets under the sea so there is no precise number of the annual volume of water that emanates from all of its outlets. Water table is about 2–3 m ASL. The water is saline ~1500 mg/l chloride and is used as raw water for desalination plants. The majority of marshes in the coastal area of Mount Carmel were converted completely to extensive fish ponds. One of the marshes that still exist, at least partially, is "Dif'le." "Dif'le" is the Arabic common name of the plant *Nerium oleander* that probably was a dominant species in the marsh at that time. Today, the name is "Dalia Marsh nature reserve," after the name of the estuary nearby.

The "Dalia Marsh" is a groundwater-fed coastal marsh. The annual discharge of the springs is ~10 million cubic meters and was calculated by mass balance. The water from the aquifer flows down and joins the water of the Dalia stream (intermittent) to crate together the estuary. The nature reserve is surrounded by an active fishery that pumps water out of it and discharges back to it the effluents. The fishery also dammed the river mouth and converted the estuary to an operational reservoir for the fishery and also to facilitate floodwater abstraction for irrigation.

6.5.2 Biotic Characteristics

The Dalia marsh was the last habitat where the largest freshwater bivalve of Israel, *Unio terminalis delicatus* used to live. This species is endemic to the Levant. In 1980, a survey recorded only adults that were concentrated around the remaining springs at the bottom of the marsh. It is now considered locally extinct.

Another flag species of the Dalia marsh is the soft-shell turtle *Trionyx triunguis*. This species is designated as vulnerable by the IUCN. The streams of the western drainage were the natural habitat for this African turtle. But, the ecological status of them was bad and deteriorated. In the early 70s, a researcher came with the idea for a plan to protect and conserve this species. Adult turtles were relocated to the Hula nature reserve in the upper Jordan that was in better ecological status from the western drainage coastal streams. The specimens that were relocated did very well and created a viable population. Today, the protection and conservation plan is to harvest hatchlings in their nest and reintroduce them to rehabilitated streams of the western drainage.

The biological records of the Dalia marsh include the fish *Aphanius mento*. This species is living and reproducing in close association with submerged plants. The record also includes the submerged species *Myriophyllum spicatum*. Submerged plants need light to penetrate the water column. Unfortunately, the Dalia marsh has and still suffers from eutrophication caused by nutrient-rich fishery effluents. The species that are sensitive to organic pollution no longer exist in the nature reserve.

The steep banks of the nature reserve do not encourage and cannot sustain the typical belt formation of the plant society. The dominant species is *Phragmites australis*. Other dominant species that create the vegetative thicket of the reserve are *Tamarix nilotica* and *Rubus sanguineus*. Despite the hard and wide disturbances, the Dalia marsh still harbors some rare plant species like *Rosa phoenicia, Ipomoea sagittata*, and *Trachomitum venetum*.

6.6 Measures of Rehabilitation and Restoration

Most of the marshes in Israel have undergone some kind of human-induced alteration. There is no pristine marsh in Israel to serve as a reference site for rehabilitation plan design. Current landscape does not tell the whole story and sometimes even masks the true nature of the habitat. The main source of information to analyze the area and its main natural characteristics is historical data and first comes the maps. For centuries, Israel (known at that time as Palestine) was a neglected province in the Ottoman empire, and no mapping efforts we invested. The first and oldest triangulation-based map of Israel was illustrated by Pierre Jacotin, for the excursions of Napoleon Bonaparte in the Levant (1799–1800). The Jacotin map gives a rough idea of the topography of Israel and gives the approximate whereabout of the marshes at that time. A more detailed and useful map was drawn by the Palestine Exploration Fund in 1884, the PEF map. This map is the most accurate and detailed map of Israel Topography and drainage systems before modern cartography. In this map, perennial streams and marshes were delineated. The local names of the aquatic habitats sometimes held clues about their nature: a pond, a marsh, a spring, a sedge meadow, and so forth. More recent and yet old enough data source of the Israeli landscape before the major development of the young established state of Israel is Arial Photos taken mostly by squadrons of the various empires that were drowned to the middle east following the collapse of the "Sick Man on the Bosphorus," the Ottoman empire in its Curtain call stage. Photos of the Bavarian, Australian, and British squadrons can be found in the state and private archives. Another rich source of information is the Jewish National Fund (JNF) and the settlement's archives. The JNF was the funder of many land acquisitions from local Arab landlords, and some of these lands were marshlands. The JND also conducted some of the operations of land reclamation and it took photos and mapped the marshes, prior to their draining (Uzan 2010).

Another source of information about marshes was a survey conducted by the antimalaria unit at the Ministry of Health of the Young State of Israel. The survey goal was to find and treat marshes infested by disease-transmitting mosquitos. The survey unintentionally delivered important data about the hydro-period, type, and the exact location and extent of the marshes, prior to their destruction. History can assist in the analysis and determination of scope of the rehabilitation project. But, to gain deeper knowledge, there is a need to understand the current status and even more important, the processes that drive it.

Before every marsh rehabilitation project, a range of surveys and investigations are carried out. The first and most important is hydrology, both surface and underground and water quality. The next is the topography, usually by means of Lidar-generated DTM, and the lithology. The next surveys are aimed to get knowledge of the fauna and flora and establish a baseline, before action.

In order to rehabilitate a marsh, there is a need to restore or at least improve the water regime. To do so, a comprehensive water plan is needed. A water plan is a pathway in which the vision, the measures and the timeline, is set. Measures can focus on elevating a dwindling aquifer water table, restoring diverted waterways, or even clogging back a canal-drained marsh. The water plan also builds alternatives for the surrounding stakeholders that drained the marsh in the first place. It can take the form of alternative water sources such as effluents, or alternative/improved land parcel, to cultivate instead of the reflooded marsh.

When water in an appropriate regime and quality is secured, it is time to act on the land itself. Many times, an impaired marsh suffers from regular and uniform shape. The common way to change that is by earthworks preferably digging, but also by filling. The bottom is preferably designed in an irregular bathymetry, thus creating distinct water bodies, with various depths and hydroperiods. The banks are the place where vegetation grows and further diversifies the landscape. Steep uniform banks support a narrow range of plants, in most cases dominant or invasive species. Creating (by excavating or by filling) a range of bank formation with different slopes and shapes allows a wider range of plant species to flourish and create the corresponding habitats. But not always a spontaneous vegetative regeneration can occur. When the marsh is in a very poor condition, when there is much earthwork to do, and when there is no near source for propagules, an active revegetation is needed. A revegetation plan is composed of a planting map (where to put what, in an attempt to match the plant with the right habitat) and the supporting system, e.g. irrigation system and maintenance to remove unwanted species that can compete with the desired plant society. Active vegetative rehabilitation is a costly project, and its success is not guaranteed.

6.7 Vernal Pools

6.7.1 Abiotic Characteristics

Vernal pools are wetlands typical of the Mediterranean climate. A vernal pool is an absolute sink, with a local drainage basin, that harvests rainfall and stores it for a period of at least four months. Vernal pools can occur in various landscape settings and lithologies and can be found on heavy soils, light soils, and even bedrock. They are present in natural, agricultural, and even urban setting, but are rarely connected to fluvial systems. Most of vernal pools are small, less than 0.5 ha, and with a depth that ranges from 30 cm to 1.5 m. If a pool is deeper than that, then it can hold its water year-round and is not considered vernal pool, but a pond.

6.7.2 Biotic Characteristics

Vernal pools attract many species from many guilds, but their uniqueness comes from the crustacean's subphylum. Vernal pools support crustaceans, like copepods, ostracods, and daphnia, that can also be found in pools with shorter than four months hydroperiod. But fairy shrimps, triops, and cyzirus can be found only in real vernal pools. These crustaceans have developed an adaptation to the wetting–drying circle of the vernal pool. They lay diapause eggs that are dropped in the mud before they vanish following the dehydration of the poll. The eggs wait the whole dry season protected by their hard shell, and emerge out it, after the pool has been filled again. Vernal pools are the home and breeding habitat of six out of Israel's seven amphibian species. The species that depend solely on vernal pools is the toad *Pelobates syriacus*. As in many species that can be found in Europe and central Asia, Israel is the southmost boundary of its distribution. Although its status out of Israel is the least concern, it is considered under a threat of local extinction. One of the reasons for that is its home range fidelity. Many vernal pools in Israel have been destroyed in times when the richness and significance of that habitat were realized. When a vernal pool is destroyed, it is likely to destroy the *Pelobates syriacus* population that used to breed in it.

References

BIOGIS (2024) Open database on biodiversity of Israel. https://biogis.huji.ac.il/ (accessed 5 May 2024).

Comair, G.F., Espinoza-Dávalos, G.E., and McKinney, D.C. (2023). Assessing water security in the Jordan River Basin: temporal changes for precipitation. Evapotranspiration and land cover. *Water* 15: 4064. https://doi.org/10.3390/w15234064.

Goren, M. (1969). Distribution and taxonomy of freshwater fishes of Israel, with emphasis on the Genera. Master thesis. University of Tel-Aviv.

Gvirzman, H. (2019). *Water Resources of Israel*. Jerusalem: Yad Ben Zvi.

Knesset (Israeli parliament) (2004). Water Law (Amendment No. 19), 5664–2004.

Milstein, D., Mienis, H., and Rittner, O. (2012). *A Guide to Israel's Mollusks*. University of Tel-Aviv. https://kineret.org.il/wp-content/uploads/2018/10/%D7%9E%D7%92%D7%93%D7%99%D7%A8-%D7%A8%D7%9B%D7%99%D7%9B%D7%95%D7%AA-ilovepdf-compressed.pdf.

Uzan, A. (2010). *The Conservation and Rehabilitation of Israels Streams*. Jerusalem: Science Division Publications. https://static.parks.org.il/wp-content/uploads/2017/08/shikumN.pdf.

Zohari, M. and Feinbrun-Dothan, N. (1966–1986). *Flora Palestina*. Jerusalem: Israeli Academy Publishing.

7

Wetlands of Angola: Locations, Biodiversity, and Conservation

Jasper Knight[1] and Mauro Lourenco[2]

[1] *School of Geography, Archaeology and Environmental Studies, University of the Witwatersrand, Johannesburg, South Africa*
[2] *National Geographic Okavango Wilderness Project, Wild Bird Trust, Johannesburg, South Africa*

7.1 Introduction

Wetlands are a common feature of semiarid areas and "wetlands in drylands" is a current research theme given that wetlands offer a good perennial water source in regions that are otherwise water scarce (Scoones 1991; Silvius et al. 2000; Tooth and McCarthy 2007; Rebelo et al. 2010; Grenfell et al. 2022). Wetlands have been mapped throughout southern Africa, focusing on their number and spatial patterns (Morant 1983; Taylor et al. 1995; Ellery et al. 2016; Sieben et al. 2018; van Deventer et al. 2020; Mandishona and Knight 2022a). Likewise, many studies in the region have examined the use of wetlands by local communities, with respect to wetland potential for agriculture and water supply, provision of different ecosystem services, and relationship to rural sustainable development (Mwakaje 2009; Dixon and Carrie 2016; Stephenson et al. 2020; Sinthumule 2021; Adeeyo et al. 2022; Mandishona and Knight 2022b). These studies highlight the critical role that wetlands play for local communities in southern Africa. There have been fewer studies, however, that have examined the topographic context and the soil and water properties of wetlands (Ellery et al. 2016; Mandishona and Knight 2022a). This is essential, however, because such an understanding can be used to predict where wetlands should be found in the landscape, and this can direct future research. This approach is particularly useful in situations where wetlands have not yet been fully identified and mapped, which includes almost all regions of southern Africa, and where there is generally a lack of field data on hydrology and soils, especially in remote rural areas.

This approach, based on an understanding of the physical landscape, is the starting point for the analysis of wetlands in Angola. Angola does not currently have any Ramsar inscribed wetlands although some have been recently proposed. The 6th National Report on Biodiversity identifies wetlands and mangroves as important ecosystems in the country (Republic of Angola 2019) but only the latter has been investigated in any consistent way and there are many thematic and spatial gaps in mangrove research in Angola (Harris et al. 2013; Cardoso et al. 2021). Apart from mangroves, there is a virtual absence of information on wetlands in the country, in English or Portuguese peer-reviewed literature, except in association with the major river systems of the Cubango, Cunene, Cuito, or on the endorheic wetlands of the Okavango and Etosha systems located outside the Angolan border (Mendelsohn 2022; Huntley 2023). There is no literature at all on wetlands

Wetlands of Tropical and Subtropical Asia and Africa: Biodiversity, Livelihoods and Conservation,
First Edition. Edited by Thammineni Pullaiah.

associated with mountain valleys, lowland rivers, river mouths, or estuaries/lagoons, although mountain peatlands have been previously identified and mapped (Lourenco et al. 2022). This absence of literature on likely Angolan wetlands is of concern because it means that they have received relatively little attention or conservation priority at the national level. It also means that it is unclear how local communities may use and value wetlands, which is important for effective resource management and conservation, and for understanding the role of wetlands in sustainable rural development.

The fourth Ramsar Strategic Plan (2016–2024) closely links the maintenance and conservation of wetlands globally to the sustainable development goals (Ramsar Convention Secretariat 2016). This is particularly relevant to situations in southern Africa where wetland ecosystem and environmental services are vital for community livelihoods and climate resilience (Ayanlade and Proske 2016; Owethu Pantshwa and Buschke 2019; Knight 2021; Dalu et al. 2022; Musasa and Marambanyika 2022a, b). The Millennium Ecosystem Assessment (MEA 2003) identifies and discusses the four ecosystem service types (supporting, provisioning, regulating, and cultural) that are present in different ecosystems and environments and their importance for human activity. Many studies in southern Africa have identified the availability, characteristics, and use of these different ecosystem service types found in wetlands (e.g. Mandishona and Knight 2022b). These services are strongly linked to the properties of the physical environment, including topographic setting, water availability (therefore precipitation patterns and water balance), soils, and climatically controlled vegetation patterns (Chatanga et al. 2019; Huntley 2019; Chatanga and Sieben 2020; Janssen et al. 2021; Salimi et al. 2021). Furthermore, the wider literature links these elements together through the concept of hydrogeomorphic systems, which is an important theme in wetland research that considers the co-relationships between wetland hydrology and its geomorphic setting (Shaffer et al. 1999; Dvorett et al. 2012; Cole 2017; Kim et al. 2020). However, this concept has not been fully applied to wetlands in southern Africa (Ollis et al. 2015; Ellery et al. 2016; van Deventer et al. 2016; Grenfell et al. 2019), but has scope to do so where regional datasets on climate, topography, soils, and vegetation are available.

The aim of this chapter is to use a hydrogeomorphic approach, based on Ramsar Convention guidelines, to identify the locations of wetlands in Angola and to describe their large-scale characteristics. In detail, this chapter (i) uses global-scale and publicly available datasets on climate, topography, soils, and ecosystems to identify the most likely locations of wetlands in Angola, including the different wetland types that are present, based on Ramsar wetland categories; (ii) describes the likely importance of these wetlands for local communities including their ecosystem services; and (iii) identifies key issues of biodiversity, threats, and conservation status of wetlands in Angola.

7.2 Methodology

This study provides an overview of (likely) wetlands that exist in Angola, based on an understanding of national-scale topography, hydrology, climate, and ecosystems derived from global datasets. This analysis provides a good guide as to where within the country wetlands are most likely to exist. This approach is used for Angolan wetlands because there is hitherto no national-scale inventory of wetlands or their locations or characteristics, or discussion of wetlands in the peer-reviewed literature. Angola became a Contracting Party to the Ramsar Convention on Wetlands of International Importance on 10 October 2021. The accession document to the Ramsar Convention, dated 10 June 2021, identified 11 potential wetland sites in Angola for designation under the Convention (UNESCO document ref LA/DEP/2021/007) but these are not yet listed on the Ramsar

Table 7.1 Proposed Ramsar sites in Angola, identified as of June 2021.

Number in Figure 7.1	Site name	Area (ha)	Coordinates	Location (province)
1	Lagunas do Mangal do Lobito	259	12°21′45″S, 13°32′43″E	Lobito (Benguela)
2	Saco dos Flamingos	1,616	09°05′03″S, 13°00′15′E	Ramiros (Luanda)
3	Lagoa do Arco	7,568	15°46′01″S, 12°03′47″E	Tômbwa (Namibe)
4	Parque Nacional da Cameia	1,445,000	11°57′32″S, 21°40′31″E	Cameia (Moxico)
5	Complexo das Zonas Húmidas da Lagoa do Carumbo	200,000	07°48′S, 19°57′E	(Lunda Norte)
6	Lagoa do Calumbo	1,000	09°10′07″S, 13°24′43″E	Icolo e Bengo (Luanda)
7	Lagoa da Quilunda	5,111	08°53′05″S, 13°36′01″E	Icolo e Bengo (Luanda)
8	Complexo de Lagunas de Santiago-Saurico	3,763	08°43′37″S, 13°24′49″E	Panguila (Bengo)
9	Lagoa do Mangal do Chiloango	3,097	05°10′37″S, 12°07′35″E	Cacongo (Cabinda)
10	Baixo Kwanza	97,200	09°20′45″S, 13°09′04″E	(Luanda)
11	Complexo das Zonas Húmidas de Kumbilo-Dirico	11,743	17°59′18″S, 20°46′53″E	Dirico (Cuando Cubango)

Source: Adapted from USDAFS (2022).

Convention website. The 11 proposed sites are given in Table 7.1 and their locations are shown in Figure 7.1. There is very little research-based information available on many of these sites (e.g. Guerreiro et al. 2012; Neto 2020), and this is a limitation on understanding of the detailed hydrogeomorphic context of the sites, or their ecological properties such as their biodiversity.

Apart from Ramsar sites, there is one UNESCO World Heritage Site in Angola, which is a cultural and not a natural site. There are no Global Biodiversity Hotspots. There is also limited national legislation that deals with biodiversity conservation (discussed by USDAFS 2022; Knight et al. 2023). The National Biodiversity Strategy and Action Plan (2019–2025) mentions wetlands very briefly with respect to different National Objectives 5.2 and 10 but states that a national wetland inventory needs to be carried out (Republic of Angola 2020) although hitherto this has not been done. This is noted across all provinces of the country.

As there is very limited available information on wetlands in Angola, this study, as a first-pass analysis, follows the hierarchical wetland inventory procedure outlined by Ramsar (Ramsar Convention Secretariat 2010) and uses global, freely available datasets and products. This methodological approach is described in Table 7.2. This procedure involves an assessment of the most likely locations where different wetlands may be found based on the properties of the physical landscape. This is therefore founded on a hydrogeomorphological approach, where wetland site geology, geomorphology, topography, soils, climate, and ecosystems are considered together and their co-relationships understood. The methodological approach taken in this study is similar to that used by van Deventer et al. (2016) and Rivers-Moore et al. (2020) for South Africa, where possible wetland types were mapped and identified at a national scale based on their hydrogeomorphic setting. It is likely, however, that wetlands will be the most common in upland areas of Angola with higher precipitation values and in constrained valley locations, similar to the hydrogeomorphic setting of peatlands (Lourenco et al. 2022).

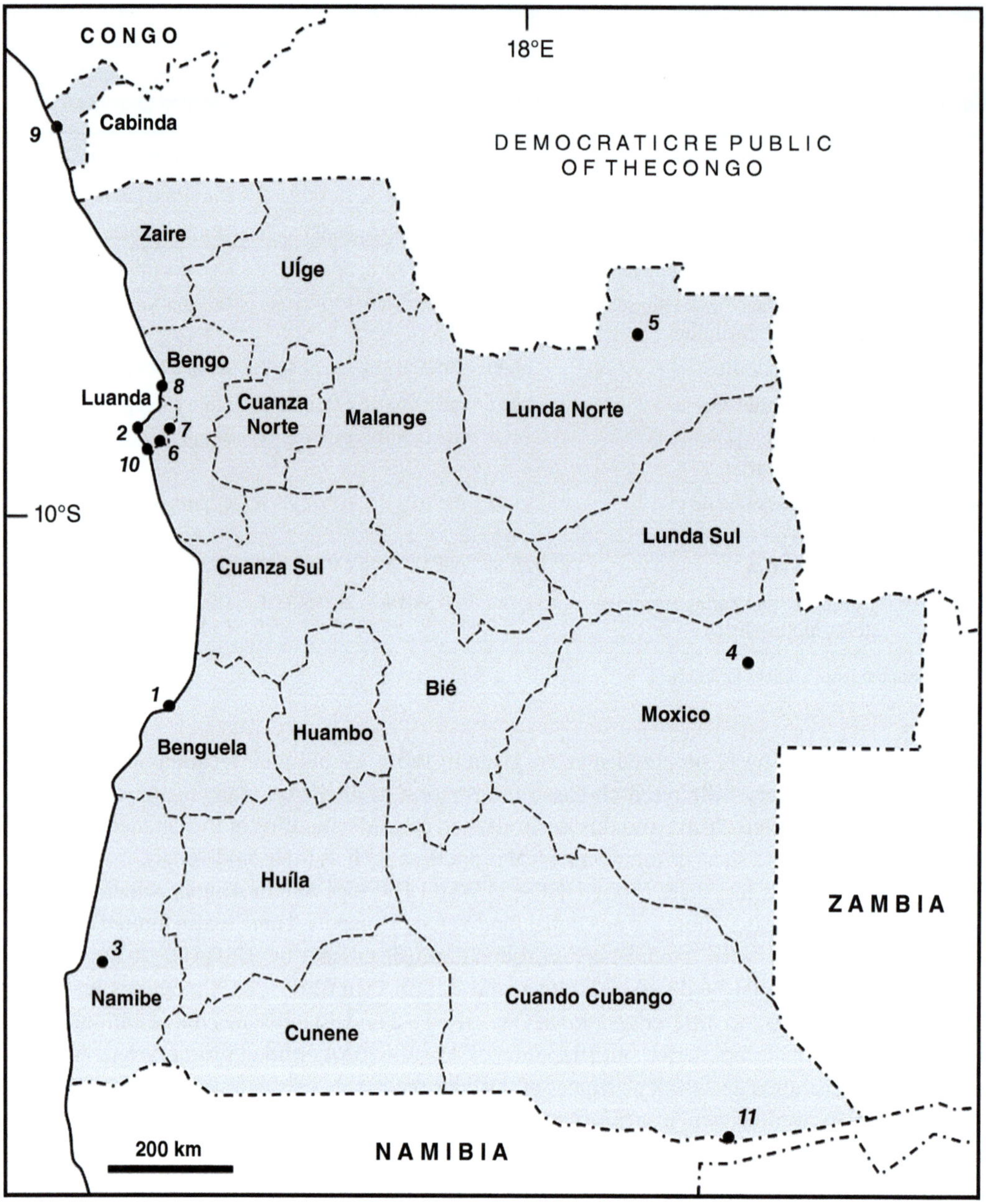

Figure 7.1 The locations of wetland sites in Angola that have been put forward as Ramsar sites (as of June 2021) (Table 7.1). Individual provinces are marked according to Table 7.1.

In the past decade, the increased accessibility of extensive open-access remote sensing data, with improvements in machine learning tools, has created new opportunities for a wider range of applications at different spatial and temporal scales within the geospatial sciences (Mahdianpari et al. 2018). Remote sensing data sources have been used in many previous studies, especially in Africa, as a first step toward mapping the locations and properties of wetlands (Ozesmi and Bauer 2002; Guo et al. 2017; Ludwig et al. 2019), and to monitor wetland degradation (Munishi and Jewitt 2019; Orimoloye et al. 2020; Steinbach et al. 2021). This has included the calculation of Normalized Difference Water Index (NDWI) and Topographic Wetness Index (TWI) and the use of different land cover classification

Table 7.2 The hierarchical approach to wetland classification inventory.

Level	Description of the spatial scale concerned	Inventory observations	Nature of data collection at this level
1	River basin or island	Geology, climate, biogeographic region	Desk study: to describe the regional geologic, climatic, and ecological features of each region using existing datasets
2	Wetland region	Location, altitude, area covered, climate, hydrology, vegetation	Desk study: to identify the wetland areas in each region, using the datasets of level 1
3	Wetland complex	Location, physical features, physico-chemical and biological features, ecosystem services	Fieldwork and analysis: to identify the major physical features, physico-chemical and biological features within each wetland complex
4	Wetland habitat	Location, physical features, physico-chemical and biological features (ecological character), land tenure, land use type and patterns, land management issues and threats	Fieldwork and analysis: to identify the detailed physical features, physico-chemical and biological features within each wetland habitat within the wetland complex

Source: Adapted from Ramsar Convention Secretariat (2010).

schemes that can identify and map areas of open water or wetlands. These approaches are based on the different spectral properties offered by different remote sensing platforms.

The emergence of robust cloud computing resources, such as Google Earth Engine, (GEE) has effectively tackled challenges related to Geo Big Data (Gorelick et al. 2017). GEE contains an extensive repository of satellite imagery and geospatial datasets, facilitating web-based algorithm development and efficient visualization of results within a reasonable processing timeframe (Mahdianpari et al. 2018). GEE has been previously used to map wetlands in an automated and spatially consistent way (e.g. Amani et al. 2019; Liu et al. 2022; Waleed et al. 2023). It has also been used for mapping wetlands and their properties (such as patterns of seasonal inundation) in some areas of Africa (e.g. Gxokwe et al. 2022; Li et al. 2022; Mashala et al. 2023). Beyond its computing and storage capabilities, GEE also incorporates several widely recognized machine learning algorithms. However, this relies on in-field training data, and in the case of Angola, these data are limited to a few locations and there is insufficient coverage and detail to produce national scale maps and inventories with confidence. Such applications have been successfully implemented (Lourenco et al. 2022), but these are constrained to local, site-specific training data.

Although the steps outlined in Table 7.2 provide a consistent framework for wetland identification and characterization, only levels 1 and 2 can be achieved using remote sensing and other spatial datasets, including through the functionality of GEE. Levels 3 and 4 require site-scale data that can only be derived through field observation and engaging with local communities. Therefore, in the absence of sufficient training data, wetland mapping was undertaken in this study using a wide range of reference maps derived from global products. This addresses levels 1 and 2 of the Ramsar Convention hierarchy (Table 7.2), nonetheless, and this is sufficient to establish the types of wetlands likely present in Angola and their major properties. The specific datasets used (Table 7.3) were obtained and extracted from GEE and ESRI's ArcGIS Living Atlas of the World, and visualizations were produced using ArcGIS Pro software.

Table 7.3 Global datasets extracted for the purpose of wetland mapping in Angola and their properties.

Dataset	Description	Reference source
SRTM DEM	The Shuttle Radar Topography Mission (SRTM) digital elevation data is an international research effort that obtained digital elevation models (DEMs) on a near-global scale. The SRTM v3 product (SRTM Plus) is provided by NASA JPL at a resolution of 1 arc-second (approximately 30 m).	Farr et al. (2007)
HydroSHEDS	The HydroSHEDS database offers a suite of global digital data layers in support of hydroecological research and applications worldwide. Its various hydrographic data products include catchment boundaries, river networks, and lakes at multiple resolutions and scales. Watersheds range from level 1 (coarse) to level 12 (detailed), using Pfastetter codes.	Lehner et al. (2008)
HydroRIVERS	HydroRIVERS represents a vectorized line network of all global rivers that have a catchment area of at least 10 km^2 or an average river flow of at least 0.1 m^3/s, or both. HydroRIVERS has been extracted from the gridded HydroSHEDS core layers at 15 arc-second resolution. The global coverage of HydroRIVERS encompasses 8.5 million individual river reaches with an average length of 4.2 km, representing a total of 35.9 million km of rivers globally.	Lehner and Grill (2013)
Soilgrids World Soil Organic Carbon 0–30 cm	This layer averages most likely interpolated carbon mass from soilgrids.org to estimate a value for the top 30 cm of soil. Soilgrids.org soil organic carbon datasets are expressed as mass in tenths of a gram (decigrams) per kilogram of soil overall. This has been converted to grams per kilogram for the Living Atlas.	Batjes et al. (2020)
Sentinel-2 NDVI	Sentinel-2 is a wide-swath, high-resolution, multispectral imaging mission supporting Copernicus Land Monitoring studies, including the monitoring of vegetation, soil and water cover, as well as observation of inland waterways and coastal areas. Normalized Difference Vegetation Index (NDVI) is one of the most widely used vegetation indices, describing the greenness, relative density, and health of vegetation. A median NDVI composite for 2021 was extracted from GEE for Angola, and images with greater than 85% cloud cover were removed from the median composite.	Rouse et al. (1974) Equation used: $\frac{(\text{Band } 8 - \text{Band } 4)}{(\text{Band } 8 + \text{Band } 4)}$
WorldClim Global Precipitation	WorldClim 2.1 provides downscaled estimates of climate variables as annual means over the period of 1970–2000 based on interpolated station measurements at 5 km resolution.	Fick and Hijmans (2017)
Sentinel-2 NDWI	The Normalized Difference Wetness Index (NDWI) as defined by McFeeters (1996) is used to monitor changes of water content in water bodies. A median NDWI composite for 2021 was extracted from GEE for Angola, and images with greater than 85% cloud cover were removed from the median composite.	McFeeters (1996) Equation used: $\frac{(\text{Band } 3 - \text{Band } 8)}{(\text{Band } 3 + \text{Band } 8)}$
JRC Water Occurrence	The Joint Research Centre (JRC) Water Occurrence dataset shows where surface water occurred between 1984 and 2021 and provides information concerning overall water dynamics. This product captures both the intra-annual and inter-annual variability and changes. The occurrence is a measurement of the water presence frequency (expressed as a percentage of the available observations over time actually identified as water). The provided occurrence accommodates for variations in data acquisition over time (i.e. temporal deepness and frequency density of the satellite observations) in order to provide a consistent characterization of the water dynamic over time.	Pekel et al. (2016)

Table 7.3 (Continued)

Dataset	Description	Reference source
JRC Water Seasonality	The JRC Water Seasonality product provides information concerning the intra-annual behavior of water surfaces. It separates "permanent" water bodies (those that are present throughout the period of observation) (nominally a year) from "seasonal" (those that are present for only part of the year); the degree of seasonality is also represented (i.e. the proportion of the total number of observed months in which water is present). For this analysis, the water seasonality for 2021 was extracted for Angola.	Pekel et al. (2016)

7.3 Results

7.3.1 Physical Properties and Hydrology

Angola exhibits diverse topography with western coastal lowlands below 200 m above sea level (asl), covering 5% of the country and extending 10–150 km inland (Figure 7.2). Areas above 1500 m asl make up 7% of Angola, while the western mountainous escarpment adjacent to the coastal lowlands rises to 1000 m asl, covering 23% of the country. Notably, the central highlands of the Bié Plateau that cover 65% of the country serve as the water tower and source of several major drainage basins, including the Cuanza (Atlantic Coast), Congo, Zambezi, and Okavango (Lourenco and Woodborne 2023) (Figure 7.2b).

According to the Soilgrids dataset (Batjes et al. 2020), Angola has a wide range of estimated soil organic carbon (SOC) contents at 0–30 cm depth. The western and southern parts of the country that are largely sand dominated and drier have lower SOC values (1.8 g/kg) while the northwestern and eastern areas have higher values (80.5 g/kg) (Figure 7.2c). The Sentinel-2-derived NDVI dataset for Angola depicts the miombo forests (green), which dominate the central plateau as well as the coastal western Namib Desert (brown) and eastern Kameia floodplains that contribute to the Zambezi River Basin (Figure 7.2d).

The precipitation, NDWI, and Joint Research Centre (JRC) water seasonality and water occurrence for Angola are presented in Figure 7.3. The WorldClim annual precipitation dataset indicates that the northern and eastern parts of Angola receive approximately 1729 mm of rainfall per year, while the western and southern parts of the country are drier, receiving as little as 25 mm/year within the Namib desert (Figure 7.3a). There is a distinct rainfall gradient along the western escarpment and the central plateau. The Sentinel-2-derived NDWI dataset (Figure 7.3b) aligns closely with river networks and JRC surface water inundation and occurrence (Figure 7.3c, d). Although the JRC water products are mapped at high (30 m) resolution, the notable surface water extents are limited to large rivers such as the Cuanza (flowing west into the Atlantic) and the Kameia floodplain (flowing east into the Zambezi) (Figure 7.3d).

7.3.2 Extent of Wetlands

The Global Wetlands Map produced by Gumbricht et al. (2017) was extracted for Angola and includes the area covered by each wetland classification type (Figure 7.4). This product spans the tropical and subtropical regions and it employs a hydrogeomorphological model based on an

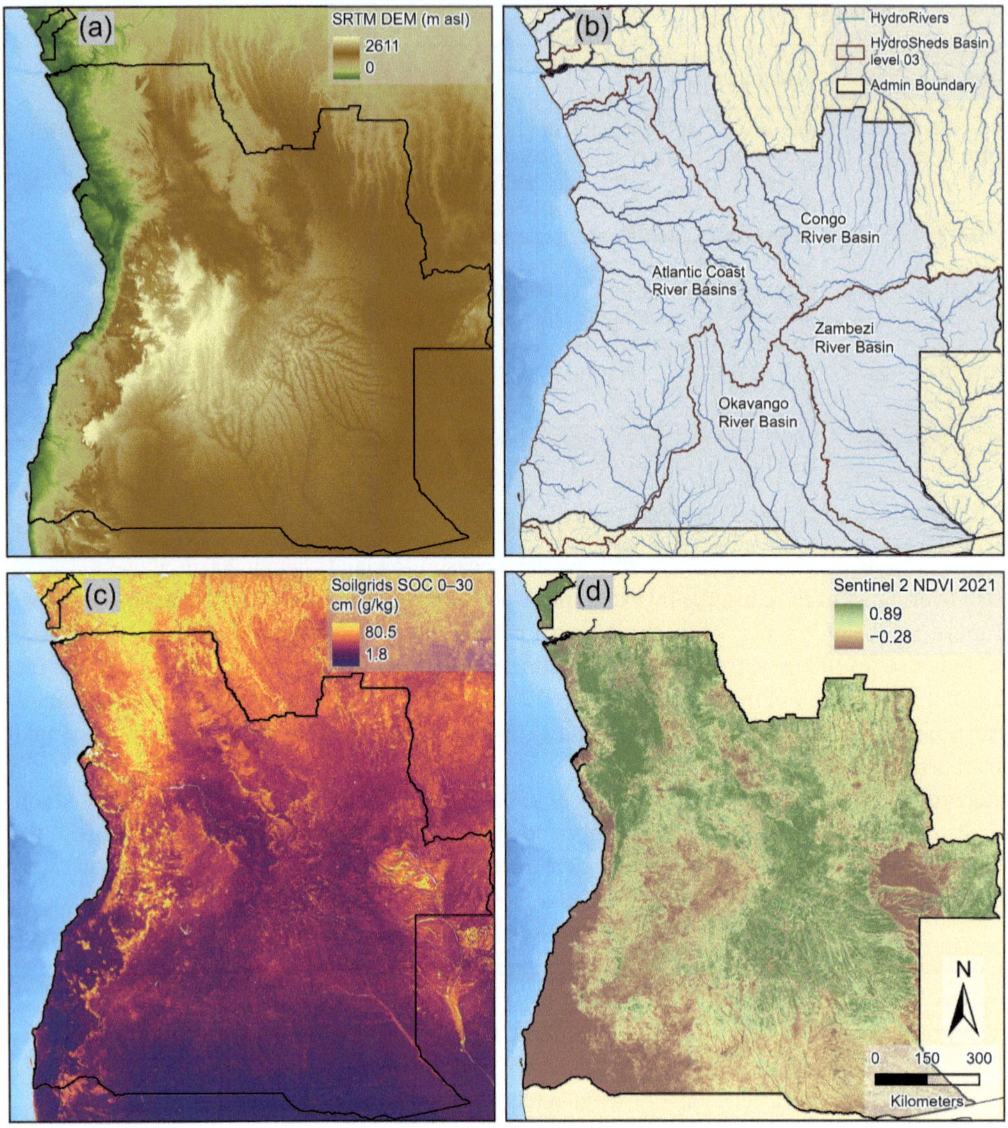

Figure 7.2 The major physical properties of Angola of relevance to the distribution of wetlands. (a) Topography (m asl), (b) major transboundary drainage basins and rivers, (c) soil organic carbon, reflecting organic content, and (d) Sentinel-2-derived NDVI.

Expert System approach, utilizing three biophysical indices associated with wetland formation: (i) sustained water supply surpassing atmospheric water demand; (ii) seasonally or annually waterlogged soils; and (iii) a geomorphological position supporting water supply and retention (Gumbricht et al. 2017). The latest version of the Global Tropical and Subtropical Wetlands Map has undergone enhancement through generalization analysis, employing tools like Majority filter and Boundary Clean. These tools effectively address minor inaccuracies by eliminating isolated pixels in the raster. The smoothing of edges involves expanding and shrinking boundaries or adjusting zones based on the values within neighboring pixels. It was mapped at 231 m resolution.

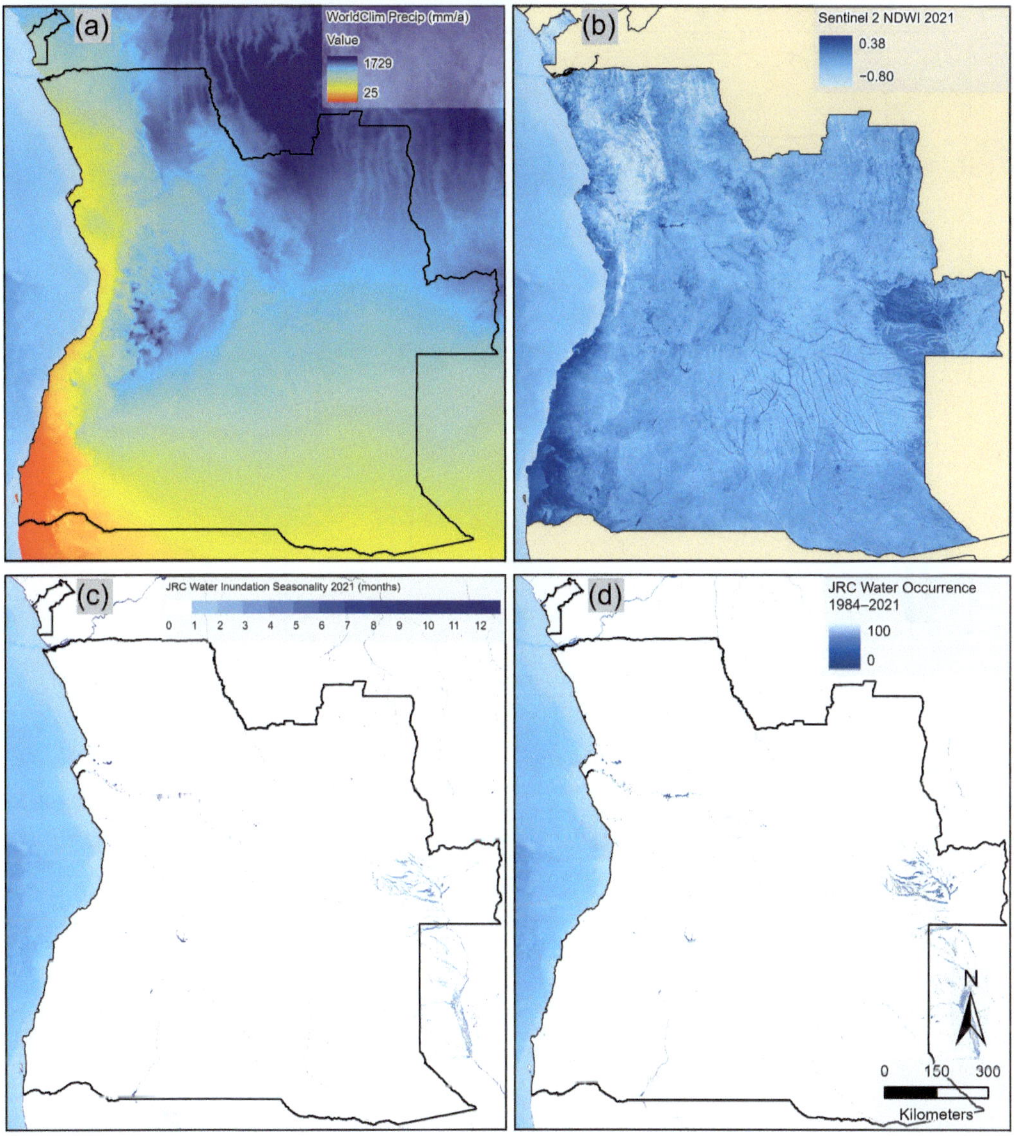

Figure 7.3 Properties of water in Angola. (a) WorldClim mean annual precipitation, (b) Sentinel-2-derived NDWI, (c) JRC water seasonality for 2021, and (d) 1984–2021 water occurrence for Angola from JRC.

According to this product, Angola contains 50,468 km^2 of wetland area with 42% classified as Marshes, 18% Floodplains and intermittent water bodies, and 17% as Floodout Swamps (Figure 7.4). In total and based upon the datasets used, wetlands make up 4.05% of the land area of the country. This compares with the National Wetland Map v5 for South Africa, which identifies that wetlands cover 3.8% of that country (van Deventer et al. 2020). As shown in the NDWI and JRC products, the Center for International Forestry Research (CIFOR) global wetland classification for Angola aligns closely with the distribution of river networks and main drainage basins (Figure 7.2b). Notable wetlands include the Kameia floodplain of Moxico province, for which there is little documentation, the Kwanza River floodplains that connect the Angolan Plateau with the capital

Legend	CIFOR Wetland V3 classes	Area (km²)
	Open Water	860
	Mangroves	13
	Swamps/Bogs	499
	Fens	24
	Riverine/Lacustrine	56
	Floodout Swamps	8,488
	Floodplains and Intermittent Water Bodies	9,006
	Marshes	21,118
	Wetland in Arid Climates	2,849
	Wet Meadow	8,415
	Total Wetland Area	50,468
	Total Area (Angola)	1,246,998

Figure 7.4 The CIFOR global wetland classification map for Angola.

Luanda on the northwest coast, and the extensive floodplains of the Cuito and Cuanavale rivers within Cuando Cubango Province. These specific examples are discussed below.

Based on the wetlands distributions shown in Figure 7.4, three hydrogeomorphic regions are identified, determined largely according to topography and water availability. (i) *Mountain wetlands* are located in higher elevation areas of the Bié Plateau region in the center of the country in Bié and Malanje provinces and are generally made up of peatlands formed in association with river terraces (Lourenco et al. 2022). As such, these are considered riparian marsh wetlands corresponding to Ramsar Convention wetland categories Tp and Ts and subordinately category U. Of the 11 proposed Ramsar sites in Angola (Table 7.1), none of these sites are mountain wetlands. (ii) *Coastal estuary and lagoonal wetlands* are found associated with river mouths, estuaries, or tidal inlets, or as lagoons located behind spits or barrier beaches. These wetlands are found at the coastline rather than inland. These are often well-known bird sites and as such there is some literature on these wetland systems, albeit focusing on elements other than the wetlands themselves (e.g. Simmons et al. 2006; Guerreiro et al. 2012; Neto 2020). These wetlands are sustained by the combination of coastal processes (waves, tides, coastal erosion, and longshore sediment transport) and fluvial processes (river water and sediment discharge) that allow for environmentally complex fresh- to brackish-water conditions to be developed. Of the proposed Ramsar sites in Angola (Table 7.1), sites 1, 2, 8, 9, and 10 are coastal wetlands. These broadly correspond to Ramsar types A, E, F, G, H, I, and J. (iii) *Alluvial, riparian, and valley-bottom wetlands* are located in mid-system river valley and floodplain settings in different provinces across Angola, in particular, in Moxico, Malanje, and Lunda Norte (Figure 7.4). The spatial pattern of these wetlands is therefore linear, following the river systems and corresponding to overbank and floodplain accumulation of mixed organic and minerogenic sediments. The most likely condition under which this type of wetland develops is where there is seasonal flooding on perennial river, where areas of the floodplain adjacent to the river channel become inundated. Low drainage capacity or high water table conditions then mean that water cannot drain away, leading to the development of hydrophilic plant species and then wetland ecosystems. This type of wetland is relatively common in southern Africa where it is associated with perennial and seasonal river systems as well as palustrine wetlands next to floodouts (van Deventer et al. 2016; Sieben et al. 2018; Grenfell et al. 2019, 2022). These wetland systems in Angola correspond to Ramsar types N, Tp, and Ts. Of the proposed Ramsar sites (Table 7.1), sites 4, 5, and 11 are of alluvial types.

7.3.3 Examples of Wetland Systems

Specific examples of river system wetlands in Angola are now described, based on CIFOR wetland classification outputs. Along the Cuito-Cuanavale river system, wetlands of different types are located along the major valleys (Figure 7.5). Floodouts are the most common geomorphic element, formed where seasonal floods affect areas outside of the main river channel. Floodouts are known to be key features of South African and Australian lowland and semiarid river systems (Ralph and Hesse 2010; Grenfell et al. 2012, 2014; Wakelin-King 2022). Swamps/bogs are found at points along the margins of the main river valley; these are likely formed in hollows or sediment depocenters, possibly associated with areas of past meander migration. It is notable that marshes are more common in the tributaries, which may reflect a narrower river valley and more restricted drainage. Field examples of these depositional environments from the Cuito-Cuanavale river system are shown in Figure 7.6.

In Moxico Province, the extensive wetland of the Kameia floodplain is similar in characteristics and behavior to wetlands of the larger Okavango, Cubango, and Etosha systems located in the surrounding regions (Milzow et al. 2009; Hamunyela et al. 2022; Mendelsohn 2022) (Figure 7.7). Here, a distributary channel system exists (Figure 7.7c) where seasonal flooding spreads water across the extent of the floodplain (Figure 7.7b). This situation allows for the development and free migration of river meanders, which is a characteristic property of rivers in this region (Mendelsohn 2022). In addition, the presence of abandoned meanders and oxbow lakes have acted as ideal depocenters for organic sediments to accumulate; hence, there is a close genetic relationship between river geomorphology and its geomorphic behavior – such as meander migration and

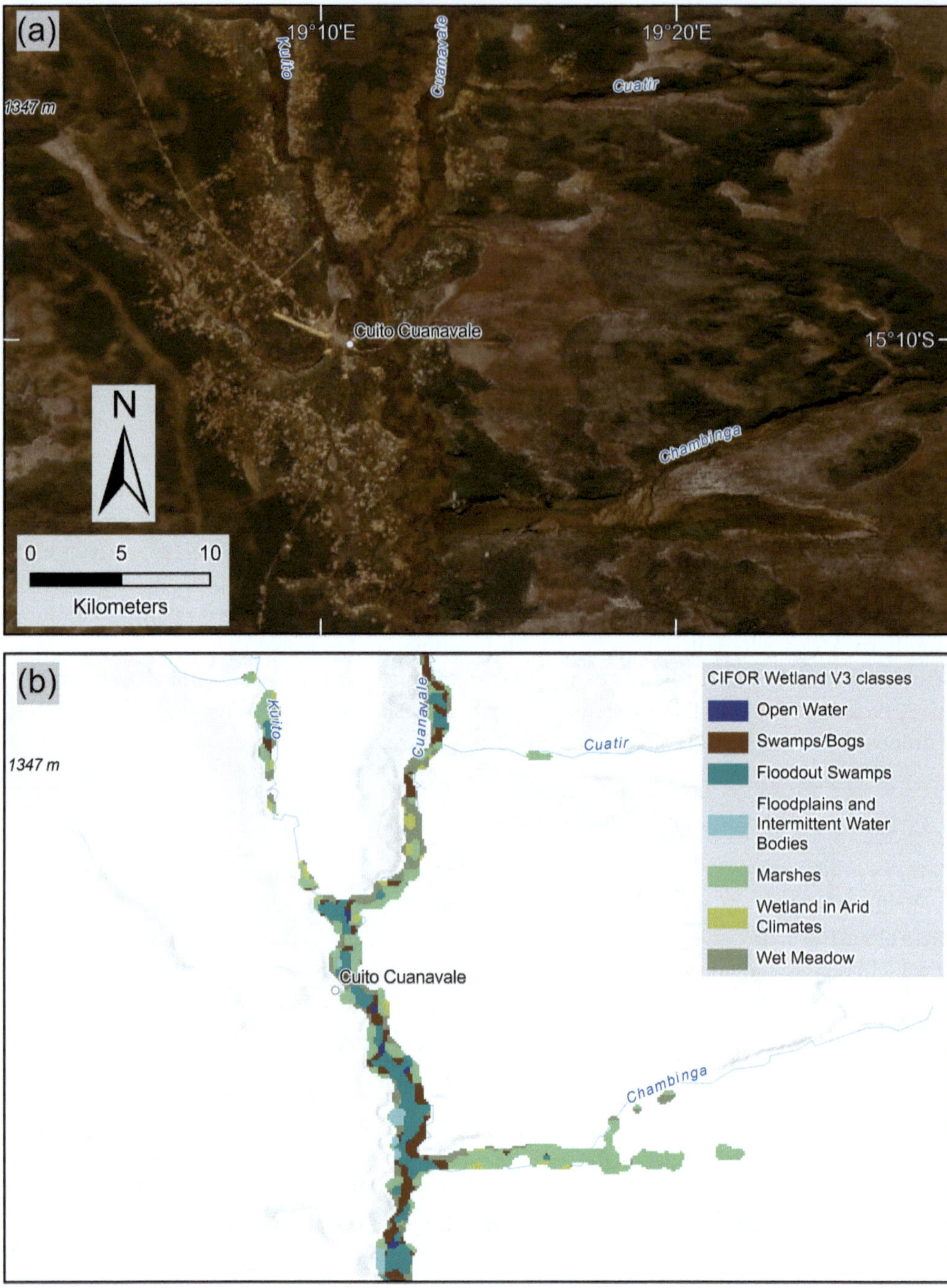

Figure 7.5 (a) Satellite view of the Cuito-Cuanavale River confluence in Cuando Cubango Province, Angola. *Source:* ESRI World Imagery. (b) CIFOR wetland classification of this region. Note: there are several different spellings of river names available.

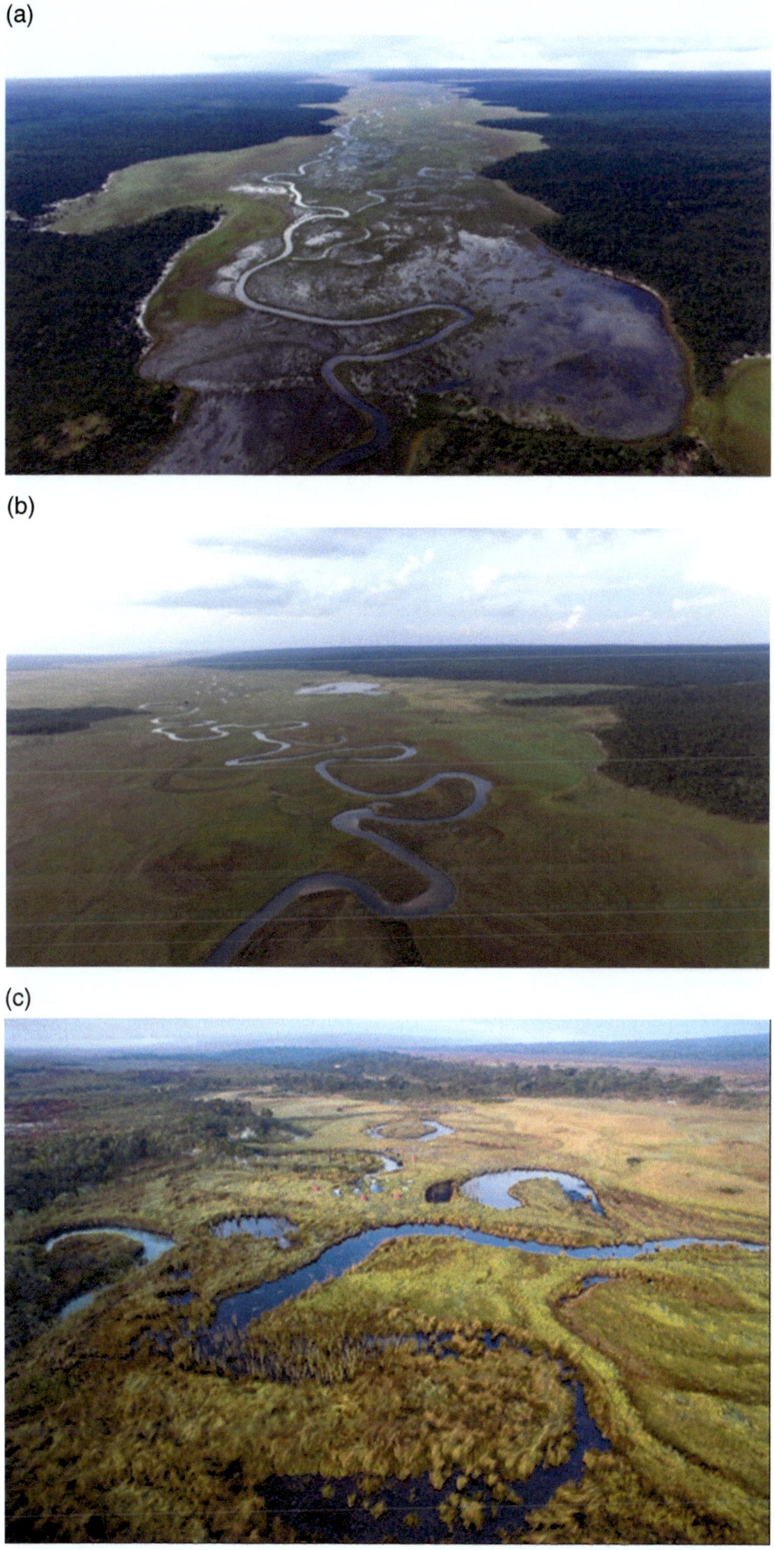

Figure 7.6 Field photos (from a drone survey) of geomorphic elements of the Cuito-Cuanavale river system, illustrating the different wetland environments present. (a) Meandering river network across a seasonally flooded floodplain, (b) active meander channels, and (c) abandoned channels and oxbow lakes with different vegetation types reflecting variations in wetland development. *Source:* Reproduced with permission from Okavango Wilderness Project/National Geographic Society.

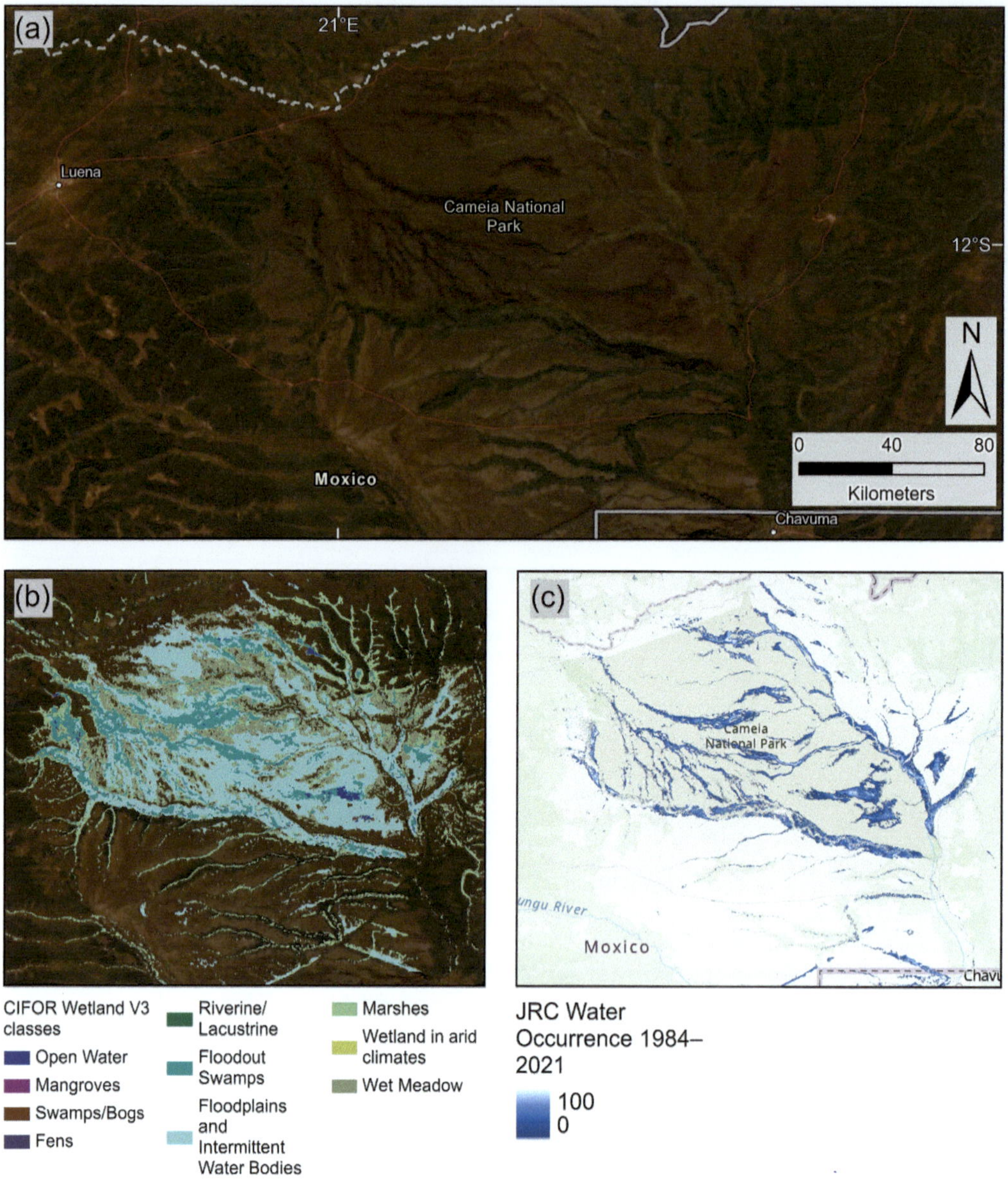

Figure 7.7 (a) Satellite view of the Kameia floodplain region, Moxico Province, Angola. *Source:* ESRI World Imagery. (b) CIFOR wetland classification. (c) JRC water occurrence 1984–2021.

abandonment – and phases of wetland development. These wetlands are therefore controlled by seasonal patterns of river discharge and overbank flooding (Awadallah and Tabet 2015; Moses et al. 2022).

Coastal wetlands are less common than other wetland types in Angola (Figure 7.4) but represent a significant proportion of the proposed Ramsar sites in the country (Table 7.1) because of their importance for birds. In the Luanda region where many of the proposed sites are located (Figure 7.1), different wetlands occur along different reaches of incoming rivers largely as a result

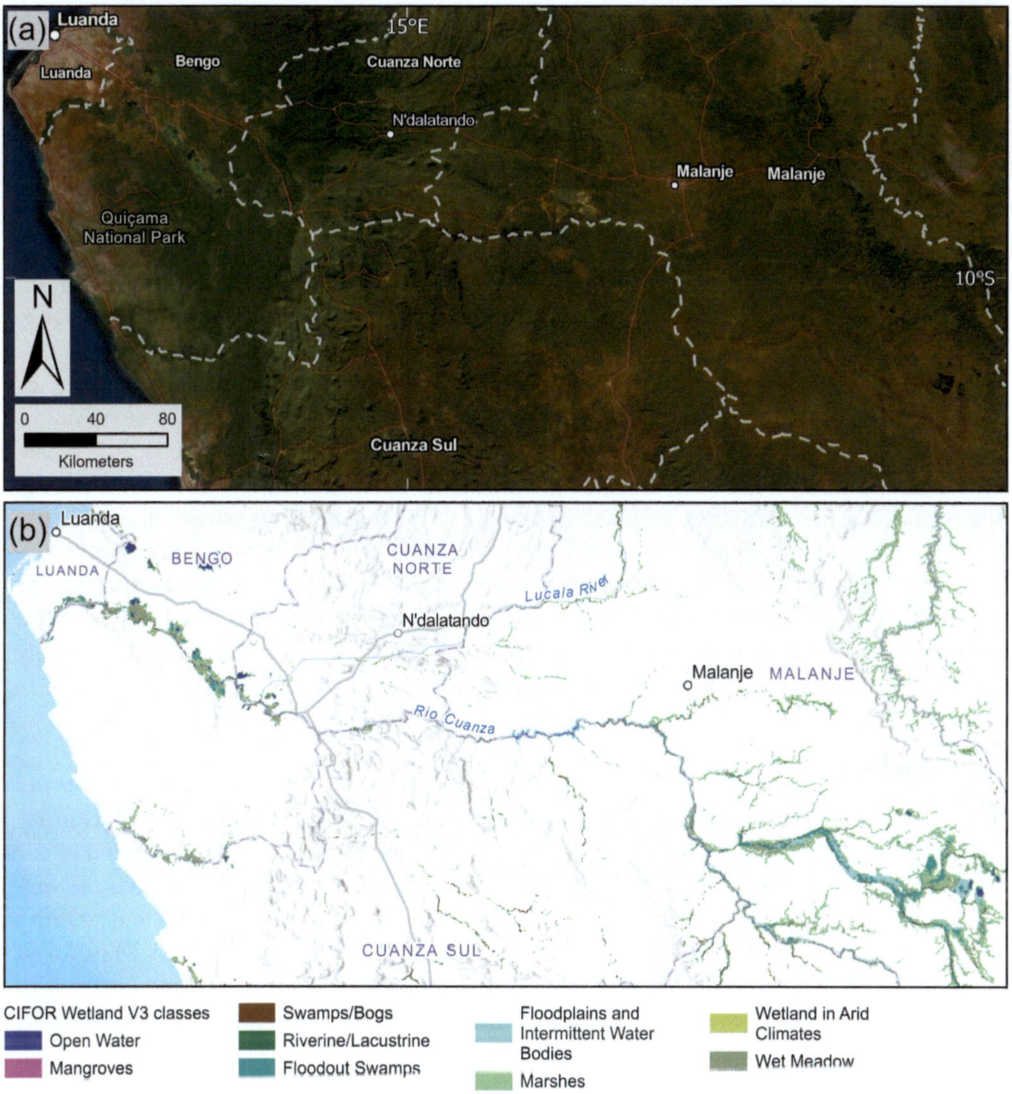

Figure 7.8 (a) Satellite view of the Kuanza River floodplain draining into Luanda city. *Source:* ESRI World Imagery. (b) CIFOR wetland classification of this region.

of local geomorphology and topography between headwater, mid-reaches, and lower reaches (Figure 7.8). Note that at the scale of Figure 7.8, any wetlands present along the coast such as mangroves are not resolved.

7.4 Discussion

This section discusses the environmental controls on wetlands in Angola, their most likely uses in contributing to the livelihoods of local communities, and their likely threats and conservation strategies. It should be noted that, as far as we are aware, there are no specific studies dealing with wetland properties and their use in Angola. Thus, in this section, we draw from evidence for wetland use and conservation elsewhere in dryland regions of southern Africa for which there is more extensive literature.

7.4.1 Wetlands in Angola: Environmental Controls

The different environmental settings in which the three main types of wetlands in Angola are found suggest that the major environmental controls are topography and water availability (Figures 7.2 and 7.3). These largely determine the spatial patterns of surface water, but poor drainage conditions then lead to the development of hydromorphic soils and wetland ecosystems (e.g. Fennessy et al. 2018; van Huyssteen and Johnson 2022). These feedbacks mean that wetland systems develop over time and lead to the formation of "wetlands in drylands," namely, wetlands developed in areas with otherwise low precipitation (Scoones 1991). This describes many valley-bottom and riparian wetlands that reflect the feedbacks between geomorphology, soils, and ecosystems. Further, this highlights that the hydrogeomorphic approach to wetland identification (e.g. Sieben et al. 2018; Rivers-Moore et al. 2020; Kotze et al. 2022) works well in practice in Angola and can be used as a predictive guide for wetland identification and mapping in the country (Figure 7.4). The analysis undertaken in this study also shows that stages 1 and 2 of the Ramsar Convention process for wetland characterization (Table 7.2) can be effectively undertaken using global and publically available datasets (Table 7.3) and that these yield good results for Angola and most likely for similar data-scarce and semiarid areas.

7.4.2 Wetland Biodiversity

Due to lack of data, wetland biodiversity in Angola cannot be evaluated in any detail. Huntley (2023, his chapter 3) identified and described 16 ecoregions in Angola and these are found in different areas according to their topography, geology, soils, and climate. It is notable that the wetlands identified from the global databases (Figure 7.4) are found scattered across different latitudes and ecoregions, as determined by their local topographic and water balance contexts. The limited studies on wetlands in Angola mean that there is very little information on wetland biodiversity despite this being examined in other Angolan biomes (Simmons et al. 2006; Tavares et al. 2012; Godlee et al. 2021; Gula et al. 2022). However, relationships between wetlands and birdlife suggest they have high biodiversity as well as high conservation status (Dean 2001; Simmons et al. 2006; Gula et al. 2022), although this has not been explored from most wetlands, including those sites that have been proposed to Ramsar (Table 7.1).

The biodiversity of different biomes in Angola, including of Red List species, has been previously identified (República de Angola 2006). The 6th National Report on Biodiversity identifies that wetlands in Serra do Pingano (Uíge province) contain 820 plant species (Republic of Angola 2019) although no species lists are given. Continuing surveys in different catchments including the Cubango and Okavango have also identified high plant biodiversity (e.g. Revermann et al. 2016) but this needs to be done systematically and in detail for all key wetlands.

7.4.3 Wetland Systems and Livelihoods

There are no specific studies in Angola that deal with community livelihoods as linked to wetland resources and services. However, many studies exist from across dryland southern Africa dealing with this issue, in particular, in South Africa and Zimbabwe, so it is envisaged that the same principles will apply in Angola. For example, many studies show how wetlands provide a range of ecosystem services of different types, but that the precise nature of these services and their degree of utilization vary according to wetland physical properties and the type of community use (Rebelo et al. 2010; Marambanyika and Beckedahl 2016; Marambanyika et al. 2017; Sieben et al. 2018; Musasa and Marambanyika 2020; Knight 2021; Mandishona and Knight 2022b). There are also trade-offs that may exist between different land uses and the types or styles of resource exploitation on wetlands, where, for example, the development of wetland agriculture may reduce the capacity of the wetland to absorb flood waters or to buffer pollution (Rodríguez et al. 2006). Although these issues have been discussed on wetlands elsewhere in the world, this has not been so widely considered in Africa (e.g. Hurford and Harou 2014; Namaalwa et al. 2020). However, the role of wetlands in southern Africa for rural community sustainable development and climate change mitigation is well established (Rebelo et al. 2010; van Deventer 2021), and this provides a framework for considering Angolan wetlands in the same light.

7.4.4 Threats to Wetland Systems

The major threats considered to impact Angolan wetlands are listed in Table 7.4. These factors have been specifically examined from wetlands elsewhere in southern Africa (e.g. Mitchell 2013; Musasa and Marambanyika 2020; Orimoloye et al. 2020; Salimi et al. 2021), and this can be used to inform on potential impacts in Angola. In addition, feedback that may amplify future threats have not been widely considered – an example is where changes in future rainfall patterns result in changes in species composition and therefore biodiversity and ecosystem services. Any future changes to the physical properties and dynamics of wetlands can potentially be mitigated (or amplified) by appropriate management strategies, highlighting the important role played by human activity in wetland systems.

Potential outcomes of climate change on Angolan biomes and ecosystems have been identified in several studies (Dean 2001; Republic of Angola 2019; Huntley 2023) but these are not specific to wetlands. Other threats also considered include deforestation, fire, erosion, illegal exploitation of natural resources, invasive species, illegal fishing, solid waste management, and human–wildlife conflict (Republic of Angola 2019). The National Biodiversity Strategy and Action Plan does not specifically mention climate change as a threat to wetlands (Republic of Angola 2020).

Table 7.4 Major ongoing and future threats to wetlands in Angola.

Threat	Likely impacts on Angolan wetlands
Climate change	Changes in rainfall regimes, including seasonality; increased temperatures resulting in higher evaporation and seasonal drying out of wetlands
Infrastructure development and urbanization	Wetland drainage; removal of vegetation and reduction in biodiversity and ecosystem services; building of fixed structures and related infrastructure (roads, pipelines); subsidence
Pollution	Water pollution coming into wetlands from industry and settlements upstream; increased water pollution from settlements and agriculture within or immediately around the wetland; increased air pollution from burning, mining
Agriculture and land use change	Removal of vegetation; burning or ploughing; reduction of existing biodiversity; increased desiccation and wind erosion; drying out and shrinkage of the wetland; changes in carbon storage potential; reduction in flood risk mitigation; potential issues for local food production

Other factors may also include changes in river properties and dynamics, in particular, given the coincidence between river valleys and the occurrence of wetlands (e.g. Figures 7.5–7.8). Several studies have examined the role of climate change and variability on river discharge patterns on the Cubango and Cuito rivers among others (Moses et al. 2022; Lourenco and Woodborne 2023) and their implications for flood frequency (Awadallah and Tabet 2015; Hamunyela et al. 2022). These processes should then be considered in terms of impacts on adjacent floodplains, sloughs, ponds, and riparian zones where wetlands are located, although this has not been fully evaluated.

7.4.5 Wetlands of International Importance

At present, there are no Ramsar sites in Angola. Those that have been proposed (Table 7.1) have not yet been fully described in the literature. However, wetlands are mentioned as important in the 6th National Report on Biodiversity (Republic of Angola 2019) although no details are given. A recent summary of protected areas in Angola (USDAFS 2022) identifies the need for a better legal and policy framework for protected areas, including guidelines for site management plans; community-based natural resources management and co-management frameworks; expansion of the protected areas network with associated infrastructure for effective management and tourism; and building institutional capacity. These national-scale issues are particularly significant for wetlands because of their importance for agriculture, water supply, and ecosystem services.

7.4.6 Wetland Conservation

Wetland conservation strategies, in Angola as elsewhere in southern Africa, need to focus on balancing the maintenance of wetland properties and ecosystem services against the use of these resources by local communities, while also ensuring the integrity and functionality of the wetland. As such, wetland management represents a balancing act between different priorities and uses in particular in the context of climate change (McCartney and Houghton-Carr 2009; Hartter and Ryan 2010). Ecological conservation in Angola also operates according to national legislation and existing protected areas including the 1998 Environment Framework Law, the Law on Forests and Wildlife (Law No. 7/17), and the Law on Environmental Conservation Areas (Law No. 8/20), as described in the National Biodiversity Strategy and Action Plan (Republic of Angola 2020; USDAFS 2022). However, different wetland uses by local communities also have implications for their sustainability and any negative environmental effects that may arise from their exploitation. An example is the role of fire in both land clearance and burning of organic wetland soils (and upland peat) as a fuel source. Previous work has shown that upland peatlands in Angola are burned more extensively than other land cover classes, and this has implications for their biodiversity and ecological sustainability (Lourenco et al. 2023).

Conservation also requires a monitoring strategy to identify any trajectory of change, but the absence of an inventory of Angolan wetlands means that a baseline for monitoring is lacking. Specific monitoring strategies mainly through remote sensing methods have been developed for southern African wetlands (van Deventer et al. 2016, 2020; Ngwenya and Marambanyika 2021). A key element in wetland conservation is the involvement of local communities as custodians of their local environments and resources (Marambanyika et al. 2017; Musasa and Marambanyika 2020; Mandishona and Knight 2022b).

7.5 Conclusions

Wetlands are a key element of Angola's ecological and hydrological environment but they have been significantly underreported and there is very little information in the peer-reviewed literature on wetlands. This study provides a first-pass analysis of wetland distributions and properties in Angola based on publicly available global datasets on topography, hydrology, and ecosystems. This analysis shows that wetlands represent 4.05% of the land area of Angola and thus make up a significant landscape and ecological element of the country, with likely associated benefits and services provided to local communities. The 11 proposed Ramsar sites in Angola represent a starting point for valuing and conserving wetlands in the country, but this should be considered an ongoing process requiring engagement with local communities and other actors.

References

Adeeyo, A.O., Ndlovu, S.S., Ngwagwe, L.M. et al. (2022). Wetland resources in South Africa: threats and metadata study. *Resources* 11: 54. https://doi.org/10.3390/resources11060054.

Amani, M., Mahdavi, S., Afshar, M. et al. (2019). Canadian Wetland Inventory using Google Earth Engine: the first map and preliminary results. *Remote Sensing* 11: 842. https://doi.org/10.3390/rs11070842.

Awadallah, A.G. and Tabet, D. (2015). Estimating flooding extent at high return period for ungauged braided systems using remote sensing: a case study of Cuvelai Basin, Angola. *Natural Hazards* 7: 255–272.

Ayanlade, A. and Proske, U. (2016). Assessing wetland degradation and loss of ecosystem services in the Niger Delta, Nigeria. *Marine and Freshwater Research* 6: 828–836.

Batjes, N.H., Ribeiro, E., and Van Oostrum, A. (2020). Standardised soil profile data to support global mapping and modelling (WoSIS snapshot 2019). *Earth System Science Data* 12: 299–320.

Cardoso, J.F., Costa, J.C., Neto, C.S. et al. (2021). Plant communities of Namibc saltmarshes (southwest of Angola). *Finisterra* 56: 99–114.

Chatanga, P. and Sieben, E.J.J. (2020). Variation in plant functional composition of the Afromontane Palustrine wetlands along an altitudinal gradient in Lesotho. *Wetlands* 40: 2539–2552.

Chatanga, P., Kotze, D.C., Janks, M., and Sieben, E.J.J. (2019). Classification, description and environmental factors of montane wetland vegetation of the Maloti-Drakensberg region and the surrounding areas. *South African Journal of Botany* 125: 221–233.

Cole, C.A. (2017). Assessment of a judgment-based hydrogeomorphic wetland classification using long-term hydrologic data. *Ecohydrology* 10: e1761. https://doi.org/10.1002/eco.1761.

Dalu, M.T.B., Mukhuwana, O., Cuthbert, R.N. et al. (2022). Understanding communities' perceptions, demographics and uses of wetlands in Vhembe Biosphere Reserve, South Africa. *Wetlands Ecology and Management* 30: 1231–1244.

Dean, W.R.J. (2001). Angola. In: *Important Bird Areas in Africa and Associated Islands: Priority Sites for Conservation* (ed. L.D.C. Fishpool and M.I. Evans), 71–91. Cambridge: Pisces Publications/Birdlife International.

Dixon, A. and Carrie, R. (2016). Creating local institutional arrangements for sustainable wetland socio-ecological systems: lessons from the 'Striking a Balance' project in Malawi. *International Journal of Sustainable Development and World Ecology* 23: 40–52.

Dvorett, D., Bidwell, J., Davis, C., and DuBois, C. (2012). Developing a hydrogeomorphic wetland inventory: reclassifying national wetlands inventory polygons in geographic information systems. *Wetlands* 32: 83–93.

Ellery, W.N., Grenfell, S.E., Grenfell, M.C. et al. (2016). Wetlands in southern Africa. In: *Quaternary Environmental Change in Southern Africa: Physical and Human Dimensions* (ed. J. Knight and S.W. Grab), 188–202. Cambridge: Cambridge University Press.

Farr, T.G., Rosen, P.A., Caro, E. et al. (2007). The Shuttle Radar Topography Mission. *Reviews of Geophysics* 45: RG2004. https://doi.org/10.1029/2005RG000183.

Fennessy, M.S., Wardrop, D.H., Moon, J.B. et al. (2018). Soil carbon sequestration in freshwater wetlands varies across a gradient of ecological condition and by ecoregion. *Ecological Engineering* 114: 129–136.

Fick, S.E. and Hijmans, R.J. (2017). WorldClim 2: new 1-km spatial resolution climate surfaces for global land areas. *International Journal of Climatology* 37: 4302–4315.

Godlee, J.L., Ryan, C.M., Bauman, D. et al. (2021). Structural diversity and tree density drives variation in the biodiversity–ecosystem function relationship of woodlands and savannas. *New Phytologist* 232: 579–594.

Gorelick, N., Hancher, M., Dixon, M. et al. (2017). Google Earth Engine: planetary-scale geospatial analysis for everyone. *Remote Sensing of Environment* 202: 18–27.

Grenfell, S.E., Rowntree, K.M., and Grenfell, M.C. (2012). Morphodynamics of a gully and floodout system in the Sneeuberg Mountains of the semi-arid Karoo, South Africa: implications for local landscape connectivity. *Catena* 89: 8–21.

Grenfell, S.E., Grenfell, M.C., Rowntree, K.M., and Ellery, W.N. (2014). Fluvial connectivity and climate: a comparison of channel pattern and process in two climatically contrasting fluvial sedimentary systems in South Africa. *Geomorphology* 205: 142–154.

Grenfell, S., Grenfell, M., Ellery, W. et al. (2019). A genetic geomorphic classification system for southern African palustrine wetlands: global implications for the management of wetlands in drylands. *Frontiers in Environmental Science* 7: 174. https://doi.org/10.3389/fenvs.2019.00174.

Grenfell, S., Grenfell, M., Tooth, S. et al. (2022). Wetlands in drylands: diverse perspectives for dynamic landscapes. *Wetlands Ecology and Management* 30: 607–622.

Guerreiro, R., Dinis, P.A., and Silva, M.M.V.G. (2012). Interações entre águas e sedimentos das Lagunas do Mangal do Lobito (Angola). In: *Para conhecer a terra. Memórias e notícias de geociências no espaço lusófono* (ed. F.C. Lopes, A.I. Andrade, M.H. Henriques, et al.), 379–387. Coimbra: Imprensa da Universidade de Coimbra.

Gula, J., Green, M.C., Fritts, S. et al. (2022). Assessments of range-wide distribution of six African storks and their relationships with protected areas. *Ostrich* 93: 34–52.

Gumbricht, T., Roman-Cuesta, R.M., Verchot, L. et al. (2017). An expert system model for mapping tropical wetlands and peatlands reveals South America as the largest contributor. *Global Change Biology* 23: 3581–3599.

Guo, M., Li, J., Sheng, C. et al. (2017). A review of wetland remote sensing. *Sensors* 17: 777. https://doi.org/10.3390/s17040777.

Gxokwe, S., Dube, T., and Mazvimavi, D. (2022). Leveraging Google Earth Engine platform to characterize and map small seasonal wetlands in the semi-arid environments of South Africa. *Science of the Total Environment* 803: 150139. https://doi.org/10.1016/j.scitotenv.2021.150139.

Hamunyela, E., Hipondoka, M., Persendt, F. et al. (2022). Spatio-temporal characterization of surface water dynamics with Landsat in endorheic Cuvelai-Etosha Basin (1990–2021). *ISPRS Journal of Photogrammetry and Remote Sensing* 191: 68–84.

Harris, L., Holness, S., Nel, R. et al. (2013). Intertidal habitat composition and regional-scale shoreline morphology along the Benguela coast. *Journal of Coastal Conservation* 17: 143–154.

Hartter, J. and Ryan, S.J. (2010). Top-down or bottom-up? Decentralization, natural resource management, and usufruct rights in the forests and wetlands of Western Uganda. *Land Use Policy* 27: 815–826.

Huntley, B.J. (2019). Angola in outline: physiography, climate and patterns of biodiversity. In: *Biodiversity of Angola* (ed. B.J. Huntley, V. Russo, F. Lages, and N. Ferrand), 15–42. Cham: Springer.

Huntley, B.J. (2023). *Ecology of Angola: Terrestrial Biomes and Ecoregions*. Cham: Springer.

Hurford, A.P. and Harou, J.J. (2014). Balancing ecosystem services with energy and food security – assessing trade-offs from reservoir operation and irrigation investments in Kenya's Tana Basin. *Hydrology and Earth System Sciences* 18: 3259–3277.

Janssen, A.B.G., Hilt, S., Kosten, S. et al. (2021). Shifting states, shifting services: linking regime shifts to changes in ecosystem services of shallow lakes. *Freshwater Biology* 66: 1–12.

Kim, J., Kim, J.G., Jung, J. et al. (2020). Modified hydrogeomorphic approach for estimating quantitative change of riverine wetland functions. *Ecological Engineering* 152: 105876. https://doi.org/10.1016/j.ecoleng.2020.105876.

Knight, J. (2021). Environmental services: a new approach toward addressing sustainable development goals in Sub-Saharan Africa. *Frontiers in Sustainable Food Systems* 5: 687863. https://doi.org/10.3389/fsufs.2021.687863.

Knight, J., Flack-Davison, E., Engelbrecht, S. et al. (2023). Compliance with the Nagoya protocol in the Southern Africa Development Community. *South African Journal of Botany* 159: 302–318.

Kotze, D.C., Rivers-Moore, N.A., Job, N., and Grenfell, M. (2022). Predicting wetland occurrence, main hydrogeomorphic type and vulnerability in the predominantly arid to semi-arid interior of the Western Cape, South Africa. *Wetlands Ecology and Management* 30: 879–898.

Lehner, B. and Grill, G. (2013). Global river hydrography and network routing: baseline data and new approaches to study the world's large river systems. *Hydrological Processes* 27: 2171–2186.

Lehner, B., Verdin, K., and Jarvis, A. (2008). New global hydrography derived from spaceborne elevation data. *Eos, Transactions American Geophysical Union* 89 (10): 93–94.

Li, A., Song, K., Chen, S. et al. (2022). Mapping African wetlands for 2020 using multiple spectral, geo-ecological features and Google Earth Engine. *ISPRS Journal of Photogrammetry and Remote Sensing* 193: 252–268.

Liu, Y., Zhang, H., Zhang, M. et al. (2022). Vietnam wetland cover map: using hydro-periods Sentinel-2 images and Google Earth Engine to explore the mapping method of tropical wetland. *International Journal of Applied Earth Observation and Geoinformation* 115: 103122. https://doi.org/10.1016/j.jag.2022.103122.

Lourenco, M. and Woodborne, S. (2023). Defining the Angolan Highlands Water Tower, a 40 plus-year precipitation budget of the headwater catchments of the Okavango Delta. *Environmental Monitoring and Assessment* 195: 859. https://doi.org/10.1007/s10661-023-11448-7.

Lourenco, M., Fitchett, J.M., and Woodborne, S. (2022). Angolan highlands peatlands: extent, age and growth dynamics. *Science of the Total Environment* 810: 152315. https://doi.org/10.1016/j.scitotenv.2021.152315.

Lourenco, M., Woodborne, S., and Fitchett, J.M. (2023). Fire regime of peatlands in the Angolan Highlands. *Environmental Monitoring and Assessment* 195: 78. https://doi.org/10.1007/s10661-022-10704-6.

Ludwig, C., Walli, A., Schleicher, C. et al. (2019). A highly automated algorithm for wetland detection using multi-temporal optical satellite data. *Remote Sensing of Environment* 224: 333–351.

Mahdianpari, M., Salehi, B., Mohammadimanesh, F. et al. (2018). The first wetland inventory map of Newfoundland at a spatial resolution of 10 m using Sentinel-1 and Sentinel-2 data on the Google Earth Engine cloud computing platform. *Remote Sensing* 11: 43. https://doi.org/10.3390/rs11010043.

Mandishona, E. and Knight, J. (2022a). Inland wetlands in Africa: a review of their typologies and ecosystem services. *Progress in Physical Geography* 46: 547–565.

Mandishona, E. and Knight, J. (2022b). Feedbacks and trade-offs in the use of wetland ecosystem services by local communities in rural Zimbabwe. *Sustainability* 14: 1789. https://doi.org/10.3390/su14031789.

Marambanyika, T. and Beckedahl, H. (2016). Wetland utilisation patterns in semi-arid communal areas of Zimbabwe between 1985 and 2013 and the associated benefits to livelihoods of the surrounding communities. *Transactions of the Royal Society of South Africa* 71: 175–186.

Marambanyika, T., Beckedahl, H., and Ngetar, N.S. (2017). Community strategies to promote sustainable wetland-based food security in rural areas of Zimbabwe. *GeoJournal* 82: 987–1003.

Mashala, M.J., Dube, T., Ayisi, K.K., and Ramudzuli, M.R. (2023). Using the Google Earth Engine cloud-computing platform to assess the long-term spatial temporal dynamics of land use and land cover within the Letaba watershed, South Africa. *Geocarto International* 38: 2252781. https://doi.org/10.1080/10106049.2023.2252781.

McCartney, M.P. and Houghton-Carr, H.A. (2009). Working wetland potential: an index to guide the sustainable development of African wetlands. *Natural Resources Forum* 33: 99–110.

McFeeters, S.K. (1996). The use of the Normalized Difference Water Index (NDWI) in the delineation of open water features. *International Journal of Remote Sensing* 17: 1425–1432.

Mendelsohn, J. (2022). The Angolan catchments of Northern Botswana's major rivers: the Cubango, Cuito, Cuando and Zambezi rivers. In: *Landscapes and Landforms of Botswana. World Geomorphological Landscapes* (ed. F.D. Eckardt), 15–35. Cham: Springer.

Millennium Ecosystem Assessment (MEA) (2003). *Ecosystems and Human Well-Being: A Framework for Assessment*. Washington, DC: Island Press.

Milzow, C., Kgotlhang, L., Bauer-Gottwein, P. et al. (2009). Regional review: the hydrology of the Okavango Delta, Botswana – processes, data and modelling. *Hydrogeology Journal* 17: 1297–1328.

Mitchell, S.A. (2013). The status of wetlands, threats and the predicted effect of global climate change: the situation in Sub-Saharan Africa. *Aquatic Sciences* 75: 95–112.

Morant, P.D. (1983). Wetland classification: towards an approach for southern Africa. *Journal of the Limnological Society of Southern Africa* 9: 76–84.

Moses, O., Blamey, R.C., and Reason, C.J.C. (2022). Relationships between NDVI, river discharge and climate in the Okavango River Basin region. *International Journal of Climatology* 42: 691–713.

Munishi, S. and Jewitt, G. (2019). Degradation of Kilombero Valley Ramsar wetlands in Tanzania. *Physics and Chemistry of the Earth* 112: 216–227.

Musasa, T. and Marambanyika, T. (2020). Threats to sustainable utilization of wetland resources in Zimbabwe: a review. *Wetlands Ecology and Management* 28: 681–696.

Musasa, T. and Marambanyika, T. (2022a). The nexus between drivers of wetland utilisation patterns, land use/land cover change and ecosystem services in two wetlands, Zimbabwe. *Transactions of the Royal Society of South Africa* 77: 119–131.

Musasa, T. and Marambanyika, T. (2022b). Assessing the sustainability of land uses in Driefontein and Intunjambili wetlands, Zimbabwe. *Scientific African* 16: e01195. https://doi.org/10.1016/j.sciaf.2022.e01195.

Mwakaje, A.G. (2009). Wetlands, livelihoods and sustainability in Tanzania. *African Journal of Ecology* 47: 179–184.

Namaalwa, S., van Dam, A.A., Gettel, G.M. et al. (2020). The impact of wastewater discharge and agriculture on water quality and nutrient retention of Namatala wetland, Eastern Uganda. *Frontiers in Environmental Science* 8: 148. https://doi.org/10.3389/fenvs.2020.00148.

Neto, C.A. (2020). Avaliação da evolução da área no Mussulo usadando geotecnologias e necessidade de protecção a luz da Ramsar. *Revista Angolana de Geociências* 1: 51–61.

Ngwenya, K. and Marambanyika, T. (2021). Trends in use of remotely sensed data in wetlands assessment and monitoring in Zimbabwe. *African Journal of Ecology* 59: 676–686.

Ollis, D.J., Ewart-Smith, J.L., Day, J.A. et al. (2015). The development of a classification system for inland aquatic ecosystems in South Africa. *Water SA* 41: 727–745.

Orimoloye, I.R., Kalumba, A.M., Mazinyo, S.P., and Nel, W. (2020). Geospatial analysis of wetland dynamics: wetland depletion and biodiversity conservation of Isimangaliso Wetland, South Africa. *Journal of King Saud University, Science* 32: 90–96.

Owethu Pantshwa, A. and Buschke, F.T. (2019). Ecosystem services and ecological degradation of communal wetlands in a South African biodiversity hotspot. *Royal Society Open Science* 6: 181770. https://doi.org/10.1098/rsos.181770.

Ozesmi, S.L. and Bauer, M.E. (2002). Satellite remote sensing of wetlands. *Wetlands Ecology and Management* 10: 381–402.

Pekel, J.F., Cottam, A., Gorelick, N., and Belward, A.S. (2016). High-resolution mapping of global surface water and its long-term changes. *Nature* 540: 418–422.

Ralph, T.J. and Hesse, P.P. (2010). Downstream hydrogeomorphic changes along the Macquarie River, southeastern Australia, leading to channel breakdown and floodplain wetlands. *Geomorphology* 118: 48–64.

Ramsar Convention Secretariat (2010). *Inventory, Assessment, and Monitoring. Ramsar Handbooks for the Wise Use of Wetlands*, 4e, vol. 13. Gland: Ramsar Convention Secretariat.

Ramsar Convention Secretariat (2016). *The Fourth Ramsar Strategic Plan 2016–2024. Ramsar Handbooks on the Wise Use of Wetlands*, 5e, vol. 2. Gland: Ramsar Convention Secretariat.

Rebelo, L.-M., McCartney, M.P., and Finlayson, C.M. (2010). Wetlands of Sub-Saharan Africa: distribution and contribution of agriculture to livelihoods. *Wetlands Ecology and Management* 18: 557–572.

Republic of Angola (2019). *6th National Report on Biodiversity in Angola and the Achievement of the AICHI Goals 2011–2020*. Luanda: Ministry of the Environment.

Republic of Angola (2020). *National Biodiversity Strategy and Action Plan 2019–2025*. Luanda: Ministry of the Environment.

República de Angola (2006). *Primeiro Relatório Nacional para a Conferência das Partes da Convenção da Diversidade Biológica*. Luanda: Ministério do Urbanismo e Ambiente, República de Angola/UNDP.

Revermann, R., Gomes, A.L., Gonçalves, R.M. et al. (2016). Vegetation database of the Okavango Basin. *Phytocoenologia* 46: 103–104.

Rivers-Moore, N.A., Kotze, D.C., Job, N., and Mohanlal, S. (2020). Prediction of wetland hydrogeomorphic type using morphometrics and landscape characteristics. *Frontiers in Environmental Science* 8: 58. https://doi.org/10.3389/fenvs.2020.00058.

Rodríguez, J.P., Beard, T.D., Bennett, E.M. et al. (2006). Trade-offs across space, time, and ecosystem services. *Ecology and Society* 11: 28. http://www.ecologyandsociety.org/vol11/iss1/art28/.

Rouse, J.W., Haas, R.H., Schell, J.A., and Deering, D.W. (1974). Monitoring vegetation systems in the Great Plains with ERTS. *NASA Special Publication* 351: 309–317.

Salimi, S., Almuktar, S.A.A.N., and Scholz, M. (2021). Impact of climate change on wetland ecosystems: a critical review of experimental wetlands. *Journal of Environmental Management* 286: 112160. https://doi.org/10.1016/j.jenvman.2021.112160.

Scoones, I. (1991). Wetlands in drylands: key resources for agricultural and pastoral production in Africa. *Ambio* 20: 366–371.

Shaffer, P.W., Kentula, M.E., and Gwin, S.E. (1999). Characterization of wetland hydrology using hydrogeomorphic classification. *Wetlands* 19: 490–504.

Sieben, E.J.J., Khubeka, S.P., Sithole, S. et al. (2018). The classification of wetlands: integration of top-down and bottom-up approaches and their significance for ecosystem service determination. *Wetlands Ecology and Management* 26: 441–458.

Silvius, M.J., Oneka, M., and Verhagen, A. (2000). Wetlands: lifeline for people at the edge. *Physics and Chemistry of the Earth, Part B* 25: 645–652.

Simmons, R.E., Sakko, A., Paterson, J., and Nzuzi, A. (2006). Birds and conservation significance of the Namib Desert's least known coastal wetlands: Baia and Ilha dos Tigres, Angola. *African Journal of Marine Science* 28: 713–717.

Sinthumule, N.I. (2021). An analysis of communities' attitudes towards wetlands and implications for sustainability. *Global Ecology and Conservation* 27: e01604. https://doi.org/10.1016/j.gecco.2021.e01604.

Steinbach, S., Cornish, N., Franke, J. et al. (2021). A new conceptual framework for integrating earth observation in large-scale wetland management in East Africa. *Wetlands* 41: 93. https://doi.org/10.1007/s13157-021-01468-9.

Stephenson, P.J., Ntiamoa-Baidu, Y., and Simaika, J.P. (2020). The use of traditional and modern tools for monitoring wetlands biodiversity in Africa: challenges and opportunities. *Frontiers in Environmental Science* 8: 61. https://doi.org/10.3389/fenvs.2020.00061.

Tavares, A.O., Máquina, M.A., and Henriques, M.H. (2012). The impact of tourism in a fragile wetland ecosystem in Angola: the Arco (Namibe) case study. *WIT Transactions on Ecology and the Environment* 161: 205–216.

Taylor, A.R.D., Howard, G.W., and Begg, G.W. (1995). Developing wetland inventories in southern Africa: a review. *Vegetatio* 118: 57–79.

Tooth, S. and McCarthy, T.S. (2007). Wetlands in drylands: geomorphological and sedimentological characteristics, with emphasis on examples from southern Africa. *Progress in Physical Geography* 31: 3–41.

US Department of Agriculture Forest Service, International Programs (USDAFS) (2022). *Angola Protected Area Network Assessment*. Washington, DC: U.S. Forest Service International Programs/USAid.

van Deventer, H. (2021). Monitoring changes in South Africa's surface water extent for reporting sustainable development goal sub-indicator 6.6.1.a. *South African Journal of Science* 117: 8806. https://doi.org/10.17159/sajs.2021/8806.

van Deventer, H., Nel, J., Mbona, N. et al. (2016). Desktop classification of inland wetlands for systematic conservation planning in data-scarce countries: mapping wetland ecosystem types, disturbance indices and threatened species associations at country-wide scale. *Aquatic Conservation: Marine and Freshwater Ecosystems* 26: 57–75.

van Deventer, H., van Niekerk, L., Adams, J. et al. (2020). National Wetland Map 5: an improved spatial extent and representation of inland aquatic and estuarine ecosystems in South Africa. *Water SA* 46: 66–79.

van Huyssteen, C.W. and Johnson, T.L. (2022). The physical, chemical, mineralogical, and hydrological properties of three different wetland types in the Kruger National Park. *Wetlands Ecology and Management* 30: 695–713.

Wakelin-King, G.A. (2022). Landscapes of the Lake Eyre Basin: the catchment-scale context that creates fluvial diversity. *Transactions of the Royal Society of South Australia* 146: 109–167.

Waleed, M., Sajjad, M., Shazil, M.S. et al. (2023). Machine learning-based spatial-temporal assessment and change transition analysis of wetlands: an application of Google Earth Engine in Sylhet, Bangladesh (1985–2022). *Ecological Informatics* 75: 102075. https://doi.org/10.1016/j.ecoinf.2023.102075.

8

Wetlands of Benin (West Africa): Biodiversity, Livelihoods, and Conservation

Salako K. Valère[1], Dassou G. Hospice[2], Gnansounou S. Constant[1], Zanvo M. G. Serge[1], Akodékou A. David[1], and Glèlè Kakaï Romain[1]

[1] *Laboratoire de Biomathématiques et d'Estimations Forestières, Faculté des Sciences Agronomiques, Université d'Abomey-Calavi, Cotonou, République du Bénin*
[2] *Département de Biologie Végétale, Laboratoire des Sciences du Végétal et Pharmacopée, Ecole Doctorale des Sciences de la Vie et de la Terre, Université d'Abomey-Calavi, Cotonou, République du Bénin*

8.1 Introduction

Wetlands are ecosystems that can be flooded either seasonally or permanently for varying durations, ranging from a few years to decades or even for short periods of time (Hu et al. 2017). The concept of wetlands is closely tied to the temporal or permanent presence of water, which imparts unique characteristics and distinct features to these ecosystems (Meng et al. 2017). Ecologically, Brinson (2011) defines wetlands as a specific segment within the continuum between the wettest ecosystems (such as rivers, lakes, estuaries, and the marine environment), and the driest environments within a physiographic region. The distribution of wetlands and their services to humanity underscores their significance in environmental protection and human well-being.

Like other vital ecosystems, wetlands play a crucial role in ecosystem service provision. Approximately, 15% of wetlands are utilized for rice production, which represents the staple food for over 40% of the world's population (Brinson 2011). Wetlands are also recognized for their contributions to biodiversity conservation and serve as carbon sinks, filters, and transformers for pollutants, nutrients, and sediments, while acting as buffer zones between aquatic ecosystems and upland areas (Wang et al. 2023). Additionally, wetlands manage hydrological processes in water bodies and contribute to tracking the ecological environment and biogeochemical cycling (Jiang et al. 2015).

Wetlands are distributed across the globe, spanning from the tropics to tundra regions, and representing 9% of the Earth's surface (Davidson et al. 2018; Sharma et al. 2021). Asia hosts the largest wetland areas, accounting for 32% of the global wetland area, followed respectively by North America (27%), Latin America and Caribbean (16%), Europe (13%), Africa (10%), and Oceania (10%) (Davidson et al. 2018). The widely recognized significance of wetlands prompted their international protection and conservation through many initiatives, including the Ramsar Convention on Wetlands of international importance (Sakané et al. 2011). As of now, 2414 wetlands sites of international importance, covering an estimated area of 543,972 ha, have been internationally recognized (Sharma et al. 2021).

Wetlands of Tropical and Subtropical Asia and Africa: Biodiversity, Livelihoods and Conservation,
First Edition. Edited by Thammineni Pullaiah.

Despite their significance to both human beings and the global environment, wetlands are facing severe threats due to both man-made and climate-related actions (Sharma et al. 2021). Factors such as rapid urbanization, agricultural expansion, and increasing pressure on water supplies contribute to the degradation of wetlands at global, regional, and local levels (Khan and Arshad 2014). According to Hu et al. (2017), there has been a loss of 33% of wetlands globally as of 2009, including 4.58 million km^2 of nonwater wetlands and 2.64 km^2 of open water. Although climate change impacts the distribution and ecological functions of wetlands, man-made actions are considered the primary contributors to wetlands degradation (Khan and Arshad 2014). It is estimated that the earth would have approximately 29.83 million km^2 of wetlands if humans did not interfere with wetlands ecosystems (Hu et al. 2017).

As observed globally, wetlands in West Africa, particularly in Benin, have undergone significant human-induced degradation and unsustainable exploitation due to factors such as demographic growth, land encroachment, and urbanization (Ajonina et al. 2014). This chapter provides a comprehensive overview on the typology, conservation, biodiversity, threats, and socioeconomic importance of wetlands in Benin. Additionally, it proposes key policy actions geared toward ensuring the sustainable use and protection of these valuable ecosystems.

8.2 Methods

8.2.1 Study Area

This chapter focuses on the Benin Republic, situated in West Africa, sharing borders with four other West-African countries: Togo, Nigeria, Niger, and Burkina Faso. The country extends between the latitudes 6°15′N and 12°25′N and longitudes 0°40′E and 3°45′E (Neuenschwander and Toko 2011), covering a total land area of 114.763 km^2 with a population of over 13.3 million inhabitants (Mensah et al. 2023).

The Benin Republic is subdivided into 12 departments and 77 municipalities, following the decentralization initiative implemented in 2003 (Mensah et al. 2023). The nation comprises various ethnic groups, which interact in diverse ways to sustain the cultural identity of the country. Ethnic groups found in the southern part of Benin include, but are not limited to, Adja, Ouémè, Toli, Sahouè, Cotafon, Gun, Holly, Watchi, Péda, Pla, Yorouba, Nago, and Mina. In the northern part of the country, the population consists of, among others, Kotokoli, Lokpa, Wama, Bariba, Peulh, and Wama, while the central region is inhabited by groups such as Idatcha, Fon, Mahi, and Tchabè (Dansi et al. 2008).

The southern and the central regions of Benin fall within the humid agroecological zones, characterized by two rainy seasons and two dry seasons annually. The mean annual rainfall of these areas ranges from 1100 to 1400 mm/year (Houessou et al. 2013). In contrast, the northern part of the country lies within arid and semiarid agroecological zones, experiencing an irregular rainfall pattern averaging between 800 and 950 mm/year, with one rainy and one dry season (Mensah et al. 2023). The mean annual temperatures of the Benin Republic range from 26 to 28 °C but can rise up to 35–40 °C under exceptional conditions (Houessou et al. 2013).

Various types of vegetation are found across the country, including semideciduous forest in the south, woodland, and savannah woodland in the central east and the northeast, dry semideciduous forest in the central west and the northwest, and tree and shrub savannahs in the far north (Dansi et al. 2008). From the southern to the northern regions, a diverse array of wetlands exists throughout the country. Being a coastal state, the south of Benin is remarkably endowed with a 120-km-long coastline with various coastal ecosystems. These wetlands scattered across the country play a crucial role in safeguarding lives and livelihoods.

8.2.2 Data Collection

This chapter was developed using various data collection approaches, including document reviews, such primary data collection, and field observation. We drew insights over 10 years of research work on wetlands in Benin to comprehensively document the distribution, conservation efforts, and ecosystem services provided by wetlands in the Benin Republic. Specific emphasis was placed on exploring their potential in biodiversity conservation, ecosystem services provision, and livelihoods supply.

8.2.2.1 Document Reviews

Initially, an extensive review of existing literature was conducted to identify the various wetlands present in Benin, elucidate their unique characteristics, and understand their sociocultural significance within local communities. This served as the foundation for our subsequent analyses and discussions on wetland ecosystems in the country. For the conduct of the review, scientific articles on wetlands in Benin were downloaded from multiple search engines, including Google Scholar, Scopus, and Web of Science. A wide range of keywords such as “mangroves”, “oceans”, “wetlands”, “lowlands”, “inland valleys”, “swamp forests”, and their corresponding in French were employed during the search process for papers downloading. Afterward, duplicate articles were identified and removed, and a screening exercise was conducted to select articles entirely focused on wetland ecosystems of Benin for further analyses. In total, 50 documents including articles, book chapters, technical reports, and handbooks were thoroughly reviewed for this study, ensuring a robust and comprehensive understanding of wetlands in Benin.

8.2.2.2 Primary Data

A comprehensive series of community engagement through focus group discussions, in-depth interviews, informal conversations, and direct observations was undertaken across different regions of the country. This enabled us to understand the functioning and the governance of wetlands as well as their roles in enhancing people’s well-being and livelihoods in Benin. For in-depth interviews, key stakeholders involved in wetlands conservation were actively engaged, including local civil society organizations ($n = 10$), state agencies ($n = 5$), and local and traditional authorities ($n = 30$). Open-ended questions were employed during interviews to facilitate in-depth discussions and gather diverse perspectives. Focus group discussions conducted were concentrated in southern Benin ($n = 63$), involving over hundred participants. This enabled to assess the distribution, value, threats, and conservation status of wetlands in Benin. These discussions provided a platform for local actors such as mangrove users, marine small-scale fishers, and smallholder farmers to share insights and experiences related to wetland ecosystems. Quantitative data were also sought from wetland users ($n = 300$) to examine the interactions between coastal populations and wetlands in terms of use patterns, livelihoods provision, and perceived threats. Participants were further asked to attribute scores to the identified threats using a five-point Likert scale, from least detrimental to wetlands (1) to highly detrimental to wetlands (5).

8.2.2.3 Secondary Data

In addition to document reviews and the primary data gathered, secondary data on plant and fish diversity collected by the authors were also used. These include primary data on plant inventory collected from mangroves from 2016 to 2023 in 600 plots, fish data collected from Agonlin-Lowé (Ouémé River, Southern Benin), Pendjari River (Northern Benin), Ahémé Lake, and Nokoué Lake in 2018 as well as participants observations made from 2016 to date in socioeconomically important wetlands such as mangroves, swamp forests, inland valley, and the ocean.

8.2.3 Data Analysis

Articles and relevant documents obtained were reviewed and information was synthesized in order to characterize the diverse wetlands present in Benin. This process involved triangulating the information obtained from the literature review with the primary data used. Frequency distribution was used to summarize quantitative data. Data on fish and plant inventory were complemented with information extracted from the literature, to illustrate the biological diversity of wetlands in Benin. The relative importance index (RII) was computed, as the local importance of each identified threat, following the formula (Aheto et al. 2016):

$$\mathrm{RII} = \frac{\sum_{i}^{n} W_i}{A \times N}$$

W_i is the score given by the respondent I, ranging from 1 to 5, A is the highest score (5 in this study), and N stands for the total number of respondents.

8.3 Results

8.3.1 Wetlands Ecosystems in Benin: Typology, Mapping, and Geographical Distribution

Document reviews and interactions with various stakeholders across the country helped identify six main types of wetlands in Benin (Figure 8.1). They include water bodies, mangroves, swamp forests, marshy meadows, riparian forests, and inland valleys and lowlands (Table 8.1 and Figure 8.2).

Water bodies represent the first category of wetlands extensively documented in Benin. The principal water bodies shaping the hydrological network in Benin can be classified into two categories: marine and coastal water bodies and inland water bodies. Marine and coastal water bodies include the Atlantic Ocean, extending approximately 120 km from Sèmè-Kpodji (Nigeria border) to Hillacondji (Togolese border), the coastal lagoon, a 65-km-long water body spanning from Togbin (Abomey-Calavi) to Grand-Popo, and the Gbaga channel, a natural border between Benin and Togo, stretching from Agbannankin to Aneho and measuring approximately 100 km. Concerning inland water bodies, five primary hydrographic basins exist in Benin. They include the basins of Niger, Volta, Couffo, Mono, and Ouémé. The Niger basin is situated in the northern part of the country and encompasses the Niger river (137 km) and its main tributaries, including Mékrou, Alibori, and Sota Rivers. The volta basin, shared between Benin and Burkina Faso, is represented in Benin by the Pendjari river (300 km), and its sole tributary, the Magou River. The Couffo basin is fed by the Couffo River (125 km), while the Mono and Ouémé basins are, respectively, fed by Mono (350 km) and Ouémé (510 km). Zou, Okpara, and Alpouro are the main tributaries of the Ouémé River. Two major lakes, namely, Lake Ahémé connected to the Aho channel and Lake Nokoué linked to the Porto-Novo lagoon, receive water from the Couffo and Ouémé rivers and connect with the Atlantic Ocean. Other important inland waters, including ancient lakes (such as Toho, Togbadji, Doukon, Dofé, Datchi, and Wozo), and other small rivers (Sazoué, Savédo, and Dévédo), also serve as valuable sources of livelihood to local —

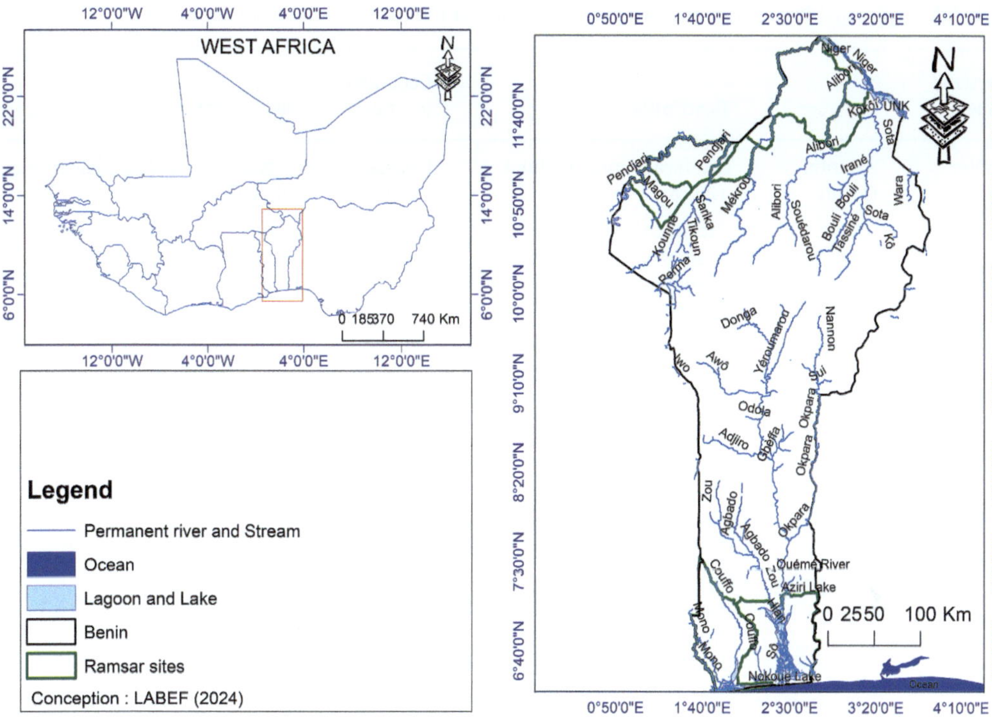

Figure 8.1 Map of the study area showing the major water areas of Benin. *Source:* IGN Benin topographic base (2018) and field data (2023).

populations. With an inundated surface of 5000 km^2, the plains of Ouémé and Sô are one of the biggest flood plains in West Africa.

The second most-documented wetlands in Benin are mangroves (Figure 8.3). They thrive along the coast and offer livelihood opportunities for thousands of coastal residents. The significant degradation of mangroves in Benin, stemming from the multitude of benefits they provide, has prompted governance and conservation efforts to curb their decline. Mangroves, characterized as coastal forests, thrive in the intertidal zones of tropical and subtropical regions. Key informants consulted described mangroves as among the most significant wetlands in terms of ecosystem services provision in Benin. They are distributed across coastal villages and are prevalent in major coastal municipalities, including Abomey-calavi, Aguégués, Sèmè-kpodji, Porto-Novo, Comè, Bopa, Grand-Popo, Ouidah, Kpomassè, and Sô-Ava. However, the municipality of Grand-Popo, located in the southeastern part of the country, has been identified by key informants and the existing literature as the hotspot of mangroves, because it reportedly hosts over 50% of mangroves in Benin.

Although less prevalent compared to mangroves and water bodies, some swamp forests also exist in Benin as wetlands. Among the most prominent and well documented are the swamp forests of Lokoli (500 ha) and Agonvè (300 ha), located in the municipalities of Zogbodomey and Zagnanado, in central Benin. These forests remain the rare Guineo-Congolese swamp forests in Benin. They are preserved through a combination of indigenous practices and legal regulations, which help mitigate their degradation, although certain human-induced threats still compromise their long-term sustainability.

Inland valleys are also well represented in Benin as wetlands (Figure 8.4). Inland valleys are mostly found landwards and are documented as pristine sites for biodiversity conservation in Benin. Existing data on inland valleys in Benin indicate that they cover an area of 23,552 ha and are primarily used for agricultural purposes, particularly rice production. Notable inland valleys in

Table 8.1 Typology, description, and geographic location of the major wetlands in Benin.

Wetland types	Subtypes	Description	Geographic locations	References
Water bodies	Atlantic Ocean	Sea facade long of about 120 km	Coastal ecosystem	Ogou et al. (2016), Deguenon et al. (2023)
	Rivers	Freshwater rivers occurring in Benin, including Niger, Mono, Couffo, Pendjari, and Ouémé rivers	Inland ecosystems	Biao (2017), Attogouinon et al. (2017), Djihouessi and Aina (2018)
	Lakes	Major lakes encountered in Benin, including Ahémé and Nokoué lakes	Inland ecosystems	Bremond et al. (2017), Olodo et al. (2020)
	Lagoons	Lagoonal systems occurring in the country, including coastal lagoon, the Gbaga channel, the Aho channel, and the Porto-Novo lagoon	Coastal and inland ecosystems	Okpeitcha et al. (2022), Deguenon et al. (2023)
	Tributaries	Tributaries of the main rivers, including Zou, Okpara, Alpouro, Mékrou, Alibori, Sota, and Magou	Inland ecosystems	Badou et al. (2017)
	Ancient lakes and lagoons	The ancient lakes located in the interior of the country, including Toho, Togbadji, Doukon, Dofé, Datchi, Wozo, Dati, Ahouangan, Todougba, and Djonou	Inland ecosystems	Achoh et al. (2021), Houssou et al. (2017)
	Small rivers	The small rivers occurring in the country, including Sazoué, Savédo, and Dévédo	Inland ecosystems	Yantikoua et al. (2023)
Mangroves	—	Salt-tolerant coastal forests	Coastal ecosystem	Gnansounou et al. (2021), Teka et al. (2019), Zanvo et al. (2021), Sinsin et al. (2022)
Swamp forests	—	Permanently flooded inland forests	Inland ecosystem	Adomou et al. (2009), Etienne et al. (2012), Dossou et al. (2012), Ahossou et al. (2017)
Marshy meadows	—	Seasonally flooded wetlands that serve primarily as buffer zones between wetter areas and dry lands	Inland and coastal ecosystems	Adjonou et al. (2020)
Riparian forests	—	Floodable forests adjacent to water bodies	Inland and coastal ecosystems	Ceperley et al. (2010), Richard et al. (2013), Natta et al. (2012), Giertz et al. (2012)
Inland valleys	—	Upper reaches of river systems where alluvial sedimentation processes are absent or low	Inland ecosystems	Danvi et al. (2017), Djagba et al. (2019)

Source: Primary data and document reviews.

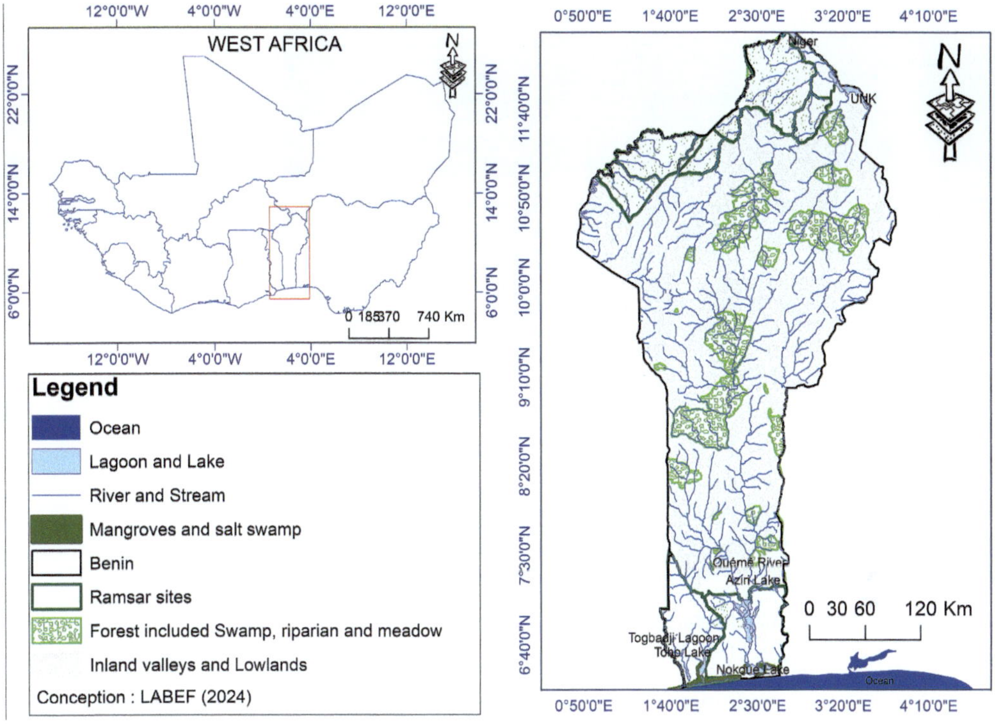

Figure 8.2 Map showing the distribution of existing wetlands in Benin. *Source:* IGN Benin topographic base (2018) and field data (2023).

Figure 8.3 Fishing in mangroves in Benin.

Figure 8.4 Rice production in inland valleys in Benin.

Benin include the lower and upper valleys of Ouémé as well as those in Salavou, Zagnanado, and Zogbodomey. Field expeditions conducted in the framework of this study facilitated the identification of three categories of inland valleys management in Benin, namely, the smart, classical, and traditional inland valleys. Smart valleys refer to structured inland valleys recognized for their capabilities in water control, whereas classical and traditional inland valleys represent those managed through dyke building and indigenous practices, respectively.

Marshy meadows and riparian forests are also present as wetlands in Benin, playing an important role in maintaining the hydrological flow of water bodies, providing essential services, and safeguarding the local environment. Marshy meadows, including salt marshes, are predominantly located along the coast of Benin, primarily functioning as buffer zones that alleviate flooding during periods of high water. Additionally, there are seasonally flooded riparian forests spread across the country, offering vital goods and services to rural communities. While two principal swamp forests are identified, over 10 riparian forests were documented in Benin. The principal riparian forests documented in Benin include Samiondji (in the south of Benin), Bètèkoukou, Toui-kilibo, and Idadjo (in central Benin) as well as Bétérou, Pénéssoulou, Pendjari Biosphere Reserve, Yarpao, and Gbèssè (in northern Benin).

8.3.2 Conservation Status of the Wetlands in Benin

In recent years, the Republic of Benin has ratified several international treaties aimed at conserving the environment, with a particular focus on wetlands. These include the 1971 Ramsar Convention on Wetlands, the 1973 Convention on International Trade in Endangered Species, the International Convention for Oil Pollution Damage (1969), the International Convention on Conservation of Migratory Species Wild Animals (1979), and the 1992 Convention on Biological Diversity. To comply with the clauses of these conventions and conserve wetlands in Benin, numerous measures have been adopted by the government, local authorities, and civil society organizations. These include the designation of certain wetlands as RAMSAR sites (RAMSAR sites 1017 and 1018 in southern Benin and RAMSAR sites 1668 and 1669 in northern Benin), the establishment of Community Biodiversity Conservation Areas, the creation of Marine Protected Areas, the creation of Biosphere Reserves and Nature Parks, the classification of certain sites as classified forests, and the sacralization of certain wetlands (Table 8.2). Despite these conservation measures,

Table 8.2 Conservation status of wetlands in Benin.

Conservation measures	Description	References
Establishment of Ramsar sites	The Ramsar convention was ratified by Benin on 24 January 2000, to protect wetlands. It compels Benin to take necessary measures to safeguard the ecological characteristics of wetlands. This led to the establishment of four Ramsar sites of international importance in Benin, including the Ramsar sites 1017 and 1018 in southern Benin and the Ramsar sites 1668 and 1669 in northern Benin	Adanguidi et al. (2020), Padonou et al. (2021), Gnansounou et al. (2024)
Creation of Community Biodiversity Conservation Areas (CBCAs)	They represent areas managed collectively by the local community and municipalities to protect natural resources. They are managed through the lens of bottom-up approaches, with technical assistance from state agencies and civil society organizations. Many CBCAs exist in Benin, namely, Bouche du Roy, Lac Toho, Chenal Gbaga, Lagune Côtière, and Togbin-Adounko	Gnansounou et al. (2024)
Marine Protected Area (MPA)	Created by presidential decree on 5 January 2022, the MPA designates a protected area set aside for the conservation of faunal, floral, historical, and cultural resources found both within the 200 nautical miles of maritime and territorial waters. Resources located in these areas are either totally or partially protected by legal frameworks and other protection measures. Two MPAs, namely, Donaten MPA and the Bouche du Roy MPA exist in Benin.	Queffelec et al. (2021)
Sacralized and sanctuary wetlands	They represent portions of wetlands set aside and managed by traditional and endogenous rules. Sacralized and sanctuary wetlands existing in Benin include but are not limited to the sacred forest of Kpékonzoun, the sacred forest of Togbozoun, the sacred forest of Lodouko, the sacred mangroves of Mitôgbodji, etc.	Gnansounou et al. (2024), CENAGREF/PAPE (2013)
Biosphere Reserves	Benin has created Biosphere Reserves to protect its natural resources. Many of these reserves are in Benin's wetlands, such as the Lower Ouémé Valley Biosphere Reserve and the Mono Transboundary Biosphere Reserve.	CENAGREF/PAPE (2013), Gnansounou et al. (2022), Lougbegnon et al. (2023)
Nature Parks	Nature Parks are generally protected areas delineated and managed for resources protection and recreational purposes. These can be community parks or national parks. Some examples include the Community Park of the Sitatunga Valley and the Pandjari Park.	CENAGREF/PAPE (2013)
Wetlands of classified forests	They represent riparian forests or inland valleys located within forests recognized as "classified" through formal legal procedure, benefiting from limited interaction with local communities or individuals.	CENAGREF/PAPE (2013)

Source: Primary data and document reviews.

the conservation and sustainable use of wetlands in Benin are still not effective. For instance, although many water bodies, including the Atlantic Ocean, lakes and rivers are located within Ramsar sites, community biodiversity areas, and marine protected areas, they still undergo significant man-made pressure. Similarly, mangroves, albeit declared as state-protected forests, also exhibit some degree of degradation. The same degradation trends are observed with swamp forests, riparian forests, inland valleys, and marshy meadows, which are also degraded, despite their presence in Ramsar sites, community parks, classified forests, and sanctuaries.

8.3.3 Biological Diversity of the Wetlands in Benin

8.3.3.1 Plant Diversity

Benin's wetlands present a large diversity of plant and animal species (Table 8.3). Our field inventory and literature review allowed us to identify 117 plant species in these vital ecosystems in the country, of which 39 were recorded for swamp forests. The commonest species to both Lokoli and Agonvè swamp forests include *Raphia hookeri, Nauclea diderrichii, Pterocarpus santalinoides, Phoenix reclinata, Xylopia rubescens, Mitragyna inermis, Alstonia congensis, Spondianthus preussii, Cola millenii, Bambusa vulgaris,* and *Antocleista vogelii*. Considering their abundance, seven species, namely, *Xylopia rubescens, Raphia hookeri, Syzygium owariense, Spondianthus preussii, Hallea ledermannii, Cleistopholis patens,* and *Nauclea diderrichii*, were dominant (Table 8.3). In riparian forests, about 77 species were recorded. The common ones include *Albizia coriaria, Albizia zygia, Borassus aethiopum, Raphia sudanica, Cola millenii, Cola laurifolia, Combretum lecardii, Ficus mucuso, Ficus* spp., *Ficus sur, Mitragyna inermis, Pterocarpus santalinoides, Crateva adansonii, Ceiba pentandra, Combretum collinnum, Combretum nigricans, Combretum molle, Vachellia sieberiana,* and *Zanthoxylum xanthoxyloides*. Mangroves are special wetlands with low plant diversity in Benin, due to the specific coastal conditions in which they thrive. In Benin, mangroves are characterized by five true mangrove species,

Table 8.3 Plant diversity in the three most important wetlands in Benin.

Species	Family	Swamp forests	Riparian forests	Mangroves	Abundance
Acacia auriculiformis	Fabaceae			x	[a]
Adansonia digitata	Bombacaceae		x		[a]
Albizia coriaria	Fabaceae		x		[b]
Albizia glaberrima	Fabaceae		x		[b]
Albizia zygia	Fabaceae		x		[b]
Alstonia congensis	Apocynaceae	x			[c]
Anarcadium occidentale	Anacardiaceae		x		[a]
Anthocleista vogelii	Gentianaceae	x			[c]
Antiaris toxicaria	Moraceae		x		[a]
Antidesma venosum	Euphorbiaceae		x		[a]
Avicennia germinans	Verbenaceae			x	[b]
Azadirachta indica	Meliaceae		x		[a]
Balanites aegyptiaca	Zygophyllaceae		x		[a]

Table 8.3 (Continued)

Species	Family	Swamp forests	Riparian forests	Mangroves	Abundance
Berlinia grandifolia	Fabaceae		x		*a*
Bligia sapida	Sapindaceae		x		*a*
Borassus aethiopum	Arecaceae		x		*b*
Burkea africana	Fabaceae		x		*b*
Brenadia salicina	Rubiaceae		x		*a*
Bridelia ferruginea	Phyllanthaceae		x		*a*
Bridelia scleroneura	Phyllanthaceae		x		*b*
Celtis integrifolia	Ulmaceae		x		*a*
Celtis toka	Celtidaceae		x		*a*
Chrysobalanus atacoriensis	Chrysobalanaceae		x		*a*
Cleistopholis patens	Annonaceae	x			*c*
Cola gigantea	Sterculiaceae		x		*c*
Cola laurifolia	Sterculiaceae		x		*c*
Cola millenii	Malvaceae	x			*c*
Combretum lecardii	Combretaceae		x		*b*
Combretum racemosum	Combretaceae	x			*a*
Connarus africanus	Connaraceae	x			*a*
Conocarpus erectus	Combretaceae			x	*a*
Crateva adansonii	Capparidaceae		x		*a*
Crinum jagus	Amaryllidaceae	x			*a*
Crossopteryx febrifuga	Rubiaceae		x		*a*
Cuviera macroura	Rubiaceae	x			*a*
Cynometra megalophylla	Fabaceae		x		*a*
Dalbergia ecastaphyllum	Fabaceae			x	*a*
Daniellia oliveri	Caesalpiniaceae		x		*a*
Detarium senegalense	Fabaceae		x		*b*
Dialium guineense	Fabaceae	x	x		*a*
Diospyros mespiliformis	Ebenaceae		x		*a*
Drepanocarpus lunatus	Fabaceae			x	*a*
Drypetes floribunda	Putranjivaceae		x		*a*
Elaeis guineensis	Arecaceae	X	x		*b*
Eriocoelum kerstingii	Sapindaceae		x		*a*
Fadogia agrestis	Rubiaceae		x		*a*
Feretia apodanthera	Rubiaceae		x		*a*
Ficus leprieuri	Moraceae	x			*b*
Ficus mucuso	Moraceae		x		*b*

(*Continued*)

Table 8.3 (Continued)

Species	Family	Swamp forests	Riparian forests	Mangroves	Abundance
Ficus ovata	Moraceae	x			[b]
Ficus sur	Moraceae		x		[b]
Ficus trichopoda	Moraceae	x			[b]
Flueggea virosa	Euphorbiaceae		x		[a]
Garcinia livingstonei	Clusiaceae		x		[a]
Grewia bicolor	Tiliaceae		x		[a]
Hallea ledermannii	Rubiaceae	x			[c]
Hallea stipulosa	Rubiaceae	x			[c]
Hexalobus crispiflorus	Annonaceae		x		[a]
Isoberlinia doka	Fabaceae		x		[b]
Khaya grandifoliola	Meliaceae		x		[a]
Khaya senegalensis	Meliaceae		x		[a]
Kigelia africana	Bignoniaceae		x		[a]
Laguncularia racemosa	Combretaceae			x	[a]
Lasimorpha senegalensis	Araceae	x			[a]
Lecaniodiscus cupanioides	Sapindaceae		x		[a]
Leptoderis brachyptera	Fabaceae	x			[a]
Lonchocarpus sericeus	Fabaceae		x		[a]
Macaranga heterophylla	Euphorbiaceae	x			[a]
Macaranga heudelotii	Euphorbiaceae	x			[a]
Manilkara multinervis	Sapotaceae		x		[a]
Macrosphyra longistyla	Rubiaceae		x		[a]
Millettia thonningii	Fabaceae		x		[a]
Mimusops andongensis	Sapotaceae		x		[a]
Mitragyna inermis	Rubiaceae	x	x		[c]
Morelia senegalensis	Rubiaceae		x		[a]
Mucuna flagellipes	Fabaceae	x			[a]
Napoleonaea vogelii	Lecythidaceae		x		[a]
Nauclea xanthoxylon	Rubiaceae	x			[c]
Newbouldia laewis	Bignoniaceae	x			[a]
Oxytenanthera abyssinica	Poaceae		x		[b]
Parinari congensis	Chrysobalanaceae	x	x		[a]
Paullinia pinnata	Sapindaceae	x			[a]
Pavetta corymbosa	Rubiaceae	x			[a]
Pentadesma butyracea	Clusiaceae		x		[a]

Table 8.3 (Continued)

Species	Family	Swamp forests	Riparian forests	Mangroves	Abundance
Pericopsis laxiflora	Leguminosae-Papilionoideae		x		[a]
Phoenix reclinata	Arecaceae	x	x	x	[a]
Pierreodendron kerstingii	Simaroubaceae	x			[a]
Pouteria alnifolia	Sapotaceae		x		[a]
Psychotria articulata	Rubiaceae	x			[a]
Psychotria calva	Rubiaceae	x			[a]
Pterocarpus santalinoides	Fabaceae	x	x	x	[b]
Raphia hookeri	Arecaceae	x			[b]
Raphia vinifera	Arecaceae	x			[b]
Raphia sudanica	Arecaceae		x		[b]
Rhizophora racemosa	Rhizophoraceae			x	[c]
Sarcocephalus latifolius	Rubiaceae		x		[a]
Spondianthus preussii	Phyllanthaceae	x			[c]
Sterculia tragacantha	Sterculiaceae		x		[a]
Synsepalum passargei	Sapotaceae		x		[a]
Syzygium guineense	Myrtaceae		x		[c]
Syzygium owariense	Myrtaceae	x			[c]
Tamarindus indica	Fabaceae		x		[a]
Tarenna bipindensis	Rubiaceae	x			[a]
Terminalia catappa	combretaceae			x	[a]
Terminalia glaucescens	Combretaceae		x		[b]
Terminalia leiocarpa	Combretaceae		x		[b]
Terminalia macroptera	Combretaceae		x		[b]
Trilepiseum madagascariense	Moraceae		x		[a]
Uapaca togoensis	Phyllanthaceae		x		[a]
Uvaria chamae	Annonaceae		x		[a]
Vitellaria paradoxa	Sapotaceae		x		[a]
Vitex chrysocarpa	Lamiaceae		x		[a]
Vitex doniana	Lamiaceae		x		[a]
Xylopia parviflora	Annonaceae	x	x		[c]
Xylopia rubescens	Annonaceae				
Zanthoxylum zanthoxyloides	Rutaceae		x		[a]

[a] Low abundance.
[b] Moderate abundance.
[c] High abundance.
Source: Primary data and document reviews.

namely, *Rhizophora racemosa*, *Avicennia germinans*, *Laguncularia racemosa*, *Acrostichum aureum*, and *Conocarpus erectus*. *R. racemosa* represents the dominant and common mangrove species in Benin. *Dalbergia ecastaphyllum* and *Drepanocarpus lunatus* are the dominant mangrove associate species encountered in Benin. In addition, three non-mangrove species, namely, *Acacia auriculiformis*, *Pterocarpus santaniloides*, and *Terminalia catappa* were recorded in mangrove environments in Benin.

8.3.3.2 Animal Diversity

Wetlands in Benin provide conducive habitats for the conservation of many animal species. The review of existing literature and field expeditions helped identify various mammal, fish, and reptile species found in the country's wetlands (Table 8.4). The most important mammal species inventoried include *Hippopotamus amphibius*, *Potamocherus porcus*, *Cercopithecus aethiops*, *Cercopithecus mona*, *Tragelafus speckii*, and *Genetta genetta*. Regarding reptile species, *Crocodilus niloticus*, *Varanus* sp., *Python sebae*, and *Python regius* were observed among others. Aquatic species observed in wetlands include marine and coastal species that accommodate saline and brackish and salty waters, and freshwater species found in inland lakes and rivers. Most represented marine and coastal species include *Sardinella maderensis*, *Sarotherodon melanotheron*, *Chrysichthys nigrodigitatus*, and *Tilapia guineensis*. Considering freshwater species, the most important include *Decapterus rhoncus*, *Pseudotolithus senegalensis*, *Selene dorsalis*, and *Chloroscombrus*.

8.3.4 Importance of Wetlands in Benin

Wetlands provide multiple goods and services to local communities. Major local uses of wetlands recorded in the field and identified in the existing literature include exploitation for food production, fuelwood collection, medicinal plant collection, and service wood collection (Table 8.5). However, the collection of medicinal plants represents an important category of wetland use in Benin. Many plant species are derived from these valuable ecosystems for the treatment of many tropical and subtropical conditions. Plant parts such as leaves, barks, roots, stems, seeds, and grains are abundantly used for the indigenous curing of diseases. Timber extraction and wood collection for domestic uses such as house construction, cooking, or fencing are also well represented in wetlands in Benin. Mangrove trees are mainly used for two major purposes in Benin: firewood and construction. In some minor cases, wetlands including mangroves serve as sanctuaries that harbor various local divinities "vodoun," where indigenous communities gather year-round for ceremonies and rituals.

8.3.5 Threats to Wetlands in Benin

Threats to Benin's wetlands are both natural and man-made. Natural threats include climate change/extreme weather conditions (severe flooding and drought), physicochemical changes in wetlands' water, and hydrological constraints. Anthropogenic threats include deforestation or excessive wood extraction, increasing urbanization, breach of traditional and legal rules, overfishing, overgrazing, illegal and unregulated fishing, overexploitation of forest resources, waste and agricultural pollution, poaching, hydrological constraints, sedimentation, agricultural expansion, and mining (Figure 8.3). Field expeditions and community engagement enabled to understand that threats to wetlands in Benin are highly dynamic and vary according to the type of wetlands. Oceans, lakes, and rivers are vulnerable to climatic pressures (RII = 0.72), sedimentation (RII = 0.60), and heavy chemical pollution (RII = 0.53) from agricultural development. For mangroves, the main threats include excessive wood extraction (RII = 0.62), bushfires (RII = 0.53),

Table 8.4 Animal diversity in the three most important wetlands in Benin.

Species	Category	Mangrove	Swamp	Ocean	Species	Category	Mangrove	Swamp	Ocean	Species	Category	Mangrove	Swamp	Ocean
Elops lacerta	Fish	x			*Cercopithecus erythrogaster*	Mammal	x	x		*Python sebae*	Reptile	x		
Illisha africana	Fish			x	*Cercopithecus mona*	Mammal		x		*Ramphotyphlops braminus*	Reptile	x	x	
Chloroscombrus chrysurus	Fish			x	*Cercopithecus tantalus*	Mammal		x		*Chelonia mydas*	Reptile		x	
Chrysichthys auratus	Fish	x			*Colobus vellerosus*	Mammal		x		*Dermochelys coriacea*	Reptile		x	
Chrysichthys nigrodigitatus	Fish	x			*Erythrocebus patas*	Mammal	x			*Eretmochelys imbricata*	Reptile		x	
Citharinus latus	Fish	x			*Galago senegalensis*	Mammal		x		*Eretmochelys imbricata*	Reptile		x	
Decapterus rhoncus	Fish			x	*Genetta genetta*	Mammal	x			*Kinixys belliana nogueyi*	Reptile		x	
Elops senegalensis	Fish	x			*Hippopotamus amphibius*	Mammal	x			*Lepidochelys olivacea*	Reptile		x	
Engraulis encrasicolis	Fish			x	*Lepus Sp*	Mammal	x			*Bitis arietans*	Reptile		x	
Ethmalosa fimbriata	Fish	x			*Lutra maculicollis*	Mammal	x			*Chelonia mydas*	Reptile		x	
Galeoides decadactylus	Fish			x	*Potamocherus porcus*	Mammal	x			*Dermochelys coriacea*	Reptile	x	x	
Labeo parvus	Fish	x			*Procolobus verus*	Mammal		x		*Eretmochelys imbricata*	Reptile	x	x	
Labeo senegalensis	Fish	x			*Thryonomys swinderianus*	Mammal	x			*Ardea purpurea*	Bird	x	x	
Lates niloticus	Fish	x			*Tragelafus speckii*	Mammal	x			*Dendrocygna viduata*	Bird	x	x	

(*Continued*)

Table 8.4 (Continued)

Species	Category	Mangrove	Swamp	Ocean	Species	Category	Mangrove	Swamp	Ocean	Species	Category	Mangrove	Swamp	Ocean
Mugil cephalus	Fish	x			*Trichechus senegalensis*	Mammal	x			*Egretta alba*	Bird	x	x	
Porogobius schlegelii	Fish	x			*Artherurus africanus*	Mammal		x		*Egretta ardesiaca*	Bird	x	x	
Pseudotolithus senegalensis	Fish			x	*Manis gigantea*	Mammal		x		*Porphyrio alleni*	Bird	x	x	
Pseudotolithus typus	Fish			x	*Leptailurus serval*	Mammal		x		*Anastomus lamelligerus*	Bird	x	x	
Sardinella maderensis	Fish	x		x	*Genetta maculata*	Mammal		x		*Ardea goliath*	Bird	x	x	
Sarotherodon galilaeus	Fish	x			*Aonyx capensis*	Mammal		x		*Calidris ferruginea*	Bird	x	x	
Sarotherodon melanotheron	Fish	x			*Bitis arietans*	Reptile	x			*Corvus albus*	Bird	x	x	
Schilbe intermedius	Fish	x			*Chamaeleo spp*	Reptile	x	x		*Dendrocygna viduata*	Bird	x	x	
Selene dorsalis	Fish			x	*Crocodilus niloticus*	Reptile	x	x		*Egretta ardesiaca*	Bird	x	x	
Tilapia guineensis	Fish	x			*Naja melanoleuca*	Reptile	x			*Ergetta alba*	Bird	x	x	
Galerella sanguinea	Mammal	x			*Python regius*	Reptile	x			*Atilax paludinosus*	Mammal	x		
Herpestes ichneumon	Mammal	x			*Cercopithecus aethiops*	Mammal	x							

Source: Primary data and document reviews.

Table 8.5 Major local uses of plant species found in wetlands in Benin.

Species	Food	Local medicine	Firewood	wood	Service wood	Species	Food	Local medicine	Firewood	wood	Service wood
Acacia sieberiana		x				*Ekebergia capensis*	x	x			
Adansonia digitata	x	x				*Elaeis guineensis*	x				x
Alafia benthamii						*Ficus ovata*					
Albizia ferruginea	x					*Ficus sycomorous*	x	x			
Albizia glaberrima						*Ficus trichopoda*		x			
Alstonia congensis	x	x	x			*Ficus vallis-choudae*			x		
Annona senegalensis	x	x	x			*Flacourtia flavescens*		x			
Anogeissus leiocarpus	x	x	x	x							
Antidesma venosum		x				*Hallea stipulosa*	x	x	x		
Antocleista vogelii		x				*Holarrhena floribunda*					x
Aphania senegalensis		x				*Irvingia gabonensis*	x	x	x		
Bambusa sp.	x	x	x			*Isoberlinia doka*	x	x	x		
Blighia sapida	x	x	x	x		*Khaya senegalensis*	x		x	x	
Borassus aethiopum	x					*Kigelia africana*		x			
Cassipourea congoense		x				*Lannea barteri*		x			
Ceiba pentandra		x	x			*Lannea nigritana*	x				
Celtis zenkeri						*Launaea taraxacifolia*	x				
Cleistopholis patens		x				*Raphia hookeri*	x	x	x		x
Cola gigantea		x	x			*Salacia longipes*					
Cola laurifolia						*Sarcocephalus latifolius*	x				
Cola millenii	x	x	x			*Spondianthus preussii*	x	x	x		x
Crateva adansonii		x				*Spondias mombin*	x	x	x		
Cynometra megalophylla	x					*Strychnos nigritana*		x			
Daniellia oliveri	x	x	x			*Synsepalum brevipes*					x
Dialium guineense	x	x	x	x		*Syzygium owariense*		x			

(Continued)

Table 8.5 (Continued)

Species	Food	Local medicine	Firewood	wood	Service wood	Species	Food	Local medicine	Firewood	wood	Service wood
Diospyros monbuttensis			x			*Tricalysia okelensis*					
Drypetes floribunda	x	x				*Uvaria chamae*	x	x			x
Lecaniodiscus cupanioides		x	x			*Vitex chrysocarpa*					
Lonchocarpus sericeus	x	x				*Vitex doniana*	x	x	x		
Malacantha alnifolia					x	*Xylopia parviflora*		x			
Manilkara multinervis					x	*Xylopia rubescens*	x	x	x		
Millettia thonningii						*Zanthoxylum xanthoxyloides*			x		
Mimusops kummel		x				*Laguncularia racemosa*			x		
Mitragyna inermis		x	x	x		*Avicennia germinans*			x	x	x
Morelia senegalensis		x				*Raphia hookeri*	x	x	x		x
Motandra guineensis		x				*Salacia longipes*					
Nauclea xanthoxylon	x	x	x			*Sarcocephalus latifolius*	x				
Oncoba spinosa					x	*Spondianthus preussii*	x	x	x		x
Pancovia bijuga		x				*Spondias mombin*	x	x	x		
Parinari congensis		x				*Strychnos nigritana*		x			
Parkia biglobosa	x	x	x			*Synsepalum brevipes*					x
Piliostigma thonningii		x				*Syzygium owariense*		x			
Premna angolensis		x				*Tricalysia okelensis*					
Pterocarpus erinaceus		x	x	x		*Uvaria chamae*	x	x			x
Rhizophora racemosa		x	x		x	*Vitex chrysocarpa*					
Conocarpus erectus		x	x			*Vitex doniana*	x	x	x		
Dalbergia ecastaphyllum			x			*Xylopia parviflora*		x			
Drepanocarpus lunatus			x			*Xylopia rubescens*	x	x	x		
Pterocarpus santalinoides	x		x			*Zanthoxylum xanthoxyloides*			x		
Avicennia germinans			x	x	x	*Laguncularia racemosa*			x		

Source: Primary data and document reviews.
x here means it is used for that purpose (say medicine).

illegal, unreported and unregulated fishing known as IUU (RII = 0.50), and climate change and climate risks (RII = 0.45). Swamp forests also face similar threats, with overcollection of wood (RII = 0.65), pronounced climate change and climate risks (RII = 0.45), and urbanization of greater concern (RII = 0.38). Meadows are particularly vulnerable to bushfires (RII = 0.69) and agricultural expansion (RII = 0.61), while riparian forests are threatened by excessive wood extraction (RII = 0.65), agricultural expansion (RII = 0.52), and climate change (RII = 0.38), while inland valleys face strong agricultural expansion (RII = 0.67), hydrological constraints (RII = 0.62), and sedimentation (RII = 0.45). Overexploitation of mangrove resources, physico-chemical modification of water parameters, and increased salt production are also relevant threats to mangrove survival, while chemical pollution from agricultural practices is of greater concern for inland valleys. In addition, mining is a significant threat to mangroves, meadows, and inland valleys alike.

8.4 Discussion

8.4.1 Spatial Distribution, Conservation, and Importance of Wetlands in Benin

In Benin, wetlands have a wide and diversified geographical distribution, ranging from rivers and water bodies to surrounding ecosystems. The typology of wetlands as reported in this study aligns with the characterization of the Convention on Wetlands of International Importance RAMSAR, which classified wetlands in various groups. Indeed, the convention categorized wetlands into five groups, namely, marine zones (coastal wetlands), estuarine zones (deltas, tidal marshes, and mangroves), lacustrine zones (lakes and associated wetlands), riverine (rivers, streams, and associated wetlands), and palustrine (marshes swamps and bogs) depending on their hydrology and salt content (Ramsar 2016; Gardner and Finlayson 2018). Groups of wetlands identified in the convention perfectly concur with those reported in our typology, especially oceans, lakes and rivers, mangroves, swamp forests, marshy meadows, riparian forests, and inland valleys. According to Davidson et al. (2018), wetland distribution can be influenced by land use types. The authors highlight three main types of wetlands, including wetlands of natural environments, anthropized wetlands for rice production and reservoirs, drawing insights from land use types and wetlands typology. In addition, the Benin Republic is home to a diversity of coastal, marine, and brackish watercourses and other wetlands such as riparian forests, swamp forests, and marshy meadows, regarded as valuable wetlands owing to their contribution in sustaining the local environment and people's well-being. According to Colombani et al. (1972), the deltaic zone stretching from the Ouémé river to the Lake Nokoué and the Porto-Novo lagoon, which is an important wetland in Benin, was established to control the overflooded water of the Ouémé and Sô rivers and provide substantial livelihood to local populations. This wetland is critical in the development and implementation of several livelihood-support activities, including fishing and aquaculture. According to Attingli et al. (2016), rivers, lakes, and their associated resources in the lower Ouémé valley provide income for over 65% of active population in the area, emphasizing the need to act for their conservation for future generations. Grizzetti et al. (2019), indicated that all the contracting parties of the International Convention on Biological Diversity (CBD) are bound to promote the Sustainable Development Goals (SDGs), including SDGs 6, 14, and 16, which advocate for the importance of maintaining, conserving, and protecting valuable wetlands such as water bodies as well as coastal and marine resources.

8.4.2 Biodiversity, Uses, and Threats to Wetlands in Benin

Tropical ecosystems including wetlands are the biologically richest ecosystem on earth (Bradshaw et al. 2009; Gibson 2011) and contain up to two-third of the earth's terrestrial biodiversity (Gardner et al. 2009). They are highly productive and provide significant benefits for millions of people (FAO 2011). Wetlands mainly composed of riparian, swamp forests, and mangroves are considered important hotspots for biodiversity conservation. In tropical regions, they represent an important refuge for animal species in fire-prone landscapes (Kellman and Meave 1997). In Benin, the global forest inventory showed about 117 tree species including swamp forests, riparian forests, mangroves, and other wetlands. Plant diversity observed in Benin's wetlands confirms their capacity to act as a hotspot for biodiversity, particularly for plant communities (Kuglerová et al. 2016). Typically, riparian vegetations along streams and rivers present a unique biological community characterized by diverse plant species and vegetation structures that function as an ecotone between terrestrial and aquatic ecosystems (Maingi and Marsh 2006). A total of 77 plant species were documented in this study, of which, several have been mentioned in previous works as principal plant components of wetlands in Benin (Natta 2003). Some plant species found in Benin's wetlands are also reported for other wetlands in West Africa. For instance, *Pterocarpus santalinoides, Syzygium guineense, Dialium guineense,* and *Elaeis guineensis,* among others, reported as occurring in wetlands in Benin were also mentioned in the Comoé riparian forests in Ivory Coast (Natta 2003; Sambare et al. 2011; Tsheboeng 2016). Besides plant species, Benin's wetlands host various animal species. Mammals, reptiles, and fish species were predominantly reported in the country's wetlands. Previous studies often mentioned these ecosystems as home to entirely unique fauna communities compared to surrounding non-riparian habitats (Sabo et al. 2005). Specifically, reptiles and mammals are recognized to rely exclusively on these fragile and moisty areas (Surasinghe and Baldwin 2015).

In terms of uses, tropical forest ecosystems including wetlands are biodiversity reservoirs that play a fundamental role in meeting the basic needs of local communities (Agbo et al. 2017). Plants are the principal resources directly used for many purposes. Food uses, medicinal plants collection, and firewood collection were the principal use categories recorded in this study. In Africa, about 80% of the human population and 90% of livestock depend on traditional medicine and abundantly utilize plant parts for curing diseases (Aragaw et al. 2020). Food and medicine are the most important ecosystem services terrestrial ecosystems and wetlands provide to local people (Ouédraogo et al. 2020). Therefore, wild food species are heavily exploited by local people to improve food security and their livelihoods causing high human pressure on moisty savanna species (Ouédraogo et al. 2017). The importance given to a species does not depend on its availability but on its ability to meet the needs of populations (Lykke et al. 2004). Woody products extracted from wetlands are specifically high in mangroves. In fact, the preference of mangrove woods for local utilization is first due to the lack of other plantations to mitigate high fuelwood demand. Second, the local population prefers to collect firewood from mangrove ecosystems due to the high calorific properties, which are believed to be well appreciated by coastal dwellers (Scales and Friess 2019). The findings of this study pointed out the socioeconomic importance of mangrove ecosystems for coastal dwellers in Benin and aligned with other studies that have also identified mangroves as a major source of income for coastal communities and are consistent with studies carried out on mangrove forests of Cameroon (Feka et al. 2009; Ajonina et al. 2014; Aheto et al. 2016).

8.4.3 Dynamics of Wetlands in Benin: Current Situation and Prospects

The dynamic of wetlands in Benin over the past decade portrays the need to safeguard these vital ecosystems. Threats to wetlands as documented by this study are various and mostly man-made, although climate actions also affect wetlands in Benin. This concurs with many other authors around the world, which pointed anthropogenic pressures as the main drivers of wetlands degradation. The evolution and the dynamics of wetlands in Benin are context dependent and are highly linked to their uses, protection status, and governance (Ajonina et al. 2014). For instance, smart inland valleys in Benin expanded rapidly from 110 ha in 2012 to 474 ha in 2014, due to their significance in rice production (Arouna et al. 2017). Apart from rice production which represents one of the major staple foods in Benin, the expansion of inland valleys has been largely associated with their high fertility, moisty soils, and their interest in smallholder farmers (Azonningbo et al. 2018). Regarding mangroves, Zanvo et al. (2021) noted a loss of 62.07% of the ecosystem in Benin from 1988 to 2001 and an increase of 18.84% from 2001 to 2019. These trends support observations made by Friess et al. (2020) who indicated a global increase in mangrove cover in recent years, following conservation initiatives in place in many countries. Although there is an increasing trend in mangroves cover in Benin, the ecosystem continues to suffer severe anthropogenic threats and needs sound conservation policies to thrive (Adanguidi et al. 2020). Like mangroves and inland valleys, the other wetlands existing in Benin are also under severe threats, which highly influence their dynamics. Lakes and rivers, coastal lagoons, and other water bodies have seen the modification of their hydrological regimes as a result of human-led impacts such as dams, land encroachment, or riverbank degradation.

With regard to threats encountered by wetlands in Benin, it is important to take adequate scientific and policy actions to safeguard them. Actions needed include but are not limited to

- Conducting more research to understand the governance, threats, and ecosystem services provided by wetlands in Benin as well as their trade-offs and synergies.
- The introduction of local ecological knowledge in wetlands management in Benin.
- The mapping of the categorization of wetlands in Benin depending on their nature, functions, and location (inland or coastal).
- The empowerment of state agencies responsible for wetlands management in Benin
- The promotion of bottom-up approaches for wetlands management.
- The implementation of alternative fuelwood supply projects, particularly along the coast to lessen pressure on mangrove plant resources.

References

Achoh, M.E., Agadjihouédé, H., Gangbè, L. et al. (2021). Dynamics of Land Use in the Basin of Ancient Lagoons (Toho, Todougba, Ahouangan, Dati, Djonou) in South Benin from 1990 to 2020 and their current vulnerability to pollution. *Journal of Geoscience and Environment Protection* 9 (11): 28–45.

Adanguidi, J., Padonou, E.A., Zannou, A. et al. (2020). Fuelwood consumption and supply strategies in mangrove forests-Insights from RAMSAR sites in Benin. *Forest Policy and Economics* 116: 102–192.

Adjonou, K., Bindaoudou, I.A.K., Segla, K.N. et al. (2020). Land use/land cover patterns and challenges to sustainable management of the Mono transboundary biosphere reserve between Togo and Benin, West Africa. *International Journal of Biological and Chemical Sciences* 14 (5): 1734–1751.

Adomou, A.C., Mama, A., Missikpode, R., and Sinsin, B. (2009). Mapping and characterization of the swamp forest flora of Lokoli (Benin). *International Journal of Biological and Chemical Sciences* 3 (3): 492–503.

Agbo, I.R., Missihoun, A.A., Vihotogbe, R. et al. (2017). Impacts des usages traditionnels sur la vulnérabilité de *Detarium microcarpum* Guill. & Perr. (Caesalpiniaceae) dans ledistrict phytogéographique Zou au Bénin (en Afrique de l'Ouest). *International Journal of Biological and Chemical Sciences* 11 (2): 730–742. https://doi.org/10.4314/ijbcs.v11i2.16.[French].

Aheto, D.W., Kankam, S., Okyere, I. et al. (2016). Community-based mangrove forest management: implications for local livelihoods and coastal resource conservation along the Volta estuary catchment area of Ghana. *Ocean and Coastal Management* 127: 43–54.

Ahossou, O.D., Fandohan, B., Stiers, I. et al. (2017). Extraction of timber and non-timber products from the swamp forest of Lokoli (Benin): use patterns, harvesting impacts and management options. *International Forestry Review* 19 (2): 133–144.

Ajonina, G.N., Ago, E.E., Amoussou, G. et al. (2014). Carbon budget as a tool for assessing mangrove forests degradation in the western, coastal wetlands complex (Ramsar Site 1017) of Southern Benin, West Africa. In: *The Land/Ocean Interactions in the Coastal Zone of West and Central Africa. Estuaries of the World* (ed. S. Diop, J.P. Barusseau, and C. Descamps). Cham: Springer https://doi.org/10.1007/978-3-319-06388-1_12.

Aragaw, T.J., Afework, D.T., and Getahun, K.A. (2020). Assessment of knowledge, attitude, and utilization of traditional medicine among the communities of Debre Tabor Town, Amhara Regional State, North Central Ethiopia: a cross-sectional study. *Evidence-Based Complementary and Alternative Medicine* 2020: 6565131. https://doi.org/10.1155/2020/6565131.

Arouna, A., Akpa, A.K.A., and Adegbola, P.Y. (2017). Impact de la technologie smart-valley pour l'amenagement des basfonds sur le revenu et le rendement des petits producteurs de riz au Benin. *Cah. du CBRST* 12: 47–66.

Attingli, A.H., Vissin, E.W., Ahouansou-Montcho, S. et al. (2016). Perception endogène de l'influence des changements climatiques sur la pêche dans la basse vallée de l'Ouémé (Sud Bénin). *International Journal of Biological and Chemical Sciences* 10: 1998–2014.

Attogouinon, A., Lawin, A.E., N'Tcha M'Po, Y., and Houngue, R. (2017). Extreme precipitation indices trend assessment over the Upper Oueme River Valley-(Benin). *Hydrology* 4 (3): 36.

Azonningbo, S.H.W., Adjakpa, J.B., Dissou, F.E. et al. (2018). Specific diversity of avifauna of wetland of international importance of Southwest Benin (Ramsar site 1017). *Journal of Entomology and Zoology Studies* 6 (6): 644–654.

Badou, D.F., Kapangaziwiri, E., Diekkrüger, B. et al. (2017). Evaluation of recent hydro-climatic changes in four tributaries of the Niger River Basin (West Africa). *Hydrological Sciences Journal 62* (5): 715–728.

Biao, E.I. (2017). Assessing the impacts of climate change on river discharge dynamics in Oueme River Basin (Benin, West Africa). *Hydrology* 4 (4): 47.

Bradshaw, C. J. A., Sodhi, N. S., Brook, B. W. (2009). Tropical turmoil – a biodiversity tragedy in progress. *Frontiers in Ecology and the Environment*, 7: 79–87. https://doi.org/10.1890/070193.

Bremond, L., Favier, C., Ficetola, G.F. et al. (2017). Five thousand years of tropical lake sediment DNA records from Benin. *Quaternary Science Reviews* 170: 203–211.

Brinson, M.M. (2011). Classification of wetlands. Wetlands: Integrating Multidisciplinary Concepts, pp. 95–113.

CENAGREF/PAPE (2013). *Catégorisation des Aires Protégées de la République du Bénin suivant la nomenclature de l ' Union Mondiale pour la Conservation de la Nature (UICN).*

Ceperley, N., Montagnini, F., and Natta, A. (2010). Significance of sacred sites for riparian forest conservation in Central Benin. *Bois et forêts des tropiques* 303: 5–23.

Colombani, J., Sircoulon, J., Monod, F., and Rodier, J. (1972). Monographie du delta de l'Ouémé (tome 1). Rapport ORSTOM, Hydrol., France, Bondy, 200 p (Ronéo).

Dansi, A., Adjatin, A., Adoukonou-Sagbadja, H. et al. (2008). Traditional leafy vegetables and their use in the Benin Republic. *Genetic Resources and Crop Evolution* 55: 1239–1256.

Danvi, A., Giertz, S., Zwart, S.J., and Diekkrüger, B. (2017). Comparing water quantity and quality in three inland valley watersheds with different levels of agricultural development in central Benin. *Agricultural Water Management* 192: 257–270.

Davidson, N.C., Fluet-Chouinard, E., and Finlayson, C.M. (2018). Global extent and distribution of wetlands: trends and issues. *Marine and Freshwater Research* 69: 620–627.

Deguenon, S.D.D.M., Hounmenou, C.G., Adade, R. et al. (2023). Simulation of the impacts of sea-level rise on coastal ecosystems in Benin using a combined approach of machine learning and the sea level affecting marshes model. *Sustainability* 15 (22): 16001.

Djagba, J.F., Zwart, S.J., Houssou, C.S. et al. (2019). Ecological sustainability and environmental risks of agricultural intensification in inland valleys in Benin. *Environment, Development and Sustainability* 21: 1869–1890.

Djihouessi, M.B. and Aina, M.P. (2018). A review of hydrodynamics and water quality of Lake Nokoué: current state of knowledge and prospects for further research. *Regional Studies in Marine Science* 18: 57–67.

Dossou, M.E., Houessou, G.L., Lougbégnon, O.T. et al. (2012). Ethnobotanical study of ligneous species in Agonvè swampy forest and surrounding area in Benin. *Tropicultura* 30 (1): 41–48.

Etienne, D.M., Toussaint, L.O., Laurent, H.G. et al. (2012). Phytoecological assessment of plant communities in the swampy forest of Agonvè and related ecosystems, Benin. *Journal of Applied Biosciences* 53: 3821–3830.

FAO (2011). *The State of the World's Forest 2011*. Rome: Food and Agriculture Organisation of the United Nations.

Feka, N.Z., Chuyong, G.B., and Ajonina, G.N. (2009). Sustainable utilization of mangroves using improved fish-smoking systems: a management perspective from the Douala-Edea wildlife reserve, Cameroon. *Tropical Conservation Science* 2 (4): 450–468.

Friess, D.A., Yando, E.S., Abuchahla, G.M. et al. (2020). Mangroves give cause for conservation optimism, for now. *Current Biology* 30 (4): R153–R154.

Gardner, R.C. and Finlayson, C. (2018). Global wetland outlook: state of the world's wetlands and their services to people. In: *Ramsar Convention Secretariat. 2018, Stetson University College of Law Research Paper No. 2020-5*. Available at SSRN: https://ssrn.com/abstract=3261606.

Gardner, T.A., Barlow, J., Chazdon, R. et al. (2009). Prospects for tropical forest biodiversity in a human-modified world. *Ecology Letters* 12: 561–582. https://doi.org/10.1111/j.1461-0248.2009.01294.x.

Gibson, L. (2011). Primary forests are irreplaceable for sustaining tropical biodiversity. *Nature* 487: 378–381. https://doi.org/10.1038/nature10425.

Giertz, S., Steup, G., and Schönbrodt, S. (2012). Use and constraints on the use of inland valley ecosystems in central Benin: results from an inland valley survey. *Erdkunde* 66: 239–253.

Gnansounou, S.C., Toyi, M., Salako, K.V. et al. (2021). Local uses of mangroves and perceived impacts of their degradation in Grand-Popo municipality, a hotspot of mangroves in Benin, West Africa. *Trees, For People* 4: 100080.

Gnansounou, S.C., Salako, K.V., Sagoe, A.A. et al. (2022). Mangrove ecosystem services, associated threats and implications for wellbeing in the Mono Transboundary Biosphere Reserve (Togo-Benin), West Africa. *Sustainability* 14: 2438.

Gnansounou, S.C., Salako, K.V., Visée, C. et al. (2024). The role of local deities and traditional beliefs in promoting the sustainable use of mangrove ecosystems. *Forest Policy and Economics* 160: 103145.

Grizzetti, B., Liquete, C., Pistocchi, A. et al. (2019). Relationship between ecological condition and ecosystem services in European rivers, lakes and coastal waters. *Science of the Total Environment* 671: 452–465.

Houessou, L.G., Teka, O., Imorou, I.T. et al. (2013). Land use and land-cover change at "W" Biosphere Reserve and its surroundings areas in Benin Republic (West Africa). *Environment and Natural Resources Research* 3 (2): 87.

Houssou, A., Ahouansou Montcho, S., Montchowui, E., and Bonou, C. (2017). Spatial and seasonal characterization of water quality in the Ouémé River Basin (Republic of Benin, West Africa). *Egyptian Journal of Chemistry* 60 (6): 1077–1090.

Hu, S., Niu, Z., Chen, Y. et al. (2017). Global wetlands: potential distribution, wetland loss, and status. *Science of the Total Environment* 586: 319–327.

Jiang, T.T., Pan, J.F., Pu, X.M. et al. (2015). Current status of coastal wetlands in China: degradation, restoration, and future management. *Estuarine, Coastal and Shelf Science* 164: 265–275.

Kellman, M. and Meave, J. (1997). Fire in the tropical gallery forests of Belize. *Journal of Biogeography* 24 (1): 23–34.

Khan, A.A. and Arshad, S.A.N.A. (2014). Wetlands of Pakistan: distribution, degradation and management. *Pakistan Geographical Review* 69 (1): 28–45.

Kuglerová, L., Dynesius, M., Laudon, H., and Jansson, R. (2016). Relationships plant assemblages and water flow across a boreal forest land-scape: a comparison of liverworts, mosses, and vascular plants. *Ecosystems* 19: 170–184.

Lougbegnon, T., Houessou, G.L., Akplo, T.M. et al. (2023). Actualisation de plan d' Aménagement et de Gestion de la Réserve de Biosphère de la Basse Vallée de l ' Ouémé Rapport Provisoire Novembre 2023.

Lykke, A., Kristensen, M., and Ganaba, S. (2004). Valuation of local use and dynamics of 56 woody species in the Sahel. *Biodiversity and Conservation* 13: 1961–1990.

Maingi, J.K. and Marsh, S.E. (2006). Composition, structure, and regeneration patterns in a gallery forest along the Tana River near Bura Kenya. *Forest Ecology and Management* 236 (2–3): 211–228. https://doi.org/10.1016/j.foreco.2006.09.006.

Meng, W., He, M., Hu, B. et al. (2017). Status of wetlands in China: a review of extent, degradation, issues and recommendations for improvement. *Ocean and Coastal Management* 146: 50–59.

Mensah, J.H.R., dos Santos, I.F.S., and Tiago Filho, G.L. (2023). A critical analysis of the energy situation in the Benin Republic and its evolution over the last decade. *Renewable Energy* 202: 634–650.

Natta, A.K. (2003). Ecological assessment of riparian forests in Benin: phytodiversity, phytoscoiology, and special distribution of tree species. PhD. thesis. Wageningen University, 215 p.

Natta, A.K., Yedomonhan, H., Adomou, A.C. et al. (2012). Structure, ecological spectra and species dominance in riparian forests from Benin (West Africa). *International Journal of Biological and Chemical Sciences* 6 (1): 37–50.

Neuenschwander and Toko 2011. Benin, its natural environment, and socioeconomic data (Chapter 2), In Neuenschwander P, Sinsin B, Goergen G (2011) Nature Conservation in West Africa: Red List for Benin. Ibadan: IITA

Ogou, F.K., Batebana, K., Ogwang, B.A. et al. (2016). Investigation of the influence of Atlantic Ocean on rainfall variability over Benin Republic, West Africa. *Ethiopian Journal of Environmental Studies and Management* 9 (1): 70–79.

Okpeitcha, O.V., Chaigneau, A., Morel, Y. et al. (2022). Seasonal and interannual variability of salinity in a large West-African lagoon (Nokoué Lagoon, Benin). *Estuarine, Coastal and Shelf Science* 264: 107689.

Olodo, I.Y., Cocquyt, C., Abou, Y., and Kokou, K. (2020). Seasonal variations and distribution of diatom flora of Lake Ahémé (Benin, West Africa). *Botany Letters* 167 (1): 160–173.

Ouédraogo, P., Bationo, B.A., Sanou, J. et al. (2017). Uses and vulnerability of ligneous species exploited by local population of northern Burkina Faso in their adaptation strategies to changing environments. *Agriculture & Food Security* 6 (15): 1–16. https://doi.org/10.1186/s40066-017-0090z.

Ouédraogo, I., Sambaré, O., Savadogo, S., and Thiombiano, A. (2020). Perceptions locales des services écosystémiques des aires protégées à l'Est du Burkina Faso. *Ethnobotany Research and Applications* 20 (13): 1–18. https://doi.org/10.32859/era.20.13.1-18.[French].

Padonou, E.A., Gbaï, N.I., Kolawolé, M.A. et al. (2021). How far are mangrove ecosystems in Benin (West Africa) conserved by the Ramsar Convention? *Land Use Policy* 108: 105583.

Queffelec, B., Bonnin, M., Ferreira, B. et al. (2021). Marine spatial planning and the risk of ocean grabbing in the tropical Atlantic. *ICES Journal of Marine Science* 78 (4): 1196–1208.

Ramsar (2016). Introduction à la Convention Sur les Zones Humides, Manuels Ramsar 5ª Édition.

Richard, A., Galle, S., Descloitres, M. et al. (2013). Interplay of riparian forest and groundwater in the hillslope hydrology of Sudanian West Africa (northern Benin). *Hydrology and Earth System Sciences* 17 (12): 5079–5096.

Sabo, J.L., Sponseller, R., Dixon, M. et al. (2005). Riparian zones increase regional species richness by harboring different, not more, species. *Ecology* 86: 56–62.

Sakané, N., Alvarez, M., Becker, M. et al. (2011). Classification, characterisation, and use of small wetlands in East Africa. *Wetlands* 31: 1103–1116.

Sambare, O., Bognounou, F., Wittig, R., and Thiombiano, A. (2011). Woody species composition, diversity and structure of riparian forests of four watercourses types in Burkina Faso. *Journal of Forest Research* 22: 145–158.

Scales, I. and Friess, D. (2019). Patterns of mangrove forest disturbance and biomass removal due to small-scale harvesting in southwestern Madagascar. *Wetlands Ecology and Management* 27: 609–625. https://doi.org/10.1007/s11273-019-09680-5.

Sharma, S., Phartiyal, M., Madhav, S., and Singh, P. (2021). Global wetlands: categorization, distribution and global scenario. In S. Sharma & P. Singh (Eds.), *Global wetlands*. Wiley. https://doi.org/10.1002/9781119692621.ch1 (Chapter 1)

Sinsin, C.B.L., Salako, K.V., Fandohan, A.B. et al. (2022). Potential climate change induced modifications in mangrove ecosystems: a case study in Benin, West Africa. *Environment, Development and Sustainability* 24: 4901–4917.

Surasinghe, T.D. and Baldwin, R.F. (2015). Importance of riparian forest buffers in conservation of stream biodiversity: responses to land uses by stream-associated salamanders across two southeastern temperate ecoregions. *Journal of Herpetology* 49 (1): 83–94.

Teka, O., Houessou, L.G., Djossa, B.A. et al. (2019). Mangroves in Benin, West Africa: threats, uses and conservation opportunities. *Environment, Development and Sustainability* 21: 1153–1169.

Tsheboeng, G. (2016). Population structure, species composition and distribution of Riparian Woody Plant Communities in the Okavango Delta, Botswana. PhD thesis. Univ. of Botswana, Botswana.

Wang, H., Zhou, Y., Wu, J. et al. (2023). Human activities dominate a staged degradation pattern of coastal tidal wetlands in Jiangsu province, China. *Ecological Indicators* 154: 110579.

Yantikoua, T.S., Kaki, C., Djara, M.B., and d'Ameida, G.A.A.F. (2023). Assessment of the vulnerability of the southwestern coast of Benin to the risk of coastal erosion and flooding. *Journal of Water Resource and Protection* 15 (1): 1–22.

Zanvo, M.S., Salako, K.V., Gnanglè, C. et al. (2021). Impacts of harvesting intensity on tree taxonomic diversity, structural diversity, population structure, and stability in a West African mangrove forest. *Wetlands Ecology and Management* 29: 433–450.

9

Wetlands of Burkina Faso: Biodiversity, Livelihoods, and Conservation

Kaboré Idrissa[1], Bancé Victor[1,2], Konaté Sidiki Roland[1], Sawadogo Yabyouré Marc-Florent[1], and Ouéda Adama[1]

[1] *Laboratoire de Biologie et Ecologie animales (LBEA), UFR/SVT, Université Joseph KI-ZERBO (Burkina Faso), Ouagadougou, Burkina Faso*
[2] *Centre Universitaire de Manga, Université Norbert ZONGO, Koudougou, Burkina Faso*

9.1 Introduction

Wetlands are important complex ecosystems including forest and watersheds with many valuable functions and services for human well-being (Fonseca et al. 2018; Kaboré et al. 2022). In tropical areas, wetlands, particularly inland ones, constitute an important natural resource with high ecological and socioeconomic values (Schuyt and Brander 2004; Lannas and Turpie 2009). In addition, many wetlands are the hotspots of a high diversity of birds, mammals, fish, amphibians, macroinvertebrates, other microorganisms and flora (Adamus et al. 2001; Fonseca et al. 2018; Ola and Benjamin 2019; Kaboré et al. 2022).

However, wetlands for a long time have been considered a driver of agricultural and rural development for a growing economy driven by a burgeoning human population argued by authors (Kaboré et al. 2022; Chakraborty et al. 2023; van Dam et al. 2023). As a result, wetland ecosystems are threatened by multiple human pressures and climate changes, making them the most vulnerable ecosystems. Indeed, wetlands are facing a number of challenges such as environmental degradation, biodiversity erosion, siltation, water systems engineering, desertification, and pollutions. The subtropical wetlands are no exception, because of a lack of understanding of their ecological and socioeconomic importance (Mitchell 2013; Ayanlade and Proske 2016; Adeeyo et al. 2022; Kaboré et al. 2022).

In West Africa, wetlands degradation from intense urbanization, mining, and agriculture, as well as climate changes, are leading to the decrease in wetland areas, which are crucial to the livelihood of its citizens (Gopal 2013; Kaboré et al. 2022; Dube et al. 2023). Many West Africans rely much more on resources from wetlands to support their livelihoods, but these wetlands areas are likely to decrease due to overexploitation of biological resources, rainfall reduction, and/or increase of evaporation, as well as the disrupt of rivers and floodplains. Many authors have demonstrated that research and inventory studies on wetlands are mostly fragmented (Kansiime et al. 2007; Landmann et al. 2010; Ola and Benjamin 2019; Kaboré et al. 2022).

In Burkina Faso, wetlands are either natural or artificial, and the latter serve as reservoirs to collect rainwater (Kaboré et al. 2024). These wetlands are endangered ecosystems because of ongoing

Wetlands of Tropical and Subtropical Asia and Africa: Biodiversity, Livelihoods and Conservation, First Edition. Edited by Thammineni Pullaiah.

human pressures. Indeed, most forests are cleared for agriculture, settlements, and other human activities using pesticides and fertilizer. These pressures pose a potential threat to wetlands' quality including overall ecosystems functioning (Kaboré et al. 2016a, 2022). Therefore, the management of wetlands is highly challenging in Burkina Faso because, when the interaction of wetland plants and animals with their environment is disrupted, many of the functions provided by wetlands will be altered or lost (Kaboré et al. 2022).

The present chapter describes the biodiversity, the livelihoods, and conservation measures of wetlands in Burkina Faso, and we discussed each good and service of wetlands independently, and finally we address long-term sustainability of wetland.

9.2 Burkina Faso and Its Wetlands

Burkina Faso is a Sub-Sahelian country located in the heart of West Africa, between 9°20′ and 15° north latitudes and 5°30′ west longitude and 2°30′ east longitude. It has an area of 274,200 km^2 and an estimated population of more than 20 million (INSD and ICF 2022). While it is landlocked, the country remains closely connected to the Gulf of Guinea via the large Volta River. It shares borders with six countries: Mali in the north and northwest, Cote d'Ivoire, Ghana and Togo in the south, Benin in the southeast, and Niger in the east and northeast (Figure 9.1).

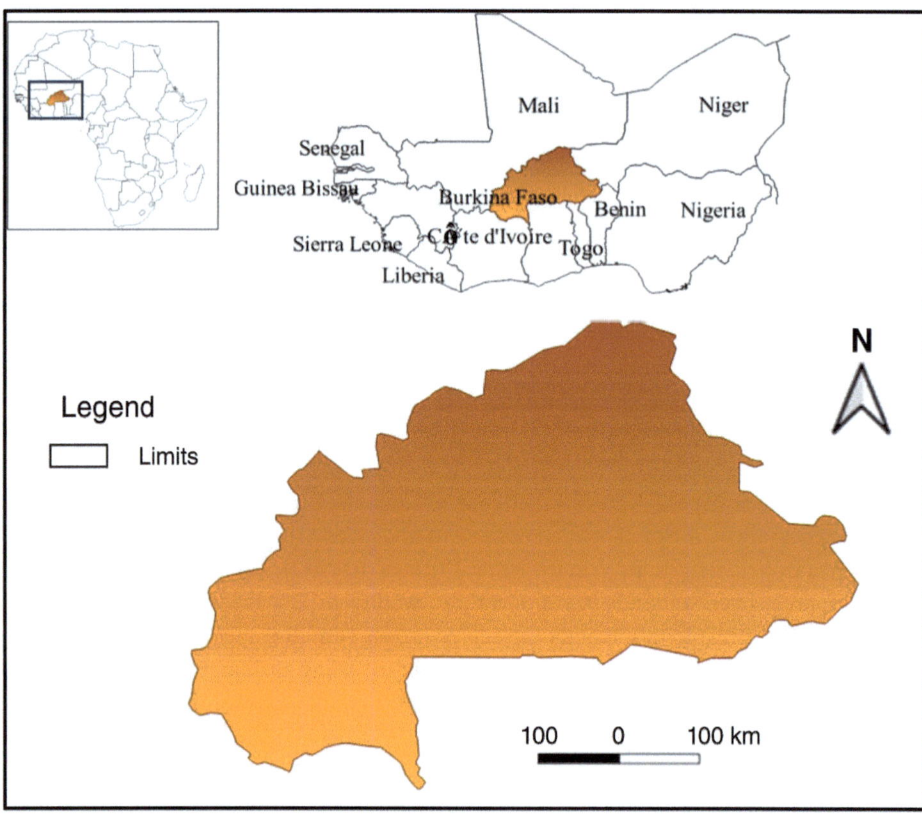

Figure 9.1 Burkina Faso location in Africa and its border.

Burkina Faso has a Sudano-Sahelian climate characterized by two seasons (dry season and rainy season). The rhythm of the seasons is determined by the Intertropical Front (ITF). It is characterized by three main bioclimatic zones; the Sahelian sector above parallel 14°N covers about 25% of the territory with an average annual rainfall between 300 and 600. The vegetation is shrubby and wooded grassland or steppe dominated by the genus *Acacia* (Nacoulma et al. 2018). The Sudano-Sahelian sector located between the parallels 11°30′ and 14°N covers about 50% of the territory with rainfall between 600 and 900 mm; the Sudanian sector located in the south of 11°30′N parallel in the southwest covers about 25% of the territory with rainfall ranging between 900 and 1200 mm. Gallery forests and savanna with trees or shrubs are mainly observed in this area (Gnoumou et al. 2011; Schmidt et al. 2013). Following the bioclimatic zones, the vegetation dry forests, deciduous forests, woody, arboreous, and shrubby savanna and steppes in the Sahelian area (Figure 9.2).

9.2.1 Wetlands in Burkina Faso

Wetlands are considered one of the most productive and diverse ecosystems. In Burkina Faso, four main watersheds with their flooded areas are encountered. The Volta catchment covers 63.1% of the nation's area and provides the majority of the surface and groundwater in Burkina Faso (see also Belemtougri et al. 2021, for more details); the Nakanbé (formerly White Volta catchment) in the central part of Burkina Faso (area of 70,000 km^2) and the Mouhoun (formerly Black Volta catchment) in the west (92,000 km^2). The Niger Basin (83,442 km^2) is drained by the tributaries of the Niger river (Banfing, Beli, Sirba, Gourouol, Gorondi and Tapoa, and Diamango). The Comoé Basin (17,590 km^2) is drained mainly by Comoé and its main tributaries (Leraba, Kodoum, Baoue, and Irngou). The hydrology in Burkina Faso is highly sensitive to and dependent on seasonal rainfall. The precipitation and evaporation are the main drivers determining the volume and dynamics of surface water and the replenishing of groundwater aquifers (Koussoube et al. 2018).

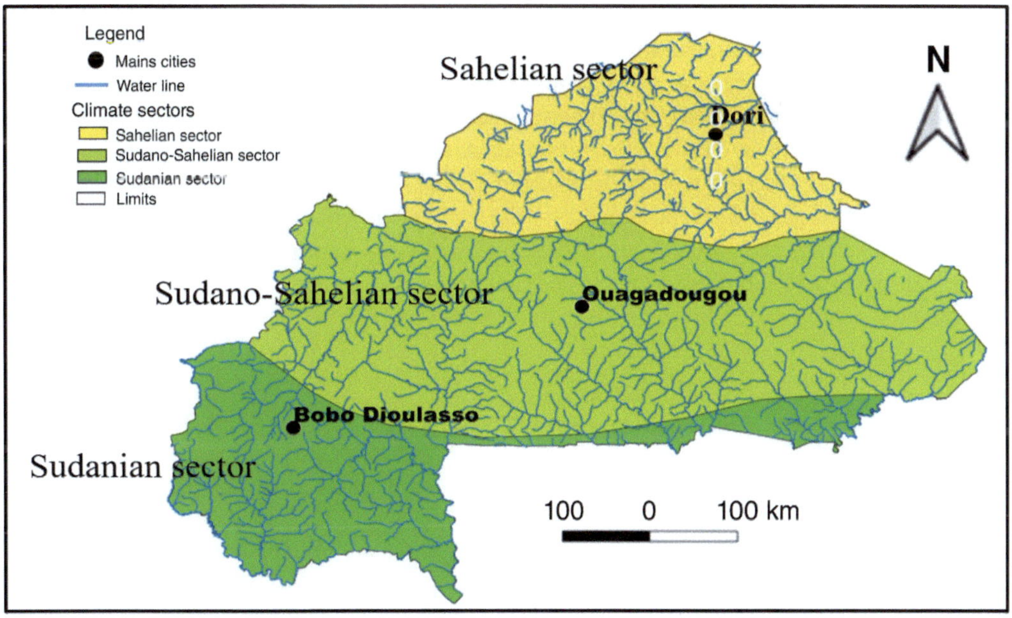

Figure 9.2 Wetlands climate sectors in Burkina Faso.

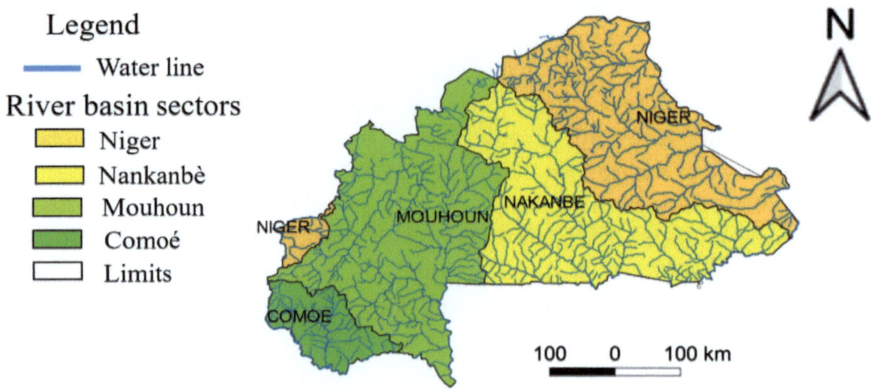

Figure 9.3 Main rivers basins and their sectors encountered in Burkina Faso.

Some rivers create a network that collects and distributes water flowing to neighboring countries: Côte d'Ivoire, Ghana, Mali, Benin, Togo, and Niger. Only the Mouhoun and Comoé are permanent watercourses with many smaller streams drying out in the dry season (Figure 9.3).

Lentic ecosystems in Burkina Faso exist as natural and artificial bodies of water either isolated in uplands or in a matrix of different wetland types or areas of development with competing interests (Melcher et al. 2020). Natural bodies of water include small natural pools on "bowe" armored plateaus, mares, natural ponds or swamps, natural lakes, such as Lake Bam or Lake Tengrela, natural ponds fed by endorheic basins and marshlands and lowlands constituting submerged wetlands in running waters: Kou and Voun Hou flood plains in the Mouhoun basin; Niena-Dionkele and Foulasso, Lelasso in the Bafing basin; plains of Kompienga, Mouhoun, and Sourou (Oueda et al. 2023).

The artificial bodies of water include boulis, excavated waterholes used for storage of water from the rainy season or filled by seasonal water streams. Dykes were put in place to impound the water (Figure 9.4). As a response to severe drought, more than 1680 reservoirs were built across Burkina Faso (Caballer Revenga 2019) making Burkina Faso the country with the highest density of reservoirs in West Africa (Cecchi et al. 2008). These range from $0.01\,km^2$ to over $10\,km^2$ and are scattered across the national territory and climatic zones (Figure 9.4). Artificial wetlands (reservoirs, dams, irrigation culverts and canals, fish farms, aquaculture ponds, and rice fields) are known to contribute significantly to the Burkina Faso wetland area, and they often provide important habitats for flora and fauna, and several benefits for the well-being of local population.

In Burkina Faso, several wetlands were identified, including Ramsar sites, biosphere, and protected sites that provide humankind with goods and services since early times (see some details in Table 9.1).

9.3 Biodiversity in Wetlands of Burkina Faso

9.3.1 Fish and Amphibians

Most of fish data from wetlands were sampled using gill nets, cast nets, long lines, and electric fishing gears, and fish were identified to species level by using morphological characters (Paugy et al. 2003), and then specimens were counted (Figures 9.5 and 9.6). Fish species occurrences in wetlands have shown that 79 fish species belonging to 41 genera grouped in 19 families were

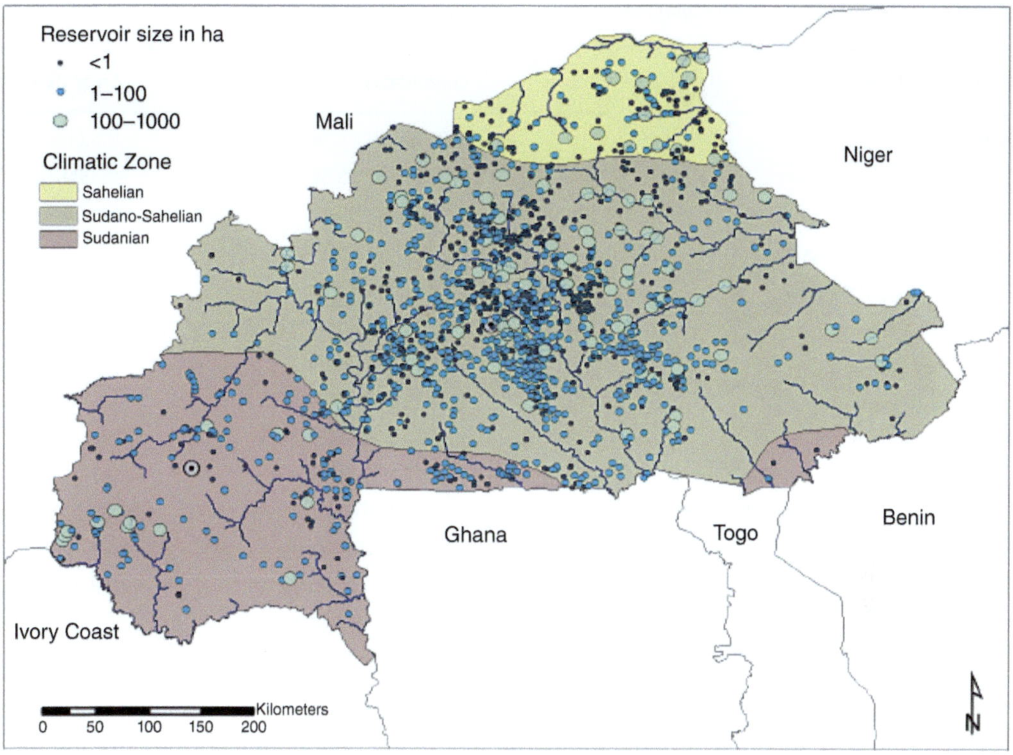

Figure 9.4 Distribution of artificial wetland in Burkina Faso. *Source:* Adapted from Caballer-Revenqa (2019).

identified (Mano et al. 2020; see details of some species in Figure 9.7). Other authors have shown that Cyprinidae, Alestidae, Mormyridae, and Mochokidae were the most represented families in terms of species and individual number. *Oreochromis niloticus*, *Coptodon zillii*, *Brycinus nurse*, *Clarias anguillaris, Sarotherodon galilaeus*, and *Enteromius macrops* were the constant species (Melcher et al. 2018). Roman (1966) has previously listed 121 species in the Volta catchment, but Fish base has mentioned that more than 123 fish species exist in Burkina Faso. So, the number of species recorded in wetlands is still unclear, because Silga (2021) have compiled the results of previous studies and has listed 128 species. This could be explained by the fact that wetland plays a crucial role in preserving fish diversity, as demonstrated by authors (Melcher et al. 2012, 2020; Mano et al. 2020), but in the recent studies, some authors have demonstrated that wetlands in Burkina Faso are extremely overexploited and impaired by many human activities leading to the decline of wetlands services and goods (Kaboré et al. 2022, see Figure 9.6). The overall wetlands biodiversity degradation, particularly fish species in Burkina Faso, was highlighted by many authors (Ouédraogo 2010; Mano 2016; Meulebroek et al. 2019; Silga 2021; Kaboré et al. 2022). In rural areas, the local population depends on wetlands resources such as fish for their daily livelihood and incomes; therefore, wetland ecosystem preservation and restoration remain major challenges for suitable socioeconomic development of local community of Burkina Faso.

Field amphibian data were gathered using visual search methods, acoustic signals, and dip netting for tadpoles (see Ayoro et al. 2020), and literature research. The amphibians constitute an important component fauna in wetland ecosystems (Ayoro et al. 2020). Because wetland amphibians are diverse

Table 9.1 List of some important wetlands among other reported in Burkina Faso.

Name of wetlands	Coordinates	Areas (ha)
Béli river	14°58′N 00°25′W	105,000
Lake Oursi–Lake Darkoye	14°42′N 00°05′W	45,000
Sourou valley	12°50′N 03°28′W	22,000
Parc Urbain Bangr-Wéoogo	12°22′N 01°30′W	230
Mare aux Hippopotames	1°33′N 04°09′W	19,200
Bagré reservoir	11°34′57″N 00°41′13″W	36,793
Lake Kompienga	11°11′N 00°36′59″E	48,000
Samandéni reservoir	11°23′12″N 04°42′58″W	68,202
Nakanbé-Mané Basin	12°58′12″N 01°25′46″W	19,477
Complex Protected areas Pô-Nazinga-Sissili	11°19′57″N 01°22′37″W	301,972.6
Corridor forestier de la Boucle du Mouhoun	12°10′09″N 02°56′07″W	134,553
Léra gellery forest	10°35′52″N 05°18′19″W	542
Lake Bam	13°24′06″N 01°31′08″W	5,300.4
Lake Dem	13°12′01″N 01°08′27″W	1,354
Lake de Tingrela	10°38′38″N 04°50′20″W	580
Lake Higa	13°36′50″N 00°43′23″E	1,514
Arli–W–Singou complex	11°33′N 01°27′E	923,000
Diéfoula–Logoniégué forest	09°45′N 04°42′W	114,000
Bérégadougou hill	10°45′N 04°43′W	15,000

Source: Adapted from Ramsar list (2023).

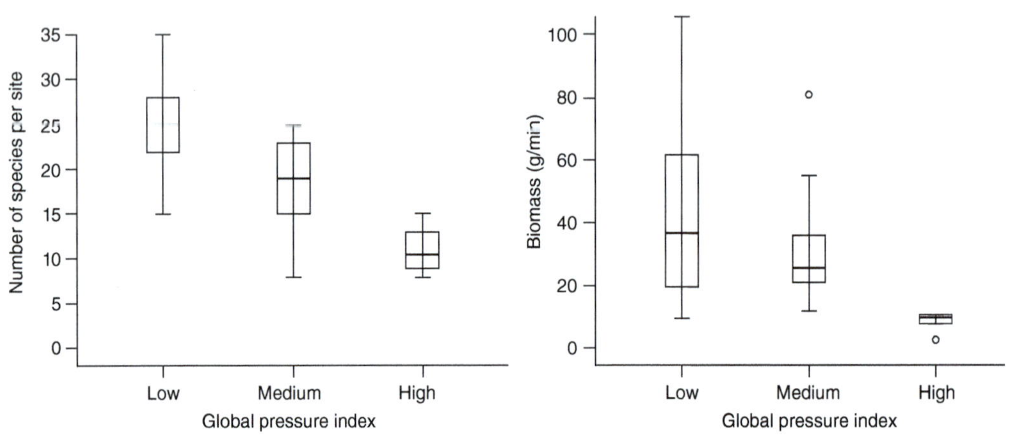

Figure 9.5 Response of fish species and biomass to different levels of human pressures.
Source: Meulebroek et al. (2019)/MDPI/CC BY 4.0.

and reflect habitat conditions, they are used as tools in wetlands bioassessment (Kaboré et al. 2022). Several amphibian species were reported in wetlands in Burkina Faso. According to Ayoro et al. (2020), 36 amphibian species belonging to 11 families were identified. Among others, the families of Arthroleptidae, Bufonidae, Dicroglossidae, Hemisotida, Hyperoliidae, Microhylidae, Phrynobatrachidae,

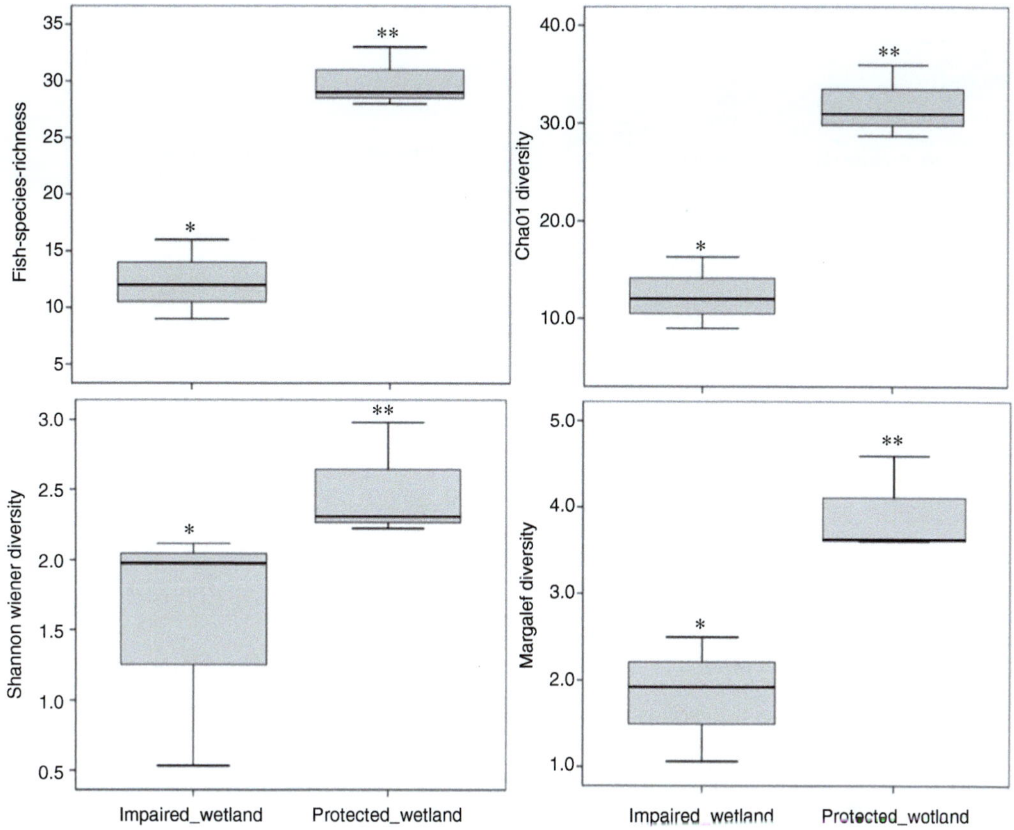

Figure 9.6 Box-plot comparing fish diversity between impaired wetlands and protected wetlands. (*) indicates statistical significance (*p*-value <0.05). *Source:* Kaboré et al. (2022)/Frontiers Media S.A./CC BY 4.0.

Pipidae, Ptychadenidae, Pyxicephalidae, and Ranidae including the genus of *Arthroleptis*, *Leptopelis*, *Sclerophrys*, *Hoplobatrachus*, *Hemisus*, *Afrixalus*, *Kassina*, *Phrynomantis*, *Hildebrandtia*, *Phrynobatrachus*, *Xenopus*, *Ptychadena*, *Pyxicephalus*, *Tomopterna*, *Sclerophrys*, and *Amnirana* have been reported in diverse wetland habitats, reflecting the high diversity of amphibians in study area (see some species in Figure 9.8). But these species are threatened by human pressures such as habitat fragmentation, introduction of predators, and the spread of toxic substances as well as climate warming (Ayoro 2021; Kaboré et al. 2022).

9.3.2 Birds and Mammals

Most of bird surveys were conducted based on local managers and experts' designs (Thiollay 2006; Daboné et al. 2016). To do so, the roads and small trails are divided into separate transects. All birds seen were counted within 100 m, and the raptors within 1 km. On either side of the trails, visual assessment and auditory cues were considered to record each species. To capture peak activity, allowing for an accurate count of species, the bird samples were taken between 9 a.m. and 14 p.m. All birds were identified with binoculars (Opticron 8×30, Super Zenith 8×40) by experts and using the field standard guide (Balança et al. 2007). Because of habitats heterogeneity, wetlands

Figure 9.7 Some species of fish recorded in wetlands. *Source:* Susfish Consortium (2015).

Figure 9.8 Some species of amphibians recorded in wetlands in Burkina Faso. *Source:* From Ayoro (2021).

provide more niches and nest site conditions for birds (Kaboré et al. 2022). Most of wetland habitats are diverse and include shrubs, trees, and grass cover, making them the hotspot of biodiversity, particularly for birds. According to Lungren et al. (2001), 455 of species birds are recorded in Burkina, whereas 318 are thought to be residents and 119 are regular seasonal migrants. Most of these species belong to families of Accipitridae, Burotidae, Musophagidae, Maloconotidae, Coraciidae, Ardeidae, Estrildidae, Scopidae, Passeridae, and Columbidae (see some species in Figure 9.9). These species include raptors and non-raptors, as well as ground birds, arboreal birds, and grass birds which are associated with habitats (Thiollay 2006; Kaboré et al. 2022). Two species: *Falco naumanni* (VU) and *Gallinago media* (NT) were classified as global conservation concern in

Figure 9.9 Some species of birds reported in Wetlands. *Source:* Photo by Sawadogo Y.M Floren (chapter author).

Burkina IUCN due to human pressures. From the literature review, most of the birds' species were reported in wetlands, but Thiollay (2006) has demonstrated that human pressures such as hunting, bush fires, timber and firewood harvesting, and agriculture with heavy pesticides spraying may affect bird species in wetlands that cause the decline or extinction or of birds' species in West Africa, particularly in Burkina Faso.

Mammal data were collected following the transect methods described by Cornélis (2007), Bouché et al. (2016), Hema et al. (2017), and literature reviews. Several mammal species are found in Burkina Faso, and some samples of species are listed in Table 9.2. Most of the species, including *Papio, Alcelaphus, Syncerus, Sylvicapra, ellipsiprymnus, Loxodonta, Acinomyx, Tragelaphus, Hippotragus, Ourebia, Phacochoerus, Kobus*, and *Redunca*, are encountered in the protected areas drained by the national rivers such as Nakanbé, Mouhoun, Nazino, and Comoé-léraba in Burkina Faso.

9.3.3 Reptiles

Herpetological surveys were undertaken in Burkina Faso by several authors (Chirio 2009; Petrozzi et al. 2017; Böhme and Heath 2018; Adje 2023). Most of reptiles were collected in wetlands. In 1966, Roman has listed 54 species of snakes in Burkina Faso, but in the recent studies 46 and 29

Table 9.2 List of some mammal species recorded in Burkina Faso.

Nom	Nom Scientifique
Babouins	*Papio anubis*
Bubale	*Alcelaphus buselaphus*
Buffle	*Syncerus caffer caffer*
Céphalophe de grimm	*Sylvicapra grimmia*
Cob de buffon	*Kobus kob bob*
Cob defassa	*Kobus ellipsiprymnus defassa*
Eléphant	*Loxodaonta africana africana*
Guib harnaché	*Tragelaphus scriptus*
Ourébi	*Ourebia ourebi*
Phacochère	*Phacochoerus africanus*
Singe vert	*Cercopithecus aethiops sabaceus*
Patas	*Cercopithecus erythrocebus patas*
Hippotrague	*Hippotracus equinus*
Redunca	*Redunca redunca*
Oryctérope	*Orycteropus afer*
Porc-épic	*Hystrix cristata*
Civette	*Viverra civetta*
Genette	*Genetta genetta*
Hyène tachetée	*Crocuta crocuta*
Hyène rayée	*Hyaena hyaena*
Chacal	*Canis adustus*
Ratel	*Mellivora capensis*

species of Ophidian snakes were, respectively, reported by Chirio (2009) and Adje (2023). Many genera among others such as *Rhamphiophis*, *Boaedon*, *Echis*, *Python*, *Naja*, *Dromophis*, *Crotaphopeltis*, *Afronatrix*, *Dasypeltis*, *Prosymna*, *Atractaspis*, *Lycophidium*, *Eryx*, *Amblyodipsas*, *Bitis*, *Mehelya*, *Telescopus*, *Philothamnus*, *Elapsoidea*, *Causus*, *Leptotyphlops*, *Rhamphiophis*, and *Typhlops* were encountered in Burkina Faso. For the tortoises, fives species including the genus of *Pelomedusa*, *Pelusios*, *Cyclanorbis*, *Centrochelys*, and *Kinixys* were identified in Burkina Faso (Chirio 2009; Trape et al. 2012; Luiselli et al. 2016; Lankoandé et al. 2020). In terms of crocodile diversity, three species were found belonging to one *Crocodylus* (Chirio 2009; Ouédraogo et al. 2020a, b). More than 16 Gekkos were encountered by Chirio (2009), Trape et al. (2012), and Dayama (2020). This means that the wetlands harbor a high diversity of reptiles, but extensive studies are still needed to build an exhaustive list of reptiles in wetlands in Burkina Faso (see some pictures in Figures 9.10 and 9.11).

9.3.4 Macro-Invertebrates

The macro-invertebrates were sampled following the multihabitat sampling approach from Moog (2007) and described in Kaboré et al. (2016b). The wetlands are the home of highly diversified invertebrate fauna. 219 species of insects and more than forty (40) species of Gastropoda and Bivalvia were recorded (Kaboré 2016; Ouédraogo 2018; Melcher et al. 2020; Oueda et al. 2023). Zooplankton including Rotifera, Copepoda, and Cladorcera and annelids composed of Oligochaeta, Hirudinea, Hydrozoa, and crustaceans belonging to Decapoda and Ostracoda are also reported in different habitats of wetlands. For the insects, 08 orders (Diptera, Hemiptera, Coleoptera, Odonata, Ephemeroptera, Trichoptera, Lepidoptera, Plecoptera, and Orthroptera) and more than 60 families were found in Wetlands. For the specific research group, seven (07) species (*Macrobrachium dux* Lenz, 1910; *Macrobrachium macrobrachion* Herklots, 1851; *Macrobrachium niloticum* P. Roux, 1823; *Macrobrachium felicium* Holthuis, 1949; *Macrobrachium thysi* (Powell 1980); *Macrobrachium raridens* Hilgendorf 1893; and *Macrobrachium* sp.) of shrimp were identified (Kaboré et al. 2023). In regards to crabs, according to Dohinnon (2023), one family of Potamidae belongs to three species

Figure 9.10 Kobus population watering. *Source:* Photo by Konaté (chapter author).

Figure 9.11 Some reptile species reported in Burkina Faso.

(*Potamonautes ecorssei* Marchand, 1902; *Potamonautes luoboensis* Rathbun, 1904; and *Geotelphusa schubotzi* Balss, 1914). Kaboré et al. (2016c) took a closer look at the diversity of the aquatic beetle families Dytiscidae, Noteridae, and Hydrophilidae (Oueda et al. 2023). In this study, a total of 60 species of water beetles was found in various habitats, most belonging to the families Dytiscidae (27), Hydrophilidae (22), and Noteridae (11), and 23 species were encountered for the first time in Burkina Faso. Some details of taxa reported in Burkina Faso wetlands are given in the Figure 9.12.

9.4 Wetland Goods and Services

An ethnoecological survey was carried (see details in Nakéna 2024). Many authors argued that wetlands provide several services including crops, livestock, fish, wild foods (nontimber forest products), fuelwood, and provision of water to humankind as demonstrated by many authors (De Groot et al. 2018; Alikhani et al. 2021; Díaz-Pinzón et al. 2022; Zekarias and Gelaw 2023). From our survey, several categories of services such as tourism (cultural services), groundwater recharge and discharge (regulating services), and biodiversity and habitat (supporting services) were listed by interviewers. For example, only the urban wetland Bangr-Wéoogo in Burkina Faso receives about 250,000 visitors per year, including researchers, students, schoolchildren, and tourists (Thiombiano and Kampmann 2010; Nakéna 2024). From our field observation, we found that grasses composed of short annuals were cut to feed animals. Others many studies have demonstrated that Bangr-Wéoogo located in the center of Ouagadougou contributes substantially to improve air condition, because of moist forest and water that provide an extensive range of direct benefits to local community (Kaboré et al. 2022). Figure 9.13 shows goods and services in the case study in specific wetlands in Burkina Faso.

Figure 9.12 Families of macro-invertebrates reported in wetlands. *Source:* From Kaboré (2016).

9.5 Wetland Conservation Measures

For suitable management of wetlands, four levels of intervention may be taken into account: (i) policymakers, (ii) researchers, (iii) local population, and (iv) partners. The close cooperation between policymakers, researchers, partners, and local community is required for sustainable development processes of wetlands. A multidisciplinary team skilled in the natural, human, and policy sciences and the involvement of environment policymakers to develop and implement more sustainable strategies are suitable to the wetlands of Burkina Faso. This approach requires new integrative networks including all stakeholders (Figure 9.14). Giving power to local community may help them to control their wetland resources, thus developing the management responsibility skills.

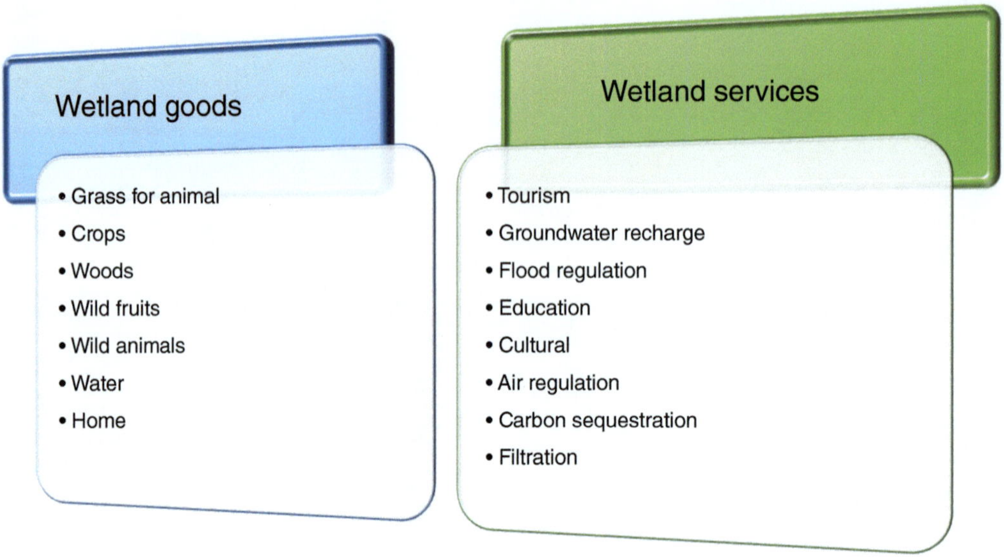

Figures 9.13 Wetland goods and services reported in the case study in Burkina Faso. *Source:* Adapted from Nakéna (2024).

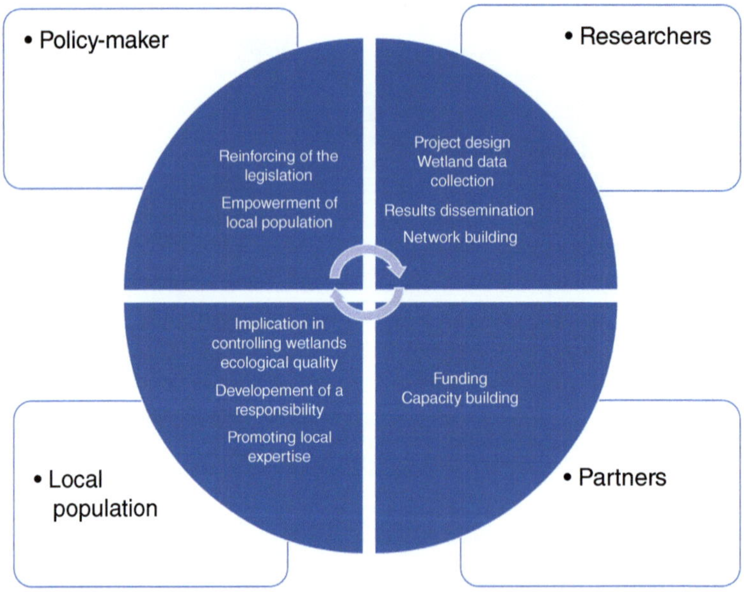

Figure 9.14 Adaptive management process of wetlands in Burkina Faso.

9.6 Conclusion

Wetland plays a crucial role in biodiversity conservation as far as demonstrated by this chapter. Many species of fauna are found in wetlands or have a style live link to wetlands, and constitute an important source of income for local population. Burkina Faso socioeconomic development relies much more on wetlands goods and services, but these wetlands are facing severe human pressures and climate warming. Urgently, many measures needed to be taken into account such as wetlands data collection, implementation legislation, promoting research project on wetlands, and close cooperation of all stakeholders for suitable management of valuable wetlands in Burkina Faso.

References

Adamus, P.R., Danielson, T.J., and Gonyaw, A. (2001). *Indicators for Monitoring Biological Integrity of Inland, Freshwater Wetlands: A Survey of North American Technical Literature, (1990–2000).* Washington, DC: U.S. Environmental Protection Agency, Office of Water, Wetlands Division.

Adeeyo, A.O., Ndlovu, S.S., Ngwagwe, L.M. et al. (2022). Wetland resources in South Africa: threats and metadata study. *Resources* 11: 54. https://doi.org/10.3390/resources11060054.

Adje, A.M.H. (2023). *Ophidiens (Serpentes) du Burkina Faso : Diversité, Régime alimentaire et Perceptions en zones urbaine et péri-urbaine de Ouagadougou*. Ouagadougou: Mémoire de Master, Université Joseph KI-ZERBO 68 p.

Alikhani, S., Nummi, P., and Ojala, A. (2021). Urban wetlands: a review on ecological and cultural values. *Water* 1: 3301. https://doi.org/10.3390/w13223301.

Ayanlade, A. and Proske, U. (2016). Assessing wetland degradation and loss of ecosystem services in the Niger delta, Nigeria. *Marine and Freshwater Research* 67: 828–836. https://doi.org/10.1071/MF15066.

Ayoro, H. J. (2021). Diversité et distribution des espèces d'amphibiens au Burkina Faso : effet des variables bioclimatiques sur la répartition de quelques espèces. Thèse de doctorat unique de l'université Joseph KI-ZERBO. Burkina Faso, 215 p.

Ayoro, H.J., Segniagbeto, G.H., Hema, E.M. et al. (2020). List of amphibian species (Vertebrata, tetrapoda) of Burkina Faso. *Zoosystema* 42 (28): 515550. https://doi.org/10.5252/zoosystema2020v42a28.

Balança, G., Cornelis, D., and Wilson, R. (2007). Les Oiseaux du complexe WAP (consortium ECOPAS). *Cirad* 35–199, ISBN 978-2-87614-645-7.

Belemtougri, A.P., Ducharne, A., Tazen, F. et al. (2021). Understanding key factors controlling the duration of river flow intermittency: case of Burkina Faso in West Africa. *Journal of Hydrology: Regional Studies* 37: 100908.

Böhme, W. and Heath, J. (2018). Amphibian and reptilian records from south-central Mali and western Burkina Faso. *Bonn Zoological Bulletin* 67 (1): 59–69.

Bouché, P., Lejeune, P., Bailly, V. et al. (2016). Conserving wildlife amongst the cotton fields. A third of a century of experience at the Nazinga Game Ranch, Burkina Faso. *Environmental Monitoring and Assessment* 188: 1–13.

Caballer Revenga, J. (2019). GIS Based Analysis of the Environmental and Societal Importance of the Reservoirs in Burkina Faso. Master Thesis. University of Natural and Life Sciences, Vienna.

Cecchi, P., Meunier-Nikiema, A., Moiroux, N., and Sanou, B. (2008). Towards an atlas of lakes and reservoirs in Burkina Faso, pp. 1–20.

Chakraborty, S.K., Sanyal, P., and Ray, R. (2023). Pollution, environmental perturbation and consequent loss of wetlands. In: *Wetlands Ecology*. Cham: Springer https://doi.org/10.1007/978-3-031-09253-4_8.

Chirio, L. et al. (2009). *Bulletin de la Société Herpétologique de France* 132: 13–41.

Cornélis, D. (2007). Le suivi écologique. In: *Nazinga* (ed. W. Delvingt and C. Vermeulen), 227–246. Gembloux and Ouagadougou: Région Wallonne, Presse Agronomique de Gembloux, Nature+, APEFE, Ministère de l'Environnement et du Cadre de Vie (Burkina Faso).

Daboné, C., Oueda, A., Adjakpa, J.B. et al. (2016). Phénologie de la reproduction du Vautour charognard Necrosyrtes monachus en zone soudano sahélienne (Garango, Burkina Faso), 2013–2015. *Malimbus: Journal of the West African Ornithological Society* 38: 38–49.

van Dam, A.A., Fennessy, M.S., and Finlayson, C.M. (2023). What's driving wetland loss and degradation? In: *Ramsar Wetlands* (ed. P.A. Gell, N.C. Davidson, and C.M. Finlayson), 259–306. Elsevier. https://doi.org/10.1016/B978-0-12-817803-4.00012-7.

Dayama, W.F. (2020). *Abondance, distribution et croyances populaires sur les lézards urbains : cas des geckos de la ville de Ouagadougou (Burkina Faso).* Mémoire de Master de l'Université Joseph KI-ZERBO 71 p.

De Groot, D., Brander, L., and Finlayson, C.M. (2018). Wetland ecosystem services. In: *The Wetland Book: I: Structure and Function, Management, and Methods*, 323–333. Netherlands: Springer.

Díaz-Pinzón, L., Sierra, L., and Trillas, F. (2022). The economic value of wetlands in urban areas: the benefits in a developing country. *Sustainability* 14 (14): 8302.

Dohinnon, Y.A.F. (2023). *Diversité et écologie des crabes d'eau douce en zone sahélienne, Burkina Faso.* Mémoire de Master de l'Université Joseph KI-ZERBO 60 p.

Dube, T., Dube, T., and Marambanyika, T. (2023). A review of wetland vulnerability assessment and monitoring in semi-arid environments of sub-Saharan Africa. *Physics and Chemistry of the Earth, Parts A/B/C* 103473. https://doi.org/10.1016/j.pce.2023.103473.

Fonseca, B.M., de Mendonça-Galvão, L., Sousa, F.D.R. et al. (2018). Biodiversity in pristine wetlands of central Brazil: a multi-taxonomic approach. *Wetlands* 38: 145–156. https://doi.org/10.1007/s13157-017-0964-7.

Gnoumou, A., Bognounou, F., Hahn, K., and Thiombiano, A. (2011). Woody plant diversity and stand structure in the ComoéLéraba reserve, southwestern Burkina Faso (West Africa). *Journal of Biological Sciences* 11: 111–123.

Gopal, B. (2013). Future of wetlands in tropical and subtropical Asia, especially in the face of climate change. *Aquatic Sciences* 75: 39–61. https://doi.org/10.1007/s00027-011-0247.

Hema, E.M., Ouattara, Y., Karama, M. et al. (2017). Population dynamics of medium and large mammals in a West African gallery forest area and the potential effects of poaching. *Journal of Threatened Taxa* 9 (5): 10151–10157.

INSD et ICF (2022). *Enquête Démographique et de Santé du Burkina Faso 2021.* Ouagadougou, Burkina Faso et Rockville, MA: INSD et ICF.

Kaboré, I. (2016). Benthic invertebrate assemblages and assessment of ecological status of water bodies in the Sahelo Soudanian area (West Africa, Burkina Faso). Doctoral thesis. University of Natural Resources and Life Sciences, Vienna, Austria, 202 p.

Kaboré, I., Moog, O., Alp, M. et al. (2016a). Using macroinvertebrates for ecosystem health assessment in semi-arid streams of Burkina Faso. *Hydrobiologia* 766: 57–74. https://doi.org/10.1007/s10750-015-2443-6.

Kaboré, I., Jäch, M.A., Ouéda, A. et al. (2016b). Dytiscidae, Noteridae and Hydrophilidae of semi-arid rivers and reservoirs in Burkina Faso: species inventory, diversity and ecological notes. *Journal of Biological and Environmental Sciences* 8 (4): 1–14.

Kaboré, I., Ouédraogo, I., Tampo, L. et al. (2016c). Composition and dynamic of benthic macroinvertebrates community in semi-arid area rivers of Burkina Faso (West Africa). *International Journal of Biological and Chemical Sciences* 10 (4): 1542–1561.

Kaboré, I., Tampo, L., Bancé, V. et al. (2022). Preliminary biological data of Sahel wetland ecosystems in Burkina Faso: implications for ecological health assessment. *Frontiers in Conservation Science* 3: 913333. https://doi.org/10.3389/fcosc.2022.913333.

Kaboré, I., Kabore, B.L., and Bance, V. (2023). Knowledge of shrimp's composition and their implication in food security in dryland area in Burkina Faso (West Africa). *International Journal of Advanced Research* 11 (05): 628–639.

Kaboré, I., Bancé, V., and Silga, R.P. (2024). Supporting of biodiversity in Sahelian arid areas: case of artificial lakes and macroinvertebrate. *World Journal of Advanced Research and Reviews* 21 (01): 2906–2912.

Kansiime, F., Saunders, M.J., and Loiselle, S.A. (2007). Functioning and dynamics of wetland vegetation of Lake Victoria: an overview. *Wetlands Ecology and Management* 15: 433–451.

Koussoube, Y., Upton, K., Dochartaigh, B., and Bellwood-Howard, I. (2018). Africa groundwater atlas: hydrogeology of Burkina Faso. *British Geological Survey*, https://www.bgs.ac.uk/africagroundwateratlas/index.cfm (accessed 5 June 2019).

Landmann, T., Schramm, M., Colditz, R.R. et al. (2010). Wide area wetland mapping in semi-arid Africa using 250-meter MODIS metrics and topographic variables. *Remote Sensing* 2 (7): 1751–1766.

Lankoandé, A., Hema, E.M., Konate, S. et al. (2020). Unleashing the potential of local captive populations for conservation in the West African savannahs – the case study of the African spurred tortoise. *Acta Oecologica* 105: 103581.

Lannas, K.S.M. and Turpie, J.K. (2009). Valuing the provisioning services of wetlands: contrasting a rural wetland in Lesotho with a Peri-Urban Wetland in South Africa. *Ecology and Society* 14 (2): 18. http://www.ecologyandsociety.org/vol14/iss2/art18.

Luiselli, L., Starita, A., Carpaneto, G.M. et al. (2016). A short review of the international trade of wild tortoises and freshwater turtles across the world and throughout two decades. *Chelonian Conservation and Biology* 15 (2): 167–172.

Lungren, C., Oueda, G.H., Walsh, F., and Belemsobgo, U. (2001). Burkina Faso. *Important Bird Areas in Africa and Associated Islands-Priority Sites for Conservation, Birdlife Conservation Series* 11: 117–125.

Mano, K. (2016). Fish assemblages and aquatic ecology integrity in Burkina Faso. Doctoral thesis, University of Natural and Life Sciences, Vienna.

Mano, K., Oueda, A., Ouedraogo, R. et al. (2020). Fish assemblages in the upper part of the Volta River, Burkina Faso: a link analysis towards fisheries management and conservation. *International Journal of Biological and Chemical Sciences* 13 (6): 2546–2559.

Melcher, A.H., Ouédraogo, R., and Schmutz, S. (2012). Spatial and seasonal fish community patterns in impacted and protected semi-arid rivers of Burkina Faso. *Ecological Engineering* 48: 117–129.

Melcher, A., Ouédraogo, R., Moog, O. et al. (2018). Healthy fisheries sustain society and ecology in Burkina Faso. In: *Riverine Ecosystem Management. Aquatic Ecology Series*, vol. 8 (ed. S. Schmutz and J. Sendzimir). Cham: Springer https://doi.org/10.1007/978-3-319-73250-3_27.

Melcher, A., Ouedraogo, R., Oueda, A. et al. (2020). *SUSFISHBook – Sustainable Fisheries and Water Management. Transformation Pathways for Burkina Faso.* SUSFISH+ Project Consortium – http://susfish.boku.ac.at/.

Meulebroek, P., Stranzl, S., Ouéda, A. et al. (2019). Fish communities, habitat use, and human pressures in the Upper Volta Basin, Burkina Faso, West Africa. *Sustainability* 11: 5444. https://doi.org/10.3390/su11195444.

Mitchell, S.A. (2013). The status of wetlands, threats and the predicted effect of global climate change: the situation in Sub-Saharan Africa. *Aquatic Sciences* 75: 95–112.

Moog, O. (2007). Deliverable 8 – part 1. manual on pro-rata multi-habitatsampling of benthic invertebrates from wadeable rivers in the HKH-region. In: *BOKU*. Vienna: University of Natural Resources and Applied Life Sciences), 28 p. www.assess-hkh.at.

Nacoulma, B.M.I., Ouedraogo, I., Ouedraogo, O. et al. (2018). Phytodiversity of Burkina Faso. In: *Global Biodiversity* (ed. T. Pullaiah), 1–33. Apple Academic Press.

Nakena, D. (2024). Services écosystémiques et diversité des macroinvertébrés dans des zones humides : Cas du Parc Urbain Bangre-Wéogo. In: *Mémoire de Master de l'Institut, de Gestion, des Risques Industriels et du Développement Durable*. Ouagadougou, Burkina Faso: Institut des gestions des risques industriels et du developpement durable 56 p.

Ola, O. and Benjamin, E. (2019). Preserving biodiversity and ecosystem services in West African forest, watersheds, and wetlands: a review of incentives. *Forests* 2019 (10): 479. https://doi.org/10.3390/f10060479.

Oueda, A., Ouedraogo, R., Somda, J. et al. (2023). *Pêche durable et gestion de l'eau, Voies de transformation pour le Burkina Faso*. Livre SUSFISH. SUSFISH-plus Project Consortium - http://susfish.boku.ac.at/.

Ouédraogo, R. (2010). Fish and fisheries prospective in arid inland waters of Burkina Faso, West Africa. Doctoral thesis. University of Natural Resources and Life Sciences, Vienna, Austria, November 2010, 231 p.

Ouédraogo, I. (2018). Biodiversité et distribution des mollusques d'eau douce au Burkina Faso. Thèse de doctorat unique. Université de Ouagadougou, 202 p.

Ouédraogo, I. (2020). Les crocodiles du Burkina Faso: diversité, abondance, distribution et relations Homme-crocodile. Thèse de doctorat unique. Université Joseph KI-ZERBO, Burkina Faso, 145 p.

Ouédraogo, I., Oueda, A., Kangoyé, N. et al. (2020a). Croyances traditionnelles et conservation du *Crocodylus Suchus* dans les mares sacrées de Bazoulé et de Sabou (Burkina Faso). *European Scientific Journal* 16: 188–205.

Ouédraogo, I., Ouedraogo, I., Kpoda, N.W. et al. (2020b). Impact of road construction on the distribution of *Crocodylus suchus* (Étienne Geoffroy Saint-Hilaire 1807) in urban park Bangr-Weoogo (Burkina Faso). *International Journal of Biological and Chemical Sciences* 14 (2): 390–401.

Paugy, D., Lévêque C., and Teugels G.G. (2003). Faune des poissons d'eaux douces et saumâtres de l'Afrique de l'Ouest. IRD/MNHN/MRAC, Collection Faune et Flore Tropicale (40), tome 2, Paris, pp 53–752.

Petrozzi, F., Hema, E.M., Sirima, D. et al. (2017). Habitat determinants of the threatened Sahel Tortoise *Centrochelys sulcata* at two spatial scales. *Herpetological Conservation and Biology* 12 (2): 402–409.

Petrozzi, F., Hema, E.M., Demaya, G.S. et al. (2020). *Centrochelys sulcata* (Miller 1779)–African spurred tortoise, grooved tortoise, Sahel tortoise, tortue sillonnée. *Chelonian Research Monographs* 5 (14): 110–111.

Ramsar List (2023). The list of wetlands of international importance, p. 1–57. https://www.ramsar.org/sites/default/files/documents/library/sitelist.pdf (accessed 14 March 2024).

Roman, F.S.C.B. (1966). Les poissons des Hauts-Bassins de la Volta. Annales – Série IN/8° / Sciences Zoologiques / N° 150. Musée Royale de l'Afrique Centrale. Tervuren. Belgique. 191 p + planches.

Schmidt, M., Traoré, S., Ouédraogo, A. et al. (2013). Geographical patterns of woody plants' functional traits in Burkina Faso. *Candollea* 68: 197–207.

Schuyt, K. and Brander, L. (2004). Living waters conserving the source of life the economic values of the World's wetlands. https://wwfeu.awsassets.panda.org/downloads/wetlandsbrochurefinal.pdf (accessed 13 March 2024).

Silga, R.P. (2021). Effets des pressions anthropiques et climatiques et perceptions locales sur la distribution des poissons au Burkina Faso. Thèse de doctorat unique de l'Université Josep KI-ZERBO, 187 p.

SUSFISH Consortium (2015). *Sustainable Management of Water and Fish Resources in Burkina Faso: A Synthetic Overview of the SUSFISH Project*. SUSFISH Consortium. Supported by the Austrian

Partnership Programme in Higher Education & Research for Development. http://susfish.boku.ac.at/download.htm.

Thiollay, J. M. (2006). Large bird declines with increasing human pressure in savanna woodlands (Burkina Faso). *Biodiversity and Conservation* 15, 2085–2108. https://doi.org/10.1007/s10531-004-6684-3

Thiombiano, A. and Kampmann, D. (eds). (2010). Areas of special importance for biodiversity conservation in Burkina Faso. In: *Biodiversity Atlas of West Africa: Burkina Faso*, vol. II, 381–401. Ouagadougou & Frankfurt/Main.

Trape, J.F., Chirio, L., and Trape, S. (2012). *Lézards, Crocodiles et Tortues d'Afrique Occidentale et du Sahara*. Paris: Institute de Recherche sur le Developpement Editions.

Zekarias, T. and Gelaw, A. (2023). Impacts of land use/land cover change on wetland ecosystem services of Lake Abaya-Chamo Wetland, Rift Valley of Ethiopia. *Geology, Ecology, and Landscapes* 1–12. https://doi.org/10.1080/24749508.2023.2202436.

10

Wetlands of Cameroon: Biodiversity, Livelihoods, and Conservation

Gordon N. Ajonina[1,2], Thomas E. Efole[2], Barthelemy Ndongo[3], and Minette Eyango Tomedi-Tabi[2]

[1] *Cameroon Wildlife Conservation Society, CWCS Coastal Forests, Mangrove & Marine Programme, Yaoundé, Cameroon*
[2] *Institute of Fisheries and Aquatic Sciences, University of Douala (Yabassi), Douala, Cameroon*
[3] *Department of Rural Engineering, Faculty of Agronomy & Agricultural Sciences, National Focal Point of Ramsar Convention, University of Dschang, Dschang, Cameroon*

10.1 Introduction

Cameroon is a Central African country (population in 2003: 15.7 million) opened to the Atlantic coast in the west with a total coastline length of 590 km. It is bounded to the west by Nigeria; northeast by Chad; south by Gabon, Congo, and Equatorial Guinea; and to the east by Central African Republic. From the Gulf of Guinea to Lake Chad, the country forms a triangle with a surface area of 475,442 km^2. The tropical rainforest covers about 45% (21.2 million ha or 212,000 km^2) of the country (MINFOF 2005). There are more than 50 protected areas with more than 10 national parks, and 5 national parks are located along the Atlantic coast. The rest of the national territory (15%) is covered by various shades of savannah, open marsh, lakes, and human space including towns, villages, and roads.

Cameroon is composed principally of western highlands which form the Cameroon range; its chain of mountain massifs orientates southwest-north from Mount Cameroon the highest point (4090 m) in West and Central Africa to the Mandara mountains (2050 m) with interspersing lowlands, the central-eastern lowlands, and the coastal plains. The climatic pattern mimics the pattern of relief from the Equatorial climate in the south with modified Cameroon type within the highlands with Debundscha the windward coastal zone of Mount Cameroon as the second wettest area in the world with annual rainfall of over 10,000 mm to the Sudan type up to lake Chad.

Cameroon has a very unique wetland diversity from its very dense hydrological network, drained from the Adamawa and from Nyong-Sanaga Dorsal along the four drainage basins: Atlantic, Congo, Niger, and Chad. The Atlantic basin consists mainly of Sanaga and coastal rivers (Nyong, Ntem, Moungo, and Wouri) with Sanaga as the largest river of over 920 km long occupying a basin area of about 140,000 km^2 (Olivry 1986). Numerous waterfalls and rapids interrupt these rivers (especially Mentchum, Abi, Nachtigal, Ekom, Mungo, and Lobe falls). There are also numerous lakes including many crater lakes (Barombi, Oku, Nyos, Wum, etc.) from volcanic activities along its highlands; tectonic lakes from large depressions especially in

Wetlands of Tropical and Subtropical Asia and Africa: Biodiversity, Livelihoods and Conservation,
First Edition. Edited by Thammineni Pullaiah.

the coastal plains (Ossa, Ejagani, etc.); basin lakes including Lake Chad and a range of artificial lakes from dams, irrigation (Bamendjin on Noun river, Mbakaou on Djerem, Lagdo on Benue, Mape on the Mbam, and Songloulou on Sanaga) (Laclavère 1980). The rocky and sandy coast has a number of lagoons, mangroves (over 200,000 ha), mudflats, and other wetland habitats very rich in biodiversity.

It is important to appraise the country's wetlands in order to ascertain its biodiversity status and evaluate their ecosystem services contribution to socioeconomic development and the fight against climate change. In this chapter, we attempt to describe wetlands biodiversity status, which includes their habitats and species and their contribution to livelihoods and conservation status using the available secondary data and information coupled with personal observations.

10.2 Potentials of Wetlands Biodiversity

10.2.1 Delimitation, Typology, and Distribution of Wetland Habitats in Cameroon

Wetlands considered originally as waterfowl habitats are defined according to the Ramsar 1971 convention as "wetlands are areas of marsh, fen, peatland or water, whether natural or artificial, permanent or temporary, with water that is static or flowing, fresh, brackish or salt, including areas of marine water the depth of which at low tide does not exceed six metres" (Article 1.1) and may incorporate riparian and coastal zones adjacent to wetlands, and islands or bodies of marine water deeper than six metres at low tide lying within wetlands (Article 2.1).

> As a result of these provisions, the coverage of the convention extends to a wide variety of habitat types including rivers, shallow coastal waters and even coral reefs but not deep sea.

Ramsar further divides wetlands into three main categories as a broad framework to aid rapid identification of wetland habitats: marine/coastal wetlands that include estuaries, intertidal marshes, brackish, saline and freshwater lagoons, mangrove swamps, as well as coral reefs and rocky marine shores such as sea cliffs; inland wetlands refer to areas such as lakes, rivers, streams and creeks, waterfalls, marshes, peat lands, and flooded meadows; and lastly, man-made wetlands include canals, aquaculture ponds, water storage areas, and even wastewater treatment areas.

For the purpose of this write-up, we classify wetlands as mapable hydrological units with vegetation that boarders water bodies and including the water bodies within the continental landscape and coastal–marine interface up to the artisanal fishing limit (5–6 nautical miles) (Folack 2024). Wetlands can therefore be represented as covering the continental landscape including coastal waters and marine waters at the limit of artisanal fishing zone.

Cameroon hydrology has been extensively studied by Olivry (1986). According to him, the rivers of Cameroon are divided into several hydrographic units. Some are tributaries of Cameroon's Atlantic coastline; others are part of larger river basins, and in particular three of Africa's five largest hydrological complexes: the Niger basin, the Lake Chad basin, and the Congo basin in the Democratic Republic of Congo. Cameroon's hydrological network contributes to the drainage of a vast area covering around a third of the African continent, shared by 23 states, i.e. almost half of all African countries. This sharing of water resources has often justified the creation of interstate bodies such as the Niger Basin Authority (A.B.N.) or the Lake Chad Basin Commission (C.B.LT.) (Olivry 1986). Cameroon has a landscape of 475,442 km^2 and marine seascape with a coastline of

Table 10.1 Cameroon hydrological basins.

No.	Hydrological basin	% Country surface	Estimated area (km^2)
1	Cross-River, draining the Mamfe region and joining the ocean at Calabar (Nigeria)	2.8	13,020
2	Coastal rivers, northwestern Sanaga (Ndian, Meme, Mungo, Wouri)	6.8	31,620
3	Sanaga, Cameroon's most important river	28.6	132,990
4	Nyong and southern coastal rivers (Lobe, Lokoundje, etc.)	8.1	37,665
5	Ntem, whose source is in Gabon	3.9	18,135
6	Ayina or upper Ivindo (Ogoou basin)	1.7	7,905
7	Kadéi and Dja, Sangha (Congo basin)	20.4	94,860
8	Lower Bénoué tributaries (Met-chum, Katsina, Donga), Niger basin	2.2	10,230
9	Bénoué (North Cameroon), which gives rise (along with the preceding tributaries) a second wind of Niger, whose resources have been exhausted in its Sahelian peregrinations Sahelian peregrinations	17.3	80,445
10	Upper Logone (North Vina and Mbéré), tributary of the Lake Chad basin	4.3	19,995
11	Mayos of North-Cameroon, Yaéré and lower Logone, Lake Chad basin	3.9	18,135
	Total	**100**	**465,000**

Source: Percentage data from Olivry (1986).

more than 490 km, a continental shelf of 10,000 km^2, and an Economic Exclusive zone (EEZ) of 15,400 km^2 (Failler and Njock 2011) broadly divided into 11 basins (Table 10.1) represented by many watersheds in Figure 10.1.

10.2.2 Estimation of the Size of Wetlands Habitats in Cameroon

Wetlands in Cameroon covering more than 95% of the country landscape (rest being dry lands) are estimated from several studies and published national forest inventory surveys in Cameroon (see Table 10.2 for estimates and the methodologies). According to this estimate, wetlands occupy a continental surface of 465,000 km^2 divided into inland wetlands of 459,632 km^2 and coastal wetlands of 3204 km^2. Aquatic forests, defined as forested[1] vegetation of wetlands of swamps and herbaceous vegetation of marshes also known as "blue forests" (Ajonina and Tomedi 2014) (for example, riverine forests, flooded forests, gallery forests, and Raffia forests.), occupies 37,066 km^2 representing inland aquatic forests (34,822 km^2) and coastal aquatic forests (2244 km^2) of mangroves, coastal gallery forests, and coastal swamp forests. The artisanal fishing areas occupy more than 4500 km^2 of the EEZ marking the limit of marine wetlands (Folack 2024). These wetland types are globally represented in Figure 10.2 as continental wetlands (459,632; 98%), coastal wetlands (3204 km^2; 0.7%), and marine wetlands (artisanal fishing area) (45,000 km^2, 1%). The artificial wetlands represent different forms of human disturbance of these wetlands categories.

1 "Forest" in Cameroon is defined as a plot of land of at least 0.5 ha covered by trees of at least 5 m in height or capable of reaching 5 m at maturity, with a crown projection on the ground of at least 10%. Land use is essentially for forestry purposes (MINFOF and FAO 2005).

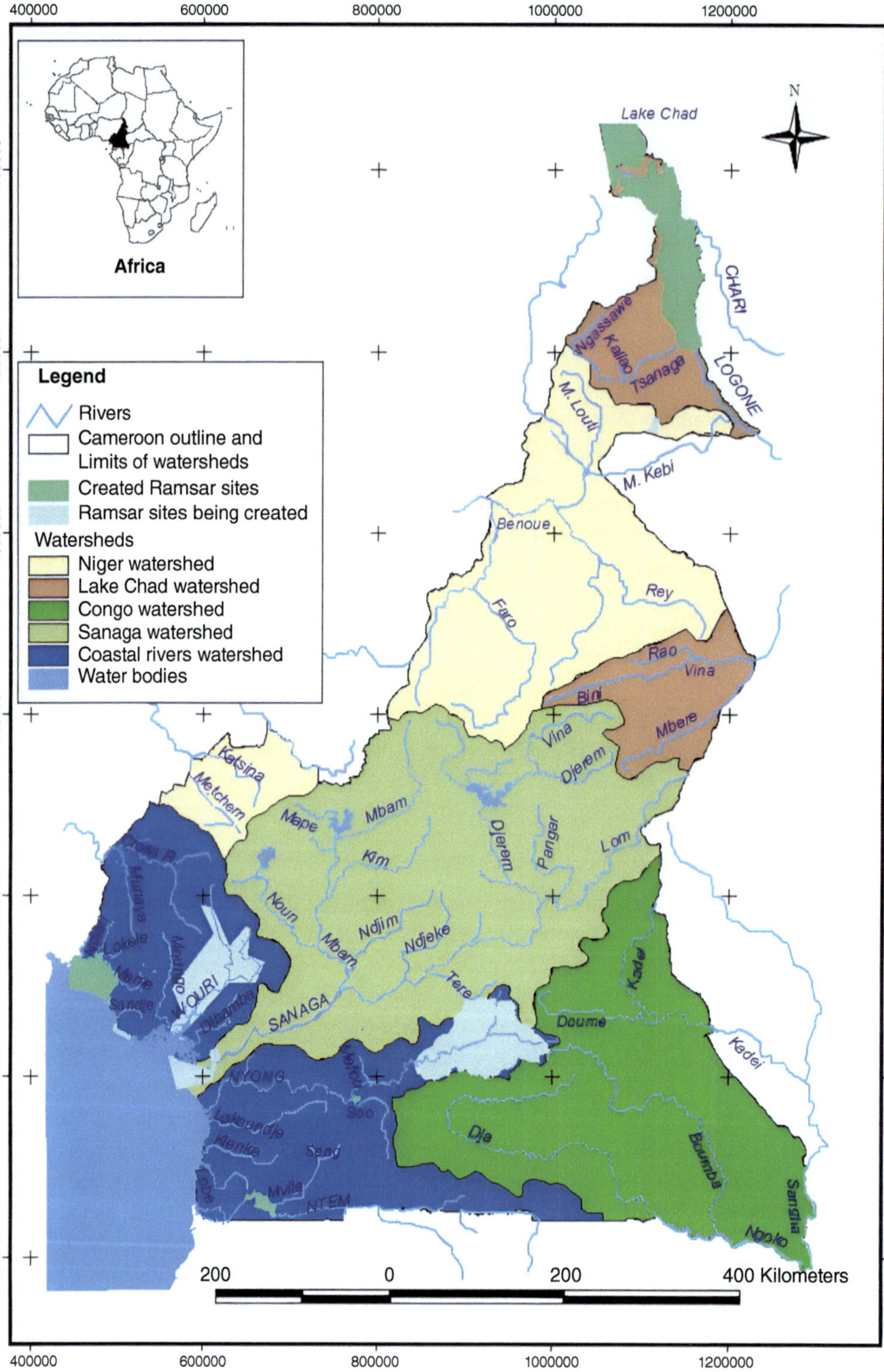

Figure 10.1 Map of Cameroon major watersheds.

Table 10.2 Estimating the size of wetland habitats in Cameroon.

No.	Description of category	Estimated area in km²	As % subtotal	As % total country's land surface	Methodological notes and references
A	Cameroon				
	Continental wetlands (inland and coastal wetlands)				
	Continental waters	2853	0.6	0.6	MINFOF and FAO 2005
	Gallery forests	17,002	3.7	3.6	MINFOF and FAO 2005
	Swamp forests	17,852	3.8	3.8	MINFOF and FAO 2005
	Marshes	427,294	91.9	89.9	By subtraction of sum of gallery forests, swamp forests, and water surfaces from the hydrological basin
	Hydrological basins (potential wetlands, i.e. aquatic forests (gallery forests and swamp forests); marshes and continental water)	465,000	100.0	97.8	Olivry (1986)
	Subtotal continental wetlands	**465,000**	**100.0**	**97.8**	
	Noncontinental wetlands				
	Dryland forests	3,247	31.1	0.7	MINFOF and FAO 2005
	Other dry lands and settlements	7,195	68.9	1.5	
	Subtotal non-wetlands (dry forests, settlements, bare land, etc.)	**10,442**	**100.0**	**2.2**	By subtraction from the area of hydrological basin from the total country land surface
	Grand total Cameroon land surface	**475,442**	**100.0**	**100.0**	
B	**Inland wetlands (2003 and 2004 assessments)**				
	Water surfaces	2,164	0.5	0.5	Total water surfaces (by MINFOF and FAO 2005) less coastal waters (by MINEPDED and RCM 2017)
	Wetlands (swamps, galleries, and marshlands)	459,632	97.4	96.7	Areas total area of hydrological basins (Olivry 1986) less coastal wetlands less inland water surfaces
	Non-wetlands (settlements, dry lands, and bareland surfaces) inland	9,977	2.1	2.1	Country surface area less total area of hydrological basins from Olivry (1986) less coastal non-wetlands
	Country's inner continental surface	**471,773**	**100.0**	**99.2**	
	Upland gallery forests				

(Continued)

Table 10.2 (Continued)

No.	Description of category	Estimated area in km^2	As % subtotal	As % total country's land surface	Methodological notes and references
	(i) Gallery forests in evergreen noncoastal forests	590	3.5	0.1	Areas of galleries in evergreen forests less areas of coastal non-mangrove galleries
	(ii) Gallery forests in semideciduous forests	1,445	8.5	0.3	MINFOF and FAO 2005
	(iii) Gallery forests in Congolian forests	—	0.0	0.0	MINFOF and FAO 2005
	(iv) Gallery forests in Northern savanna	14,962	88.0	3.1	MINFOF and FAO 2005
	Subtotal gallery forests	**16,997**	**100.0**	**3.6**	
	Upland swamp forests				
	(i) Swamp forests in evergreen noncoastal forests	3,668	20.6	0.8	Areas of swamps in evergreen forests less areas of coastal forests non-mangrove swamps
	(ii) Swamp forests in semideciduous forests	3,088	17.3	0.6	MINFOF and FAO 2005
	(iii) Swamp forests in Congolian forests	9,319	52.3	2.0	MINFOF and FAO 2005
	(iv) Swamp forests in Northern savanna	1,749	9.8	0.4	MINFOF and FAO 2005
	Subtotal total (inland swamp forests)	**17,825**	**100.0**	**3.7**	
C	**Coastal wetlands in 2015**				
	Coastal wetlands				
	Water	689	18.8	0.1	MINEPDED and RCM 2017
	Sedimentation areas	272	7.4	0.1	MINEPDED and RCM 2017
	Mangroves	2,212	60.3	0.5	MINEPDED and RCM 2017
	Coastal forest galleries	5	0.1	0.0	Area of coastal forests (MINEPDED and RCM) multiplied by % galleries in evergreen swamps (MINFOF and FAO 2005)
	Coastal forest swamps	27	0.7	0.0	Area of coastal forests (MINEPDED and RCM) multiplied by % swamps in evergreen swamps (MINFOF and FAO 2005)
	Subtotal coastal wetlands	**3,204**	**87.3**	**0.7**	
	Noncoastal wetlands				
	Coastal forests (less galleries and swamps)	100	2.7	0.0	
	Plantations	346	9.4	0.1	MINEPDED and RCM 2017
	Bare surface	20	0.5	0.0	MINEPDED and RCM 2017
	Subtotal noncoastal wetlands	**465**	**12.7**	**0.1**	
	Total coastal land surface	**3,669**	**100.0**	**0.8**	

Figure 10.2 Wetland habitats in Cameroon.

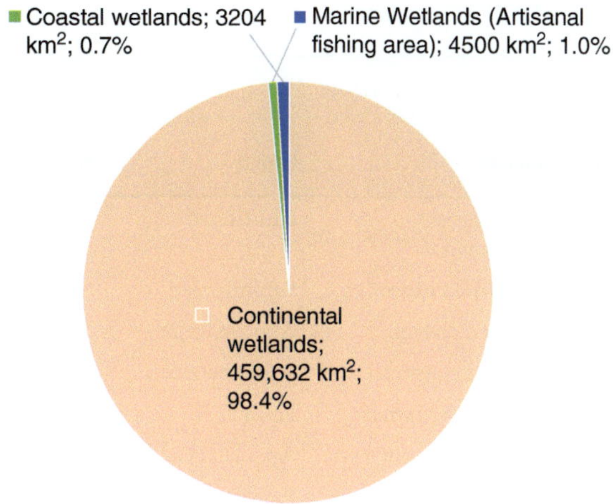

10.3 Wetland Species

These various wetland habitats harbor hundreds of plant and animal species (see Annexes 10.1–10.3) for tree (167), fish (104), and waterbird (68) species of some 40,000 individuals in coastal shores in land as far as Dschang artificial lake from 2024 annual coastal water bird census of an NGO Cameroon Wildlife Conservation Society (CWCS; www.camerooncws.org) (Ajonina and Tomedi 2014). Pure mangrove species are red mangrove tree species: *Rhizophiora mangle*, *R. racemosa*, and *R. harosonii*, white mangrove tree species: *Avicennia germinans*, mangrove shrub: *Conocorpus erectus*, mangrove fern: *Acrostichum aureum*, and mangrove exotic palm: *Nypa fruticans* (Ajonina 2008). Ayissi et al. (2014) identified four species of sea turtles common along Cameroon estuaries: *Dermochelys coriacea, Lepidochelys olivacea*, *Chelonia mydas*, and *Eretmochelys imbricata* for nesting and feeding activities. Eight cetaceans (*Sousa teuszii, Delphinus capensis, Delphinus* sp., *Tursiops truncatus, Stenella attenuata* or *S. frontalis, S. coeruleoalba, Megaptera novaeangliae, Physeter macrocephalus*) and one sirenian species (*Trichechus senegalensis*) found to be common, seasonal, or rare.

Annex 10.1 Typical aquaforest woody species of west central Africa.

No.	Species	Family	Habit	Life forms[a]	Geo affinity[a]	Forest zone A	Forest zone B	Forest zone C	Savana zone A	Savana zone B	Savana zone C
1	*Acacia ataxacantha*	Mimosaceae	Tree	mph	S				x		
2	*Aeschynomene elaphroxylon*	Papilionaceae	Tree	mph	SG	x	x		x	x	
3	*Afzelia africana*	Caesalpiniaceae	Tree	mPh	SG	x			x		
4	*Albizia malacophylla*	Mimosaceae	Tree	mPh	SG				x		
5	*Albizia zygia*	Mimosaceae	Tree	mPh	SG				x		
6	*Alchornea cordifolia*	Euphorbiaceae	Tree	mph	TA	x	x		x	x	

(Continued)

Annex 10.1 (Continued)

No.	Species	Family	Habit	Life forms[a]	Geo affinity[a]	Forest zone A	Forest zone B	Forest zone C	Savana zone A	Savana zone B	Savana zone C
7	*Allanblackia floribunda*	Guittiferae	Tree	mPh	GC	x					
8	*Allophylus africanus*	Sapindaceae	Shrub	nph	Pan	x	x	x	x	x	x
9	*Alstonia boonei*	Apocynaceae	Tree	mPh	GC	x	x				
10	*Alstonia congensis*	Apocynaceae	Tree	MPh	GC	x	x				
11	*Ancistrophyllum* sp.	Palmae	Liana	mph	GC		x				
12	*Ancylobotrys amoena*	Apocynaceae	Liana	nph	SG				x		
13	*Andira inermis*	Papilionaceae	Tree	mph	SG				x	x	
14	*Annona senegalensis*	Annonaceae	Shrub	nph	TA	x			x		
15	*Anogeissus leiocarpus*	Combretaceae	Tree	mph	SG	x			x		
16	*Anthocleista nobilis*	Loganiaceae	Tree	mPh	GC		x				
17	*Anthocleista schweinfurthii*	Loganiaceae	Tree	mPh	GC		x				
18	*Anthocleista vogelii*	Loganiaceae	Tree	mPh	SG				x	x	
19	*Antiaris africana*	Moraceae	Tree	mPh	SG				x		
20	*Antidesma venosum*	Euphorbiaceae	Tree	mph	TA	x	x		x	x	
21	*Aphania sengalensis*	Sapindaceae	Tree	mph	TA	x	x	x	x	x	x
22	*Avicennia germinans*	Avicenniaceae	Tree	mPh	GC			x			x
23	*Berlinia grandiflora*	Caesalpiniaceae	Tree	mPh	GC	x			x		
24	*Borassus aethiopum*	Palmae	Palm	mPh	TA				x		
25	*Bosqueia angolense*	Moraceae	Tree	MPh	GC	x			x		
26	*Brachystegia eurycoma*	Caesalpiniaceae	Tree	MPh	GC	x					
27	*Breonadia salicina*	Rubiaceae	Tree	mph	SG				x		
28	*Bridelia micrantha*	Euphorbiaceae	Tree	mPh	SG				x		
29	*Calamus* sp.	Palmae	Liana	mph	GC		x				
30	*Carapa procera*	Meliaceae	Tree	mPh	GC	x					
31	*Cassia sieberiana*	Caesalpiniaceae	Tree	mph	SG				x		
32	*Ceiba pentandra*	Bombacaceae	Tree	mPh	SG				x		
33	*Celtis integrifolia*	Ulmaceae	Tree	mPh	SG				x		
34	*Clausena anisata*	Rutaceae	Shrub	nph	SG				x		
35	*Cleistopholis glauca*	Annonaceae	Tree	mPh	GC	x					
36	*Cleistopholis patens*	Annonaceae	Tree	mPh	GC	x	x				
37	*Clerodendrum thyrsoideum*	Verbenaceae	Liana	mph	SG						
38	*Cola laurifolia*	Sterculiaceae	Tree	mPh	GC	x			x		
39	*Combretum paniculatum*	Combretaceae	Shrub	nph	TA	x			x		
40	*Combretum tomentosum*	Combretaceae	Tree	mPh	SG	x			x		

Annex 10.1 (Continued)

No.	Species	Family	Habit	Life forms[a]	Geo affinity[a]	Forest zone A	Forest zone B	Forest zone C	Savana zone A	Savana zone B	Savana zone C
41	*Cordia myxa*	Boraginaceae	Tree	mph	SG				x		
42	*Cordia sinensis*	Boraginaceae	Tree	mph	Pan	x	x		x	x	
43	*Crataeva adansonii*	Capparaceae	Tree	nph	TA	x	x		x	x	
44	*Cynometra megalophylla*	Caesalpiniaceae	Tree	mPh	GC	x	x				
45	*Cynometra sanagaensis*	Caesalpiniaceae	Tree	mPh	GC	x	x				
46	*Cynometra vogelii*	Caesalpiniaceae	Tree	mph	SG				x		
47	*Dalbergia melanoxylon*	Papilionaceae	Tree	mph	Pan	x			x		
48	*Dalbergia sissoo*	Papilionaceae	Tree	mph	Pan	x			x		
49	*Detarium senegalense*	Caesalpiniaceae	Tree	mph	SG				x		
50	*Dialium dingklagei*	Caesalpiniaceae	Tree	mPh	GC	x	x				
51	*Dialium guineense*	Caesalpiniaceae	Tree	mph	SG	x			x		
52	*Diospyros mespiliformis*	Ebenaceae	Tree	mPh	SG				x		
53	*Dracaena arborea*	Agavaceae	Tree	mPh	GC	x	x				
54	*Endodesmia calophylloides*	Hypericaceae	Tree	mPh	GC		x				
55	*Entada manii*	Mimosaceae	Liana	mph	GC	x					
56	*Erythrophleum suaveolens*	Caesalpiniaceae	Tree	mPh	SG	x			x		
57	*Eugenia nigerina*	Myrtaceae	Shrub	nph	SG				x		
58	*Ficus abutilifolia*	Moraceae	Shrub	nph	GC				x		
59	*Ficus asperifolia*	Moraceae	Shrub	nph	GC	x					
60	*Ficus capreifolia*	Moraceae	Shrub	nph	GC				x		
61	*Ficus dicranostyla*	Moraceae	Shrub	nph	GC				x		
62	*Ficus exasperata*	Moraceae	Shrub	nph	GC				x		
63	*Ficus glumosa*	Moraceae	Shrub	nph	GC				x		
64	*Ficus natalensis*	Moraceae	Shrub	nph	GC				x		
65	*Ficus ovata*	Moraceae	Shrub	nph	GC				x		
66	*Ficus polita*	Moraceae	Shrub	nph	GC				x		
67	*Ficus sur*	Moraceae	Shrub	nph	GC				x		
68	*Ficus sycomorus*	Moraceae	Shrub	nph	GC				x		
69	*Ficus thonningii*	Moraceae	Shrub	nph	GC				x		
70	*Ficus trichopoda*	Moraceae	Tree	mPh	GC				x	x	
71	*Ficus vallis-choudae*	Moraceae	Tree	mPh	GC				x		
72	*Ficus vogelii*	Moraceae	Tree	mPh	GC				x		
73	*Flacourtia indica*	Flacourtiaceae	Tree	mph	SG				x		
74	*Gambeya africana*	Sapotaceae	Tree	MPh	GC	x					

(*Continued*)

Annex 10.1 (Continued)

						Forest zone			Savana zone		
No.	**Species**	**Family**	**Habit**	**Life forms[a]**	**Geo affinity[a]**	**A**	**B**	**C**	**A**	**B**	**C**
75	*Garcinia livingstonei*	Guittiferae	Tree	mph	SG				x		
76	*Garcinia ovalifolia*	Guittiferae	Tree	mph	SG				x		
77	*Gardenia imperialis*	Rubiaceae	Tree	mph	GC	x	x				
78	*Guibourtia capallifera*	Caesalpiniaceae	Tree	mph	SG				x		
79	*Guibourtia demeusei*	Caesalpiniaceae	Tree	MPh	GC	x	x				
80	*Hippocratea africana*	Hippocrateaceae	Liana	mph	SG				x		
81	*Holarrhena floribunda*	Apocynaceae	Tree	mph	GC	x			x		
82	*Ipomoea carnea*	Convolvulaceae	Shrub	nph	Pan	x	x		x	x	
83	*Irvingia smithii*	Simaroubaceae	Tree	mPh	PRA	x					
84	*Jasminum dichotomum*	Oleaceae	Liana	mph	SG				x		
85	*Kaya senegalensis*	Meliaceae	Tree	MPh	SG				x		
86	*Keayodendron bridelioides*	Euphorbiaceae	Tree	mPh	GC	x	x				
87	*Keetia cornelia*	Rubiaceae	Shrub	nph	SG				x		
88	*Laguncularia racemosa*	Combretaceae	Tree	mPh	G			x			x
89	*Lecaniodiscus cupanioides*	Sapindaceae	Shrub	nph	SG				x		
90	*Leptoderris brachyptera*	Papilionaceae	Liana	LmPh	GC	x					
91	*Loeseneriella africana*	Hippocrateaceae	Liana	mPh	SG				x		
92	*Lonchocarpus cyanescens*	Papilionaceae	Shrub	mph	SG				x		
93	*Lonchocarpus griffonianus*	Papilionaceae	Tree	mPh	GC		x				
94	*Lonchocarpus sericeus*	Papilionaceae	Tree	mPh	SG				x		
95	*Lophira alata*	Ochnaceae	Tree	MPh	SG		x				
96	*Macrosphyra longistyla*	Rubiaceae	Shrub	nph	SG				x		
97	*Malacantha alnifolia*	Sapotaceae	Tree	nph	TA				x		
98	*Manilkara multinervis*	Sapotaceae	Tree	mPh	TA	x					
99	*Militia excelsa*	Moraceae	Tree	mPh	SG				x		
100	*Mimosa pigra*	Mimosaceae	Shrub	nph	Pan	x					
101	*Mitragyna ciliata*	Rubiaceae	Tree	MPh	GC	x	x				
102	*Mitragyna inermis*	Rubiaceae	Shrub	nph	SG				x		
103	*Mitragyna stipulosa*	Rubiaceae	Tree	mPh	SG		x				
104	*Morelia senegalensis*	Rubiaceae	Shrub	mph	SG	x	x		x		
105	*Napoleonaea vogelii*	Lecythidaceae	Tree	mph	G	x					
106	*Nauclea pobeguinii*	Rubiaceae	Tree	mph	G		x				
107	*Neocarya macrophylla*	Chrysobalanaceae	Tree	nph	SG				x		
108	*Olax subsorpioides*	Olacacea	Shrub	nph	SG				x		
109	*Oncoba spinosa*	Flacourtiaceae	Tree	mph	SG				x		

Annex 10.1 (Continued)

No.	Species	Family	Habit	Life forms[a]	Geo affinity[a]	Forest zone A	Forest zone B	Forest zone C	Savana zone A	Savana zone B	Savana zone C
110	*Ouratea glaberrima*	Ochnaceae	Shrub	nph	GC	x					
111	*Oxytenanthera abyssinica*	Poaceae	Bamboo	mPh	SG				x		
112	*Pachystela pobeguiniana*	Sapotaceae	Shrub	nph	TA				x		
113	*Parinari congensis*	Chrysobalanaceae	Tree	MPh	GC	x					
114	*Paullinia pinnata*	Sapindaceae	Liana	nph	TA				x		
115	*Pericopsis laxiflora*	Papilionaceae	Tree	mph	SG				x		
116	*Phoenix reclinata*	Palmae	Palm	mPh	SG	x	x		x	x	
117	*Phyllanthus muellerianus*	Euphorbiaceae	Shrub	nph	SG				x		
118	*Phyllanthus reticulatus*	Euphorbiaceae	Liana	nph	SG	x			x		
119	*Phyllanthus welwitschianus*	Euphorbiaceae	Shrub	nph	SG	x			x		
120	*Plagiosiphon longitubus*	Caesalpiniaceae	Tree	mPh	GC	x	x				
121	*Psychotria calva*	Rubiaceae	Shrub	nph	GC	x					
122	*Pterocarpus santalinoides*	Papilionaceae	Tree	mPh	PRA	x					
123	*Raphia hockeri/vinifera*	Palmae	Palm	mPh	SG	x	x		x	x	
124	*Raphia sudanica*	Palmae	Palm	mPh	SG				x	x	
125	*Rhizophora harrisonii*	Rhizophoraceae	Tree	mPh	TA			x			x
126	*Rhizophora mangle*	Rhizophoraceae	Shrub	nph	AA			x			x
127	*Rhizophora racemosa*	Rhizophoraceae	Tree	mPh	TA			x			x
128	*Ricinodendron heudelotii*	Euphorbiaceae	Tree	mPh	GC	x					
129	*Rinorea* spp.	Violaceae	Liana	mPh	SG				x		
130	*Rotula aquatica*	Boraginaceae	Shrub	nph	Pan	x	x	x	x	x	x
131	*Rourea thomsonii*	Connaraceae	Shrub	mph	SG				x	x	
132	*Rytigynia senegalensis*	Rubiaceae	Shrub	nph	SG				x		
133	*Saba comorensis*	Apocynaceae	Liana	mPh	SG	x			x		
134	*Saba senegalensis*	Apocynaceae	Liana	mPh	SG	x					
135	*Sacoglottis gabonensis*	Humiriaceae	Tree	mPh	GC		x				
136	*Salix subserrata*	Salicaceae	Shrub	nph	Pan	x	x	x	x	x	x
137	*Santaloides afzelii*	Connaraceae	Shrub	nph	SG				x		
138	*Scyphocephalium mannii*	Myristericacee	Tree	mPh	GC		x				
139	*Sesbania sesban*	Papilionaceae	Shrub	nph	Pan	x	x		x	x	
140	*Sorindeia grandifolia*	Anacardiaceae	Tree	mPh	GC	x					

(*Continued*)

Annex 10.1 (Continued)

No.	Species	Family	Habit	Life forms[a]	Geo affinity[a]	Forest zone			Savana zone		
						A	B	C	A	B	C
141	*Spathodea campanulata*	Bignoniaceae	Tree	mPh	GC	x					
142	*Spathodea tomentosa*	Bignoniaceae	Tree	mPh	SG				x		
143	*Spondianthus preussii*	Euphorbiaceae	Tree	mPh	GC	x	x				
144	*Spondias mombin*	Anacardiaceae	Tree	mPh	GC	x			x		
145	*Sterculia tragacantha*	Sterculiaceae	Tree	mPh	SG	x	x		x	x	
146	*Strombosia grandifolia*	Olacacea	Tree	mPh	GC	x					
147	*Strophanthus sarmentosus*	Apocynaceae	Liana	mPh	SG	x			x		
148	*Symphonia globulifera*	Guittiferae	Tree	mPh	GC		x				
149	*Synsepalum brevipes*	Sapotaceae	Tree	mph	GC	x					
150	*Syzygium guineense*	Myrtaceae	Tree	mPh	TA	x	x				
151	*Taccazea apiculata*	Aslepiadaceae	Liana	mPh	GC	x					
152	*Talbotiella batesii*	Caesalpiniaceae	Tree	mPh	GC	x	x				
153	*Terminalia glaucescens*	Combretaceae	Tree	mPh	SG				x		
154	*Terminalia laxiflora*	Combretaceae	Tree	mPh	SG				x		
155	*Tetracera alnifolia*	Dilleniaceae	Liana	mph	SG				x		
156	*Uapaca guineensis*	Euphorbiaceae	Tree	mPh	GC	x					
157	*Uapaca heudelotii*	Euphorbiaceae	Tree	mPh	GC	x					
158	*Uapaca togoensis*	Euphorbiaceae	Tree	mPh	SG				x		
159	*Uvaria chamae*	Annonaceae	Shrub	nph	SG				x		
160	*Vitex chrysocarpa*	Verbenaceae	Tree	mPh	SG				x		
161	*Vitex doniana*	Verbenaceae	Tree	mPh	SG				x		
162	*Vitex madiensis*	Verbenaceae	Tree	mPh	SG				x		
163	*Voacanga africana*	Apocynaceae	Tree	mPh	SG				x		
164	*Xylopia aethiopica*	Annonaceae	Tree	mPh	GC	x			x		
165	*Xylopia aurantiiodora*	Annonaceae	Tree	mph	GC	x	x				
166	*Zanthoxylum zanthoxyloides*	Rutaceae	Shrub	nph	SG				x		
167	*Ziziphus spina-christi*	Rhamnaceae	Shrub	nph	SG				x		
	Total = 167					**75**	**43**	**9**	**108**	**19**	**9**
	As % of total					**44.9**	**25.7**	**5.4**	**64.7**	**11.4**	**5.4**

A. River banks/fringing forests; B. Freshwater swamps; C. Tidal swamps/mangroves.
[a] ***(I) Life forms (LF)*** following Raunkaier (1934), Schnell (1971) and Keay and Hepper (1954–1972): Phanerophytes (Ph): megaphanerophytes (MPh: >30 m), mesophanerophytes (mPh: 8–29 m), microphanerophytes (mph: 2–7 m), and nanophanerophytes (nph <2 m); and **(II) Phytogeographic types (PTs)** following White (1986) and Keay and Hepper (1954–1972) are as follows: (i) Species widely distributed in the tropics (Cosmopolitan – cosmo, pantropical – pan, Afro-American – AA, and Paleotropical – Paleo); (ii) species widely distributed in Africa (Tropical Africa – TA and Pluri Regional in Africa – PRA); (iii) regional species in Africa (Sudanian – S, Guinean – G, Sudano-Guinean – SG, Sudano-Zambesian – SZ, and Guineo Congolian – GC).
Source: Ajonina and Tomedi (2014)/Taylor & Francis.

Annex 10.2 Typical aquaforest fish species of west Central Africa.

No.	Scientific name	Family	Common name	Type of aquaforests		
				Fringing forest	Freshwater	Mangroves/ coastal
1	*Acanthurus monroviae*	Acanthuridae	Surgeonfish	0	1	1
2	*Albula vulpes*	Albulidae	Sea banana	0	1	1
3	*Alectis alexandrinus*	Carangidae	Carangue	0	0	1
4	*Alestes* sp.	Alestidae	Alestes	0	1	1
5	*Alopochen aegyptiacus*	Anantidae	Ouette of Egypt	0	1	0
6	*Antennarius striatus*	Antenarridae	Frog fish	0	1	1
7	*Aplocheilichthys spilauchen*	Poeciliidae	Poeciliids	1	0	0
8	*Aristichthys nobilis*	Cyprinidae	Big carp head	0	1	0
9	*Arius* spp.	Bagridae	Cat fish	0	1	1
10	*Barbus* sp.	Gobeiidae	River barber	1	1	1
11	*Batrachoides liberiensis*	Batrechoididae	Toadfish	0	0	1
12	*Brachydeuterus auritus*	Haemulidae	Grunds	0	0	1
13	*Brycinus longipinnus*	Alestidae	Tétras africains	1	1	1
14	*Bufo regularis*	*Bufonidae*	Toad	1	0	0
15	*Carassius auratus*	Cyprinidae	Gilded(bronzed) carasin	0	1	0
16	*Carcharhinus leucas*	Carcharhinidae	Shark	0	0	1
17	*Chrysichtys nigrodigitatus*	Bagridae	Cat fish	0	1	0
18	*Cirrhina molitorella*	Cyprinidae	Carp of the vase	0	1	0
19	*Clarias gariepinus*	Claridae	Cat fish	1	1	0
20	*Clupea harengus*	Clupeidae	Herring	0	0	1
21	*Coryphaena hippurus*	Coryphaenidae	Sea bream	0	0	1
22	*Crocodilus fuscus*	Crocodylidae	Caiman	1	1	0
23	*Crocodylus* sp.	Crocodylidae	Crocodile	0	1	1
24	*Ctenopharyngodon idella*	Cyprinidae	Herbivorous carp	0	1	0
25	*Cynoglosis* spp.	Soléidae	Sole	0	1	1
26	*Cynoglossus senegalensis*	Soléidae	Sole	0	0	1
27	*Cyprinus carpio*	Cyprinidae	Common carp	0	1	0
28	*Dasyatis* spp.	Rhinobatidae	Skate	0	1	1
29	*Dermocherys coriacea*	Dermochelyidae	Tortoise	0	1	1
30	*Dicentractchus labrax*	Serranidae	Common bar	0	0	1
31	*Dorminator lebretonis*	Eleotridae	Sleeper	0	1	1
32	*Dreissena polymorpha*	Dreissenidae	Streaked mussel	0	1	0
33	*Drepana africana*	Drepaneidae	Disk	0	0	1
34	*Elops lacerta*	Elopidae	Ladyfish	0	1	1
35	*Engraulis encrasicolus*	Engraulidae	Anchovies	0	0	1

(*Continued*)

Annex 10.2 (Continued)

				Type of aquaforests		
No.	**Scientific name**	**Family**	**Common name**	**Fringing forest**	**Freshwater**	**Mangroves/ coastal**
36	*Ephippion guttifer*	Tetraodontidae	Puffer	1	0	0
37	*Epinephelus* sp.	Serranidae	Grouper	0	0	1
38	*Ethmalosa fimbriata*	Clupeidae	Wadding	0	1	1
39	*Fluviatilis* sp.	Fluviatilis	Ecrevisse	0	1	1
40	*Fodiator acutes*	Exocotidae	Wheel fish	0	0	1
41	*Galeoides decadactylus*	Polynemidae	Threadfins	0	0	1
42	*Gymnarcus niloticus*	Gymnarchidae	Gymnarque of the nil	0	0	1
43	*Gymnura micrura*	Gymnuridae	Butterfly-rays	0	0	1
44	*Hemichromis faciatus*	Cichlidae	Panther fish	0	1	0
45	*Hemiramphus balao*	Hemiramphidae	Halfbeaks	0	0	1
46	*Hepsetus odoe*	Hepsetidae	Pikes	0	1	1
47	*Hetérobrancus longifilis*	Claridae	Cat fish	1	1	0
48	*Heterotis niloticus*	Ostéoglossidae	Kanga	0	1	0
49	*Hippoglossus Hippoglossus*	Pleuronectidae	Halibut	0	0	1
50	*Hippoglossus stenolepis*	Pleuronectidae	Halibut	0	0	1
51	*Hydrochoeris hydrochoeris*	Hydrochoeridae	Water pig	0	1	0
52	*Hydrocynus forskalii*	Alestidae	Tiger fish	0	0	1
53	*Hypleurochilus langi*	Blennidae	Blennie	0	1	1
54	*Hypophtalmichthys molitrix*	Cyprinidae	Silver carp	0	1	0
55	*Inia geoffrensis*	Platanistidae	Fresh	0	0	1
56	*Labeo* sp.	Cyprinidae	Labeo	0	1	0
57	*Lates niloticus*	Centropomidae	Captain	0	1	0
58	*Lepomis gibbosus*	Centrachidae	Perche sun	0	1	0
59	*Lisha africana*	Clupeidae	Clupeids	0	1	1
60	*Litjanus agennes*	Lutjaidae	Snapper	1	1	1
61	*Liza* sp.	Mugilidae	Mullet	0	1	1
62	*Malapterus electricus*	Malaptéruridae	Electricial fish	0	1	1
63	*Mugil cephalus*	Mugilidae	Mullet	0	1	1
64	*Mylopharyngodon piceus*	Cichlidae	Black carp	0	0	1
65	*Myocastor coypus*	Myocastoridae	Coypu	0	0	1
66	*Myrophis plumbeus*	Ophichthyidae	Snake eel	0	1	1
67	*Oncorhynchus mykiss*	Salmonidae	Rainbow trout	0	1	0
68	*Ondatra zibethicus*	Muridae	Muskrat	0	1	0
69	*Oreochromis mosambicus*	Cichlidae	Tilapia	0	1	0
70	*Oreochromis niloticus*	Cichlidae	Tilapia of the nil	0	1	0

Annex 10.2 (Continued)

				Type of aquaforests		
No.	**Scientific name**	**Family**	**Common name**	**Fringing forest**	**Freshwater**	**Mangroves/ coastal**
71	*Oxyura jamaicensis*	Anantidae	Fuzz Erismature	0	1	0
72	*Pangasius gigas*	Pangasiidés	Mekong's fish	1	1	1
73	*Parachanna abscura*	Channidae	Viper fish	1	1	0
74	*Paraconger arisona*	Ophichthyidae	Conger	0	1	1
75	*Penaeus* sp.	Peneidae	Shrimp	0	1	1
76	*Periphtalmus* sp.	Gobeiidae	Periophtalm	0	0	0
77	*Poecilia latipina*	Poeciliidae	Molly	0	0	1
78	*Poecilia reticulata*	Poeciliidae	Guppy	0	0	1
79	*Polydacrylus* sp.	Polynemidae	Small African threadfin	0	0	1
80	*Polypterus senegalus*	Polyotéridae	Polyptére of sénégal	0	0	1
81	*Pomedasys jubilini*	Sparidae	Sompat grunt	0	0	1
82	*Procyon lotor*	Procyonidae	Racoon	0	1	0
83	*Psettodes belcheri*	Psettodiae	Psettodis	0	0	1
84	*Psettus sebas*	Drepaneidae	African moony	0	0	1
85	*Pseudotolithus brachygnathus*	Sciaenidae	Drums	0	1	1
86	*Pseudotolithus elongatus*	Sciaenidae	Bobo croaker	0	1	1
87	*Pseudotolithus senegalensis*	Sciaenidae	Bar	0	0	1
88	*Pteroscion peli*	Sciaenidae	fried fish	0	1	1
89	*Rana* sp.	Ranidae	Frog	0	1	0
90	*Rhinobatos rhinobatos*	Rhinobatidae	Guitar skate	0	0	1
91	*Salmo salar*	Salmonidae	Salmon	0	0	1
92	*Sardinella maderensis*	Clupeidae	Clupeids	0	1	1
93	*Sarotherodon galilaeus*	Cichlidae	Tilapia	0	1	0
94	*Schilber mystus*	Schilbedae	Cat fish	0	1	0
96	*Scomber scombrus*	Scombridae	Mackerel	0	0	1
97	*Sphyraena barracuda*	Sphyraenidae	Barracuda	0	0	1
98	*Sphyraena dubia*	Hepsetidae	Brochet	0	0	1
99	*Sphyraena piscatorium*	Spyraenidae	Guinean barracuda	0	0	1
100	*Strongylura senegalensis*	Belonidae	Orphies	1	1	1
101	*Synodontis* sp.	Mochokidae	Synodontis	0	1	0
102	*Tilapia guineensis*	Cichlidae	Tilapia	0	1	0
103	*Trichechus manatus*	Trichechidae	Manatee	1	1	1
104	*Trichiurus lepturus*	Trichiuridae	Cutlassfish	0	0	1

Source: Ajonina and Tomedi (2014)/Taylor & Francis.

Annex 10.3 Coastal waterbirds census in Cameroon.

Species English name	Species scientific name	Wouri Estuary	Douala-Edea National Park	Sanaga Estuary & Coast	Upper Sanaga River	Mount Cameroon coast	Ndian Basin	Dschang Artificial Lake	Total	As % total waterbirds	As % Grand total
I-Large waders (Cormorants to Ibises)											
African Openbill Stork	*Anastomus lamelligerus*		472	120	1				593	**1.54**	**1.53**
African Darter	*Anhinga rufa*	1	28	2	1				32	**0.08**	**0.08**
Grey Heron	*Ardea cinerea*	15	33	7	1		252	1	309	**0.80**	**0.80**
Goliath Heron	*Ardea goliath*		3	1					4	**0.01**	**0.01**
Purple Heron	*Ardea purpurea*				1				1	**0.00**	**0.00**
Squacco Heron	*Ardeola ralloides*	17							17	**0.04**	**0.04**
Cattle Egret	*Bubulcus ibis*	109	7			335		30	481	**1.25**	**1.24**
Green-backed Heron	*Butorides striatus*	1	59	6		1	121		188	**0.49**	**0.49**
Woolly-necked Stork	*Ciconia episcopus*	4	29		47		326		406	**1.05**	**1.05**
Great White Egret	*Casmerodius albus*	10	16				2,954		2,980	**7.72**	**7.69**
Black Heron	*Egretta ardesiaca*						4		4	**0.01**	**0.01**
Little Egret	*Egretta garzetta*	55	144	33	16		121		369	**0.96**	**0.95**
Western Reef Egret	*Egretta gularis*	22	89	10		6	1,518		1,645	**4.26**	**4.24**
Intermediate Egret	*Egretta intermedia*	7	12	1	3				23	**0.06**	**0.06**
	Mesophyx intermedia						50		50	**0.13**	**0.13**
Hadada Ibis	*Bostrychia hagedash*		2						2	**0.01**	**0.01**
Yellow-billed Stork	*Mycteria ibis*	1	52				357		410	**1.06**	**1.06**
Black-crowned Night-heron	*Nycticorax nycticorax*		3				150		153	**0.40**	**0.39**
Great White Pelican	*Pelecanus onocrotalus*		8				500		508	**1.32**	**1.31**
Pink-backed Pelican	*Pelecanus rufescens*	2	48				1,502		1,552	**4.02**	**4.00**
Long-tailed Cormorant	*Phalacrocorax africanus*	5	20	1		51	813	14	904	**2.34**	**2.33**
Lesser Flamingo	*Phoeniconaias minor*						550		550	**1.43**	**1.42**

Annex 10.3 (Continued)

Species English name	Species scientific name	Wouri Estuary	Douala-Edea National Park	Sanaga Estuary & Coast	Upper Sanaga River	Mount Cameroon coast	Ndian Basin	Dschang Artificial Lake	Total	As % total waterbirds	As % Grand total
Hammerkop	*Scopus umbretta*	5	20						25	**0.06**	**0.06**
Little Grebe	*Tachybaptus ruficollis*							2	2	**0.01**	**0.01**
Sacred Ibis	*Threskiornis aethiopicus*	5	8				12		25	**0.06**	**0.06**
Subtotal I-Large waders (Cormorants to Ibises) 25 species:		**259**	**1053**	**181**	**70**	**393**	**9,230**	**47**	**11,233**	**29.11**	**28.98**
II-Ducks, rails, and finfoots										**0.00**	
Yellow-billed Duck	*Anas undulata*							40	40	**0.10**	**0.10**
White-faced Whistling Duck	*Dendrocygna viduata*	51							51	**0.13**	**0.13**
Black Crake	*Amaurornis flavirostra*					1		1	2	**0.01**	**0.01**
Purple Swamphen	*Porphyrio porphyrio*							6	6	**0.02**	**0.02**
Hartlaub's Duck	*Pteronetta hartlaubi*		3						3	**0.01**	**0.01**
Subtotal II-Ducks, rails and finfoots: 5 species		**51**	**3**			**1**		**47**	**102**	**0.26**	**0.26**
III-Small waders										**0.00**	
African Jacana	*Actophilornis africana*	23	4		6	3			36	**0.09**	**0.09**
Common Sandpiper	*Tringa hypoleucos*	6	116	14	20	64	76		296	**0.77**	**0.76**
Ruddy Turnstone	*Arenaria interpres*						4		4	**0.01**	**0.01**
Senegal Thick-knee	*Burhinus senegalensis*		15	1					16	**0.04**	**0.04**
Curlew Sandpiper	*Calidris ferruginea*		20						20	**0.05**	**0.05**
Little Stint	*Calidris minuta*	12					2		14	**0.04**	**0.04**
Ringed Plover	*Charadrius hiaticula*	2	267		20		4,021		4310	**11.17**	**11.12**
White-fronted Sandplover	*Charadrius marginatus*		11	7	36		28		82	**0.21**	**0.21**
Grey Pratincole	*Glareola cinerea*		252	2	401				655	**1.70**	**1.69**

(Continued)

Annex 10.3 (Continued)

Species English name	Species scientific name	Wouri Estuary	Douala-Edea National Park	Sanaga Estuary & Coast	Upper Sanaga River	Mount Cameroon coast	Ndian Basin	Dschang Artificial Lake	Total	As % total waterbirds	As % Grand total
Eurasian Oystercatcher	*Haematopus ostralegus*			1					1	**0.00**	**0.00**
Black-winged Stilt	*Himantopus himantopus*	1							1	**0.00**	**0.00**
Bar-tailed Godwit	*Limosa lapponica*		26						26	**0.07**	**0.07**
Eurasian Curlew	*Numenius arquata*		15			4	12		31	**0.08**	**0.08**
Whimbrel	*Numenius phaeopus*	7	92	15		16	55		185	**0.48**	**0.48**
Ruff	*Philomachus pugnax*						2		2	**0.01**	**0.01**
Grey Plover	*Pluvialis squatarola*		459	22			406		887	**2.30**	**2.29**
Avocet	*Recurvirostra avosetta*		4		1		1,500		1505	**3.90**	**3.88**
Wood Sandpiper	*Tringa glareola*	7					500		507	**1.31**	**1.31**
Greenshank	*Tringa nebularia*	34	233	21	28	8	1,873		2,197	**5.69**	**5.67**
Marsh Sandpiper	*Tringa stagnatilis*	21	204				20		245	**0.63**	**0.63**
Redshank	*Tringa totanus*		68				6		74	**0.19**	**0.19**
White-headed Lapwing	*Vanellus albiceps*		15		24				39	**0.10**	**0.10**
Spur-winged Plover	*Vanellus spinosus*	17							17	**0.04**	**0.04**
Black-headed Plover	*Vanellus tectus*				7				7	**0.02**	**0.02**
Terek Sandpiper	*Tringa cinereus*		1						1	**0.00**	**0.00**
Subtotal III-Small waders: 25 species		**130**	**1802**	**83**	**543**	**95**	**8505**		**11,158**	**28.92**	**28.79**
IV-Gulls, Terns, and African Skimmers											
Black Tern	*Chlidonias niger*						5,050		5,050	**13.09**	**13.03**
Moorhen	*Gallinula chloropus*					1		15	16	**0.04**	**0.04**
Lesser Black-backed Gull	*Larus fuscus*						549		549	**1.42**	**1.42**

Annex 10.3 (Continued)

Species English name	Species scientific name	Wouri Estuary	Douala-Edea National Park	Sanaga Estuary & Coast	Upper Sanaga River	Mount Cameroon coast	Ndian Basin	Dschang Artificial Lake	Total	As % total waterbirds	As % Grand total
Black-headed Gull	*Larus ridibundus*	2					1		3	**0.01**	**0.01**
African Skimmer	*Rynchops flavirostris*		1		567		220		788	**2.04**	**2.03**
Little Tern	*Sterna albifrons*		183				5,130		5313	**13.77**	**13.71**
Caspian Tern	*Sterna caspia*			2					2	**0.01**	**0.01**
Royal Tern	*Sterna maxima*	22	1565	55		14	2,688		4344	**11.26**	**11.21**
Sandwich Tern	*Sterna sandvicensis*	2	20	1			5		28	**0.07**	**0.07**
Subtotal IV-Gulls, Terns, and African Skimmers: 9 species		**26**	**1769**	**58**	**567**	**15**	**13,643**	**15**	**16,093**	**41.71**	**41.52**
Total waterbird species: 54		**466**	**4627**	**322**	**1180**	**504**	**31,378**	**109**	**38,586**	**100.00**	**99.55**
As % total waterbird species		**1.21**	**11.99**	**0.83**	**3.06**	**1.31**	**81.32**	**0.28**	**100**		
V. Kingfishers and other birds											
Unidentified Harrier	*Circus* spp.		1						1		**0.00**
Black Kite	*Milvus migrans*				69		29		98		**0.25**
Osprey	*Pandion haliaetus*		5		2	1			8		**0.02**
Peters' Finfoot	*Podica senegalensis*		6	1					7		**0.02**
African Pied Hornbill	*Lophoceros fasciatus*		1						1		**0.00**
Blue-breasted Kingfisher	*Halcyon malimbica*	3	3						6		**0.02**
Giant Kingfisher	*Megaceryle maxima*					2			2		**0.01**
Malachite Kingfisher	*Corythornis cristatus*	1			1				2		**0.01**
Palm-nut Vulture	*Gypohierax angolensis*		13	6	2				21		**0.05**
Pied Kingfisher	*Ceryle rudis*		2		2	2	10		16		**0.04**
Shining-blue Kingfisher	*Corythornis nais*		2						2		**0.01**

(Continued)

Annex 10.3 (Continued)

Species English name	Species scientific name	Wouri Estuary	Douala-Edea National Park	Sanaga Estuary & Coast	Upper Sanaga River	Mount Cameroon coast	Ndian Basin	Dschang Artificial Lake	Total	As % total waterbirds	As % Grand total
White-crested Hornbill	*Horizocerus albocristatus*		1						1		**0.00**
Woodland Kingfisher	*Halcyon senegalensis*	3				6			9		**0.02**
Yellow Wagtail	*Motacilla flava*		1						1		**0.00**
Subtotal V. Kingfishers and other birds: 14 species		**7**	**35**	**7**	**76**	**11**	**39**		**175**		**0.45**
Grand total		**473**	**4662**	**329**	**1256**	**515**	**31,417**	**109**	**38,761**		**100.00**
As % grand total		**1.22**	**12.03**	**0.85**	**3.24**	**1.33**	**81.05**	**0.28**	**100**		

Source: Ajonina et al., 2023/Taylor & Francis.

10.4 Importance and Values of Wetlands for Livelihoods and Development in Cameroon

Wetlands are complex interactive ecosystems with intrinsic and extrinsic attributes (immense diversity in wetlands and within wetlands) providing a wide range of ecological services ranging from provisioning, regulating, cultural, and supporting services important for the welfare of both people (Table 10.3).

Cameroon has 11 dams (5 completed, 2 under construction, and 4 planned, with a total capacity of 3700 MW (Table 10.4). They supply the country's hydroelectric energy and fisheries resources and water for irrigation upstream. Despite these advantages upstream, there are many disadvantages downstream including changes in stream flow and biodiversity (Ajonina 2023).

It is known, however, that the economic and social impacts of large dams on African communities living on the floodplains have been mostly adverse (Adams 1996).

10.5 Threats to Wetlands Degradation and Loss in Cameroon

Wetlands being a resource with different interests and conflicting uses (Figure 10.3) are being threatened. Considerable wetlands have been lost and the underlying root causes are associated with population pressure, poor governance, economic pressure in the rural and urban centers, poverty status of local communities, and unequal distribution of resources. In addition, climate change-related factors such as sea level rise and increased sedimentation have affected the fringing communities.

Yemele et al. (2023) in threat analysis based on a questionnaire survey of 277 individuals interviewed in 6 wetlands including Ramsar sites from the Rio del Rey, Barombi, (southwest

Table 10.3 Classification of wetlands ecosystem services.

Provisioning services (environmental goods: the products/goods obtained from ecosystems)	
Food	Production of fish, wild game, fruits, and grains
Freshwater	Storage and retention of water for domestic, industrial, and agricultural use
Fiber and fuel	Production of logs, fuelwood, peat, fodder
Biochemical	Extraction of medicines and other materials from biota
Genetic materials	Genes for resistance to plant pathogens, ornamental species, and so on
Regulating services (benefits obtained from the regulation of ecosystem processes)	
Climate regulation	Source of and sink for greenhouse gases; influence local and regional temperature, precipitation, and other climatic processes
Water regulation (hydrological flows)	Groundwater recharge/discharge
Water purification and waste treatment	Retention, recovery, and removal of excess nutrients and other pollutants
Erosion regulation	Retention of soils and sediments
Natural hazard regulation	Flood control, storm protection
Pollination	Habitats for pollinators
Cultural services (the nonmaterial benefits that people obtain from ecosystems through spiritual enrichment, cognitive development, reflection, recreation, and aesthetic experiences) that directly affect people	
Spiritual and inspirational	Source of inspiration; many religions attach spiritual and religious values to aspects of wetland ecosystems
Recreational	Opportunities for recreational activities
Aesthetic	Many people find beauty or aesthetic value in aspects of wetland ecosystems
Educational	Opportunities for formal and informal education and training
Supporting services (services needed to maintain other services)	
Soil formation	Sediment retention and accumulation of organic matter
Nutrient cycling	Storage, recycling, processing, and acquisition on nutrients
Nursery grounds	For fish, molluscs, and shellfish

Source: Adapted from Millennium Ecosystem Assessment (2005).

region), Ebogo (Centre Region), the Cameroonian portion of Ntem wetlands (south region), Dschang's municipal lake (west region), and Douala-Edea (littoral regions) revealed that overexploitation (37%), urbanization (23%), agriculture (21%), and pollution (12%) were the main threats exerting the greatest pressure on wetlands in relatively all the sites studied. This distribution varies from one region to another, probably because of the different challenges they face and their needs. The littoral region was dominated by overexploitation, urbanization, fisheries, agriculture, invasive species, and cultures. The same causes were identified in the west, but with less significance. The southwest region was dominated by insecurity. Overall, the littoral region was the one with the most threatened wetlands with heavy exploitation of mangrove wood for fish preservation, followed by the west region and the south. The mangrove wood is preferred from

Table 10.4 Hydroelectrical dams in Cameroon.

No	Hydroelectric station	Community	Coordinates	Type	River	Reservoir	Capacity (MW)	Year completed
1	Grand Eweng				Sanaga River		1,000	Planned
2	Chollet Hydroelectric Power Station	Border with the Republic of the Congo	02°00′48″N; 15°00′07″E		Dja River		600	Planned
3	Nachtigal Amont						420	Planned
4	Song Loulou Power Station		04°04′41″N; 10°27′54″E	Run-of-the-river	Sanaga River	Song Loulou Reservoir	384	1981 and 1988 [1]
5	Mouséré (Mbakaou)				Djérem River		330	Planned
6	Edea Power Station [2]	Edéa	03°48′46″N; 10°07′40″E	Run-of-the-river	Sanaga River	Edea Reservoir	264	1953
7	Memve'ele Power Station	Nyabessan	02°23′47″N; 10°23′56″E	Reservoir	Ntem River	Memve'ele Reservoir	201	Under construction
8	Lagdo Power Station	Lagdo	09°03′36″N; 13°41′17″E	Reservoir	Benue River	Lagdo Reservoir	72	1982
9	Lom Pangar Dam		05°23′01″N; 13°30′09″E	Reservoir	Lom River		30	2016
10	Mekin Power Station				Dja River		15	2016
11	Song Loulou Power Station		04°04′41″N; 10°27′54″E	Run-of-the-river	Sanaga River	Song Loulou Reservoir	384	1981 and 1988

Source: Adapted from Wikipedia (2011).

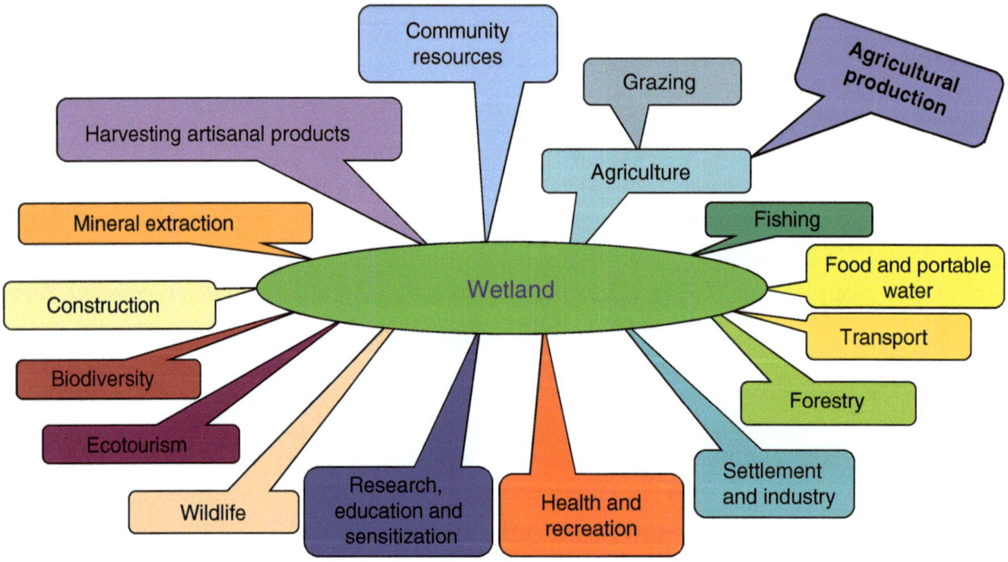

Figure 10.3 Wetland being a resource with different interests with conflicting uses.

other species for its high calorific value and good burning characteristics under wet conditions thus reducing unnecessary wood processing cost and time especially drying before use. Traditional low energy serving (just up to 30% efficient) open-type smoking rafts implanted in kitchens are used. Health problems such as breathing and sight difficulties, headache, general fatigue, and inhalation of poisonous wood gases as well as occasional fire accidents occur during smoking operations (Feka et al. 2009; Ajonina 2023). Invasive plant species also have raided more watercourses and wetlands in Cameroon especially Water Hyacinth (*Eichhornia crassipes*) in waterbodies (Nwamo et al. 2022) and nypa palms (*Nypa fruticans*) in mangrove areas (Moudingo et al. 2019). Mangroves have been heavily lost in Cameroon till remaining less than 200 000 ha (Ajonina 2022).

Current mangrove forest degradation activities seriously threaten this ecosystem and reduce its resilience to mitigate climate change effects (Ajonina 2022). Frequent impacts of invasive species, sea surges, inundations and natural disasters, and other climate change impacts recorded in recent years in the coastal areas are evidence of increasing vulnerability of this system due to increasing human pressures with more than 30% of the people living within the coastal areas.

10.6 Overview of Evolving Strategies for Mainstreaming Wetlands Conservation and Wise Use into Livelihoods Development and Poverty Reduction Efforts in Cameroon

Facing threats, the importance of their protection for the survival of local people is increasingly recognized. Maintaining a balance between the needs of the local communities and the ecological integrity of wetlands ecosystems has been a challenging national and international cause for concern for governments especially as wetland issues transcend national boundaries. Figure 10.4 presents key conservation and mainstreaming elements that are extensively discussed by Ajonina (2016) including linkages to poverty reduction with an imminent march toward wetlands policy development. In certain cases, the level of achievements, shortcomings, and the way forward are further highlighted.

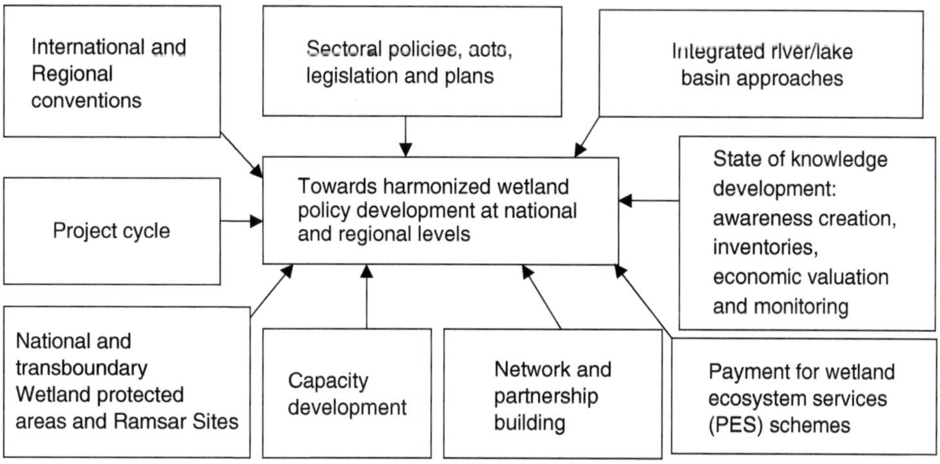

Figure 10.4 Key conservation and mainstreaming elements in African wetlands. *Source:* Ajonina (2009) and Ajonina (2016).

10.6.1 Impact of International and Regional Conventions on Wetlands Conservation

Regulatory instruments at the international level relevance to wetlands have been put in place including: the Bonn Convention on Migratory Species (CMS); the 1971 Ramsar Conventions on wetlands; the "Washington" Convention on International Trade in Endangered Species of Wild Fauna and Flora (CITES) in 1975; the 1991 "Caring for the Earth" document; the Agenda 21 document of the 1992 Rio de Janeiro Earth; the 1997 New York United Nations Convention on the Non-Navigable Uses of International Watercourses. As with other regions of the world, most African countries including Cameroon have manifested direct serious concerns over wetlands in particular, but all have concerns over water resources with implications on wetlands. There are several international agreements that deal with Shared Watercourse Systems on the Protection and Sustainable Utilization of the Water Resources and on the Development of the Lake Basins, while important basin initiatives are being developed, e.g. Lake Chad and Congo basin Commissions.

10.6.2 Status of Mainstreaming Within Sectoral Policies, Plans, Programs, and Projects

Wetlands issues are generally not explicitly stated in many sectoral policies, acts, and legislations (Tata and Ajonina 2007a, b). Many were drawn not originally targeting wetlands but issues related to water management. Wetland issues tend to be dispersed in several ministries and departments including agriculture, fisheries, and forestry. These sectoral policies, programs, and projects usually have inadequate provisions for linkages and coordination and do not incorporate clearly defined collaborative framework within the projects – a situation that at times results to conflicts, duplication of efforts, and waste of resources and funds during the implementation stages. At the project level, except projects implemented under the different integrated river/lake basins and other wetland-specific initiatives, for example, various Integrated Coastal Zone Management (ICZM) projects, e.g. Gulf of Guinea Large Marine Ecosystem Project, and national and transboundary basin initiatives, e.g. Chad Basin and Congo basin, wetlands issues are generally not taken into consideration in other development projects. These sector-oriented projects are mostly directed for general rural development, but may directly or indirectly contribute to wetlands degradation – a situation that can be effectively addressed by the incorporation of Integrated Wetlands Management (IWM) approaches into the project cycle. There is an increasing need to incorporate wetland conservation concerns into these development project cycles especially through Environmental Impact Assessment (EIA) as well as the growing concern that IWM processes ought to be adapted according to the requirements of each project cycle, the type of project, the specific characteristics of the wetland zone under consideration, and the country's general environmental situation.

10.6.3 Integrated River/Lake Basin and Transboundary Approaches

Many integrated river/lake basin initiatives due to these shared watercourses within and between nations have provided opportunities to mainstream wetland conservation issues now coming under different names of IWM, Integrated Water Resources Management (IWRM), Integrated Water Catchment Management (ICM), Integrated River Basin Management (IRBM), and ICZM. Examples include various ICZM projects, e.g. Gulf of Guinea Large Marine Ecosystem Project, and national and transboundary basin initiatives, e.g. River Sanaga/Nyong basin in Cameroon and

Congo basin. Some of the countries within these River/Lake Basins have created Basin Commissions and Conventions that are forums where they can discuss, elaborate, and implement projects of common interest within the basin, e.g. the Lake Chad Basin and Niger Commissions. There is a growing recognition of the benefits derived from taking a river/lake basin approach because it is the only way to identify the most critical needs of each basin. It is a good framework for building partnerships and strengthening teamwork between national institutions and between countries sharing common challenges and common wetlands resources. This basin approach is making many African subregions, national and subregional organizations, including lake/river basin organizations, shifting from immediate and short-term planning to much longer-term vision. A part from problems that usually arise because some countries tend to put their national interests above overall riparian countries' interests, managing lakes and rivers for their long-term sustainable use at the basin scale is the most challenging undertaking because it is costly, complex, and lengthy (Tata and Ajonina 2007a, b).

10.6.4 National, Transboundary Wetland-Protected Areas and Ramsar Sites

10.6.4.1 National Wetlands Conservation

It should be recalled that Cameroon was one of the few countries in the world to have achieved Target 17 of the Aichi Targets for the 2011–2020 Strategic Plan, which stipulated that: "by 2020, at least 17% of terrestrial and inland water areas and 10% of marine and coastal areas, including areas that are particularly important for biological diversity and ecosystem services, are conserved through ecologically representative and well-connected networks of effectively and equitably managed protected areas and other effective area-based conservation measures, and integrated across the terrestrial and marine landscape."

Cameroon has already set a target of 30% protection in Article 22 of the January 1994 Forestry, Wildlife and Fisheries Law, which has guided the 30% rate of protection currently set by global environmental policy. The following articles of the aforementioned law set out the details:

a) Article 21. – (i) Permanent forests or classified forests are those located on the permanent forest estate. (ii) The following are considered permanent forests: State forests; communal forests;
b) Article 22. – Permanent forests must cover at least 30% of the total surface area of the national territory and represent the country's ecological diversity. Each permanent forest must be the subject of a management plan approved by the competent administration;
c) Article 24. – (i) For the purposes of the present law, the following are considered State forests: protected wildlife areas such as national parks, wildlife reserves, areas of hunting interest, state-owned game ranges, state-owned zoos, wildlife sanctuaries and buffer zones. Forest reserves such as integral ecological reserves, production forests, protection forests, recreation forests, teaching and research forests, flora sanctuaries, botanical gardens, and reforestation areas. (ii) The definition, rules, and conditions the use the various types of state-owned forests are laid down by decree.

Cameroon-protected areas that cut across several wetland types (Tomedi Eyango 2017) taking into account various systems of classification including international: International Union for Conservation of Nature (IUCN), Birdlife, Ramsar, CBD, National, and traditional summarized in numbers in Cameroon per type in Table 10.5. All these protected areas based on IUCN model representative of Cameroon's ecosystems as a whole cover more than 20% of the national territory (see map in Figure 10.5). To promote conservation especially critical sites outside protected areas

Table 10.5 International, national, and traditional conservation systems and numbers in Cameroon.

Conservation system or model	Promoter (s)	Category	Numbers in Cameroon	Comments
International conservation models				
Classic protected areas	International Union for Conservation of Nature (IUCN)	National parks and wilderness areas	22	
		Nature reserves (forests and wildlife)	90	5 wildlife reserves, 5 sanctuaries, 5 zoological, 3 botanical gardens, and 72 hunting zones and forest reserves
		Community areas	267	Area of 1 million ha since 1994
		National Monuments	Unknown	
		Marine seascape	3	One in process (23.9% EEZ)
Protection within and may be outside classic protected area	Birdlife International (BL)	Key biodiversity areas (KBAs)	33	2056 km^2, 8.8% national surface
Wetlands of International Importance	Wetlands International, IUCN, BL	RAMSAR sites	7	Six in the process of creation (Base Sanaga, Lake Ossa, Nkam-Wouri, Ma Mbed, Mbed, Upper Nyong, and Batoke-Bakinguele Ramsar sites)
High Conservation Value Areas (IHCVAs)	Forest Steward-ship Council, SFC	Aquatic forests	Many	Found on all habitats especially areas of forest concession areas – Forest Management Units (FMUs)
Ocean conservation	Convention of Biological Diversity (CBD)	Ecologically and Biologically Significant Areas (EBSAs) in the ocean	3	Yet to be fully mapped (Campo, Douala-Edea, Ndongore)
National classification				
Cameroon	Ministry in charge of forests and wildlife	Category A	10	At least 150 ha
		Category B	5	100,000–15,000 ha
		Category C	7	Less than 10,000 ha
		Community and communal forests	701	Legal entities (community-based organizations) and councils for community and communal forests, respectively, but community forests should be at most 5000 ha while the communal forests may take any surface
Traditional classification				
Traditional (sacred sites)	Traditional cultures and customs	Lakes, forests, other sites, etc.	Unknown	Particularly in West Region

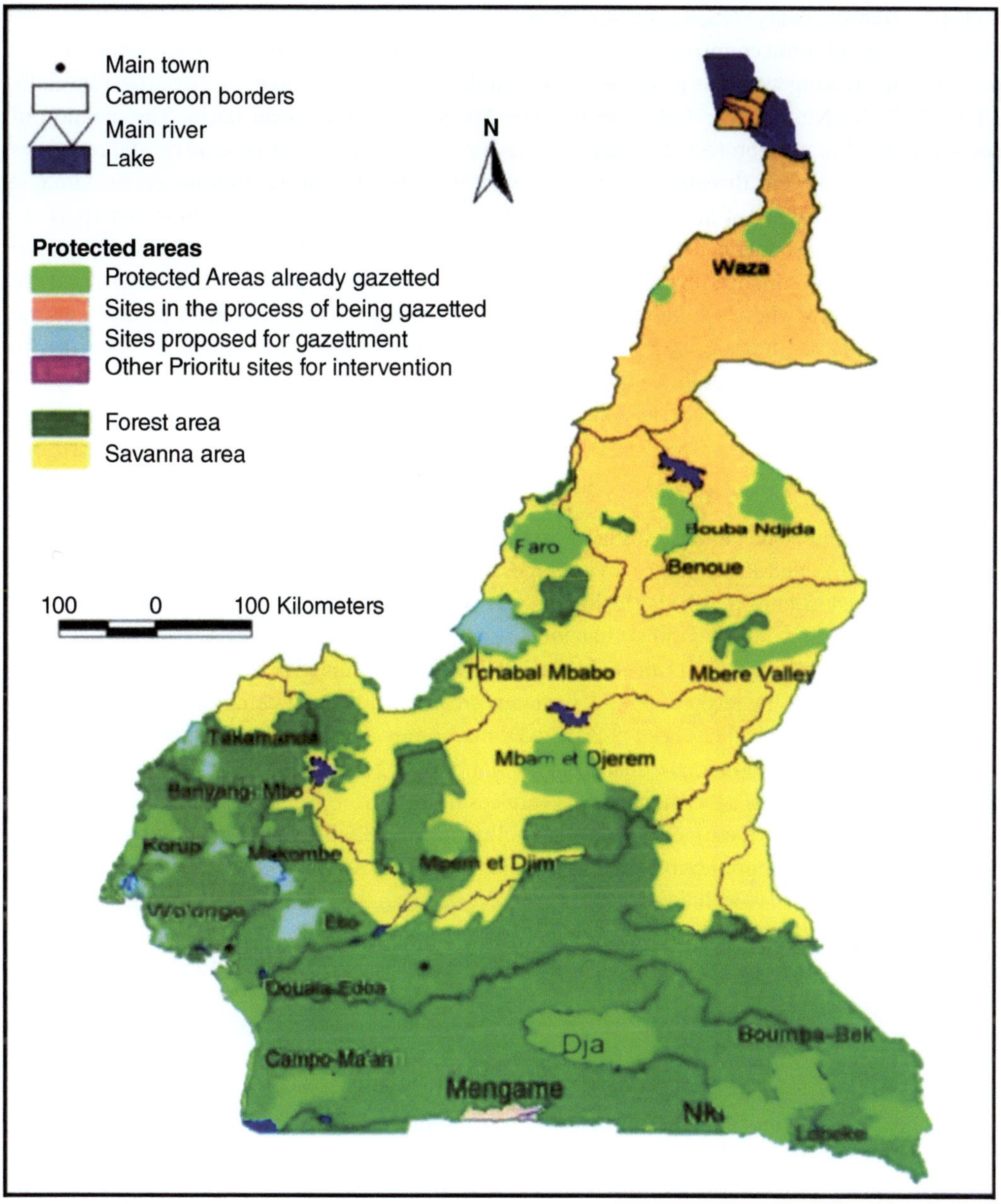

Figure 10.5 Network of continental-protected areas in Cameroon.

belonging to various land-sea holders, Cameroon is currently embarking on the creation of sites known as Other Effective Area-Based Conservation Measures (OECMs) of IUCN 2018 (Jonas and Sandwith 2019) beginning with marine zone where over 15 marine-coastal critical sites have been identified outside marine-protected areas (MINFOF-CWCS 2024).

There are currently three marine national parks all covering 23% of the EEZ of 1540 km^2: Douala Edéa National Park (both terrestrial and marine) (less than 50%) created in 2018; Manyange na Elombo-Campo Marine Park in 2021 (entire marine) created in 2021, and Ndongere National Park (being created).

10.6.4.2 Transboundary Wetlands Conservation

Cameroon neighboring countries especially Chad, Congo, and Nigeria have set up commendable initiatives in creating wetlands protected areas and Ramsar sites in shared watercourses especially Lake Chad, Rio Ntem, Niger Delta-Rio Del Rey Mangroves, and Congo Basin. The creation of coastal wetland/marine protected areas (MPAs) is now a global priority to rescue coastal resources under extremely severe threats, as agreed unanimously by the World Summit on Sustainable Development, 26 August–4 September 2002, Johannesburg, given that these protected areas are now proven positive benefits for fisheries and coastal communities. These initiatives are also seen as a mechanism to enhance ownership to local communities through a clear zoning strategy to protect community fisheries zones often evaded by big fishing interests or simply regarded as "no man's land." These transboundary initiatives could be important vehicles and mechanisms to resolve sectoral and cross-border conflicts and promote subregional peace and stability especially if the necessary policies are developed and the political will to implement them.

10.6.5 Wetland Inventories and Economic Valuation

A significant number of national wetland inventories and mapping such as MINFOF and FAO (2005), MINEDED, and RCM (2017) have been established within the past few years, even though the data collected in some may be fairly limited, with an emphasis on identifying wetlands of importance for conservation purposes rather than for their importance for direct uses and for functions and services, the reasons being mainly due to inadequate capacity in trained personnel and lack of funds. Given the ever-increasing recognition that wetlands are important ecosystems and that they require sound management, a need has been recognized for a more systematic, comprehensive approach to wetland inventory. Few economic valuations of wetlands have been undertaken notwithstanding the nonmarketability of most other nonphysical services derived from wetlands using various market and nonmarket methods (Noumeyi 2015). Such evaluations especially when expressed on cost per ha basis enumerate the importance of wetlands for local people in Africa, and what the costs will be to society if these wetlands should disappear or when comparing the value of the wetland with values of alternative wetland uses. These help decision-makers to take these costs into account in plans to convert wetlands' lands for other purposes, including agriculture and urban or industrial expansions.

10.6.6 Building Networks and Partnerships

There is growing enlargement of the various partnerships/networks at various levels starting from Communty Based Organisations (CBOs), Non Governmental Organisations (NGOs), Governments, Multilateral Development Banks (MDBs), and international/bilateral organizations involved in promoting IWM in particular and sustainable environmental management in general. Very significant is the need for effective collaboration and coordination of activities between partners, in order to minimize the possibilities of unnecessary and costly duplication and overlapping of interventions. The main actors in the area of wetlands in Africa can be classified into the following broad categories (Tata and Ajonina, 2007a):

a) Those that finance projects but are not directly involved in implementation – these include: The ADB, The World Bank, IUCN, UNEP, EU, and Ramsar Secretariat.
b) Those who look for finances to add to their own resources and get involved in direct project implementation in the field – these include WWF and Wetlands International.
c) Bilateral organizations that provide funds and are sometimes involved in implementation: GTZ, DFID, etc.

d) Special projects, such as the Gulf of Guinea Large Marine Ecosystems Management, include important components that deal with coastal wetlands.
e) Various regional and national NGOs that receive external and internal funding to directly implement projects at the local level including NGOs within the Cameroon Mangrove Network (CMN) (CWCS 2024).
f) National governments, some of which are still lagging behind in wetlands management issues within their countries.
g) The riparian local communities that live within and around wetland areas and whose activities impact wetland resources.

There is still a very low level of private sector participation especially extractive industries in wetlands issues though blamed for most destruction of wetlands habitats. There are often conflicts with local communities over issues of compensation. There is a need to encourage a climate of mutual trust and move toward sustainable exploitation of resources with these actors especially as they constitute potential sources of funding for environmental conservation and wise use issues especially in various payments of ecosystems service (PES) schemes.

10.6.7 Payment for Wetland Ecosystem Service Schemes and Carbon Markets

New concepts and approaches are emerging of voluntary material or monetary compensations or rewards by the so-called buyers of ecosystems services to owners or sellers of ecosystem services usually close to ecosystems or wetland habitats for maintaining healthy ecosystems through good land or water management practices that enhance ecosystem service flows especially those that produce externalities (environmental services, ES). The "Payment for Environmental Services" (PES) is part of economic instruments based on market mechanisms in the context of pollution prevention and ecosystem conservation. These are new approaches that promote positive environmental externalities through the transfer of financial resources among the beneficiaries of ES and their suppliers or managers of environmental resources. The PES is based on the principle of beneficiary – payer. It is defined according to Wunder (2005) as: *(a) a voluntary transaction where (b) a well-defined environmental service (or a use of land likely to secure that service), (c) is "bought" by a buyer of service (minimum one), (d) from a service provider (minimum one), (e) if and only if service provider secures the service provision (conditionality).*

The PES schemes focus on the ES provided by forest conservation, reforestation, and sustainable forestry and some practices of agroforestry or silvopastoral farming. The use of these PES schemes for watershed conservation, biodiversity, landscape beauty, and carbon sequestration has become more common over the past decade (Mayrand and Paquin 2004). The concept behind PES being simple and strives to put environmental protection within the perspective that can more closely resemble contracts between service providers and service beneficiaries than the object of governmental regulations.

The first insight of an ecosystem services perspective is that investing in natural capital can prove more efficient than using built capital to deliver key services. For example, floodwaters can be addressed through built capital, such as engineered works (e.g. construction and maintenance of dykes and levees) or through natural capital, such as landscape management (e.g. restoration of wetlands in flood plains). In some instances, perhaps many, landscape management may prove a better public and private investment strategy for providing flood control once one accounts for the positive externalities of improved water quality, wildlife habitat, and recreational amenities. Some PES programs involve contracts between consumers of ecosystem services and the suppliers of these services. However, the majority of the PES programs is funded by governments and involves intermediaries, such as nongovernment organizations. The party supplying the ES normally holds the property rights over an environmental good that provides a flow of benefits to the demanding party in return for compensation. In the case of private contracts, the beneficiaries of the ecosystem services are willing to pay a price that can be expected to be lower than their welfare gain due to the services. The providers of the ecosystem services can be expected to be willing to accept a payment that is greater than the cost of providing the services. PES schemes are now being explored as voluntary schemes by corporate bodies to promote conflict-free exploitation of natural resources in wetland areas with due compensations to resource owners especially local communities.

10.6.8 Toward National Wetland Policies, Legal and Institutional Framework Development

Although the government has elaborated far-reaching biodiversity conservation policies, there is still not adequate provision for wetlands. Wetland issues tend to be treated in a rather scanty and scattered manner in various sectoral policies (Tata and Ajonina 2007a, b). The are many ministries in Cameroon that deal with wetland issues or have something to deal with wetlands, for example, the Ministry of Environment, Nature Protection and Sustainable Development (MINEPDED), Ministry of Forestry & Wildlife (MINFOF), Ministry of Livestock and Animal Industries (MINEPIA), and Ministry of Agriculture and Rural Development (MINADER), though in most cases there are interministerial (intersectoral) coordination committees or by Prime Ministry.

There is an imminent evolution toward wetlands policy development into which wetland conservation issues are mainstreamed. These processes are gradually being co-ordinated harmonized, culminating in the elaboration of wetland policy based on various sectoral policies, action plans, and legal and institutional mechanisms to define conservation priorities at national, regional, and international levels. For example, many partners and network partners have developed mechanisms to establish National Policies, Legislation and Institutional Frameworks that should be strengthened and harmonized to address cross-border wetlands, in order to achieve more effective IWM. The starting point in most cases is to revisit the country's environmental management plans, sustainable development strategies, and Poverty Reduction Papers to ensure that wetland issues are inadequately addressed therein with appropriate legal and institutional backing mechanisms before the approval of loans.

10.6.9 Wetland Conservation and Poverty Reduction Efforts

Good progress has been made in the field on poverty reduction by working with local communities in wetland areas, if success in mainstreaming mangrove and wetland biodiversity conservation effort is to be achieved and maintained sustainably. Linkages between poverty and wetlands are complex because of the nature or type of definition attached to poverty, the extent of poverty, and very importantly the causes of poverty.

Conventionally, poverty was defined by income per household and macro-economic indicators. Within this narrow definition of poverty, the value of nature and the role natural resources and ecosystem services play is difficult to deduce. Today, poverty is better understood broadly in many dimensions in terms of: access to and quality of physical and social infrastructure (roads, housing, water, and sanitation); social support structures and social progress (distribution of benefits of resource exploitation, social relations and networks including gender equality and children rights, empowerment, and cultural identity); capacity development (knowledge, skills, education, and health); rights or legal position (land tenure including demarcation and elimination of overlapping land rights, laws, and policies – specifically those which govern community rights and access to natural and other resources, benefit sharing); and financial conditions (income, access to credits, cash flow).

Nature and natural resources form the foundation of economies in the region. Ecosystem products (food, fodder, timber, fish, etc.) and services (regulation of climate and clean air, water flow, coastal protection, etc.) support these dimensions. When the ecological systems fail and natural resources are depleted, quality of life decreases and poverty (as defined above) is exacerbated. Nature conservation and environmental interventions are therefore unsustainable if poverty is not addressed. Consequently, poverty reduction efforts have been hinged around addressing sustainable resource use. This has been through contributions to improved livelihood conditions of local communities (increased income levels, employment, enterprise development, ecotourism development, etc.) by creating opportunities and mechanisms by many organizations for added value and rates of return based on secured access and good governance; rational methods of resource extraction, processing, storage, and use; wetland restoration; etc. Mechanisms of poverty reduction efforts can be through: capacity building; secured access to resources, good governance, participation, and changing ownership patterns; improved techniques of resource extraction, processing, storage, and use; wetland restoration; marine protected areas; development of alternatives; mainstreaming wetland conservation into country strategy and poverty reduction strategy papers; development of ecotourism; etc.

10.7 Conclusion

The scope of wetlands in Cameroon in terms of habitat and species diversity and distribution, their importance given their environmental and ecological services, and their socioeconomic value to human populations and as well as for wildlife conservation have been highlighted. Cameroon wetlands are already threatened, degraded, and sometimes destructed as a consequence of both natural and anthropogenic causes. This dismal situation has been recognized by the donor and development agencies and the government, but measures to address the situation so far have been inadequate. There is an evolving march toward filling policy gaps by developing national wetlands policy within which wetlands conservation and poverty reduction issues are mainstreamed. It has also been demonstrated that site programs largely employing the ecosystem approach can contribute to strengthening mainstreaming at different levels be it national and international levels.

Acknowledgements

This chapter is a fruit of work done by experts and key stakeholders' consultations and interactions at all levels (Government Services, NGOs, Private Sector, and local communities including elected representatives, traditional authorities, and opinion leaders) including mangrove villages.

References

Adams, W.M. (1996). Economics and hydrological management of African floodplains. In: *Water Management and Wetlands in Sub-Saharan Africa* (ed. M.C. Acreman and G.E. Hollis). Gland: IUCN.

Ajonina, G.N. (2008). Inventory and modelling mangrove forest stand dynamics following different levels of wood exploitation pressures in the Douala-Edea Atlantic coast of Cameroon, Central

Africa. *Mitteilungen der Abteilungen für Forstliche Biometrie, Albert-Ludwigs- Universität Freiburg.* 2: 215 p.

Ajonina, G.N. (2009). Mainstreaming coastal wetland biodiversity conservation in African mangroves, Cameroon. In: *Mainstreaming Wetland Biodiversity Conservation: Experience and Lessons Learned in Practical Applications of Mainstreaming* (ed. E.D. Ongley and M.H. Allard). Beijing: UNDP and the State Forestry Administration of China 314 pp.

Ajonina, G.N. (2016). Keynote address: mangroves and wetlands of sub-saharan Africa: potential for sustainable livelihoods and development. *Proceedings of the 38th Annual Conference of Forestry Association of Nigeria* (7–11 March 2016), Portharcourt, Rivers States, pp. xxii–xLix.

Ajonina, G.N. (2022). Cameroon mangroves: current status, uses, challenges, and management perspectives. In: Das, S.C., Pullaiah,T., Ashton, E.C. (eds) Mangroves: Biodiversity, Livelihoods and Conservation. Springer, Singapore. https://doi.org/10.1007/978-981-19-0519-3_21.

Ajonina, G.N. (2023). *Programme de suivi des impacts avals du barrage de Lom Pangar: Synthèse bibliographique et contribution au plan d'action «estuaire» Sanaga.* France: Rapport de consultation BRLi/JMN 102 pp.

Ajonina, G.N. and Tomedi Eyango, M. (2014). Aquaforests and aquaforestry: Africa. In: *Encyclopedia of Natural Resources: Land*, 16–38. New York: Taylor and Francis. (Published online: 21 October 2014). http://www.crcnetbase.com/doi/10.1081/E-ENRL-120047504.

Ajonina, G.N., Timba, M., Njie, F., and Ngnintedem, J.M. (2023). Dénombrement International des Oiseaux d'eau de janvier 2023 au Cameroun. In: *Simultaneous January 2023 Waterbird Census Along the East Atlantic Flyway: National Reports* (ed. G. Agblonon, A.Y. Diallo, K. Gueye, et al.). Wilhelmshaven, Wageningen, Cambridge: Wadden Sea Flyway Initiative p/a Common Adden Sea Secretariat, Wetlands International, BirdLife International.

Ayissi, I., Ajonina, G.N., and Angoni, H. (2014). Status of large marine flagship faunal diversity within Cameroon estuaries of Central African Coast. In: *The Land/Ocean Interactions in the Coastal Zone of West and Central Africa, Estuaries of the World* (ed. S. Diop et al.), 97–107. Springer International Publishing Switzerland 2014 https://doi.org/10.1007/978-3-319-06388-1_9.

CWCS (2024). *CWCS Annual Report 2023 /Rapport Annuel 2023.* Cameroon Wildlife Conservation Society.

Failler, P. and Njock, J. (2011). *Stratégie de recherche halieutique et aquacole du Cameroun.* Institut de Recherche Agricole pour le Developpement (IRAD) 49 pp.

Feka, N.Z., Chuyong, G.B., and Ajonina, G.N. (2009). Sustainable utilization of mangroves using improved fish-smoking systems: a management perspective from the Douala-Edea wildlife reserve, Cameroon. *Tropical Conservation Science* 2 (4): 450–468.

Folack, J. (2024). *Lignes directives national des Aires protégées marines et autres measures de conservation au Cameroun.* Cameroun: Ministere des Forets et de la Faune.

Jonas, H. and Sandwith, T. (ed.) (2019). *Towards Recognising, Reporting and Supporting OECMs: Report of the Fourth Expert Meeting of the IUCN-WCPA Task Force on Other Effective Area-based Conservation Measures.* Gland: IUCN.

Keay, R.W.J. and Hepper, F.N. (1954–1972). *Flora of West Tropical Africa*, 2e, vol. I–III. London: Millbank.

Laclavere, G. (1980). *Atlas of the United Republic of Cameroon.* Paris: Edition Jeune Afrique 72 pp.

Mayrand, K. and Paquin, M. (2004). *Le paiement pour les services environnementaux: Etude et évaluation des systèmes actuels.* Montréal: UNISFERA 59 p.

Millennium Ecosystem Assessment (MEA) (2005). *Ecosystems and Human Well-being: Synthesis.* Washington, DC: Island Press.

MINEP-RCM (2017). *Etats des lieux et Atlas sur les mangroves du Cameroun*. Ministère de l'Environnement et Protection de la Nature (MINEPDED) et Réseaux Camerounais de Mangroves (RCM).

MINFOF and FAO (2005). *Évaluation des ressources forestières nationales du Cameroun 2003-2004: Résumé des résultats de l'Inventaire Forestier National*. Ministère des Forêts et de la Faune (MINFOF) et Organisation des Nations Unies pour l'Alimentation et l'Agriculture, FAO 26 pp.

MINFOF-CWCS (2024). *Inventory of Coastal-Marine Other Effective Conservation Area-based Measures (OECMs) in Cameroon*. Yaounde: Ministry of Forestry & Wildlife and Cameroon Wildlife Conservation Society 52 pp.

Moudingo, J.-H., Ajonina, G.N., Dibong, D., and Tomedi, M. (2019). Introduction, distribution and drivers of non-native mangrove palm *Nypa fruticans* Van Wurmb (Arecaceae) in Cameroon, Gulf of Guinea. *Advances in Ecological and Environmental Research* 1: 1–13.

Noumeyi, S.M.J. (2015). *Evaluation économique des biens et services écosystémiques: Cas de la mangrove et forêts associées du paysage côtier Douala-Edéa au Cameroun*. Diplôme Master, Université Senghor: Egypt 93 pp.

Nwamo, R.D., Ajonina, G.N., Ngwasiri, P.N. et al. (2022). Problems of invasive species of water hyacinth (*Eichhornia crassipes* [Mart.] Solms) in Cameroon with special reference to its eradication and valorization: a bibliographical review. *Energy and Environment Research* 12 (1): 2022. https://doi.org/10.5539/eer.v12n1p56.

Olivry (1986). *Fleuves et Rivières du Cameroun*, 733. Paris: MESRES-ORSTOM.

Raunkaier, C. (1934). *The Life Forms of Plants and Statistical Plant Geography*. Clarendon Press: London.

Schnell, R. (1971). *Introduction à la Phytogéographie des Pays Tropicaux. Milieux, les groupements Végétaux*, vol. II, 503–951. Paris: Edition Gauthier-Villars.

Tata, F.T. and Ajonina, G.N. (2007a). *Wetlands assessment, planning, management and monitoring in Africa. Desk Study*. African Development Bank Publication 123 pp.

Tata, F.T. and Ajonina, G.N. (2007b). *Wetlands Assessment, Planning, Management and Monitoring in Africa. Selected Case Studies*. African Development Bank Publication 97 pp.

Tomedi Eyango, M. (2017). *Bassin versant du Nkam-Wouri: prototype de la biodiversité des zones humides du Cameroun*, 262p. L'harmattan.

White, F. (1986). *La végétation de l'Afrique. Mémoire accompagnant la carte de végétation de l'Afrique*. UNESCO/AET FAT/UNSO.

Wunder, S. (2005). *Payments for Environmental Services: Some Nuts and Bolts*. CIFOR Occasional Paper 42. Bogor: Center for International Forestry Research.

Yemele, O.M., Ngaba, M.J.Y., and Ajonina, G.N. (2023). Stakeholder perceptions of wetlands management effectiveness in Cameroon. *Journal of Ecology and the Natural Environment* 15 (3): 50–60.

11

Wetlands of Ghana: Biodiversity, Community Livelihoods, and Conservation

Yaa Ntiamoa-Baidu[1,2] *and Emmanuel Nii Attram Taye*[1,2]

[1] *Centre for Biodiversity Conservation Research, University of Ghana, Accra, Ghana*
[2] *Department of Animal Biology and Conservation Science, University of Ghana, Accra, Ghana*

11.1 Introduction

Globally, wetlands constitute one of the most productive ecosystems, supporting a wide array of fauna and flora and providing a whole range of ecosystem services that are essential for human health and development as well as diverse resources and livelihoods for communities that live in and around wetlands. In Africa, wetlands are critical for the survival of its people (Thieme et al. 2005; Gardner et al. 2015; Okonkwo et al. 2015; IPBES 2018; Ramsar Convention on Wetlands 2018; Stephenson et al. 2020). Wetlands constitute the backbone of agricultural productivity in Africa. In many rural communities, small wetland formations, such as streams, irrigation dams, water reservoirs, and excavations, provide a lifeline on which people depend extensively for crop farming and animal husbandry. Wetland ecosystems also provide basic needs such as water for consumption and domestic use. It is estimated that more than 50% of the human population live within 3 km of a freshwater body, although freshwater ecosystems cover less than 1% of the earth's surface (Kummu et al. 2011; WWF 2022). Wetlands also play an important role in the sociocultural lives of rural communities.

Wetlands provide essential ecosystem services, including, maintenance of water cycle, water storage and purification, storm protection and groundwater recharge, and support-rich biodiversity, comprising a wide diversity of invertebrates, amphibians, reptiles, fish, and waterbirds, which often congregate in high concentrations. It is estimated that more than 10% of known animals and about one-third of known vertebrate species are associated with freshwater ecosystems (Balian et al. 2008; WWF 2018). A diverse range of mammalian species, including rare species such as the Manatee *Trichecus senegalensis* and the Sitatunga *Tragelphus spekii*, depend on wetlands. Brackish water lagoons and inland wetlands support marine and freshwater fisheries, which contribute a significant proportion of protein intake in Africa. The 2020 report on the State of World Fisheries and Aquaculture by FAO estimates that fish contributed 17% of animal proteins consumed worldwide and 7% of all proteins consumed, a figure that rises to 50% or more in some countries, including Ghana, the Gambia, and Sierra Leone (FAO 2020).

Despite the great importance of wetlands to human survival, these ecosystems are among the most threatened ecosystems in the world. It is estimated that the coverage of wetlands globally

Wetlands of Tropical and Subtropical Asia and Africa: Biodiversity, Livelihoods and Conservation,
First Edition. Edited by Thammineni Pullaiah.

may have declined by as much as 87% since 1700 (Davidson 2014). The losses in wetland ecosystems have been attributed to the numerous threats that they face, including habitat destruction, fragmentation, and modification through drainage and conversion, pollution, and climate change (Stayer and Dudgeon 2010; Ramsar Convention on Wetlands 2018; WWF 2018).

Ghana is endowed with diverse wetlands, ranging from the vast expanse of brackish water lagoons, marshlands, and floodplains associated with the Volta delta to numerous small freshwater impoundments scattered in the northern parts of the country (Gordon 2000; Ntiamoa-Baidu and Gordon 1991). In this chapter, we present the wetlands of Ghana. We highlight the types of wetlands in the country, their status, changes in the extent of coverage over the past years, their biodiversity richness and importance for people's livelihoods, threats faced by the Ghanaian wetlands, and the various actions and initiatives taken by the country to protect its wetlands.

11.2 Types of Wetlands in Ghana

Wetlands have been defined in various ways, the key characteristic being the presence of water, permanent or temporary; halfway between terrestrial and aquatic systems and supporting distinctive flora and fauna (Cowardin et al. 1979; O'Keefe et al. 2000; Miller 2004; Wagner 2004; Mitsch and Gosselink 2015). The Convention on Wetlands (Ramsar Convention) defines wetlands as "areas of marsh, fen, peatland or water, whether natural or artificial, permanent or temporary, with water that is static or flowing, fresh, brackish or salt, including areas of marine water the depth of which at low tide does not exceed six metres." The convention groups all wetlands into three categories: Marine and Coastal Wetlands, Inland Wetlands, and Human-made Wetlands (Ramsar Secretariat 2016). The convention recognizes five major natural wetland types: marine (coastal wetlands including coastal lagoons, rocky shores, seagrass beds, and coral reefs); estuarine (including deltas, tidal marshes and mudflats, and mangrove swamps); lacustrine (wetlands associated with lakes); riverine (wetlands along rivers and streams); and palustrine ("marshy" – marshes, swamps, and bogs). The convention also recognizes human-made wetlands such as irrigated agricultural lands, fishponds, salt pans, reservoirs and small dams, gravel pits, and wastewater treatment ponds.

All the different categories and types of wetlands recognized by the Ramsar Convention can be found in Ghana. The Volta River and its tributaries, the Black Volta and the White Volta, that traverse the country from the coast to the northern boundaries of neighboring countries, create a huge network of open water bodies, marshlands, and floodplains that are critical for biodiversity and community livelihoods. Other river basins in the country with important wetland associations include the Amansuri, Ankobra, Densu, Oti, Pra, and Tano.

11.2.1 Coastal Wetlands

A survey of the Ghana coast in the early 1980s aimed at identifying key habitats for waterbirds documented over 50 coastal wetlands along the 550 km stretch of coastline (Ntiamoa-Baidu and Grieve 1987; Ntiamoa-Baidu and Hepburn 1988; Ntiamoa-Baidu and Gordon 1991). The majority of these were small estuarine habitats, but some were significantly large expanse of lagoons (Figure 11.1). Lagoons along the Ghana coast have been classified as open or closed lagoons. The open lagoons have a permanent connection to the sea and are associated with rivers. The closed lagoons are usually formed behind sand bars and have no permanent connection to the sea, so seawater replenishment occurs by seepage through sand dunes (Kwei 1977; Mensah 1979; Gordon 1987; Gordon et al. 2000).

Figure 11.1 Map of Ghana showing the location of main coastal lagoons.

Wetlands along the Ghana coast fall into three main categories, estuaries and brackish water lagoon habitats (e.g. Keta and Songor lagoons, Amisa, Narkwa, and Amansuri estuaries), open and closed lagoon systems (e.g. Korle, Sakumo, and Muni lagoons), estuarine and salt pan complexes (e.g. Densu delta, Lalui, and Ningo). The most significant wetlands on the Ghana coast in terms of size, biodiversity, and livelihoods are the Keta and Songor lagoon complexes associated with the Volta delta, Sakumo lagoon, Densu delta, and Muni lagoon (Ntiamoa-Baidu and Gordon 1991). All five wetlands have been designated as Ramsar sites. Other small, but significant wetlands on the Ghana coast include the Jange, Kpeshi, Mukwe lagoons, Amisa, Narkwa, and Amansuri (Ntiamoa-Baidu 1991b; Gordon 2000; Gbogbo 2007; Owusu 2007; Nukpezah et al. 2022).

11.2.1.1 Keta Lagoon

The Keta Lagoon is by far the largest wetland along the Ghana coast, lying to the east of the Volta River estuary (Lat. 5.914201, Long. 0.986607). The wetland comprises an open water body, a large expanse of surrounding floodplain consisting of marsh, scrub and farmlands, mangrove swamps, and a sandy beach. The lagoon is an extensive brackish water body, with an estimated area of 300 km^2, stretching over ca. 40 km along the coast and separated from the sea by a narrow ridge of 2.5 km at its widest portion. Although an open lagoon, the only direct outlet into the sea is at Atiteti, near Anyanui. Freshwater inflow into the Keta lagoon is from the Todzie River, which flows into the Avu lagoon and overflows into the Keta lagoon, and from the Aka and Belikpa rivers in the north. The inflow from the Volta River into the Keta Lagoon has been greatly minimized by the regulation of the flow as a result of the construction and operationalization of the Akosombo dam (Piersma and Ntiamoa-Baidu 1995; Finlayson et al. 2000; Ntiamoa-Baidu et al. 2017). The Keta lagoon and surrounding floodplain support numerous settlements that heavily exploit the resources from the wetland, primarily fisheries, mangrove stands for fuelwood, and other wetland plants for building construction and production of a range of household items. Sand and salt mining are also common in some parts of the wetland. The greatest socioecological threats to the wetland and communities who live in and around the wetland are as follows: (i) the severe coastal erosion that has led to loss of homes, (ii) periodic flooding during the rainy season leading to destruction of houses, farms, and property, (iii) insufficient land, that is high enough for settlements and farming, resulting in that part of coast having some of the highest population densities in the country (Piersma and Ntiamoa-Baidu 1995). The Keta wetland was designated as a wetland of international importance (Ramsar site) based on the total waterbird population the site regularly supports and the occurrence of internationally important populations of eight species of waders (Ntiamoa-Baidu and Gordon 1991). However, like all the other coastal wetland sites in the country, the biodiversity importance of the site has declined over the years (Lamptey and Ofori Danson 2014; EPA 2017; Mattah et al. 2024).

11.2.1.2 Songor Lagoon

The Songor wetland is the second of the two lagoons associated with the Volta delta and lies to the west of the estuary (Lat. 5.797843, Long. 0.434680). The wetland comprises a brackish water lagoon, extensive mudflats, islands and salt flats, a floodplain with degraded mangroves, and coastal savannah vegetation consisting of waterlogged grasslands, riverine woodlands and scattered thickets to the north, and a broad sandy beach, which is important for nesting turtles. The lagoon extends some 20 km along the coast and ca. 8 km inland, covering an area up to ca. 115 km^2. Songor is a very shallow lagoon with a depth of only 50 cm in several places (Piersma and Ntiamoa-Baidu 1995) and it is closed, so seawater inflow occurs by seepage through the sand dune. Freshwater inflow occurs through small streams, notably Sege and Zano. The channels, which in the past provided direct connections with the Volta River, have been mostly blocked by construction of the Akosombo dam. Small villages, occupied primarily by fishermen, are found along the narrow sand dune, while larger towns are concentrated along the northern edges of the lagoon. The Songor area is important for both commercial and traditional salt production. Significant stretches of the lagoon dry up during the dry season, at which time the communities collect salt from the mudflats. The Songor wetland faces similar threats as the Keta lagoon, the area is low lying, and the potential and real-time impacts of climate change and sea level rise are obvious (Agyeman-Bonsu et al. 2008). Boakye-Boaten (2010) projected that villages, such as Pute, Totokpoe, and Lolonya, situated on the narrow sand dune that separates the lagoon from the sea could disappear by 2100. The Songor wetland was designated as a Ramsar site based on the internationally significant total numbers of waterbirds and population of seven species of waterbirds that the site regularly supported.

11.2.1.3 Sakumo Lagoon

The Sakumo wetland is situated on the coastal road between the cities of Accra and Tema (Lat. 5.619430, Long. −0.038873) and comprises an open lagoon, a floodplain, and a freshwater marsh. The brackish water lagoon covers an area of 1–3 km^2 while the floodplain extends over ca. 7 km^2 depending on the season (Ntiamoa-Baidu and Gordon 1991). A narrow sand dune on which the Accra-Tema coastal road was built separates the lagoon from the sea. The lagoon was fed by the Gbagbla-Ankonu and Mamahuma streams, but with the intensive peri-urban farming taking over the floodplain and marshlands, the lagoon receives very little freshwater inflow. Sakumo lagoon was fished intensively in the 1970s and 1980s by communities in Sakumono, Tema New Town and Teshi-Nungua, with the lagoon providing significant livelihoods for the local communities (Pauly 1975; Mensah 1979; Ntiamoa-Baidu 1991b). The Sakumo wetland was the third most important wetland on the Ghana coast for waterbirds at the time of its designation as a Ramsar site, holding internationally important populations of six species of waders. The lagoon is presently a mere shadow of what it was at the time of designation as a Ramsar site, with serious adverse changes in its ecological character. This has been caused by the many housing and industrial developments in the catchment area, severe pollution from domestic and industrial waste and effluents, resulting in high nutrients, particularly phosphate and pesticide pollution from farming in the catchment area (Nartey et al. 2011; Nonterah et al. 2015).

11.2.1.4 Densu Delta

The Densu delta wetland (also known as Sakumo I or Panbros Salt Pans) is one of the two wetlands along the Ghana coast that are situated close to the capital city of Accra (Lat. 5.536430, Long. −0.288623). The wetland is ca. 11 km west of Accra and lies in the valley formed by the Weija-McCarthy and Aplaku-Takuse Hills. It encompasses the Densu river estuary, a commercial salt industry, a network of natural brackish water bodies that are heavily fished by local communities, saltwater reservoirs and salt pans, scrub, and marshlands (Ntiamoa-Baidu and Hollis 1988). The bunds of the salt pans are used by large populations of terns for roosting. The wetland receives freshwater mainly from the Densu river, but the hydrology of the area has been impacted heavily by the construction and operation of the Weija dam that supplies water to parts of Accra. The Densu delta wetland is inundated periodically during the rainy season, and especially when the Water Company releases water to protect the dam. The Densu delta wetland was designated as an internationally important wetland based on the populations of waterbirds it supported, but over the years demand for land in the urban area for housing and industrial developments has eroded more than 60% of the land originally designated for protection. Although the site is still important for waterbirds, it is obvious that the entire wetland would have been lost without the private ownership and protection given by the Panbros Salt company. In addition to the threat posed by the housing and industrial developments, the site is also exposed to pollution, overexploitation of resources, and the impacts of climate change and sea level rise (Appeaning-Addo et al. 2011; Appeaning-Addo and Adeyemi 2013).

11.2.1.5 Muni Lagoon

The Muni lagoon lies on the outskirts of Winneba town (Lat. 5.325450, Long. −0.652746), about 55 km west of Accra. The lagoon normally covers an area of 3 km^2 but expands considerably over the ca. 6 km^2 floodplain during the rainy season (Ntiamoa-Baidu and Hollis 1988). The Muni wetland constitutes the catchment area of three streams, Pratu, Boaku, and Muni, which are seasonal and drain into the lagoon. The wetland encompasses the open water body, two protected areas, the Yenku A and Yenku B forest reserves, the traditional hunting area of the Efutu people and a broad sandy beach. Muni lagoon lies behind a sand bar and is effectively a closed lagoon, cut off from the sea for most of the year, creating hypersaline conditions. There are reports that the lagoon and surrounding communities get flooded occasionally during periods of heavy rains. At this time, the sand bar may be breached naturally or by the people in the communities who dig a canal to let the flood waters drain into the sea to prevent flooding of people's homes (Gordon 1994; Tumbulto and Bannerman 1995; Gordon et al. 2000; Ryan and Ntiamoa-Baidu 2000). The Muni wetland was designated as a wetland of international importance on the basis of the waterbird species and populations that the site supported (Ntiamoa-Baidu et al. 2000; Ryan and Ntiamoa-Baidu 2000).

11.2.2 Inland Wetlands

Information on the inland wetlands of Ghana is limited. The first wetland designated as a Ramsar site in Ghana was the small inland wetland, the Owabi Wildlife Sanctuary, in the Kumasi Metropolitan Area (KMA). This site is a rather unique nature reserve, comprising a bird sanctuary, a plantation with exotic plants, and a lake formed through the construction of a dam across the Owabi River to supply potable water to people living in the Kumasi metropolis. The Owabi Wildlife Sanctuary is notable for its riverine forests and the diversity of aquatic plants and butterfly species it supports. The reserve has a significant appeal for visitors but has come under severe pressure from intensive farming and demand for land for human settlements. Other notable wetlands in the Ashanti region include the Barekese dam, also in the KMA, and Lake Bosumtwe.

A survey of inland wetlands in the northern half of the country by the Ghana Wildlife Society in 1999, aimed at documenting their ornithological importance, identified 100 inland wetlands comprising 13 natural wetlands and 87 human-made wetlands (irrigation dams, water reservoirs, and excavations) (Nyame 2000). A more recent survey of the northern wetlands by the Centre for African Wetlands (now Centre for Biodiversity Conservation Research), with the primary aim of documenting community use of wetlands and how communities could be supported to manage their wetlands to enhance their livelihoods, provides information on 34 wetlands associated with rivers and dams (Quartey et al. 2017). These included the Yapei, Bontanga dam, Nasia river, and Janga wetland in the northern region; the Binaba, Vea, and Tono dams in the Upper East region; the Wechiau and Kulpawn wetlands, the Katua, Busa, and Sankana dams in the Upper West regions. Although most of these wetlands are very small, they are critical for the livelihood of communities living around them, providing the main source of water for domestic use, livestock production, and for farming, especially during the dry periods of the year. Several of the northern small wetlands have also been studied in relation to the initiatives by the Ghana government to encourage lowland rice farming (Nsor et al. 2014). Other general studies on the inland wetlands include Nyarko et al. (2015); Kuusaana et al. (2021).

11.3 Biodiversity Richness of Ghanaian Wetlands

Wetlands have been described as "cradles of biological diversity" and are known to be among the world's most productive environments, with many species of plants and animals depending on wetlands for survival (Ramsar Convention 2016; IPBES 2018; WWF 2018). Wetlands are critical as reservoirs for plant genetic material; they provide habitats for high numbers of invertebrate species, amphibians, reptiles, fish, birds, and mammals.

The few Ghanaian wetlands that have been inventoried and assessed for the fauna and flora they support demonstrate the biodiversity richness of the country's wetlands. While a few taxonomic groups such as fish and waterbirds have been the subjects of many studies and are by far the most abundant groups on the Ghanaian wetlands, data on others such as invertebrates, amphibians, and reptiles are limited.

11.3.1 Waterbirds

The importance of the Ghanaian coastal wetlands for waterbirds is very well documented (Grimes 1974, 1987; Macdonald 1978; Ntiamoa-Baidu and Grieve 1987; Ntiamoa-Baidu 1988, 1991a, 1993; Ntiamoa-Baidu and Hepburn 1988; Ntiamoa-Baidu et al. 1998, 2000, 2014). Over 70 species of waterbirds have been recorded on the Ghana coast, with 9 species of shorebirds occurring in internationally important numbers at various points in time (Appendix 11.1). These are the Curlew sandpiper *Calidris ferruginea*, Sanderling *Calidris alba*, Little stint *Calidris minuta*, Greenshank *Tringa nebularia*, Spotted redshank *Tringa erythropus*, Ringed plover *Charadrius hiaticula*, Black-tailed godwit *Limosa limosa*, Avocet *Recurvirostra avosetta*, and the Black-winged stilt *Himantopus himantopus*. The Avocet, which in past occurred in large aggregations at Songor, now occurs in small numbers of less than 100. The earlier surveys of the Ghana coast for waterbirds identified at least eight coastal wetland sites (Keta lagoon, Songor lagoon, Sakumo lagoon, Korle lagoon, Densu delta, Muni lagoon, Elmina Salt pans, and Amansure/Esiama Beach) that supported significant populations of waterbirds. The total bird populations using these sites (ranging from 23 000 to 111 000) and the occurrence of several species of waders in internationally significant numbers provided the justification for the designation of five of coastal wetlands as Ramsar sites in 1992 (Ntiamoa-Baidu and Gordon 1991). The waterbird population threshold for qualification for designation as a Ramsar site at the time was 20 000 (Smit and Piersma 1989).

Apart from waders, the wetlands along the Ghana coast also support large numbers of terns, herons, and ducks, with six species of terns, Sandwich tern *Sterna sandvicensis*, Royal tern *S. maxima*, Little tern *S. albifrons*, Common tern *S. hirundo,* Roseate tern *S. dougallii*, and Black tern *Chlidonias*, occurring in internationally important populations.

While the ecological character of some of the wetlands has suffered adverse changes as a result of habitat destruction and pollution, leading to steep declines in the waterbird populations using the sites (Lamptey and Ofori-Danson 2014), the Ghanaian coastal sites still support significant waterbird populations (Appendix 11.1) and remain important for both resident and migratory birds using the East Atlantic Flyway, as staging, feeding, and roosting sites (Wetlands International 2002) and as nesting sites for resident species (Taye et al. 2022). The populations of a few of the waterbird species using the Ghana coast have seen increases, e.g. the Sanderling. Ntiamoa-Baidu et al. (2014) reported that the sanderling population along the Ghana coast had increased fourfold over a 20-year period.

11.3.2 Fish Diversity

Coastal wetland habitats, including the open water bodies of coastal lagoons, provide important fisheries as well as spawning and nursery grounds for some commercial marine fish species, such as *Mugil* spp. *Caranx* spp. and *Ethmalosa fimbriata* (Koranteng et al. 2000; Asmah et al. 2008; Gbogbo et al. 2008; Kumi et al. 2015). The studies of Koranteng (1995) and Koranteng et al. (1998) under the Ghana Coastal Wetlands Management Project (CWMP) provided detailed information on the fish diversity of the five coastal Ramsar sites. The dominant species of fish in the coastal lagoons are *Sarotherodon melanotheron, Hemichromis fasciatus, Tilapia zillii,* and *Ethmalosa fimbriata.* Earlier studies in the 1970s (e.g. Pauly 1976) reported four species of fish that were abundant or fairly common in the Sakumo lagoon, *Saratherodon melanotheron*; *Ethmalosa fimbriata, Elops senegalensis, Oxyurichtys occidentalis,* with a further five species, *Lutjanus* sp., *Synaptura punctissima*, *Caranx hippos, Mugil* sp., and *Pomadasys jubelini* being uncommon or occurring in small numbers. Koranteng et al. (2000) reported three finfish belonging to three genera and two families in the Muni lagoon, in addition to eight species of shell-fish belonging to eight families and eight genera. The most abundant mollusc at Muni was *Tympanotonus fuscatus*, a species that occurred abundantly on the shores of other small coastal lagoons (Ntiamoa-Baidu 1991a). *Sarotherodon melanotheron* was the dominant species in all the coastal lagoons, accounting for 68% of the total catch in the Densu delta (Enstuah-Mensah 1998) and 97% of the total catch at Sakumo lagoon (Koranteng 1995). A more recent study of the fisheries of the Sakumo lagoon by Sawyerr et al. (2022) reported *Oreochromis niloticus* to be the most abundant species by weight, accounting for 47.6% of samples collected, while *Sarotherodon melanotheron* accounted for 35.6%.

A comparison of the trends in fisheries of the coastal wetlands indicates that there have been significant decreases in the diversity, relative abundance, and sizes of fish species caught in the coastal lagoons. While *Sarotherodon melanotheron* remains the most abundant fish species in the lagoons, other species reported in the 1970s are either completely absent or occur in very small numbers, and the size of the *Sarotherodon melanotheron* has decreased from a maximum length of 19.0 cm and a modal size class of 10–11 cm to a maximum length of 12 cm and a modal class of less than 7 cm (Pauly 1976; Ntiamoa-Baidu 1991b; Koranteng et al. 2000; Dankwa et al. 2004; Sawyerr et al. 2022). The decline in the fisheries of the coastal lagoons has been attributed to overexploitation and pollution of the lagoon waters.

Ghana's freshwater fish includes 28 families, 73 genera, and 157 species (MES 2002). About 124 fish species from 62 genera and 26 families occur in the major rivers. A study of the fisheries of the Bosumtwe lake basin by Whyte (1975) recorded 11 species of fish and provided baseline data for the fisheries of inland wetlands. Nunoo et al. (2012) reported 19 species of fish belonging to 15 families from the Owabi Ramsar site, with the three species *Hemichromis fusciatus*, *Sarotherodon galilaeaes*, and *Hepsetus odoe* being the most abundant. Studies of the fisheries of small wetlands in the northern parts of the country include Obodai and Nsor (2009) on fish species diversity and production of the Kukobila wetland in the Savelugu-Nanton District. Twenty-three species of fish belonging to 19 families were reported in this study, with an estimated fish productivity of 75 000 kg in the wet season and 45 000 kg in the dry season.

11.3.3 Herpetofauna

Raxworthy and Attuquayefio (2000) reported 13 species of amphibians and 19 species of reptiles from the Muni wetland. The grassland thickets associated with the lagoon yielded the most diverse community of herpetofauna, a total of 26 species, with only 3 species of frogs recorded in the brackish water lagoon. The species recorded regularly in more than one of the five habitat types sampled in the wetland included *Bufo regularis*, *Bufo maculatus* (Bufonidae), *Rana occipitalis*, *Rana galamensis* (Ranidae), *Pelomedusa subrufa* (Pelomedusidae), and *Python regius* (Boidae). Species of herpetofauna recorded in inland wetlands in the northern parts of Ghana include tortoise, monitor lizards (*Varanus niloticus* and *V. exanthematicus*), crocodile, Agama lizards, skinks, black house snakes, sand snakes, spitting cobras, puff adder, night adder, carpet viper, green member, royal python, and the African python.

11.3.4 Invertebrates

Assessments of the invertebrate fauna in coastal wetlands include the work carried out by Finlayson et al. (2000) at Songor and Keta Lagoon, and Gordon (2000) at the Muni lagoon. The most common invertebrates recorded at Songor and Keta included annelids, molluscs, and crustaceans (Finlayson et al. 2000). Gordon (2000) found the invertebrate fauna of Muni to be impoverished, especially during the period when the lagoon was closed, with the situation improving when the lagoon was opened to the sea. The species identified included the Crustacia, *Callinectes latimanus*, *Cardiosoma armatum* and the molluscs *Anadara senilis*, *Macoma cumana*, and *Crassostera tulipa*. A follow-up study of the Muni wetland in 2010 recorded a total of 1723 individual invertebrates belonging to 10 groups. Of these, polychaetes (Annelida) constituted 90.71%, bivalves (Mollusca) 7.84%, and crustaceans (Arthropoda) 1.45% (Abdourahamane 2010). The study confirmed the stressed conditions of Muni lagoon environment, with fluctuating high temperature and salinity and a benthic population dominated by a single taxon (capitellid worms). A recent assessment of insects associated with selected northern wetlands recorded 156 families from 20 insect orders, including 36 species of butterflies belonging to 5 families (Kyerematen 2018). The study found Dipterans to be the most diverse group in the samples collected.

Other fauna of conservation interest occurring in the Ghanaian wetlands include marine turtles, manatees, sitatunga, and hippos. Four species of marine turtles (the Loggerhead *Caretta caretta*, the Green Turtle *Chelonia mydas*, the leatherback *Dermochelys coriacea*, and olive ridley *Lepidochelys olivacea*), have been confirmed to nest on Ghanaian beaches, with the most important nesting sites occurring on the sandy beach of the Songor Ramsar site in the east and the beaches bordering the Bortianor, Fete, and Elmina settlements in the west (Toth and Toth 1976; Allman et al. 2015; Agyekumhene et al. 2020). Historical records indicate that a fifth species, the hawksbill *Eretmochelys imbricata*, nested and foraged along the Ghana coast (Irvine 1947), but the hawksbill has not been confirmed to nest in Ghana for at least 30 years (Amiteye 2001; Allman et al. 2015), although local fishermen consistently recount encounters,(Agyekumhene et al. 2014; Alexander et al. 2017). Some communities along the coast, for example, the Ningo people, have taboos against killing and eating sea turtles, a traditional conservation strategy that is enforced by the chiefs and people of the area (Ntiamoa-Baidu 1991b, 1995), and which has been reported to be more revered and a stronger deterrent than the wildlife conservation laws of the country (Agyekumhene and Allman 2018). Threats to marine turtles on the Ghana coast come from adults getting entangled in fishing nets and being killed by fishermen, killing of nesting females, harvesting of eggs, and egg predation by dogs and pigs. Nesting habitats are being destroyed by sand mining and increasing beach erosion caused by sea level rise (Agyekumhene and Allman 2018).

Manatees are found in the Lower Volta estuary and in wetlands associated with several arms of the Volta River (Ofori Danson and Agbogah 1995). A study of the fauna of selected national parks by the Centre for Biodiversity Conservation Research in 2020–2023 reported a significant population of manatees in the Digya National Park (CBCR unpublished data). Manatees are also reported to occur in the Tano river and the Tano/Abbey/Ehy Lagoon complex in western Ghana (Entsua-Mensah and Ankudey 1996). de Graft-Johnson et al. (2010) reported that the Nzema (the ethnic group along the western coast of Ghana) regard the manatee as a deity, but they nevertheless exploit it for food.

The only known population of the Sitatunga in Ghana occurs in the Avu Lagoon, which is part of the Keta Lagoon Ramsar site. Only small populations of Hippos *Hippopotaus amphibius* are known to exist in Ghana (Figure 11.2), and these occur in wetlands associated with the Volta River within the Digya National Park, Bui National Park, and the Wechiau Wildlife Sanctuary.

11.3.5 Wetland Vegetation

There have been very few studies on the flora of wetlands in Ghana. The vegetation studies undertaken under the Ghana CWMP, as part of the effort to describe the ecological character of the five wetlands selected for designation as Ramsar sites, detailed the flora of the sites (e.g. Oteng Yeboah 1994, 1999; Finlayson et al. 2000). Oteng Yeboah (1999) described the key vegetation types found at the Sakumo, Densu delta, and Muni wetlands to comprise salt marshes, mangrove swamps, freshwater swamp, scrub, and grassland. Plant species characteristic of the salt marsh and freshwater swamps included *Sesuvium portulacastrum*, *Paspalum vaginatum*, and *Sporobolus virginicus*, while the dominant species in the mangrove swamps was *Avicennia africana*, with various woody plant species characterizing the scrub vegetation.

Finlayson et al. (2000) provided a list of aquatic and wetland plants identified in the Keta and Songor wetlands and associated swamps and mapped out the dominant species found at various sampling sites located within the wetlands. The study identified 126 species of macrophytes, with *Typha domingensis*, *Scirpus littoralis*, and *Paspalum vaginatum* being the most dominant species in

Figure 11.2 Hippos at the Wechiau Community Hippo Sanctuary in northern Ghana. *Source:* Courtesy: Dr Kofi Amponsah-Mensah, Centre for Biodiversity Conservation Research.

the freshwater areas of the wetland, while *Sesuvium portulacastrum* and *Sporobolus pyramidialis* were the dominant species in the drier, more saline areas around the lagoons.

Pockets of mangrove stands are found in the Keta and Songor wetlands on the eastern coast and the Amansuri wetland on the west, comprising three main species of mangroves, *Avicennia germinans*, *Rhizophora racemosa*, and *Laguncularia racemosa* (Gordon and Ayivor 2003; Nunoo and Agyekumhene 2022). All three are listed as of "Least Concern" according to the IUCN Red Data list (Ellison et al. 2010); however, mangroves on the Ghanaian coast have been badly degraded through heavy exploitation for fuelwood.

A study by Nsor et al. (2019) of the flora of six small wetlands in northern Ghana (Wuntori, Tugu, Kukobila, Nabogo, Adayili, and Bunglung) reported 46 plant species from 18 families, comprising 42.2%, grasses, 42.2% herbs, and 15.5% woody species. These were further grouped into obligate wetland species (30.4%), facultative species (47.8%), and obligate upland species (26.1%). Nsor et al. (2014) also detailed factors influencing the range shift of plant species in these northern wetlands.

11.4 Wetland Resources and Community Livelihoods

The sociocultural lives and livelihoods of many communities living within and around wetlands are interlinked with wetland ecosystems and wetland resources (Willoughby et al. 2001; Danso et al. 2021). Coastal wetlands are critical for fishing and salt production, while inland wetlands are the backbone of crop production and animal husbandry.

Artisanal fishing in the coastal lagoons constitutes an important source of protein and livelihoods source for coastal communities and plays a crucial role in the economy of these communities, especially during the off-season for marine fishing (Ntiamoa-Baidu 1991b; Entsua-Mensah 1998; Dankwa et al. 2004). There are no figures on the total fish productivity from the coastal lagoons, but the results of a few site-based studies give an indication of fisheries productivity of the coastal lagoons. Ntiamoa-Baidu (1991b) estimated the average income from fishing in the Sakumo lagoon to be three to four times higher than the minimum wage paid to government workers for an eight-hour day of work. Fishermen worked individually using cast-nets or in groups of 4–12 men using draw nets, averaging 3.5 man-hours per day and obtaining a catch per unit effort of 1.0–6.1 kg per man hour. A similar assessment of the Sakumo fisheries carried out in 2014–2015 reported a much-reduced fishing effort of about an hour and a catch per unit effort of 2.82 kg and 5.39 kg/man-h for the open and closed seasons, respectively (Sawyerr et al. 2022). These results would suggest that fishing in the Sakumo lagoon was as productive in the 2010s as it was in the 1980s. Koranteng (1995) estimated annual catches to be 75, 114, and 2701 tonnes for Muni, Sakumo, and Densu delta, respectively. Dankwa et al. (2004) estimated a daily income earned by fishermen of the Keta and Songor lagoons to be USD 8.7, based on a landing price of USD 0.4 per kg for tilapia.

The "acadja" method of fishing, a traditional fishing practice where twigs and branches of trees are placed in the water in sheltered areas as a means of attracting fish to aggregate for feeding and spawning (Mensah 1979; Entsua-Mensah 1998), is still practiced in the Densu delta. Women traders are seen regularly on the site waiting to purchase and descale the fish, mainly tilapias (Figure 11.3a) ready for resale to consumers and street food vendors.

Salt production is one of the major economic activities of several communities living in the coastal areas of the country. It takes place within and on the edges of coastal lagoons and is a major industry, especially during the drier periods of the year. Wetland sites where salt production occurs extensively include Keta, Songor, Djange, Old Ningo, Prampram, Lalui, Densu delta, and Iture. Salt production at Songor and Densu delta is a major industry for salt companies (Figure 11.3b).

(a) (b)

Figure 11.3 (a) A woman trader descaling tilapia fish at Densu delta wetland. (b) Salt works at Panbros Salt Pans.

A census in 1989 reported that about 3500 families engaged in salt mining in the Ketu district of the Anlo-Keta lagoon complex (Statistical Services 1989).

Three different methods are used for salt production in the coastal areas of Ghana:

1) The traditional method is where brackish water in the lagoons evaporates during the dry season (September to March) leaving the salt for collection by the communities.
2) The use of salt pans constructed mostly with lagoon mud, but also, occasionally, with more permanent material such as concrete. The pans are filled with water from the lagoon or tidal channels at high tide when more saline water is available and allowed to evaporate for the salt to be collected.
3) The pumping of seawater using motorized pumps into large concrete concentration pans or reservoirs and then into pans to evaporate is the method used by commercial salt production companies, as it is more capital intensive.

11.5 Threats to Ghanaian Wetlands

Almost all wetlands in Ghana face threats and pressures that have impacted and continue to impact their ecological character and ability to sustain the ecosystem functions and the resources they provide. The pressures and threats on the Ghanaian wetlands have led to losses in the extent of wetland ecosystems, the degradation of wetland habitats, and loss of wetland quality.

All coastal wetlands, including the designated Ramsar sites, are threatened by habitat destruction arising from increasing demand for land and encroachment for the development of industries and human settlements. Wetlands in urban areas such as the Sakumo and the Densu delta Ramsar sites near Accra and Tema and the Owabi and Barekese reservoirs near Kumasi, in particular, have come under intense pressure from demand for land for housing and industrial developments. As a result, over a third of the land area originally designated as the Densu delta and the Sakumo Ramsar sites have been lost to housing and industrial developments. Satellite imageries of these wetlands show significant losses in wetland areas, with increases in the extent of bare and built-up areas in the wetlands (see, for example, the Densu delta Ramsar site in Figure 11.4).

Ekumah et al. (2020) employed geospatial techniques to assess ecosystem health of the Densu delta, Sakumo, and Muni-Pomadze Ramsar sites over a 32-year period using structure, function, and resilience indicators. The findings of the study indicated that the ecosystem health of the

2000 Classified satellite image map of Densu Ramsar Site

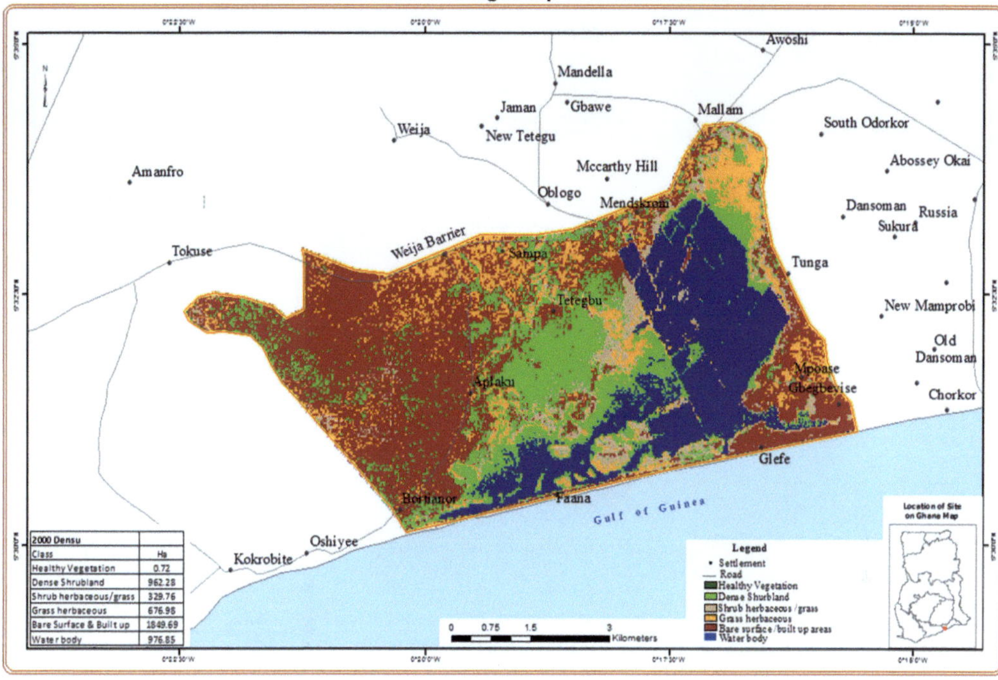

2014 Classified satellite image map of Densu Ramsar Site

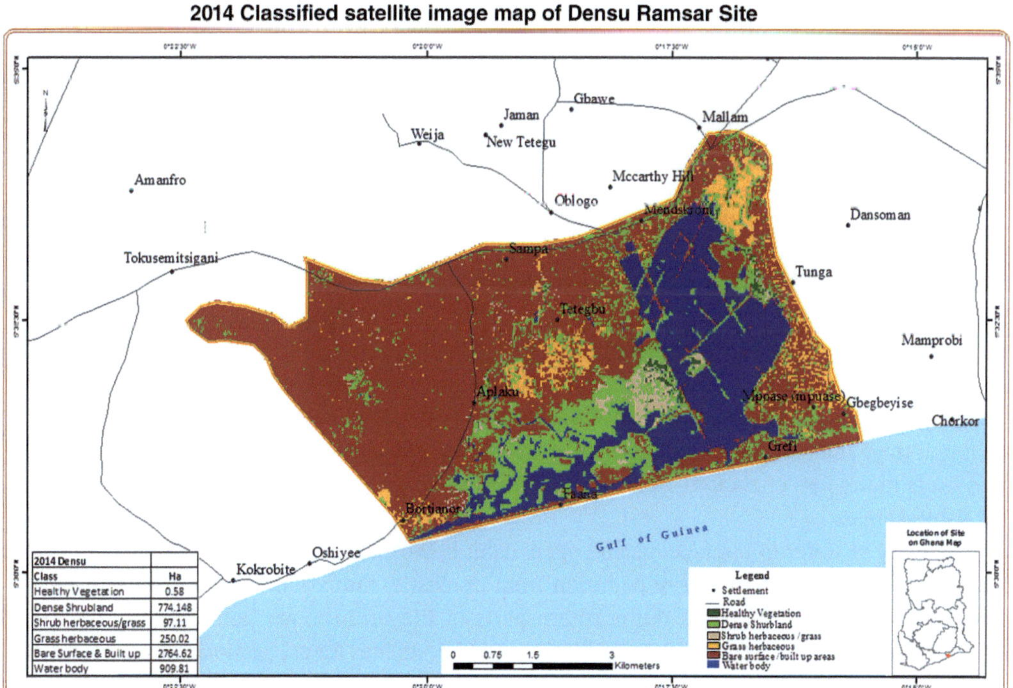

Figure 11.4 Changes in the coverage of the Densu delta Ramsar site from 2000 to 2020. (Note the increase in built-up area coverage from 1849.7 ha in 2000 to 3202.9 ha in 2020 and a decrease in size of water body by 100 ha over the same period.)

2020 Classified satellite image map of Densu Ramsar Site

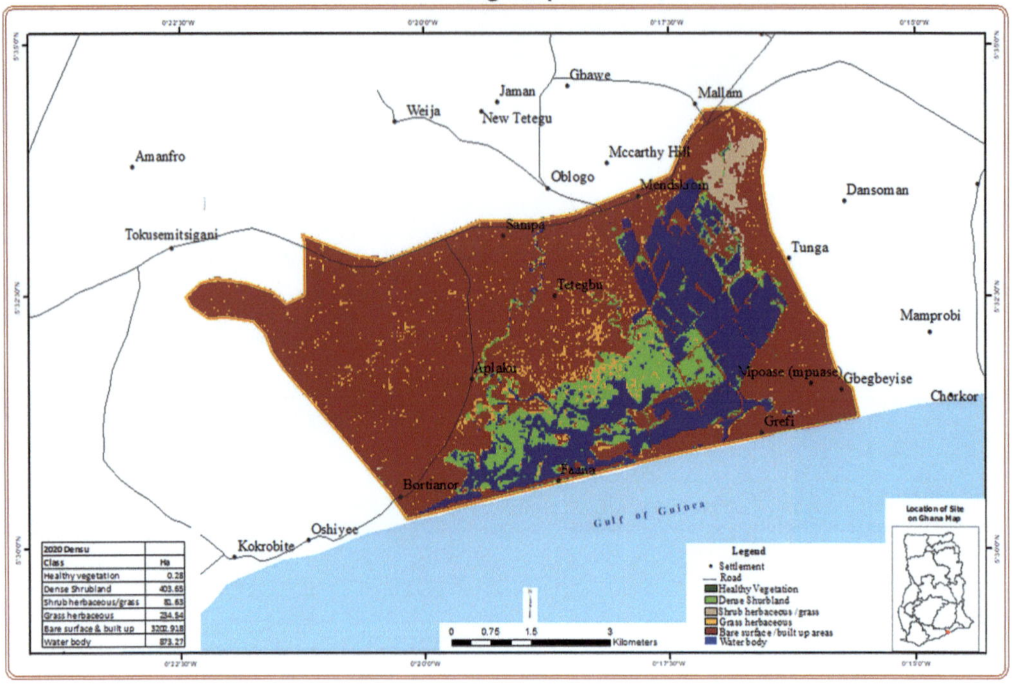

Figure 11.4 (Continued)

wetlands progressively deteriorated in 2002 and 2017 compared to the reference year of 1985. Wetland areas in the Keta municipality were reported to have reduced from 182.33 to 135.11 km^2 from 1991 to 2018 (Peters 2019).

Campion and Owusu-Boateng (2013) highlight the problem of people building in wetland areas and floodplains in their study of the political ecology of small urban wetland areas associated with streams within the KMA. The main cause of wetland habitat losses in the northern parts of the country comes from the demand for land for cultivation, a development that is increasingly aggravated by the promotion of lowland rice farming.

Several of the wetlands along the coast, such as Songor, Keta, Lalui, and Densu delta, have experienced expansions in existing salt pans and/or establishment of new salt works in recent times. Although the bunds of the pans offer increased roosting sites for waterbirds, the creation of more pans leads to the loss of natural wetland habitats, resulting in the loss of shorebird feeding sites and fishing areas.

The quality of water in the lagoons and open water bodies within most Ghanaian wetlands has been adversely impacted by heavy pollution from domestic, industrial, and agricultural wastes. Inland wetlands, particularly in the northern parts of the country, are polluted from domestic waste and agricultural practices, including heavy use of pesticides for cultivation of crops. Wetlands continue to be treated as wastelands where household refuse and industrial waste are dumped freely (Biney 1982; Osei et al. 2010). Small wetlands in urban areas such as the Fosu lagoon (Cape Coast), Chemu Lagoon (Tema), and Korle lagoon (Accra) that in the past provided lagoon fisheries for the indigenous people living around these waterbodies have been destroyed through the dumping of industrial waste and effluent (EPA 2017).

Ghana's coastal wetlands are low-lying areas; most of them are separated from the sea by a very narrow ridge on which settlements are built. This makes them particularly vulnerable to the threats of climate change, global warming, and sea level rise (Agyeman-Bonsu et al. 2008; Appeaning-Addo et al. 2011; Aheto et al. 2011; Appeaning-Addo and Adeyemi 2013; Avornyo et al. 2023). Projections of the potential impacts of climate change on the coastal wetlands present a disturbing picture. For example, studies by Aheto et al. (2011) on the River Whin estuary, Boakye-Boaten (2010) on the Songor Ramsar site, and Appeaning-Addo et al. (2011) on the Densu delta, all predict coastline recession and inundation of coastal villages. Using 1970–1990 as the baseline, Appeaning-Addo et al. (2011) predict that some 650 000 people and 926 buildings in three communities (Panbros, Grefi, and Gbegbeyise) within the Densu delta Ramsar site could be permanently inundated by 2100. Appeaning-Addo and Adeyemi (2013) also conclude that the sea level rise predictions by the IPCC will result in inundation of the Densu delta wetland, destroying homes and property. The projected inundation of the wetlands will lead to the destruction of feeding, roosting, and nesting habitats of waterbirds and marine turtles, but this could also create additional fishing areas that could positively impact community livelihoods.

11.6 Conservation Actions and Initiatives Taken by Ghana to Manage Its Wetlands

The concerted action taken by the Ghanaian government to protect Ghanaian wetlands and waterbirds can be traced back to 1985, when the Government of Ghana entered into an agreement with the Royal Society for the Protection of Birds (UK) and BirdLife International to protect sea and shorebirds, targeting specifically the Roseate Tern and their coastal habitats in Ghana. This partnership had been initiated out of concern for the dwindling population of European Roseate terns and the discovery that terns were intensely hunted on the Ghana coast (Ntiamoa-Baidu and Grieve 1987; Ntiamoa-Baidu and Hepburn 1988; Avery et al. 1995). To implement the agreement, a project, the Save the Seashore Birds Project-Ghana (SSBP-G), was set up to monitor bird populations and carry out conservation education in schools and communities along the Ghana coast (Ntiamoa-Baidu 1990).

The SSBP-G established the importance of several coastal wetlands for migratory waterbirds (Ntiamoa-Baidu and Gordon 1991). This information was fed into the development of a proposal for managing Ghana's coastal wetlands and funding was obtained from the Global Environment Facility and the World Bank to implement the Ghana CWMP. Prior to accessing the funds, Ghana had ratified the Ramsar Convention in 1988 with the designation of its first Ramsar site, the Owabi Wildlife Sanctuary. Subsequently, Ghana designated five of the key wetland sites identified under the SSBP-G (the Keta lagoon complex, Songor, Sakumo, Densu delta, and Muni-Pomadze) as Ramsar sites. The goal of the CWMP was to manage the five coastal wetland sites in order to maintain their ecological integrity and enhance the benefits derived from the wetlands by local communities (Finlayson et al. 2000).

Ghana was one of the first countries in Africa to formulate a National Wetlands Strategy (Ministry of Lands and Forestry 1999). This was followed by a National Wetlands Action Plan covering a 10-year period, 2007–2016 (Wildlife Division 2005). The Action Plan outlined six broad issues for focus, including poverty reduction and health, and nine priority actions, including wetland inventory and monitoring, wetland rehabilitation and restoration, networking, and international cooperation. The expectation was that the National Wetlands Strategy and the corresponding action plan would promote implementation of specific actions and coordination of interventions to ensure the management and conservation of the nation's wetlands.

Ghana also developed a Buffer Zone Policy, which advocated for the designation and sustainable management of buffer zones around rivers, streams, lakes, reservoirs, and other water bodies (Ministry of Water Resources, Works and Housing 2013). The expectation was that the implementation of this policy would complement the National Wetlands Action Plan and support wetland conservation in Ghana.

At the international level, Ghana is signatory to a number of international and regional treaties that promote conservation and management of biological resources, habitats, and ecosystems. A number of these treaties have direct relevance to the management of wetlands and wetland resources, namely, the Convention on Wetlands (Ramsar Convention), the Convention on Migratory Species (CMS), the Convention on Biological Diversity (CBD), and the Convention on Trade in Endangered Species (CITES). Despite these concrete actions, the implementation of Ghana's obligations under the conventions to ensure sustainable management of the country's natural habitats and biodiversity has not been adequate. The result is that critical ecosystems in the country, such as wetlands, continue to be severely degraded and lost.

11.7 Conclusion

Ghana is endowed with significant wetlands, ranging from the large expanse of the Keta and Songor lagoons along the coast to numerous small wetlands in the northern parts of the country. The coastal wetlands are important for biodiversity, notably waterbirds, fish species, and marine turtles. The inland wetlands provide the backbone for crop production and animal husbandry and are critical for household nutrition and food security. Wetlands provide important sources of livelihood for communities living in wetland catchments, with fisheries and salt production being key. Ghana has taken important steps nationally by designating its key wetlands as Ramsar sites to promote wise use of the wetlands, while maintaining their ecological integrity. Internationally, the country has acceded to all the important conventions that seek to promote the conservation of ecosystems and biodiversity. Unfortunately, the pressures and challenges presented by the growing human population and associated demands for land for housing and industrial developments have led to significant losses in the areas originally designated for protection. The quality of wetlands also continues to be adversely impacted by serious pollution from domestic and industrial waste and effluents as well as pesticides from farms within wetland catchments. All these threats are compounded by climate change and sea level rise that are projected to cause inundation of the coastal wetlands, with destruction of habitats and loss of livelihood. Unless the country takes concerted action to improve waste management and treatment, enforce approved land-use plans to prevent developments in wetland catchment areas, and allocate adequate financial resources for conservation actions, the nation's wetlands will lose their ability to provide the functions and services that are so critical for community and national welfare.

Appendix 11.1 Importance of the Ghanaian coastal Ramsar sites for waterbirds.

	Peak counts recorded in the past 10 years at the Ramsar sites				
Species	**Densu Delta**	**Keta lagoon**	**Muni Pomadze**	**Sakumo lagoon**	**Songor lagoon**
Waders					
Common sandpiper (*Actitis hypoleucos*)	114	63	132	63	41
African jacana (*Actophilornis africana*)	284	102	71	570	26
Turnstone (*Arenaria interpres*)	127	19	43	10	34
Senegal thicknee (*Burhinus senegalensis*)	14	3	11	50	13
Curlew sandpiper (*Calidris ferruginea*)[a]	7112	7666	676	742	10029
Sanderling (*Calidric alba*)[b]	523	498	416	123	1096
Little stint (*Calidris minuta*)[a]	3964	1396	411	231	870
Red knot (*Calidris canutus*)	199	56	38	1	194
Ruff (*Calidris pugnax*)	27	4		54	
Ringed plover (*Charadrius hiaticula*)[a]	4400	1992	959	464	1664
Kittlitzs sandplover (Charadrius pecuarius)	192	336	8	48	104
White fronted sandplover (*Charadrius marginatus*)	6	10	27		23
Collared pratincole (Glareola pratincola)	662	292	640	540	394
Black-winged stilt (*Himantopus himantopus*)[a]	2947	2493	520	1799	3964
Avocet (*Recurvirostra avosetta*)[c]	1	4		3	73
Black-tailed godwit (*Limosa limosa*)[b]	182	256	129	288	204
Bar-tailed godwit (*Limosa lapponica*)	81	25	48	8	141
Golden plover (*Pluvialis apricaria*)					1
Grey plover (*Pluvialis squatarola*)	247	88	263	26	377
Curlew (*Numenius arquata*)	174	60	7		26
Whimbrel (*Numenius phaeopus*)	342	54	169	14	87
Terek sandpiper (*Tringa cinereus*)		1	1		1
Spotted redshank (*Tringa erythropus*)[a]	149	612	250		1127
Wood sandpiper (*Tringa glareola*)	397	149	147	344	48
Greenshank (*Tringa nebularia*)[a]	2991	3478	357	160	4838
Marsh sandpiper (*Tringa stagnatilis*)	117	129	138	69	207
Redshank (*Tringa totanus*)	43	47	45	10	34
Senegal wattled plover (*Vanellus senegallus*)	22	17	54	114	8
Spur-winged plover (*Vanellus spinosus*)	58	84	311	233	73
Red-necked phalarope (*Phalaropus lobatus*)	1				
Oystercatcher (*Haematopus ostralegus*)	1		3		
Painted snipe (*Rostratula benghalensis*)	2			3	

(Continued)

Appendix 11.1 (Continued)

	Peak counts recorded in the past 10 years at the Ramsar sites				
Species	**Densu Delta**	**Keta lagoon**	**Muni Pomadze**	**Sakumo lagoon**	**Songor lagoon**
Terns					
Little tern (*Sterna albifrons*)[a]	1129	2104	1375	450	4373
Roseate tern (*Sterna dougallii*)[a]	171	210	28		232
Common tern (*Sterna hirundo*)[a]	5629	10 300	1199	693	19 295
Royal tern (*Sterna maxima*)[a]	4063	1204	1070	278	8703
Sandwich tern (*Sterna sandvicensis*)[a]	3050	1254	1569	272	11 368
Black tern (*Chlidonias niger*)[a]	1109	2967	292	610	11 788
Gull-billed tern (*Gelochelidon nilotica*)		156			56
Caspian tern (*Hydroprogne caspia*)	17	222	2		85
Herons/Ducks/Gulls					
Grey heron (*Ardea cinerea*)	649	316	30	98	124
Black-headed heron (*Ardea melanocephala*)		1	1	1	
Purple heron (*Ardea purpurea*)	13	18	7	22	5
Squacco heron (*Ardeola ralloides*)	97	35	16	150	6
Green=backed heron (*Butorides striatus*)	14	11	13	5	9
Great white egret (*Casmerodius albus*)	779	979	189	237	239
Black heron (*Egretta ardesiaca*)	327	710	449	88	496
Little egret (*Egretta garzetta*)	943	2477	221	280	2258
Western reef heron (*Egretta gularis*)	696	860	435	739	7778
Yellow-billed egret (*Egretta intermedia*)	52	60	11	116	31
Night heron (*Nycticorax nycticorax*)	11		1	1	2
Little grebe (*Tachybaptus ruficollis*)	19			14	
Cattle egret (*Bubulcus ibis*)	50	577	83	157	6
Little bittern (*Ixobrychus minutus*)	8	3	5	2	9
Purple swamphen (*Porphyrio porphyrio*)	8	3	5		
Garganey (*Anas querquedula*)	1			1	
Fulvous tree duck (*Dendrocygna bicolor*)	18	1230	1	21	13
Long-tailed cormorant (*Phalacrocorax africanus*)	4421	5765	1740	126	1725
African darter (*Anhinga rufa*)			1		
Lesser black-backed gull (*Larus argentatus*)	7	168	2		215
Others					
Black crake (*Amaurornis flavirostra*)	32		5		1
African Openbill stork (*Anastomus lamelligerus*)		1127			
Northern pintail (*Anas acuta*)				20	
Black stork (*Ciconia nigra*)		103			

Appendix 11.1 (Continued)

Species	Peak counts recorded in the past 10 years at the Ramsar sites: Densu Delta	Keta lagoon	Muni Pomadze	Sakumo lagoon	Songor lagoon
Pied crow (*Corvus albus*)	28				
White-faced tree duck (*Dendrocygna viduata*)	1814	4705	376	1424	349
Common moorhen (*Gallinula chloropus*)	170	4	6	20	1
Black-headed gull (*Larus ridibundus*)		16			27
Pink-backed pelican (*Pelecanus rufescens*)		5			146
African spoonbill (*Platalea alba*)				1	
Glossy ibis (*Plegadis falcinellus*)	1	6		9	
African skimmer (*Rynchops flavirostris*)	278	53			
Hammerkop (*Scopus umbretta*)			1		
Arctic skua (*Stercorarius parasiticus*)		1			
Sacred ibis (*Threskiornis aethiopicus*)					4

[a] Species occurring in internationally important populations.
[b] Species occurring in internationally important populations, but not on a Ramsar site.
[c] Species that occurred in internationally important populations on the Ramsar sites in the past.

References

Abdourahamane, I.S. (2010). Waterbird as Bioindicators of Wetland Quality: Case Study of the Muni-Pomadze Ramsar Site, Ghana. MPhil thesis. University of Ghana.

Agyekumhene, A. and Allman, P. (2018). Ghana. In: *Sea Turtles in the West Africa and East Atlantic Region: MTSG Annual Regional Report 2018* (ed. A. Agyekumhene and C.K. Kouerey Oliwina), 2018. Ghana: Draft Report of the IUCN-SSC Marine Turtle Specialist Group.

Agyekumhene, A., Agyekumhene, A., Akwoviah, J., and Allman, P. (2014). Perceptions from fishing communities on sea turtle status and conservation in central Ghana. *African Sea Turtle Newsletter* 2: 11–16.

Agyekumhene, A., Dondo, L.C., and Allman, P. (2020). Encounter of Hawksbill Sea Turtles in the coastal waters of Ghana. *African Sea Turtle Newsletter, No. 13* 2: 11–12.

Agyeman Bonsu, W., Minia, Z., Dontwi, J. et al. (2008). *Ghana Climate Change Impacts, Vulnerability and Adaptation Assessments* (ed. J. Allotey and L. Mensah). Accra: Environmental Protection Agency.

Aheto, D., Mensah, E., Aggrey-Fynn, J. et al. (2011). Spatio-temporal analysis of two coastal wetland systems in Ghana: Addressing ecosystem vulnerability and implications for fisheries development in the context of climate and land use changes. *Archives of Applied Science Research* 3 (3): 499–513.

Alexander, L., Agyekumhene, A., and Allman, P. (2017). The role of taboos in the protection and recovery of sea turtles. *Frontiers in Marine Science* 4: 237.

Allman, P., Barbour, D., and Agyekumhene, A. (2015). Loggerhead sea turtle nesting activity in Ghana. *African Sea Turtle Newsletter* 3: 13–15.

Amiteye, B.T. (2001). Distribution and ecology of sea turtles in Ghana. M. Phil thesis. University of Ghana, Legon.

Appeaning Addo, K. and Adeyemi, M. (2013). Assessing the impact of sea-level rise on a vulnerable coastal community in Accra, Ghana. *Jàmbá: Journal of Disaster Risk Studies* 5 (1): 8.

Appeaning Addo, K., Larbi, L., Amisigo, B., and Ofori-Danson, P.K. (2011). Impacts of coastal inundation due to climate change in a cluster of urban coastal communities in Ghana, West Africa. *Remote Sensing* 3 (9): 2029–2050.

Asmah, R., Dankwa, H., Biney, C.A., and Amankwah, C.C. (2008). Trends analysis relating to pollution in Sakumo Lagoon, Ghana. *African Journal of Aquatic Science* 33 (1): 87–93.

Avery, M.I., Coulthard, N.D., Del Nevo, A.J. et al. (1995). A recovery plan for Roseate Terns in the east Atlantic: an international programme. *Bird Conservation International* 5 (4): 441–453.

Avornyo, S.Y., Addo, K.A., Teatini, P. et al. (2023). A scoping review of coastal vulnerability, subsidence and sea level rise in Ghana: assessments, knowledge gaps and management implications. *Quaternary Science Advances* 100108.

Balian, E.V., Segers, H., Martens, K., and Lévéque, C. (2008). *The Freshwater Animal Diversity Assessment: An Overview of the Results*, 627–637. The Netherlands: Springer.

Biney, C.A. (1982). Preliminary survey of the state of pollution of the coastal environment of Ghana. *Oceanlogica Acta* Special volume: 39–43.

Boakye-Boaten, A.B. (2010). The potential impacts of climate change on coastal wetlands in Ghana: a case study of the Songor Ramsar Site. MPhil thesis. University of Ghana.

Campion, B.B. and Owusu-Boateng, G. (2013). The political ecology of wetlands in Kumasi, Ghana. *International Journal of Environment and Bioenergy* 7 (2): 108–128.

Cowardin, L.M., Carter, V., Golet, F.C., and LaRoe, E.T. (1979). *Classification of Wetlands and Deepwater Habitats of the United States. FGDC-STD-004-2013*, 2e, 79. USGS (December 1979). https://doi.org/FWS/OBS-79/31.

Dankwa, H., Shenker, J., Lin, J. et al. (2004). Fisheries of two tropical lagoons in Ghana, West Africa. *Fisheries Management and Ecology* 11 (6): 379–386.

Danso, G.K., Takyi, S.A., Amponsah, O. et al. (2021). Exploring the effects of rapid urbanization on wetlands: insights from the Greater Accra Metropolitan Area, Ghana. *SN Social Sciences* 1: 1–21.

Davidson, N.C. (2014). How much wetland has the world lost? Long-term and recent trends in global wetland area. *Marine and Freshwater Research* 65 (10): 934–941.

Ekumah, B., Armah, F.A., Afrifa, E.K. et al. (2020). Geospatial assessment of ecosystem health of coastal urban wetlands in Ghana. *Ocean and Coastal Management* 193: 105226.

Ellison, A., Farnsworth, E., Moore, G. (2010). *Avicennia germinans The IUCN Red List of Threatened Species* 2010: e.T178811A7613866. https://doi.org/10.2305/IUCN.UK.2010-2.RLTS.T178811A7613866.en

Entsua-Mensah, R.E.M. (1998). Comparative studies of the dynamics and management of fish populations in an open and closed lagoon in Ghana. Doctoral dissertation. University of Ghana.

Entsua-Mensah, M. and Ankudey, N. (1996). *The West African Manatee, Trichechus senegalensis*. Bionews, Newsletter of Ghana Institute of Biology.

Environmental Protection Agency (2017). Ghana State of the Environment Report, 2016. Ghana: EPA.

Finlayson, C.M., Gordon, C., Ntiamoa-Baidu, Y. et al. (2000). The Hydrobiology of the Songor and Keta Lagoons: Implications for Wetland Management in Ghana. Supervising Scientist Report 152, Australia 16, Supervising Scientist, Darwin.

Food and Agriculture Organisation (2020). The State of World Fisheries and Aquaculture (SOFIA 2020) Report.

Gardner, R.C., Barchiesi, S., Beltrame, C. et al. (2015). *State of the World's Wetlands and Their Services to People: A Compilation of Recent Analyses. Ramsar Briefing Note No. 7*. Gland: Ramsar Convention Secretariat.

Gbogbo, F. (2007). The importance of unmanaged coastal wetlands to waterbirds at coastal Ghana. *African Journal of Ecology*, 45 (4), 599–606. https://doi.org/10.1111/j.1365-2028.2007.00778.x

Gbogbo, F., Oduro, W., and Oppong, S.K. (2008). Nature and pattern of lagoon fisheries resource utilisation and their implications for waterbird management in coastal Ghana. *African Journal of Aquatic Science* 33 (3): 211–222.

Gordon, C. (1987). Coastal lagoons of Ghana. In: *African Wetlands and Shallow Water Bodies* (ed. M.J. Burgis and J.J. Symoens). Paris: ORSTOM. 211.

Gordon, C. (1994). *Reconnaissance Report on Aquatic Ecology. Ghana Coastal Wetland Management Project*. Department of Game and Wildlife, Government of Ghana.

Gordon, C. (2000). Hypersaline lagoons as conservation habitats: macro-invertebrates at Muni Lagoon, Ghana. *Biodiversity and Conservation* 9: 465–478.

Gordon, C. and Ayivor, J.S. (2003). Key Issues: Ghana. In: *Report on the Africa Regional Workshop on the Sustainable Management of Mangrove Forest Ecosystems* (ed. D.J. Macintosh and E.C. Ashton). Centre for African Wetlands, University of Ghana: Legon, Accra 66 p.

Gordon, C., Ntiamoa-Baidu, Y., and Ryan, J.M. (2000). The Muni-Pomadze Ramsar site. *Biodiversity and Conservation* 9: 447–464.

de Graft-Johnson, K.A.A., Blay, J., Nunoo, F.K.E., and Amankwah, C.C. (2010). *Biodiversity Threats Assessment of the Western Region of Ghana*. Ghana: The Integrated Coastal and Fisheries Governance (ICFG) Initiative.

Grimes, L.G. (1974). Radar tracks of Palaearctic waders departing from the coast of Ghana in spring. *IBIS* 116: 165–171.

Grimes, L.G. (1987). The birds of Ghana, an annotated check-list. B.O.U. Check-list No.9, 276 p

IPBES (2018). *The IPBES Regional Assessment Report on Biodiversity and Ecosystem Services for Africa* (ed. E. Archer, L. Dziba, K.J. Mulongoy, et al.). Bonn: Secretariat of the Intergovernmental Science-Policy Platform on Biodiversity and Ecosystem Services.

Irvine, F.R. (1947). *The Fishes and Fisheries of the Gold Coast*. UK: Government of the Gold Coast. The Crown Agents for the Colonies 352 pp.

Koranteng, K.A. (1995). Ghana Coastal Wetlands Management Project Environmental Baseline Studies (a: Densu -delta, b: Sakumo and c: Muni RAMSAR sites): Fisheries. A Report Prepared for the Ghana Wildlife Department (GW/A.285/SF2/31) Government of Ghana.

Koranteng, K.A., Ofori-Danson, P.K., and Entsua-Mensah, M. (1998). Comparative study of the fish and fisheries of three coastal lagoons in West Africa. *International Journal of Ecology and Environmental Sciences* 24: 371–382.

Koranteng, K.A., Ofori-Danson, P.K., and Entsua-Mensah, M. (2000). Fish and fisheries of the Muni lagoon in Ghana, West Africa. *Biodiversity and Conservation* 9 (4): 487–499.

Kumi, J., Kumi, M., and Apraku, A. (2015). Threats to the conservation of wetlands in Ghana: the case of Songor Ramsar site. *Journal of Scientific Research and Reports* 6 (1): 13–25.

Kummu, M., de Moel, H., Ward, P.J. et al. (2011). How close do we live to water? A global analysis of population distance to freshwater bodies. *PLoS ONE* 6 (6): e20578. https://doi.org/10.1371/journal.pone.0020578.

Kuusaana, E.D., Ahmed, A., Campion, B.B. et al. (2021). Characterisation and typology of urban wetlands in Ghana: implications for the governance of urban commons in secondary cities in Africa. *Urban Governance* 1 (1): 38–50. https://doi.org/10.1016/j.ugj.2021.09.002.

Kwei, E.A. (1977). Biological, chemical and hydrological characters of coastal lagoons of Ghana, West Africa. *Hydrobiologia* 56 (2): 157–174.

Kyerematen, R. (2018). *Insect Biodiversity Assessment of wetlands in Northern Ghana*. Centre for African Wetlands. 18 pp.

Lamptey, A.M. and Ofori-Danson, P.K. (2014). Review of the distribution of waterbirds in two tropical coastal Ramsar lagoons in Ghana, West Africa. *West Africa Journal of Applied Ecology* 22 (1): 77–91.

Macdonald, M.A. (1978). Seasonal changes in numbers of waders at Cape Coast, Ghana. *Bulletin – Nigerian Ornithologists' Society* 14 (45): 28–35.

Mattah, P. A. D, Akwetey, M. F. A., Abrokwah,S. et al. (2024). Perspectives on drivers of biodiversity and environmental changes in the Keta Lagoon Ramsar Site of Ghana, *Sustainability*, 16 (2): 666; https://doi.org/10.3390/su16020666

Mensah, M.A. (1979). *The Hydrology and Fisheries of the Lagoons and Estuaries in Ghana. Marine Fishery Report* 7, 14. Tema: Fisheries Research and Utilization Branch.

Miller, T.G.J. (2004). *Living in the Environment*, 13e. Jack Carey.

Ministry of Environment and Science, Ghana (2002). National Biodiversity Strategy for Ghana. Ministry of Environment and Science, Accra.

Ministry of Lands and Natural Resources (1999). Managing Ghana's Wetlands: A National Wetlands Conservation Strategy, Ghana.

Ministry of Water Resources, Works and Housing, (2013). Riparian Buffer Zone Policy For Managing Freshwater Bodies in Ghana.

Mitsch, W.J. and Gosselink, J.G. (2015). *Wetlands*, 5e. Hoboken, NJ: Wiley.

Nartey, V.K., Edor, K.A., Doamekpor, L.K. et al. (2011). Nutrient load of the Sakumo lagoon at the Sakumo Ramsar site Tema, Ghana. *West Africa Journal of Applied Ecology* 19: 93–105.

Nonterah, C., Xu, Y., Osae, S. et al. (2015). A review of the ecohydrology of the Sakumo wetland in Ghana. *Environmental Monitoring and Assessment* 187: 1–14.

Nsor, C.A., Obodai, E.A., and Blay, J. (2014). Factors influencing the range shift of plant species in wetlands, Northern Region (GHANA). *Annals of Experimental Biology* 2 (2): 23–33.

Nsor, C.A., Antobre, O.O., Mohammed, A.S. et al. (2019). Modelling the effect of environmental disturbance on community structure and diversity of wetland vegetation in Northern Region of Ghana. *Aquatic Ecology* 53: 119–136 (2019). https://doi.org/10.1007/s10452-019-09677-5.

Ntiamoa-Baidu, Y. (1988). *Three Years of Saving Seashore Birds in Ghana*. SSBP-G Publ. No. 2.

Ntiamoa-Baidu, Y. (1990). Coastal wetlands conservation: the save the seashore birds project. In: *Living with Wildlife* (ed. A. Kiss), 91–95. World Bank Technical Paper 130.

Ntiamoa-Baidu, Y. (1991a). Seasonal changes in the importance of coastal wetlands in Ghana for Wading Birds. *Biological Conservation* 57: 139–115.

Ntiamoa-Baidu, Y. (1991b). Conservation of coastal lagoons in Ghana: the traditional approach. *Landscape and Urban Planning* 20: 41–46.

Ntiamoa-Baidu, Y. (1993). Trends in the use of Ghanaian coastal wetlands by migrating knots. *Calidris canutus*. *Ardea* 81: 71–79.

Ntiamoa-Baidu, Y. 1995. Indigenous vs. Introduced Biodiversity Conservation Strategies: the Case of Protected Area Systems in Ghana. African Biodiversity Series No. 1, BSP, Washington.

Ntiamoa-Baidu, Y. and Gordon, C. (1991). *Coastal Wetlands Management Plans*. Accra: Ghana World Bank/EPC 149 pp.

Ntiamoa-Baidu, Y. and Grieve, A. (1987). Palaearctic Waders in Coastal Ghana in 1985/86. *Wader Study Group Bull* 49, Suppl./IWRB Special Publication 7: 76–78.

Ntiamoa-Baidu, Y. and Hepburn, I. (1988). Wintering waders in Coastal Ghana. *R.S.P.B Conservation Review* 2: 85–88.

Ntiamoa-Baidu, Y. and Hollis, G.E. (1988). Planning the management of Coastal Lagoons in Ghana. In: *Conservation and Development: The Sustainable Use of Wetland Resources* (ed. E. Maltby et al.), 113–121. IUCN.

Ntiamoa-Baidu, Y., Piersma, T., Wiersma, P. et al. (1998). Water depth selection, daily feeding routines and diets of waterbirds in coastal lagoons in Ghana. *IBIS*, 140 : 89–103. https://doi.org/10.1111/j.1474-919X.1998.tb04545.x

Ntiamoa-Baidu, Y., Nyame, S.K., and Nuoh, A.A. (2000). Trends in the use of a small coastal lagoon by waterbirds: Muni Lagoon (Ghana). *Biodiversity and Conservation* 9 (4): 527–539.

Ntiamoa-Baidu, Y., Nuoh, A.A., Reneerkens, J. et al. (2014). Population increases in non-breeding Sanderlings in Ghana indicate site preference. *Ardea* 102: 131–137. https://doi.org/10.5253/arde.v102i2.a3.

Ntiamoa-Baidu, Y., Ampomah, B.Y., and Ofosu, E.A. (ed.) (2017). *Dams, Development and Downstream Communities: Implications for Re-optimising the Operations of the Akosombo and Kpong Dams in Ghana*. Tema: Digibooks Gh. Ltd.

Nukpezah, D., Quarshie, J.T., Nyarko, E. et al. (2022). Characterisation of litter and their deposition at the banks of coastal lagoons in Ghana. *Heliyon* 8 (3): https://doi.org/10.1016/j.heliyon.2022.e08997.

Nunoo, F.K.E. and Agyekumhene, A. (2022). Mangrove degradation and management practices along the coast of Ghana. *Agricultural Sciences* 13: 1057–1079. https://doi.org/10.4236/as.2022.1310065.

Nunoo, J., Agbo, N., and Ackah, M. (2012). Fish fauna of the Owabi Dam Reservoir in Ghana. *Proceedings of the International Academy of Ecology and Environmental Sciences* 2 (1): 21.

Nyame, K.S. (2000). *Survey of inland wetlands in*. Accra: Ghana Wildlife Society.

Nyarko, B.K., Diekkrüger, B., Van De Giesen, N.C. et al. (2015). Floodplain wetland mapping in the White Volta River Basin of Ghana. *GIScience & Remote Sensing* 52 (3): 374–395. https://doi.org/10.1080/15481603.2015.1026555.

O'Keefe, T.C., Elliott, S.R., and Naiman, R.J. (2000). Wetland functions and values. http://www.epa.gov/watertrain.

Obodai, E.A. and Nsor, C.A. (2009). Aspects of biodiversity and fish production in the Kukobila wetland in the Savelugu-Nanton district of the northern region, Ghana. *Ethiopian Journal of Environmental Studies and Management* 2 (3): https://doi.org/10.4314/ejesm.v2i3.48266.

Ofori-Danson, P.K. and Agbogah, K. (1995). *Survey of Aquatic Mammals of Ghana. A Report Presented to the Oceans and Coastal Areas, Programme Activity Centre (OCA/PAC) of the United Nations Environment Programme*. Institute of Aquatic Research. Tech. Rep, 143.

Okonkwo, C.N.P., Kumar, L., and Taylor, S. (2015). The Niger Delta wetland ecosystem: what threatens it and why should we protect it? *African Journal of Environmental Science and Technology* 9 (5): 451–463.

Osei, J., Nyame, F.K., Armah, K.T. et al. (2010). Application of multivariate analysis for identification of pollution sources in the Densu delta wetland in the vicinity of a landfill site in Ghana. *Journal of Water Resource and Protection* 2 (12): 1020–1029.

Oteng Yeboah, A. (1999). Biodiversity studies in three coastal wetands in Ghana, West Africa. *Journal of the Ghana Science Association* 1 (3): 147–149.

Oteng-Yeboah, A.A. (1994). DNA Bank depositories in developing countries: the way forward in Sub-Saharan Africa. *Conservation of Plant Genes II: Utilization of Ancient and Modern DNA. Missouri Botanical Garden, St Louis* 194: 175–186.

Owusu, E.H. (2007). The ornithological importance of the Amansuri Community Nature Reserve in the Western Region of Ghana. *Journal of Science and Technology (Ghana)* 27 (3): 72–85.

Pauly, D. (1975). On the ecology of a small West Africa lagoon. *Ber. Dt. Wiss. Komm. Meeresforsch* 24: 46–62.

Pauly, D. (1976). The biology, fishery and potential for aquaculture of *Tilapia melanotherodon* in a small West African lagoon. *Aquaculture* 7: 33–49.

Peters, M.K. (2019). Assessing land use and land cover change in the Keta municipality of Ghana using remote sensing. Master's thesis. University of Ghana.

Piersma, T. and Ntiamoa-Baidu, Y. (1995). Waterbird ecology and the management of coastal wetlands in Ghana. *NIOZ Report* 1995–6: 1–105.

Quartey, J.K., Nuoh, A.A., and Mort, K. (2017). *Report on Reconnaissance Study Trip to Northern Ghana Under the Project "Enhancing Rural Livelihoods and Food Security through Wise Use of Wetlands in Northern Ghana"*. Centre for African Wetlands.

Ramsar Convention on Wetlands (2018). *Global Wetland Outlook: State of the World's Wetlands and their Services to People*. Gland: Ramsar Convention Secretariat.

Ramsar Convention Secretariat (2016). *Importance of Wetlands*. Gland: Ramsar Convention.

Raxworthy, C.J. and Attuquayefio, D.K. (2000). Herpetofaunal communities at Muni lagoon in Ghana. *Biodiversity and Conservation* 9: 501–510.

Ryan, J.M. and Ntiamoa-Baidu, Y. (2000). Biodiversity and ecology of coastal wetlands in Ghana. *Biodiversity and Conservation* 9 (4): 445–446.

Sawyerr, L.M., Quartey, J.K., and Ntiamoa-Baidu, Y. (2022). Impact of indigenous natural resources regulatory systems of fishing on the fisheries of a small coastal lagoon in Ghana, West Africa. *Journal of Applied Ecology* 30 (2): 13–23.

Smit, C. J. and Piersma, T. (1989). Numbers, midwinter distribution and migration of wader populations using the East Atlantic Flyway, pp 24–63. In H Boyd & J.-Y. Priot. *Flyways and reserve networks for waterbirds*. IWRB Spec. Publ. 9, Slimbridge.

Statistical Services (1989). *1984 Population Census of Ghana*. Accra: Report on Localities by Local Authorities. Statistical Services.

Stephenson, P.J., Ntiamoa-Baidu, Y., and Simaika, J.P. (2020). The use of traditional and modern tools for monitoring wetlands biodiversity in Africa: challenges and opportunities. *Frontiers in Environmental Science* 8: 61. https://doi.org/10.3389/fenvs.2020.00061.

Strayer, D.L. and Dudgeon, D. (2010). Freshwater biodiversity conservation: recent progress and future challenges. *Journal of the North American Benthological Society* 29: 344–358. https://doi.org/10.1899/08-171.1.

Taye, E.N.A., Quartey, J.K., and Ntiamoa-Baidu, Y. (2022). Nesting ecology of back-winged stilts *Himantopus himantopus* in Ghana. *Wader Study* 129 (1): 22–30. https://doi.org/10.18194/ws.00267.

Thieme, M.L., Abell, R.A., Stiassny, M.L.J. et al. (2005). *Freshwater Ecoregions of Africa and Madagascar: A Conservation Assessment*. Washington, DC: Island Press.

Toth, E.F. and Toth, K.A. (1976). *Coastal National Park Site Selection Survey*. Mimeo: Department of Game and Wildlife.

Tumbulto, J. and Bannerman, R. (1995). *Ghana Coastal Wetlands Management Project.Environmental Baseline Studies Sakumo Ramsar Site – Hydrology*. Gov of Ghana.

Wagner, M. (2004). *Managing Riparian Habitats For Wildlife*. Texas Parks and Wildlife.

Wetlands International (2002). *Waterbird Population Estimates*, 3e, vol. 12. Wageningen: Wetlands International Global Series.

Whyte, S. (1975). Distribution, trophic relationships and breeding habits of the fish populations in a tropical lake basin (Lake Bosumtwi-Ghana). *Journal of Zoology* 177 (1): 25–56.

Wildlife Division (2005). *Managing Ghana's Wetlands: A National Wetlands Conservation Strategy and Action Plan 2007–2016*. Wildlife Division, Accra.

Willoughby, N., Grimble, R., Ellenbroek, W. et al. (2001). The wise use of wetlands: identifying development options for Ghana's coastal Ramsar sites. *Hydrobiologia* 458: 221–234. https://doi.org/10.1023/A:1013158329107.

WWF (2018). *Living Planet Report – 2018: Aiming Higher* (ed. M. Grooten and R.E.A. Almond), 143. Gland: WWF.

WWF (2022). *Living Planet Report 2022 – Building a Nature Positive Society* (ed. R.E.A. Almond, M. Grooten, D. Juffe Bignoli, and T. Petersen), 118. Gland: WWF.

12

Biodiversity, Change, and Use of Moroccan Wetlands

Abdeslam Ennabili[1], Mohamed Libiad[2], Mohamed El Haissoufi[3], Abdelmajid Khabbach[4], and Nard Bennas[2]

[1] *Superior School of Technology, Sidi Mohamed Ben Abdellah University, Fez, Morocco*
[2] *Faculty of Sciences, Abdelmalek Essaâdi University, Tetouan, Morocco*
[3] *Polydisciplinary Faculty of Taza, Sidi Mohamed Ben Abdellah University, Fez, Morocco*
[4] *Faculty of Sciences, Sidi Mohamed Ben Abdellah University, Fez, Morocco*

12.1 Introduction

Wetlands include all areas of highly variable extent, whose main physical component is water; either above or very close to the surface, not exceeding six meters for marine water bodies at low tide. Ramsar-listed wetlands of international importance, particularly as waterfowl habitat, may include areas of shoreline or coastline adjacent to the wetland and islands or bodies of marine water deeper than 6 m by low tide, surrounded by the wetland (RCW 2016). The often seasonal nature of wetlands and their functional link (their interactions) with hydrological systems make their delimitation a problematic issue (Shine and de Klemm 1999; Ennabili et al. 2021).

Wetlands come in various forms, with varying degrees of artificialization: lakes, lagoons, marshes, sebkhas (oglet, guelta, or chotts), graras (wet depressions with temporary immersion), springs (aïn, onsar, or aghbalou) or diffuse springs (indefinite water resurgences), water seepage surfaces, ponds, peat bogs, wadis (or assif, temporary or permanent watercourses), river mouths and backwaters, riparian forests, seagrass beds, coral reefs and other marine areas not exceeding six meters in depth at low tide, natural or artificial dams, pond dams (silted-up dams), flood plains, wastewater effluent discharge areas, extensive wastewater treatment plants, and irrigation zones (Ennabili 1999; RCW 2016; Dakki et al. 2020; Ennabili et al. 2021).

The surface area of wetlands worldwide is around 4–6% of the Earth's land surface (RCW 2016). Artificial wetlands, mainly rice paddies and reservoirs, make up 12% of wetlands, and their surface area is increasing based on available data between 1970 and 2015. However, this increase cannot compensate for the losses recorded in marine, coastal, and inland wetlands, estimated at nearly 35% for the same period, i.e. a rate of regression three times greater than that of forests (RCW 2018), although there are no large-scale standardized and systematic assessments (Leberger et al. 2020).

Morocco's wetlands are far more renowned for their abundance and diversity than for their extent. The approximately 300 sites delineated as part of the national inventory of natural and artificial wetlands occupy around 0.6% of the national territory (Dakki et al. 2015; Ennabili et al. 2021). In Morocco, this rate of wetland loss is much higher, with around 50% of Moroccan

Wetlands of Tropical and Subtropical Asia and Africa: Biodiversity, Livelihoods and Conservation,
First Edition. Edited by Thammineni Pullaiah.

wetlands having disappeared over the past 50 years, due in particular to human intervention and climate change (Libiad et al. 2021). However, it is not easy to provide a reliable figure in this respect, given that coastal and continental wetlands include countless wetland sites that are difficult to map (ponds, seeps, narrow riparian forests and islets, graras, artificial wetlands, etc.) on the one hand, and seasonally flooded areas or dam reservoirs dependent on highly variable rainfall from one year to the next and/or siltation on the other hand.

Biological diversity, or biodiversity, of a genetic, specific, or ecosystemic nature, is recognized to be higher in warm and/or wet habitats than in cold and/or dry habitats. Wetlands are among the most biologically productive ecosystems, play an important role in the water cycle, regulate nutrient cycles including the filtration of pollutants, and contain the essential carbon reserves of the planet's soil (Ennabili 1999; RCW 2018; Ennabili et al. 2019, 2021; Ennabili and Radoux 2020). Wetland ecosystems are home to diverse and unique species assemblages, with variable composition and physiognomy depending on site conditions, particularly hydrodynamics and physicochemical characteristics of the water, and the season influencing plant development. This results in simple forms or successions (zoning) characteristic of hygroseries (hydroseries, haloseries) or complete natural series of fresh, brackish, or marine waters (Ennabili 1999; Lahrouz et al. 2013; Dakki et al. 2015; Leberger et al. 2020; Nogueira et al. 2022). Together with the anthropozoic impact, these environmental variables condition the development of planktonic and benthic communities, and animals in general, in aquatic ecosystems, playing an important role in maintaining their quality (Ramdani et al. 2001a, b, 2009; Mars et al. 2023a; Sammoudi et al. 2023).

Assessment of the biodiversity of Morocco's wetlands is far from complete, due to the inadequacy or even absence of coordinated and integrated research programs on the one hand, and technical and financial difficulties on the other hand. Some summary publications on the inventory of vascular plants in wetlands have recently been compiled on the basis of available data from various works and propose further studies to ensure a more in-depth coverage and representation of national wetlands (Khabbach et al. 2020; Ennabili et al. 2021; Libiad et al. 2021). The richness and diversity of marine macroalgae have been studied extensively (e.g. Riadi and Kazzaz 1998; Riadi et al. 2000; Hassoun et al. 2014, 2023), while studies on the diversity of freshwater microalgae are still scarce (e.g. Fqih Berrada et al. 1999; Rachiq et al. 2002; Rachiq 2003; Mhamdi et al. 2018, 2020). Studies on the bryophytic flora of Moroccan wetlands are rare (e.g. Chtaibi et al. 2017). Assessing the diversity of the fauna of national wetlands comes up against a number of difficulties, including the diversity of taxonomic groups with life cycles that depend entirely or temporarily on these environments, the difficulty of qualifying the habitat preferences of the species observed, and the mobility of certain species during their life cycle, which sometimes leads them to change habitat depending on the season. In fact, the descriptions, taxonomic notes, and distribution of most Moroccan species are scattered in the old literature that is difficult to access (e.g. El Haissoufi 2012; Bennas et al. 2016; Mabrouki et al. 2020c; Wakass et al. 2023), in addition to the absence of exhaustive inventories of observed species of certain taxonomic groups.

Wetland conservation and sustainable use are crucial to people's livelihoods thanks to the various ecosystem services they provide, although they are frequently underestimated by policymakers (RCW 2018). In a context of international commitment and partnership, and national legislation, Morocco is increasingly deploying its efforts to conserve wetlands through various actions and programs, aiming to implement decisive measures to preserve human health and global biodiversity by 2030 (Ennabili et al. 2021). Various nonspecific strategic plans have been adopted, such as the Strategic Plan 2011–2020 of the Convention on Biological Diversity and the United Nations Convention to Combat Desertification, or dealing with wetlands, such as the Strategic Plan 2016–2024 of the Ramsar Convention and its subsidiary conventions (Dakki et al. 2015). However, in terms of achievements, the actors involved in wetland management have no standardized, integrated scientific tools at their disposal, apart from sectoral and fragmentary local work, which does not enable them to develop a sustainable policy in this respect. This work consists of a global synthesis of previous work relating to Moroccan wetlands, in terms of biological diversity, livelihoods, and conservation. This would assist in attempts to revise the criteria for their classification by providing more information and in the development of practical sustainable planning and operational tools, useful for wetland managers, policy-makers, and the interested public.

12.2 Wetlands Biodiversity

12.2.1 Plantae

A total of 1526 species, subspecies, varieties, and formats of specific, facultative, or opportunistic vascular plants grow in Moroccan wetlands, representing around 36.3% of the national vascular flora, despite the very small surface area of wetlands compared with the national territory. Most vascular plant families are under-represented in Moroccan wetlands, i.e. more than half the species, and intraspecies correspond to only 8.2% of botanical families (Ennabili et al. 2021).

The specific vascular flora of Morocco's wetlands comprises 672 species, subspecies, varieties, and forms, corresponding to 91 botanical families, 298 genera, 578 species, 89 subspecies, 4 varieties, and 1 form, i.e. 15.9% of Morocco's vascular flora. Nearly half of all wetland plant taxa are represented by just eight botanical families: Cyperaceae, Poaceae, Asteraceae, Fabaceae, Juncaceae, Ranunculaceae, Apiaceae, and Caryophyllaceae (Khabbach et al. 2020). In addition, the facultative vascular flora of these habitats comprises 338 taxa, including 301 species, 37 subspecies, 220 genera, and 72 botanical families, i.e. 8.05% of the vascular flora of Morocco. Poaceae and Asteraceae account for 26.9% of facultative taxa, followed by Fabaceae, Amaranthaceae, Apiaceae, Caryophyllaceae, and Brassicacaeae (Libiad et al. 2021). Around 13.5% of Morocco's endemic vascular flora is represented by specific taxa (16.2%) and facultative ones (19.2%). Asteraceae, Brassicaceae, Apiaceae, Scrophulariaceae, Poaceae, Fabaceae, Caryophyllaceae, and/or Plumbaginaceae include more than half of these endemic taxa (Ennabili et al. 2021). Around 50.5% of the endemic-specific vascular flora of Moroccan wetlands is common to the Iberian Peninsula, Algeria, Tunisia, Mauritania, and/or the Canary Islands (Khabbach et al. 2020).

Bryophytes, characterizing a wide range of habitats including wetlands, include 507 mosses, 107 liverworts, and 5 anthocerotes (Ahayoun et al. 2013). Two studies focusing on wetlands revealed 26 and 46 bryophyte taxa, mostly moss-based, respectively, in the Ramsar wetland of Merja Zerga – Kenitra and the Cascades d'Ouzoud – Azilal (Chtaibi et al. 2017; Fadel et al. 2021). Charophytes number 26 species and varieties; the two characeae, *Chara strigosa* A.Br. and *Lamprothamnium succinctum* (A.Br. in Asch.) Wood, are of particular heritage interest in Morocco (Muller et al. 2017).

Moroccan Atlantic and Mediterranean coastlines are home to 540 seaweeds (5% of the world seaweeds), belonging to 229 genera, 93 families, and 7 classes. The class of Florideophycideae is the richest in Morocco with 317 seaweed taxa, followed by Pheophyceae (105), Ulvophyceae (97), Compsopogonophyceae (9), Cyanophyceae (5), Bangiophyceae (4), and Stylonematophyceae (3). No more than 10% of botanical families include 39% of Morocco's seaweeds (Table 12.1). The seaweed genera most represented are *Ulva* (19 taxa), *Gelidium* (16), *Ceramium* (14), *Cladophora* (13), *Polysiphonia* (10), *Bryopsis* (9), *Gongolaria* (9), *Lithophyllum* (9), *Codium* (8), and *Jania* (8). The high diversity of the Moroccan seaweeds could be a result of belonging to three biogeographical provinces: the western Mediterranean and the temperate and tropical Atlantics (Benhissoune et al. 2001).

Table 12.1 Specific richness (%) of algae families from Morocco.

Seaweeds		Freshwater microalgae		Marine microalgae	
Rhodomelaceae	**9.81**	**Desmidiaceae**	**7.59**	**Bacillariophyceae**	**52.5**
Ceramiaceae	6.11	Scenedesmaceae	7.24	Dinophyceae	37.0
Sargassaceae	5.74	Euglenaceae	5.52	Coccolithophyceae	4.97
Wrangeliaceae	4.63	Phacaceae	4.48	Trebouxiophyceae	1.66
Ulvaceae	4.44	Naviculaceae	4.48	Dictyochophyceae	1.10
Delesseriaceae	4.26	Spirogyraceae	4.48	Euglenophyceae	1.10
Cladophoraceae	4.07	Bacillariaceae	3.45	Eustigmatophyceae	1.10
Other 86 families	60.9	Other 72 families	62.8	Raphidophyceae	0.55

Source: González García and Conde Poyales (1994), Franchimont and Saadaoui (2001), Riadi (1998), Riadi and Kazzaz (1998), Fqih Berrada et al. (1999), Riadi et al. (2000), Benhissoune et al. (2001, 2002a, b, 2003), Rachiq et al. (2002), Rachiq (2003), Farid et al. (2009, 2012), Hanif et al. (2014b, c), Hassoun et al. (2014, 2023), Mhamdi et al. (2018, 2020), Moussa et al. (2018), Bahammou et al. (2021), Sabri et al. (2021).

12.2.2 Freshwater Animalia

An assessment of the animal diversity of Morocco's continental aquatic ecosystems revealed a total of 2883 species, 90.32% of which are invertebrates, mainly arthropods (Table 12.2). These ecosystems, representing only some 0.56% of the country's total surface area, are home to 11.71% of the total number of recognized animal species in Morocco, adopted by the "Plan d'Actions National pour la DiversitéBiologique du Maroc, 2016–2020." Arthropods and insects constitute 12.73% and 15.2% of their nationally reported equivalents in the same order. In addition, 16.3% of the total aquatic species recorded are endemic, higher than their equivalent for the global fauna.

Diptera and Coleoptera are the main representatives of both insects (85.29%) and all freshwater animal species (60.53%). A total of 3057 species, belonging to 93 families and 949 genera, were listed by Kettani et al. (2022). 22 families of this order of insects have aquatic or semiaquatic larvae, representing 50.26% of Morocco's continental aquatic fauna and 25.74% of its endemics. With an average rate of endemism among insects (Table 12.2), aquatic beetles, an important part of the macrozoobenthos in freshwater habitats, are one of the few insect groups where adults and larvae are entirely aquatic (Jäch and Balke 2008; Millán et al. 2014). They are also recognized as good indicators of ecosystem biodiversity (Bennas et al. 2005; Bilton et al. 2006; Sánchez-Fernández et al. 2008).

Other insect orders are relatively less represented, from 1.37% (Plecoptera) to Trichoptera (4.15%), but with significantly higher rates of endemism (Table 12.2). Aquatic hemipterans are hardly used as "indicators," as they are considered ubiquitous organisms with a high dispersal capacity and a wide range of occupied habitats (Taybi et al. 2018). Plecopterans are an important ecological component of running waters, particularly in unpolluted streams and rivers (Fochetti and Tierno de Figueroa 2008).

In Morocco, the diversity of Moroccan freshwater crustaceans is estimated at almost 10% of arthropods; the Gammaridae family features prominently, with a high rate of endemism. The number of species of continental aquatic molluscs, the second group of interest, is constantly increasing, thanks to new records and descriptions of new species. It currently corresponds to 4.84% of

Table 12.2 Species diversity of Moroccan freshwater Animalia.

Freshwater Animalia group	No. of families	No. of species	No. of M + MI	% Endemism	% IUCN threatened species
Vertebrates	65	279			
Mammals	1	1			
Waterbirds	36	208			3.365
Native fishes	17	52	21 + 6	51.92	15.38
Reptiles	3	4			
Turtles	2	2			
Snakes	1	2			
Amphibians	8	14	4 + 0	28.57	28.57
Invertebrates	344	2604			
Plathelmintes	4	6			
Turbellarians	4	6	2 + 0	33.33	
Arthropods	148	2325			
Crustaceans	51	232			
Decapod crustaceans	3	3			
Crustaceans Gammaridae	1	11	5 + 0	45.45	
Other Crustaceans	47	218			
Arachnids	12	47	9 + 0	19.15	
Insecta	85	2046			
Ephemeroptera	10	57	19 + 16	61.40	
Plecoptera	7	28	9 + 9	64.29	
Trichoptera	14	85	17 + 21	44.71	
Odonata	9	63	2 + 7	14.29	19.05
Coleoptera	11	296	38 + 53	30.74	

(*Continued*)

Table 12.2 (Continued)

Freshwater Animalia group	No. of families	No. of species	No. of M + MI	% Endemism	% IUCN threatened species
Hemiptera	12	68	5 + 9	20.59	
Diptera	22	1449	76 + 45	8.350	
Annelida	14	84			
Oligochaetes	9	66	1 + 0	1.515	
Hirudines	5	18	0 + 3	16.67	
Molluscs	10	126	78 + 15	73.81	45.24
Rotifers	14	63			
Total	219	2883	286 + 184	16.30	3.052

M + MI, Moroccan + Maghrebian and Ibero-Magrebian Endemic species. *Source:* Maarouf and Chemseddine (1995), Dakki et al. (1997), Neveu et al. (1997), Ghamizi et al. (1999), Ramdani et al. (2001a), Bennas(2002), El Alami (2002), Azeroual (2003), Tuzovskij and Gerecke (2003), Gerecke(2004), Delling and Doadrio (2005), Fadil and Dakki (2001, 2003, 2006), Pârvu et al. (2006), Darwall and Smith (2006), Bennas and Sáinz-Cantero (2001, 2006, 2007), Bouamama et al. (2007), Fadil et al. (2009), Abdaoui et al. (2010), Badsi et al. (2010), L'Mohdi et al. (2010), Samraoui et al. (2010), Van Damme et al. (2010), Vikhrev (2011), El Haissoufi (2012), Harrath et al. (2012), Martin and Aït Boughrous (2012), Qninba et al. (2011, 2012), Bahid and Kettani (2013), Hajji et al. (2012, 2013), El Haouari and Kettani (2014), El Haouari et al. (2014), Errochdi et al. (2014a, b), Ferreira et al. (2014), Nechad et al. (2014), Rewicz et al. (2014), Benamar (2015), Casal-Lopez et al. (2015), El Haissoufi et al. (2008, 2010, 2015), Errochdi (2015), Zerrouk et al. (2021), Afzan and Belqat (2016), Bennas et al. (1992, 2001, 2016), Doadrio et al. (2011, 2015, 2016a, b), Driauach and Belqat (2015, 2016), Froufe et al. (2016), Kettani and Negrobov (2016), Kment et al. (2016), L'Mohdi (2016), Aulagnier et al. (2017), Clavero et al. (2014a, b, 2017a, b), Corona-Santiago et al. (2016), Millán et al. (2016), Ghamizi and Bouhal (2017), Hajji (2017), Khadri et al. (2017), Trari et al. (2004, 2017), Yimlahi et al. (2017), Adghir et al. (2018), Agnès et al. (2018), Bahid et al. (2017, 2018), Belqat et al. (2011, 2018), Brahimi et al. (2018), Kettani and Moubayed-Breil (2018), Bourquia et al. (2019), Ebejer et al. (2019), Faraj and Himmi (2019), Gomes-dos-Santosa et al. (2019), Nourti et al. (2019a, b), Splendiani et al. (2019), Yahyaoui et al. (2020), Ford et al. (2020), Ghamizi (2020), Glöer et al. (2015, 2020a, b), Grichanov et al. (2020), Jong et al. (2020), Radi et al. (2020), Rihane and El Hamoumi (2020), Sahib et al. (2020), Villastrigo et al. (2020), Bouazza et al. (2021), Boulaassafer et al. (2018, 2020, 2021), Mansouri et al. (2021), Ouassou et al. (2021), Achiban et al. (2022), Benamar et al. (2021a, b, 2022a, b), Bergier et al. (2022a, b), El Hamoumi et al. (2022), Kaltenbach et al. (2022), Kettani et al. (2022), Squalli et al. (2020, 2022), Benka et al. (2023), Benlasri et al. (2023), Boudarfa et al. (2023), El Alami et al. (2000, 2022, 2023a, b), El Yaagoubi et al. (2023), Khaffou et al. (2023), Mabrouki et al. (2017, 2019a, b, 2020a, 2020b, c, 2021a, b, 2022a, b, c, 2023a, b, c, d, e), Mars et al. (2023a, b), Taybi et al. (2019, 2020, 2021a, 2021b, 2022a, 2022b, 2022c, 2023a, 2023b), Wakass et al. (2023), Zahidi et al. (2023), IUCN (2024).

invertebrates in this habitat type, with a high rate of endemism. The number of species in other freshwater invertebrate groups is comparatively low, with moderate to weak endemism (Table 12.2).

Representing 9.68% of Morocco's freshwater animals, freshwater vertebrates account for 16.24% of the country's terrestrial vertebrates and are dominated by wetland-dependent birds (74.55%), most of which are migratory. The most represented families are Scolopacidae, Anatidae, and Laridae, together grouping 52.4% of freshwater bird species. Of the 75 species of fish inhabiting Morocco's freshwater ecosystems, 52 species are indigenous and fairly highly endemic (Table 12.2). There are also two semiaquatic turtles, *Emys orbicularis* (Linnaeus, 1758) and *Mauremys leprosa* (Schweigger, 1812), and two snakes, *Natrix maura* (Linnaeus, 1758) and *N. astreptophora* (Seoane, 1984). *E. orbicularis* is considered the rarest chelonian in the Maghreb countries (Farid et al. 2009). More than a quarter of Morocco's amphibians are endemic (Table 12.2); *Salamandra algira* Bedriaga, 1883 is currently represented by four endemic subspecies (Hernandez and Escoriza 2019).

12.2.3 Microalgae

A few studies were carried out on freshwater microalgae in Northern Morocco dams, and Middle Atlas, Pré-Rif, and Saïs plain wetlands, leading to 290 taxa, belonging to 141 genera, 79 families, and 16 classes. Overall, 256 taxa are belonging to six classes: Bacillariophyceae (71 taxa), Chlorophyceae (54), Zygnematophyceae (44), Cyanophyceae (40), Euglenophyceae (29), and Trebouxiophyceae (18); the 10 other classes comprise only 34 taxa (Fqih Berrada et al. 1999; Rachiq et al. 2002; Rachiq 2003; Mhamdi et al. 2018, 2020). The Bacillariophyceae (diatoms), Chlorophyceae (green algae), and Cyanophyceae (blue-green algae) are also the richest classes worldwide according to Hassi et al. (2020). About 37% of all taxa recorded are represented by Desmidiaceae, Scenedesmaceae, Euglenaceae, Phacaceae, Naviculaceae, Spirogyraceae, and Bacillariaceae (Table 12.1). One taxon each was recorded for a total of 88 genera, while *Cosmarium, Spirogyra, Euglena, Nitzschia, Navicula*, and *Phacus*consist of 15, 13, 9, 8, 8, and 8 taxa in the same order.

So far, the Moroccan freshwater microalgae inventoried represent only 0.58% of the world's described microalgae (Hassi et al. 2020), and significant efforts are required to further explore this microalgal diversity. What's more, some wetlands are threatened by toxic cyanobacteria blooms (Belokda et al. 2018). In addition, Riadi (1998) and Elghrib et al. (2012) reported the presence of 181 taxa of marine microalgae along Morocco's southern Atlantic coast, largely dominated by Bacillariophyceae (Diatoms) and Dinophyceae (Dinoflagellates) (Table 12.1).

This results in Morocco's wetlands harbor at least 5311 taxa of plants, animals, and protists, of which 470 are endemic. However, this diversity falls far short of the reality; many taxa have yet to be discovered in most of the lower taxonomic groups, in particular protozoa and bacteria. Studies of local diversity in bryophytes, charophytes, and microalgae are relatively fragmentary. Fungi deserve particular attention here, as they are present in wetlands as saprophytic parasites of aquatic and semiaquatic organisms, in silt or drained from surrounding soils. Lichens, like fungi, can grow indirectly in wetlands on tree trunks and branches and other inert supports. It is also not easy to estimate the diversity of aquatic animals along the coast, since the literature gives only a rough idea of the marine fauna, which extends beyond the wetlands as defined in accordance with the Ramsar Convention. Like the world's marine fauna, Morocco's one, although incompletely catalogd, numbers 7136 known species, with a predominance of Arthropods (26.98%), Molluscs (22.37%), and Vertebrates (16.05%). Arthropods are mainly represented by Crustaceans, Molluscs by Gastropods and Lamellibranchs, and Vertebrates by Fishes (Franchimont and Saadaoui 2001).

12.3 Conservation Status

Wetland-specific or facultative vascular plants include 359 taxa on the IUCN Red List, 45.13% of which are near-threatened due to the lack of a specific conservation program for the taxon concerned. Critically endangered, endangered, threatened, near-threatened, and vulnerable taxa represent around 7.03% of this local wetland flora. However, more than a dozen taxa specific to wetlands, strictly endemic to Morocco and of limited distribution, have not been classified as "at risk" (Khabbach et al. 2020; Ennabili et al. 2021; Libiad et al. 2021). Moreover, Morocco is home to 50 threatened seaweeds, largely due to overexploitation, invasive species, pollution, intense grubbing, overlooked biological rest period, and/or naval traffic (Riadi 1998; Franchimont and Saadaoui 2001; Hanif et al. 2014a, b).

As for freshwater animals, a quarter of wetland-dependent species such as fish, waterbirds, or turtles are now threatened with extinction worldwide, particularly in tropical regions (RCW 2018). Birds, amphibians, reptiles, and the *Lutra lutra* (Linnaeus, 1758), as well as some Odonata species, have been assessed by the IUCN on a global scale. In the Mediterranean region, the assessment focused on fish and odonates (Darwall and Smith 2006; Riservato et al. 2009). However, the extinction risk of over 94% of invertebrate species has not yet been assessed by the IUCN. In North Africa, 31 fish species and only 5.38% of invertebrate species recorded have been assessed for conservation status, including odonates, molluscs, and crab (García et al. 2010).

In Morocco, the range of the Eurasian otter *Lutra lutra* is both declining and fragmenting (Aulagnier et al. 2017). It is globally classified as near-threatened (IUCN 2024). In terms of bird species conservation, the reference list currently used in Morocco is that of species of heritage interest (El Agbani et al. 2011), comprising 33 species considered threatened with extinction nationally. Seven species breeding and/or wintering in Moroccan wetlands are on the IUCN Red List of Threatened Species (Table 12.2). Nine other species present in Moroccan wetlands are classified as near-threatened.

In terms of native fish, 8 species are threatened and classified as endangered and vulnerable following the IUCN's assessment of the conservation status of 26 species in North Africa and the Mediterranean (Darwall and Smith 2006; Garcia et al. 2010). There has also been the extinction of the Salmonidae *Salmo pallaryi* (Pellegrin, 1924) and the regional extinction of the Clupeidae *Alosa alosa* (Linnaeus, 1758) and *A. fallax* (Lacépède, 1803). The shad is the only freshwater fish listed in Appendix IV of law n° 29-05 on the protection of species of wild fauna and flora and the control of their trade.

Listed on the IUCN Red List as a near-threatened taxon, the Moroccan distribution of *Emys orbicularis* (Linnaeus, 1758) is highly fragmented and its populations are declining (Velo-Antón et al. 2015). It is considered vulnerable (Pleguezuelos et al. 2010) and listed in Appendix IV of Moroccan law n° 29-05. *Mauremys leprosa* (Schweigger, 1812), on the other hand, has a wide distribution in Morocco and is classified as globally vulnerable (IUCN 2024) but of lesser concern in Morocco (Pleguezuelos et al. 2010).

Three of Morocco's amphibians are threatened with extinction: the Moroccan endemic *Pelobates varaldii* (Pasteur & Bons, 1959), *Salamandra algira* (Bedriaga, 1883), classified as vulnerable, and *Alytes maurus* (Pasteur & Bons, 1962). *Barbarophryne brongersmai* (Hoogmoed, 1972) is classified as being of minor concern (Table 12.2). Six amphibian species are listed in category IV of Moroccan law n° 29-05, which may be revised as often as necessary by decision of the "Haut Commissaire aux Eaux et Forêts et à la Lutte contre la Désertification." Among the crustaceans reported in Morocco, the freshwater crab *Potamon algeriense* (Bott, 1967) is the only species in the group to have been classified as Least Concern by the IUCN (Cumberlidge 2010).

All odonate species are considered in category IV of Moroccan law n° 29-05. In North Africa, 12 of these species are threatened (Table 12.2). Worldwide, only two species are considered threatened: *Calopteryx exul* (Sélys, 1853) and *Onychogomphus boudoti*. Ferreira (Ferreira et al. 2014;

IUCN 2024). *Acilius (Homoeolytrus) duvergeri* (Gobert, 1874) is the lone beetle listed as vulnerable by the IUCN (2024). Although their conservation status has not been globally assessed by the IUCN, 41% of endemic beetles in the broad sense are highly vulnerable nationwide and 70% of Morocco's endemic beetles are highly vulnerable internationally (Bennas et al. 2007, 2008, 2009; Benamar 2015; Benamar et al. 2021c). The same applies to hemipterans; 10% of endemics in the broad sense are highly vulnerable nationally (L'Mohdi 2016). The conservation status has been assessed for 66.67% of the freshwater mollusc species reported in Morocco (Van Damme et al. 2010), resulting in 57 being threatened with extinction (Table 12.2). In total, only 7.34% of the freshwater animals and vascular plants surveyed are included in the IUCN Red List; assessment approaches in this respect carried out locally either by some researchers and/or officially often lack methodology and, as a result, remain ineffectively exploited.

12.4 Wetland Services

Wetlands obviously have many functions and values, common from an ecological and scientific perspective, and more specific on the geographical, socioeconomic, and cultural side (e.g. Ennabili 1999; RCW 2018; Khabbach et al. 2020). In addition to being nesting sites for several species, Moroccan wetlands are essential resting and refueling stops, or wintering sites for migratory waterbirds. Of the 46 areas classified as "Important Bird Areas," 35% are freshwater wetlands, while the remainders are coastal areas, including estuarine wetlands (Fishpool and Evans 2001).

Wetlands, especially marshes and estuaries, outperform other ecosystem types in terms of primary productivity. Under natural and/or polluted conditions, the most productive local wetland plants are *Arundo donax* L. (118–134 tonnes of exportable dry matter·ha/year), *Phragmites australis* (Cav.) Trin. ex Steud. (23–54.3 t dw/ha/year) and *Typha angustifolia* L. (21.6–41 dw ha/year). They are followed by *Salix purpurea* L., *Tamarix gallica* L., *Populus euphratica* Olivier, *Sparganium erectum* L., *Scirpus litoralis* Schrader and *S. maritimus* L., *Juncus acutus* L., and *J. maritimus* Lam. (Ennabili 1999; Ennabili and Radoux 2020, 2021). With regular removal of cultivated biomass in urban wastewater treatment, *Lemna gibba* L. produces 28.39 t dw ha/year (Ennabili et al. 2019).

Public administrations' and local populations' perceptions of ecosystem services often have a utilitarian connotation; ecological and cultural dimensions are of little interest in this respect. Stakeholders' visions are partial, not always shared, and sometimes antagonistic or conjunctural (Bouahim et al. 2011; Choukrani et al. 2023). The habitat of "contact" with local wetlands, including nomadism, enables human communities to survive by carrying out agro-pastoral and hunting activities. In addition, other occupations are conducted here, such as harboring, collecting wood and raw materials for handicrafts and construction, sand and salt extraction, waterfowl food foraging, recreation, research, and education (AEFCS 1996; Ennabili 1999; Dakki et al. 2015; Chadli et al. 2022). Most of these activities fall within the traditional or informal economic sector (Figure 12.1), with the exception of royalties or user fees (Ennabili et al. 1996; Franchimont and Saadaoui 2001).

The most widely used native wetland plants are *Arundo donax*, *Nerium oleander* L., *Typha angustifolia*, *Juncus maritimus*, *Scirpus litoralis*, *Puccinellia distans* (Jacq.) Parl, *Juncus acutus*, and *Salix alba* L. These plant species can generate a potential financial income of a few to several tens of thousands of dirhams per hectare, occupying a modest workforce during the summer period. The limited extent of wetlands in relation to the national territory means that they make a particular contribution, which argues in favor of safeguarding this type of environment (Ennabili 1999). Using macrophytes for fencing, fixing dunes and riverbanks, as a source of wood, as livestock feed, in cooking, and as medicinal and/or aromatic plants needs to be assessed economically according to the "cost-benefit" principle.

Figure 12.1 (a) Cutting and in situ drying of *Juncus maritimus* biomass, M'diq, 1996. (b) Sale of *Juncus* mats, Souk Sebt Lalla Mimouna-Kenitra, 1996. (c) *Arundo donax* sunshades, Oued Laou beach, 2004. (d) Sale of *Salix* handicrafts, Oued Laou, 2004.

Numerous natural or artificial substitutes complement or even replace certain uses of local wetland plants, without however recording a trend toward potential and competitive utilization of these resources. However, *Azolla*, a genus introduced in Morocco, is currently marketed online from some twenty sales outlets throughout the country at an average price of around one hundred dirhams per kilo of fresh material, qualified as "free fodder." Considering its productivity in a Mediterranean climate, *Azolla filiculoides* Lam. can generate a financial product very similar to that of the native macrophytes mentioned earlier (Costa et al. 1999; Ennabili 1999; 2AM 2014).

The use of constructed wetlands based on local macrophytes in wastewater treatment has helped to optimize urban wastewater treatment processes adapted to the local socioeconomic and climatic context (Ennabili 1999; Radoux et al. 2003; Ezzahri et al. 2006; Ennabili et al. 2019; Ennabili and Radoux 2020, 2021). The option of using these natural technologies on a full-scale basis, by creating new wetland ecosystems or rehabilitating existing wetlands for wastewater treatment, could help maintain the biodiversity of local wetlands and produce valuable biomass (Ennabili et al. 2021).

Morocco's contribution to global seaweed production (over 35.76 Mt of fresh weight, 2019) is only 0.05%, but it is the leader in the global *Gelidium* market, holding over 64% of the global share, i.e. over 22.21 kt in 2020. Morocco's *Gracilaria* crop produced 273 tons wet weight in 2019 or only 0.01% of the global share (Cai et al. 2021; Mouga and Fernandes 2022). Annual exports of agar, mainly produced from *Gelidium corneum* from Morocco, ranged from 0.905 to 1.066 kt in 2012–2016, less than 1.247 kt/year permitted export quota of locally produced agar (FAO 2018; DGC 2019). A further form of electronic trade of Moroccan algae (*Gelidium* spp., *Gracilaria* spp., *Gegantina* spp.,

Laminaria spp., and *Ulva*) has been identified on "https://www.morocco-seaweed.com/about-us/," but no mention whatsoever is made of the quantities traded. This mode of exchange is apparently less controlled and could therefore threaten Morocco's algal resources.

With a view to sustaining natural resources for agar production in Morocco, suitable natural sites have recently been identified for *Gracilaria* spp. farming by the "Agence Nationale pour le Développement de l'Aquaculture (NAAD)" in selected localities on the Moroccan coast, such as the Sidi Rahal coast, Nador lagoon, and Dakhla Bay. Since 2013, fishermen have begun exploring the possibility of cultivating seaweed (*Gracilaria gracilis*) in the Nador lagoon, supervised by NAAD, as part of the "Gestion Intégrée des Zones Côtières" approach. The project is part of the "Sustainable Med" program implemented by the "Ministère de l'Environnement." The project plans to achieve an annual production of 1.3 kt dry weight of *G. gracilis* from the Bouareg, Nador lagoon (Kassila et al. 2019).

In addition, the microalgae sector is still underdeveloped in Morocco, even though it harbors interesting resources; only five companies are currently producing spirulina and its derivatives. Uses for microalgae in the food industry, agriculture, medicine, aquaculture, and wastewater treatment are in their pilot stages, and more incentives and funding are needed to develop this field in Morocco (Hassi et al. 2020).

At least 291 seaweed species are used worldwide in 43 countries (White and Wilson 2015). In Morocco, the potential uses of 53 seaweed taxa were reported, including food and hydrocolloid production (agar, alginate, ...), Pharmacopoeia, antimicrobial activities, biotechnology (plant seeds germination and plant growth), paper manufacturing, and/or wastewater treatment (Rhimou et al. 2010; Ainane 2011; Farid et al. 2009, 2012; El Atouani et al. 2016; Boujaber et al. 2016; Mzibra et al. 2018; Kassila et al. 2019; Belattmania et al. 2021).

Morocco's seaweed industry is mainly based on *Gelidium* spp. agar extraction and ranks the country third worldwide in agar production. This dynamism will result in the commercialization of new products, such as "Quick Soluble Agar" or agaroses. There are other seaweeds abundant in Morocco and potentially suitable for alginate production, such as *Gracilaria gracilis*, *G. multipartita* (Clemente) Harvey, and the invasive *Sargassum muticum* (Hanif et al. 2014d; El Atouani et al. 2016; Kassila et al. 2019).

On the other hand, fish consumed by human beings are mainly dependent on coastal wetlands. For example, the Merja Zerga – Kenitra wetland, a Ramsar-listed area covering 7300 ha, is mainly used for fishing mullet, eel, sole, sea bass, and shellfish. Total annual production varies between 70 and 300 t/year for fish and between 100 and 150 t/year for clams (Kraïem et al. 2001). Inland fishery production can reach 30 kt of fish. The fish sought by sport fishermen in rivers classified as Salmonidae are native and introduced trout, and secondarily pike, zander, black-bass, and perch. Carp, barbel and, to a lesser extent, black bass and pike-perch are sought for commercial purposes along certain unclassified rivers or in the reservoirs of large dams. Eels are fished and exported, or artificially fertilized in aquaculture units, notably in Kenitra and Nador (Franchimont and Saadaoui 2001).

12.5 Drivers of Changes

As a variant of the subtropics, Morocco's climate is predominantly Mediterranean, with almost half of the territory under Saharan, arid, and semiarid bioclimates. The impact of climate change on wetland habitats is highlighted at regional and local scales, particularly affecting precipitation and temperature (Khabbach et al. 2019; Leberger et al. 2020; Nogueira et al. 2022; Taybi et al. 2023c).

A decrease in precipitation of 10–20% and an increase in temperature of 2–3 °C are forecast for North Africa by 2050 (Schilling et al. 2012). In addition, high and accelerated lake siltation rates and/or increased sediment transport within lagoons in exchange with the ocean have been recorded (e.g. Appleby et al. 2001; Nogueira et al. 2022).

The main causes of degradation of local natural wetlands include the climatic drought cycles experienced in Morocco, and drying induced by proximal activities such as arboriculture, *Eucalyptus* plantations, overexploitation of river and groundwater, and dam construction (Ennabili 1999; Ennabili et al., 2021; Ichen et al. 2021; Sadki et al. 2022). The physico-chemical characteristics of the water obviously condition the nature of the wetland ecosystem and its living communities, well linked to the height and duration of submersion, evapotranspiration, geological feature, seawater intrusion, and pollution including discharges from water desalination plants (Ennabili 1999; Ramdani et al. 2001a; RCW 2018; Elkhiati et al. 2013; Elazzaoui et al. 2022). Excessive drainage or mobilization of water directly for utilitarian purposes (Figure 12.2) are among the main causes of local wetlands drying up (Ennabili 1999; Flower et al. 2001; Khabbach et al. 2019). To combat winter flooding in certain agricultural regions of the Maghreb in the middle of the last century, rivers were channeled from drained to coastal lakes (Ramdani et al. 2001c).

On the other hand, the development works undertaken sometimes cause real ecological scourges, namely, the construction of dykes in wetlands, artificial openings at the seashore and/or their filling in general, riprap and concreting of the banks have caused in many places modifications of living communities by indirectly influencing the microtopography and the state of static submersion.

Figure 12.2 (a) Managed ponds, Amghas-Azrou, 2002. (b) Partially managed peatland, Khmiss Sahel-Larache, 1997. (c) Typha wetland resulting from irrigation losses, Oued Laou, 1994. (d) Drained wetland under cultivation, M'diq-Tétouan, 1994.

These development actions are often linked to tourism development and urbanization work (Ennabili and Gharnit 2003; El Falah et al. 2021; Ennabili et al. 2021; Mars et al. 2023b).

The Moroccan dams policy is strategically justifiable and allows the creation of new wetlands called "artificial." But intense erosion in most watersheds, coupled with declining vegetation, would challenge the long-term profitability of these dams because of their siltation (Ennabili 1999). Other anthropogenic factors threaten wetland surfaces and their vital components, such as sand and crushed stone extraction, conversion to agricultural land (Figure 12.2), livestock, grazing, and macrophyte cutting (Ennabili 1999; Cherkaoui and Hanane 2011; El Falah et al. 2021). However, some of these traditional activities could be ecologically important and sustainable, subject of course to a reasonable exploitation plan (Ennabili et al. 2021).

Among the specific plants introduced in Moroccan wetlands, three are qualified as invasive by some authors, probably due to a progressive constitution of niches similar to continental, subtropical, tropical, and desert climates (Khabbach et al. 2020; Chadli et al. 2022). But with the exception of *Cotula coronopifolia* L., naturalized in Morocco, none of the other highly invasive species in other countries has reached naturalized and/or invasive status. Although *Pistia stratiotes* L. has invaded wetlands in several regions, its invasion is currently limited in Morocco due to environmental and management factors (Chadli et al. 2024).

Biological invasion could pose a threat to Moroccan algae stocks insofar as eight invasive species have been reported in Moroccan marine waters (Franchimont and Saadaoui 2001; Hanif et al. 2014a). For example, the invasion of the Mediterranean Nador lagoon by *Caulerpa prolifera* (Forsskål) J.V. Lamouroux and the Atlantic coasts from Oualidia to Azemmour by *Sargassum muticum* (Yendo) Fensholt could be responsible for the gradual disappearance of *Posidonia oceanica* (Linnaeus) Delile and *Gelidium corneum* (Hudson) J.V. Lamouroux, respectively (Franchimont and Saadaoui 2001; Hanif et al. 2014b). Another particularly striking case is the gradual elimination of fields of large kelp (*Laminaria*, *Saccorhiza*, and *Phyllariopsis* genera) on both the Atlantic and Mediterranean coasts; these species seem to escape surface waters to go deeper up to 30 m (Franchimont and Saadaoui 2001).

On the other hand, Morocco is home to 41 invasive animal species, including 21 fish species and 7 species of molluscs and arthropods; some of these introduced species are among the hundred most invasive species in the world, such as *Stegomyia albopictus* (Skuse, 1894), *Micropterus salmoides* (Lacépède, 1802), *Cyprinus carpio* (Linnaeus, 1758), *Gambusia holbrooki* (Girard, 1859), *Oncorhynchus mykiss* (Walbaum, 1792), and *Trachemys scripta elegans* (Wied-Neuwied, 1839). These exotic species were mainly introduced as a result of restocking initiatives, posing a significant risk to local biodiversity through predation, competition, and hybridization; trade in aquarium and ornamental species would have a secondary role in this regard (Taybi et al. 2023c). It should be noted that *Procambarus clarkii* (Girard, 1852), one of the most invasive aquatic species in the world, has been introduced in Morocco since 2008 to test its degree of acclimatization and has spread in most of the river systems of the north-eastwest of the country (Saguem et al. 2020).

12.6 Planning and Management

The Ramsar Convention on Wetlands, which Morocco ratified in 1980, provides a framework for actions and programs aimed at wetland conservation and resource use. Currently, 38 Moroccan sites are on the Ramsar list, and most of the 154 "Sites of Biological and Ecological Interest" correspond to wetlands and coastal areas (Figure 12.3). In this respect, it is also important to note the law relating to environmental impact studies, whose methodological guide classifies wetlands as sensitive areas, and the law relating to the coastline as a public domain (Ennabili et al. 2021).

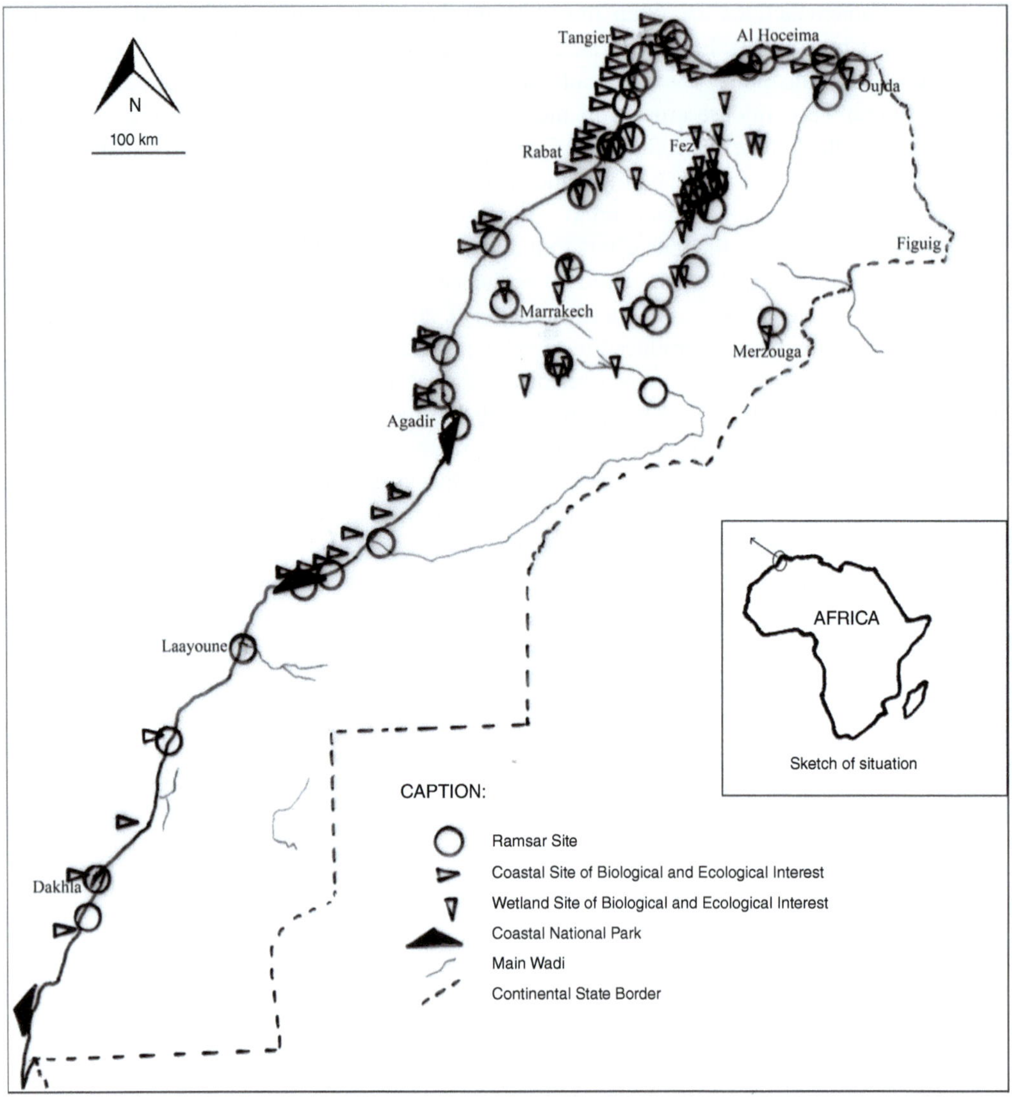

Figure 12.3 Situation of wetlands with current or planned conservation status. *Source:* Ramsar List (2023) and AEFCS (1996).

Further legislative and operational measures can only benefit wetland protection, namely, the creation of National Parks, Biological Reserves, and Biosphere Reserves, which promote solutions reconciling the conservation of biodiversity and its sustainable use. There is also legislation on water, hunting, fishing, the protection of endangered species of flora and fauna, and the "Charte Nationale de l'Environnement et du Développement Durable" (Dakki et al. 2015), in addition to the ratification of international conventions dealing with natural and cultural heritage.

Within this framework, a number of actions have been carried out on a local scale. Conservation of endangered Moroccan algae requires more and prompt work to implement coastal spatial planning, particularly marine site selection, initiated since 2000 in Morocco (Sanchez-Jerez et al. 2016), and to explore population trends and list these species on the IUCN Red Register. With a view to implementing emergency measures to reduce overexploitation and rebuild seaweed resource stocks and their traceability, Morocco's "Direction des Pêches Maritimes" has set a seaweed harvesting quota for all species, of around 9.746 kt of dried weight or 30.459 kt of wet one (DPM 2021).

Reducing harvesting pressure on *Gelidium corneum* by exploiting *Gracilaria* spp. is a promising alternative for agar production (Benhissoun et al. 2002b; Hanif et al. 2014d; Kassila et al. 2019), and past management mistakes must be prevented. When collecting biomass, the disc fixing the species and the perennial part of the thallus ensuring the annual production of new fronds must be preserved, and advanced regulations must also be worked out in this regard involving the local population (Hanif et al. 2014d). Seaweed cultivation, in particular *Gracilaria gracilis* (Stackhouse) Steentoft, L.M. Irvine & Farnham (Kassila et al. 2019), could provide a local substitute and relieve the *G. corneum* operation; incidentally, worldwide seaweed cultivation concerns nearly 30 species (FAO 2021).

Chinese selected carp have been introduced into dam reservoirs to improve freshwater fish production and combat eutrophication, and artificial propagation of these carp has been carried out in Morocco, in addition to the production of trout used to restock rivers (Franchimont and Saadaoui 2001). However, the legal and practical measures taken to conserve wetlands need to be continuously improved through the drafting of implementing legislation and specific development and management plans, in consultation with the stakeholders concerned. National or international conservation status does not necessarily guarantee a wetland's good environmental status (e.g. MedWet/Com 2016; Rifai et al. 2018; El Falah et al. 2021). Most of the 29 freshwater "Key Biodiversity Areas" identified in Morocco (Darwall et al. 2015) are located outside protected areas. Actions by local authorities to reconstitute and conserve poplar-based riparian wetlands are sometimes not subsequently capitalized on and aborted in barely a generation, despite the fact that the production of poplar seedlings, including *Populus euphratica*, was encouraged as early as the middle of the last century (Ennabili 1999; Ennabili et al. 2021).

Tourism infrastructure development and urban planning programs in Morocco have inevitably led to the massive dysfunctioning of wetlands and/or their conversion into buildable areas, apparently without taking traditional know-how into account or carrying out reliable environmental impact studies (Ennabili et al. 2021). Water and land conservation practices were often driven by the need to manage difficult environments. Despite the water deficit, the country's agricultural policy has always favored regions with high production potential, particularly irrigation schemes, which also have a negative impact on water and soil (Laouina 2010; El Falah et al. 2021). The mountain environment is subject to programs and actions to combat the silting-up of dams, and traditional mountain agriculture has been abandoned in favor of water-draining crops, calling into question traditional management methods and techniques. New agro-forestry approaches were unable to break away from water-intensive crops and plantations (Ennabili et al. 2021).

Lack of coordination, particularly between authorities and users, is occasionally observed when carrying out programs and actions relating to wetland conservation and management (Dakki et al. 2015). Among the key players in wetland conservation and use, scientists are inevitably insufficiently involved due to political, funding, and coordination challenges; the result is a lack of sustainability for projects and actions carried out (Ennabili 2012; Garnaud and Rochette 2012; Ennabili et al. 2021). In this respect, public involvement is generally utilitarian, while the private sector is apparently absent.

12.7 Concluding Remarks

This work presents a summary and updated status of Morocco's wetlands, according to the availability of data in terms of biodiversity, conservation, and management. This knowledge, which is extremely useful for both scientists and decision-makers, is intended to be updated regularly. Wetland biodiversity is thought to considerably underestimate actual diversity, for reasons of coverage, technical, and/or scientific difficulties. Different taxonomic groups interest researchers in different ways. Birds are one of the most studied and monitored animal groups, while the minor invertebrate phyla, Protista, and Fungi are less or not studied in this respect.

Many new and/or endemic taxa are part of the national heritage and should be given special attention in assessing their conservation status; those not on the IUCN Red List may also be under threat. Many introduced species are currently poorly known, and their status could change. To limit the expansion of these exotic taxa, particularly those with invasive status, a multicriteria approach should be devised, in both the short and long term, and implemented on a case-by-case basis, developing appropriate environmental and socioeconomic measures.

Considering the drivers of change in wetlands would go a long way toward resolving the current discrepancies between stakeholders' visions with regard to ecosystem services. The sustainability of subsistence activities carried out by the local population, such as inland fishing and the exploitation of macrophyte biomass, requires their support and integration into large-scale integrated development programs.

It is particularly expected that Morocco's commitment to global biodiversity conservation will lead to the development of a sustainable policy for the conservation of local wetlands, despite the many cases of inadequate conservation and disturbance recorded in many places. Involving further actors concerned with wetland conservation and management, notably researchers through their universities and research centers, would undoubtedly have a positive impact in terms of training and benefiting from all available knowledge and experience.

References

2AM (2014). Annoncesagricoles, Maroc. https://flaha.ma/fr/maroc/a_vendre/azolla?p=1 (accessed 3 February 2024).

Abdaoui, A., El Alami, M.E., and Ghamizi, M. (2010). Diversité et distribution des Baetidae (Insecta, Ephemeroptera) du Parc National de Toubkal (Haut Atlas, Maroc). *Actes de la CIFE VI, Travaux de l'Institut Scientifique, Série Zoologie, Rabat* 47 (1): 1–5.

Achiban, H., Mansouri, I., Squalli, W. et al. (2022). Avifauna of the Oued Bouhellou Valley (Morocco): remarkable diversity, five new breeding cases and mapping of nesting sites. *Zoology and Ecology* 32 (1): 36–48. https://doi.org/10.35513/21658005.2022.1.5.

Adghir, A., Herman, D.J., and Kettani, K. (2018). The Tipulidae (Diptera) of northern Morocco with a focus on the Rif region, including the description of a new species of *Tipula luna tipula* and an updated checklist for Morocco. *Annales de la Societe Entomologique de France* 54 (6): 522–538. https://doi.org/10.1080/00379271.2018.1530949.

AEFCS, Administration des eaux et Forêts et de la Conservation des Sols (1996). *Plan directeur des Aires Protégées du Maroc*, vol. 2 and 3. Maroc: Ministère de l'Agriculture et de la Mise en Valeur Agricole.

Afzan, H. and Belqat, B. (2016). Faunistic and bibliographical inventory of the Psychodinae moth-flies of North Africa (Diptera, Psychodidae). *ZooKeys* 558: 119–145. https://doi.org/10.3897/zookeys.558.6593.

Agnès, J.F., Louizi, H., Gilles, A. et al. (2018). A euryhaline fish, lost in the desert: the unexpected metapopulation structure of *Coptodon guineensis* (Günther, 1862) in the Sebkha of Imlili. *Comptes Rendus Biologies* 341 (2): 75–84. https://doi.org/10.1016/j.crvi.2018.01.002.

Ahayoun, K., Ouazzani, T.A., Benkirane, R. et al. (2013). Catalogue bibliographique des Bryophytes du Maroc (1913–2011). *Journal of Animal & Plant Sciences* 17 (1): 2433–2513.

Ainane, T. (2011). Valorisation de la biomasse algale du Maroc: Potentialités pharmacologiques et Applications environnementales, cas des algues brunes *Cystoseira tamariscifolia* et *Bifurcaria bifurcata*. PhD thesis. University of Hassan II of Casablanca.

Appleby, P.G., Birks, H.H., Flower, R.J. et al. (2001). Radiometrically determined dates and sedimentation rates for recent sediments in nine North African wetland lakes (the CASSARINA Project). *Aquatic Ecology* 35: 347–367.

Aulagnier, S., Cuzin, F., and Thevenot, M. (2017). *Mammifères sauvages du Maroc: peuplement, répartition, écologie*. France: Société Française pour l'Etude et la Protection des Mammifères.

Azeroual, A. (2003). Monographie des Poissons des eaux continentales du Maroc: systématique, distribution et écologie. PhD thesis. University of Mohammed V of Rabat.

Badsi, H., Ali, H.O., Loudiki, M. et al. (2010). Ecological factors affecting the distribution of zooplankton community in the Massa Lagoon (Southern Morocco). *African Journal of Environmental Science and Technology* 4 (11): 751–762.

Bahammou, N., Cherifi, O., Bouamama, H. et al. (2021). Checklist of Rhodophyceae and the first report of *Aglaothamnion tripinnatum* and *Gaillona gallica* in the Moroccan coastline. *Egyptian Journal of Aquatic Research* 47: 101–107.

Bahid, F.Z. And Kettani, K. (2013). Données génériques préliminaires sur les Empididae (Diptera, Empidoidea) du Rif méditerranéen (Maroc). *Travaux de l'Institut Scientifique, Rabat, Série Zoologie* 49: 59–69.

Bahid, F.Z., Kettani, K., and Daugeron, C. (2017). Two new species of the *Empis (Leptempis) rustica* group of the Mediterranean basin (Diptera: Empididae: Empidinae). *Annales de la Societe Entomologique de France* 53 (6): 422–427.

Bahid, F.Z., Kettani, K., and Daugeron, C. (2018). A new species of *Empis* (*Xanthempis*) from Morocco (Diptera, Empididae, Empidinae). *Bulletin de la Société Entomologique de France* 123 (1): 119–123.

Belattmania, Z., Bentiss, F., Jama, C. et al. (2021). Spectroscopic characterization and gel properties of agar from two *Gelidium* species from the Atlantic Coast of Morocco. *Biointerface Research in Applied Chemistry* 11 (5): 12642–12652.

Belokda, W., Khalil, K., Loudiki, M. et al. (2018). Moroccan reservoirs water quality: a review. *Journal of Materials and Environmental Science* 9 (2): 631–640.

Belqat, B., Adler, P.H., and Crosskey, R.W. (2011). Faunistic and bibliographical inventory of the blackflies (Diptera: Simuliidae) of Morocco. *Zootaxa* 2829: 46–58.

Belqat, B., Adler, P.H., Cherairia, M. et al. (2018). Inventory of the Black Flies (Diptera: Simuliidae) of North Africa. *Zootaxa* 4442 (2): 201–220. https://doi.org/10.11646/zootaxa.4442.2.1.

Benamar, L. (2015). Les coléoptères aquatiques du Maroc. Atlas, Biogéographie et degré de vulnérabilité. PhD thesis. Abdelmalek Essaadi University of Tetouan.

Benamar, L., Millán, A., Sáinz-Cantero, C.E. et al. (2021a). Annotated checklist of water scavenger beetles (Coleoptera: Polyphaga: Hydrophilidae) of Morocco. *Aquatic Insects* 42 (2): 91–159. https://doi.org/10.1080/01650424.2021.1874422.

Benamar, L., Bennas, N., Belhaj, A. et al. (2021b). An updated checklist of Haliplidae, Gyrinidae, Noteridae, Hygrobiidae and Dytiscidae (Coleoptera: Adephaga) of Morocco, with notes on chorology. *Aquatic Insects* 42 (3–4): 247–371. https://doi.org/10.1080/01650424.2021.1939884.

Benamar, L., Bennas, N., Hassoun, M. et al. (2021c). Threatened endemic water beetles from Morocco. *Insect Conservation and Diversity* 25: 465–477. https://doi.org/10.1007/s10841-021-00314-x.

Benamar, L., Bennas, N., Hassoun, M. et al. (2022a). Checklist of Moroccan aquatic beetles (Coleoptera: Hydraenidae Mulsant, 1844). New records and updates. *Zootaxa* 5129 (4): 451–504. https://doi.org/10.11646/zootaxa.5129.4.1.

Benamar, L., Bennas, N., Hassoun, M. et al. (2022b). Updating the presence, distribution and chorology of Moroccan Dryopoidea (Coleoptera: Elmidae and Dryopidae). *Aquatic Insects* 43 (4): 335–389. https://doi.org/10.1080/01650424.2022.2063337.

Benhissoune, S., Boudouresque, C.F., and Verlaque, M. (2001). A checklist of the seaweeds of the Mediterranean and Atlantic coasts of Morocco. I. Chlorophyceae Wille *s l. Botanica Marina* 44: 171–182.

Benhissoune, S., Boudouresque, C.F., and Verlaque, M. (2002a). A checklist of the seaweeds of the Mediterranean and Atlantic coasts of Morocco. II. Phaeophyceae. *Botanica Marina* 45: 217–230.

Benhissoune, S., Boudouresque, C.F., Perret-Boudouresque, M. et al. (2002b). A checklist of the seaweeds of the Mediterranean and Atlantic Coasts of Morocco. III. Rhodophyceae (excluding ceramiales). *Botanica Marina* 45: 391–412.

Benhissoune, S., Boudouresque, C.F., Perret-Boudouresque, M. et al. (2003). A checklist of the seaweeds of the Mediterranean and Atlantic coasts of Morocco. IV. Rhodophyceae – Ceramiales. *Botanica Marina* 46: 55–68.

Benka, E., Dakki, M., Ouibimah, A. et al. (2023). First annoted checklist of aquatic Diptera (Insecta) of two Ramsar sites (Ahançal and Aït Bouguemaz Rivers) at the Central High Atlas (Morocco): families Ceratopogonidae, Chironomidae, Tipulidae, Empididae, and Tabanidae. *International Journal of Zoology* 2023: https://doi.org/10.1155/2023/5581863.

Benlasri, M., Vuataz, L., Gattolliat, J.L. et al. (2023). First report of *Cloeon vanharteni* Gattolliat & Sartori, 2008 (Baetidae, Ephemeroptera) in the Maghreb. *Alpine Entomology* 7: 143–152. https://doi.org/10.3897/alpento.7.109562.

Bennas, N. (2002). Coléoptères aquatiques Polyphaga du Rif (Nord du Maroc): Faunistique, Ecologie et Biogéographie. PhD thesis. Abdelmalek Essaadi University of Tetouan.

Bennas, N. and Sáinz-Cantero, C.E. (2001). *Hydrochus aljibensis* Castro & Delgado et *H. tariqi* Ribera, Hernando & Aguilera, deux nouveaux Hydrochidae pour le nord de l'Afrique. *Latissimus* 13: 10–11.

Bennas, N. and Sáinz-Cantero, C.E. (2006). Coléoptères aquatiques Adephaga de la chaîne rifaine marocaine (Coleoptera Gyrinidae, Haliplidae, Noteridae, Hygrobiidae, Dytiscidae). *Memorie della Società Entomologica Italiana* 85: 31–73.

Bennas, N. and Sáinz-Cantero, C.E. (2007). Nouvelles données sur les Coléoptères aquatiques du Maroc: les Elmidae Curtis, 1830 et les Dryopidae Billberg, 1820 du Rif (Coleoptera). *Nouvelle Revue d'Entomologie* 24 (1): 61–79.

Bennas, N., Sáinz-Cantero, C.E., and Alba-Tercedor, J. (1992). Datos preliminares para un estudio biogeográfico del Macizo Bético- Rifeño basado en Coleópteros acuáticos. *Zoologica Baetica* 3: 167–180.

Bennas, N., Sáinz-Cantero, C.E., Ouarour, A. et al. (2001). Nouvelles données sur les Coléoptères aquatiques du Maroc: Les Hydraenidae Mulsant, 1844 du Rif Faunistique & Biogéographie. *Zoologica Baetica* 12: 135–168.

Bennas, N., Abellán, P., and Millán, A. (2005). Ecosystèmes aquatiques à intérêt particulier de conservation dans le Rif (Nord du Maroc): résultats préliminaires. *Proceedings of the 2nd International Congress WATMED*, Marrakech, Morocco (14–17 November 2005).

Bennas, N., Sánchez-Fernández, D., Abellán, P. et al. (2007). *Hydrochus obtusicollis* Fairmaire, 1877 (Coleoptera, Hydrochidae), un coléoptère aquatique du Rif marocain hautement menacé. *Boletín Sociedad Entomológica Aragonesa* 40: 491–495.

Bennas, N., Sánchez-Fernández, D., Abellán, P. et al. (2008). *Ochthebius* (*Ochthebius*) *lanarotis* Ferro 1985 (Coleoptera, Hydraenidae), un coléoptère endémique marocain spécifique des milieux aquatiques hypersalins. *Boletín Sociedad Entomológica Aragonesa* 43: 361–366.

Bennas, N., Sánchez-Fernández, D., Abellán, P. et al. (2009). Analyse de la vulnérabilité des coléoptères aquatiques dans la rive sud Méditerranéenne: cas du Rif Marocain. *Annales de la Societe Entomologique de France* 45: 309–320.

Bennas, N., Himmi, O., Benamar, L. et al. (2016). Premières données sur les coléoptères et les hémiptères aquatiques de la Sebkha d'Imlily (Région de Dakhla-Oued Ed Dahab, Sud-Ouest du Maroc), avec *Berosus guilielmi* Knisch, 1924 nouvelle pour le Maroc (Coleoptera, Hydrophilidae). *Boletín Sociedad Entomológica Aragonesa* 59: 117–126.

Bergier, P., Thévenot, M., Qninba, A. et al. (2022a). *Birds of Morocco/Oiseaux du Maroc.* Société d'Etudes Ornithologiques de France.

Bergier, P., Sadak, M., and CHM (2022b). *Les oiseaux rares au Maroc.* 27ème Rapport de la Commission d'Homologation Marocaine. Eléments d'Ornithologie Marocaine: eom22031.

Bilton, D.T., McAbendroth, L., Bedford, A. et al. (2006). How wide to cast the net? Cross-taxon congruence of species richness, community similarity and indicator taxa in ponds. *Freshwater Biology* 51: 578–590.

Bouahim, S.R.,L., Mathevet R. et al. (2011). Analysis of perception of temporary pools in western of Morocco by the local stakeholders and the interest of sustainable development. *Journal of Materials and Environmental Science* 2: 451–454.

Bouamama, L., Lebbadi, M., and Aarab, A. (2007). Bacteriological analysis of *Periplaneta americana* L. (Dictyoptera; Blattidae) and *Musca domestica* L. (Diptera; Muscidae) in ten districts of Tangier, Morocco. *African Journal of Biotechnology* 6 (17): 2038–2042.

Bouazza, A., Mouden, E.H., and Rihane, A. (2021). Checklist of amphibians and reptiles of Morocco: a taxonomic update and standard Arabic names. *Herpetology Notes* 14: 1–14.

Boudarfa, F., Gallouli, E.M., El Hafa, M. et al. (2023). Impact of water stress on the planktonic biodiversity of the Youssef Ben Tachafine Dam (Souss Massa, Morocco). *Ecological Engineering & Environmental Technology* 24 (8): 97–106. https://doi.org/10.12912/27197050/171665.

Boujaber, N., Oumaskour, K., Hassou, N. et al. (2016). Antimicrobial effect of two marine algae *Gelidium sesquipedale* and *Laminaria ochroleuca* collected from the Coast of El Jadida-Morocco. *Journal of Bio Innovation* 5 (1): 16–23.

Boulaassafer, K., Ghamizi, M., and Delicado, D. (2018). The genus *Mercuria* Boeters, 1971 in Morocco: first molecular phylogeny of the genus and description of two new species (Caenogastropoda, Truncatelloidea, Hydrobiidae). *ZooKeys* 782: 95–128. https://doi.org/10.3897/zookeys.782.26797.

Boulaassafer, K., Ghamizi, M., Machordom, A. et al. (2020). Phylogenetic relationships within *Pseudamnicola* Paulucci, 1878 (Caenogastropoda: Truncatelloidea) indicate two independent dispersal events from different continents to the Balearic Islands. *Systematics and Biodiversity* 18 (4): 396–416. https://doi.org/10.1080/14772000.2020.1771466.

Boulaassafer, K., Ghamizi, M., Machordom, A. et al. (2021). Hidden species diversity of *Corrosella* Boeters, 1970 (Caenogastropoda: Truncatelloidea) in the Moroccan Atlas reveals the ancient biogeographic link between North Africa and Iberia. *Organisms Diversity & Evolution* 21: 393–420. https://doi.org/10.1007/s13127-021-00490-3.

Bourquia, M., Garros, C., Rakotoarivony, I. et al. (2019). Update of the species checklist of *Culicoides* Latreille, 1809 biting midges (Diptera: Ceratopogonidae) of Morocco. *Parasites and Vectors* 12: https://doi.org/10.1186/s13071-019-3720-4.

Brahimi, A., Libois, R., Henrard, A. et al. (2018). *Luciobarbus lanigarensis* and *L. numidiensis*, two new species of barbels from the Mediterranean Sea basin in North Africa (Teostei [sic Teleostei]: Cyprinidae). *Zootaxa* 4433 (3): 542–560. https://doi.org/10.11646/zootaxa.4433.3.9.

Cai, J., Lovatelli, A., Aguilar-Manjarrez, J. et al. (2021). *Seaweeds and Microalgae: An Overview for Unlocking Their Potential in Global Aquaculture Development.* 1229th FAO Fisheries and Aquaculture Circular. https://doi.org/10.4060/cb5670en.

Casal-Lopez, M., Perea, S., Yahyaoui, A. et al. (2015). Taxonomic review of the genus *Luciobarbus* Heckel, 1843 (Actinopterygii, Cyprinidae) from northwestern Morocco with the description of three new species. *Graellsia* https://doi.org/10.3989/graellsia.2015.v71.135.

Chadli, C., Libiad, M., Khabbach, A. et al. (2022). Functional aspects of *Pistia stratiotes*, an invasive plant of Al Jawahir (Fez) wadi lentic habitats. *Botanica Complutensis* 46: https://doi.org/10.5209/bocm.80494.

Chadli, C., Bouslamti, R., and Ennabili, A. (2024). Do wetland vascular plants introduced in Morocco also become invasive? *Russian Journal of Biological Invasions.*

Cherkaoui, I. and Hanane, S. (2011). Status and breeding biology of Northern Lapwings *Vanellus vanellus* in the Gharb coastal wetlands of northern Morocco. *Wader Study Group Bulletin* 118 (1): 49–54.

Choukrani, H., Kuper, M., Hammani, A. et al. (2023). Visions contrastées des services écosystémiques des zones humides saisonnières du Gharb, Maroc. *Cahiers Agricultures* 32: https://doi.org/10.1051/cagri/2022031.

Chtaibi, N., Magri, N., Belahbib, N. et al. (2017). Study of bryophytic flora in the Ramsar Wetland of Merja Zerga (North-west of Morocco). *International Journal of Environment, Agriculture and Biotechnology* 2 (3): 1032–1037. https://doi.org/10.22161/ijeab/2.3.4.

Clavero, M., Adrados, B., Calzada, J. et al. (2014a). On the presence of *Petromyzon marinus* in Oued Moulouya (Morocco). *Cybium* 38 (4): 307–308.

Clavero, M., Esquivias, J., Qninba, A. et al. (2014b). Fish invading deserts: non-native species in arid Moroccan rivers. *Aquatic Conservation: Marine and Freshwater Ecosystems* 25 (1): 49–60. https://doi.org/10.1002/aqc.2487.

Clavero, M., Qninba, A., Riesco, M. et al. (2017a). Fish in Moroccan desert rives: the arid extreme of Mediterranean streams. *Fishes in Mediterranean Environments* https://doi.org/10.29094/fishmed.2017.003.

Clavero, M., Calzada, J., Esquivias, J. et al. (2017b). Nowhere to swim to: climate change and conservation of the relict Dades trout *Salmo multipunctata* in the high atlas mountains, Morocco. *Oryx* 52 (4): 627–635. https://doi.org/10.1017/S0030605316001551.

Corona-Santiago, D.K., Casal-López, M., Perea, S. et al. (2016). The complete mitogenome of the Moroccan *Luciobarbus rifensis* Doadrio, Casal-López &Yahyaoui, 2015 (Actinopterygii: Cyprinidae). *Mitochondrial DNA Part B Resources* 1 (1): 931–933. https://doi.org/10.1080/23802359.2016.1250137.

Costa, M.L., Santos, M.C., and Carrapiço, F. (1999). Biomass characterization of *Azolla filiculoides* grown in natural ecosystems and wastewater. In: *Biology, Ecology and Management of Aquatic Plants. Developments in Hydrobiology*, vol. 147 (ed. J. Caffrey, P.R.F. Barrett, M.T. Ferreira, et al.). Dordrecht: Springer https://doi.org/10.1007/978-94-017-0922-4_48.

Cumberlidge, N. (2010). The status and distribution of freshwater crabs. In: *The Status and Distribution of Freshwater Biodiversity in Northern Africa*. Gland, Cambridge, and Malaga: IUCN.

Dakki, M., Himmi, O., Qninba, A. et al. (1997). *Étude Nationale sur la Biodiversité: Faune aquatique continentale. Rapp. inedit.* PNUE & Ministère Environnement Maroc.

Dakki, M., Menioui, M., and Amhaouch, Z. (2015). Stratégie nationale et plan d'action 2015–2024 pour les zones humides du Maroc. Haut-Commissariat aux Eaux et Forêts et à la lutte Contre la Désertification & Deutshe Gesellschaft fur Internationale Zusammenarbeit (GIZ), Rabat. https://medwet.org/wpcontent/uploads/2017/03/strategie-nationale_-ZH_Maroc.pdf (accessed January 2021).

Dakki, M., El Fellah, B., and Qninba, A. (2020). Rivers' natural reservoirs: new inputs to the classification of Mediterranean and Saharan wetlands. *Bulletin de l'Institut Scientifique, Rabat, Section Science de la Vie* 42: 1–14.

Darwall, W.R.T. and Smith, K.G. (2006). *The Status and Distribution of Freshwater Fish Endemic to the Mediterranean Basin*. Gland: IUCN. https://doi.org/10.2305/iucn.ch.2006.mra.1.en.

Darwall, W., Carrizo, S., Numa, C. et al. (2015). *Les Zones clés pour la biodiversité d'eau douce dans le hotspot du bassin méditerranéen. Un éclairage pour la conservation des espèces et la planification du développement dans les écosystèmes d'eau douce*. Cambridge, Royaume-Uni et Malaga, Espagne.

Delling, B. and Doadrio, I. (2005). Systematics of the trouts endemic to Moroccan lakes, with description of a new species (Teleostei: Salmonidae). *Ichthyological Exploration of Freshwaters* 16: 49–64.

DGC (Direction Générale du Commerce) – Ministre de l'Industrie, du Commerce, de l'Investissement et de l'Economie Numérique (2019). Avis no 16/19 aux exportateurs des algues brutes et d'agar agar. www.mcinet.gov.ma/sites/default/files/avis_16-19.pdf (accessed 3 October 2023).

Doadrio, I., Perea, S., and Yahyaoui, A. (2011). Morphological and molecular analyses of freshwater Blennids: a new species of the genus *Salaria* Forsskal, 1775 (Actinopterygii, Blenniidae) in Morrocco. *Graellsia* 67 (2): 151–173.

Doadrio, I., Perea, S., and Yahyaoui, A. (2015). Two new species of Atlantic trout (Actinopterygii, Salmonidae) from Morocco. *Graellsia* 71 (2): https://doi.org/10.3989/graellsia.2015.v71.142.

Doadrio, I., Casal-López, M., and Perea, S. (2016a). Taxonomic remarks on *Barbus moulouyensis* Pellegrin, 1924 (Actinopterygii, Cyprinidae) with the description of a new species of *Luciobarbus* Heckel, 1843 from Morocco. *Graellsia* 72 (2): https://doi.org/10.3989/graellsia.2016.v72.174.

Doadrio, I., Casal-López, M., Perea, S. et al. (2016b). Taxonomy of rheophilic *Luciobarbus* Heckel, 1842 (Actinopterygii, Cyprinidae) from Morocco with the description of two new species. *Graellsia* 72: 1–17. https://doi.org/10.3989/graellsia.2016.v72.153.

DPM (Département de la Pêche Maritime). (2021). *Rapport d'Activité, Ministère de l'Agriculture, de la Pêche Maritime*. Maroc: du Développement Rural et des Eaux et Forêts.

Driauach, O. and Belqat, B. (2015). A new species of the genus *Baeoura* from Morocco, with a key to the West Palaearctic species (Diptera, Tipuloidea, Limoniidae). *ZooKeys* 532: 99–105.

Driauach, O. and Belqat, B. (2016). Additions to the Limoniidae and Pediciidae fauna of Morocco, with an updated checklist (Diptera, Tipuloidea). *ZooKeys* 563: 129–146.

Ebejer, M.J., Kettani, K., and Gatt, P. (2019). First records of families and species of Diptera (Insecta) from Morocco. *Boletín de la Sociedad EntomológicaAragonesa* 64: 143–153.

El Agbani, M.A., Qninba, A., Radi, M. et al. (2011). *Les oiseauxd'intérêt patrimonial du Maroc*. Maroc, GREPOM, 3, Adams Graphic.

El Alami, M. (2002). Taxonomie, écologie et biogéographie des Éphéméroptères du Rif (Nord du Maroc). PhD thesis. Abdelmalek Essaadi University of Tetouan.

El Alami, M., Dakki, M., Errami, M. et al. (2000). Nouvelles données sur les Baetidae du Maroc (Insecta, Ephemeroptera). *Zoologica Baetica* 11: 105–113.

El Alami, M., El Yaagoubi, S., Gattolliat, J.L. et al. (2022). Diversity and distribution of Mayflies from Morocco (Ephemeroptera, Insecta). *Diversity* 14 (6): https://doi.org/10.3390/d14060498.

El Alami, M., Vuataz, L., El Yaagoubi, S. et al. (2023a). A new species of the genus *Alainites* Waltz & McCafferty, 1994 (Ephemeroptera, Baetidae) from the north of Morocco. *ZooKeys* 1176: 221–241. https://doi.org/10.3897/zookeys.1176.107829.

El Alami, M., Vuataz, L., El Yaagoubi, S. et al. (2023b). Another new species of the genus *Habrophlebia* Eaton, 1881 (Ephemeroptera, Leptophlebiidae) from the Maghreb. *ZooKeys* 1186: 47–70. https://doi.org/10.3897/zookeys.1186.112796.

El Atouani, S., Bentiss, F., Reani, A. et al. (2016). The invasive brown seaweed *Sargassum muticumas* new resource for alginate in Morocco: spectroscopic and rheological characterization. *Phycological Research* 64: 185–193.

El Falah, S., Dakki, M., and Mansouri, I. (2021). Mapping analysis of the wetland loss in Loukkos (Morocco) under agricultural managements. *Bulgarian Journal of Agricultural Science* 27 (1): 186–193.

El Haissoufi, M. (2012). Les odonates du Maroc: Atlas, état de conservation et outils de gestion. PhD thesis. Abdelmalek Essaadi University of Tetouan.

El Haissoufi, M., Lmohdi, O., Bennas, N. et al. (2008). Les Odonates du bassin versant Laou (Rif occidental, Maroc). *Travaux de l'Institut ScientifiqueRabat* 5: 47–59.

El Haissoufi, M., Bennas, N., El Mohdi, O. et al. (2010). Analyse préliminaire de la vulnérabilité des Odonates (Odonata) du rif occidental (Nord du Maroc). *Boletin Sociedad Entomológica Aragonesa* 46: 345–354.

El Haissoufi, M., De Knijf, G., Bosch, V.J. et al. (2015). Contribution to the knowledge of the Moroccan Odonata, with first records of *Orthetrum sabina*, and an overview of first and last dates for all species. *Odonatologica* 44 (3): 225–254.

El Hamoumi, R., El Malki, S., Rihane, A. et al. (2022). The Sidi Moussa-Oualidia wetland complex a bird paradise between land and sea. *Frontiers in Science and Engineering* 12 (1): 65–75.

El Haouari, H. and Kettani, K. (2014). Premier inventaire des Tabanidés (Diptera: Tabanidae) du Rif occidental (Maroc). *Bulletin de l'Institut Scientifique, Rabat, Section Sciences de la Vie* 36: 77–88.

El Haouari, H., Kettani, K., and Ghamizi, M. (2014). Les Tabanidae (Insecta Diptera) du Maroc. *Bulletin de la Société Zoologique de France* 139 (1–4): 91–105.

El Yaagoubi, S., Vuataz, L., El Alami, M. et al. (2023). A new species of the *Baetis fuscatus* group (Ephemeroptera, Baetidae) from Morocco. *ZooKeys* 1180: 27–50. https://doi.org/10.3897/zookeys.1180.109298.

Elazzaoui, A., Kalla, Y., Lefrere, L. et al. (2022). Using Mediterranean mussel bio-indicators to assess the initial health state of ecosystems receiving desalination plant effluents in Agadir Bay (Morocco). *Journal of Coastal Research* 38 (1): 154–167. https://doi.org/10.2112/jcoastres-d-21-00035.1.

Elghrib, H., Somoue, L., Elkhiati, N. et al. (2012). Distribution du phytoplancton dans les zones d'upwelling de la côte atlantique marocaine située entre les latitudes 328300N et 248N. *Comptes Rendus Biologies* https://doi.org/10.1016/j.crvi.2012.07.002.

Elkhiati, N., Ramdani, M., Espinar,J.L. et al. (2013). Ecological similarities between two Mediterranean wetlands: Sidi Boughaba (North-West Morocco) and the Doñana National Park (South-West Spain). *Journal of Limnology* 72 (2): 301–312. https://doi.org/10.4081/jlimnol.2013.e24.

Ennabili, A. (1999). Végétation hygrophile du Maroc Méditerranéen: écologie, socio-économie et rôlepotentiel dans l'épuration des eaux usées. PhD thesis. UniversityofLiège. https://doi.org/10.13140/RG.2.2.15553.79209.

Ennabili, A. (2012). Centre expérimental MHEA® de M'Diq (Maroc) "Epuration des eaux usées urbaines": problème de communicationou incapacité d'intégration. *Revue de l'Association Forum Nord du Maroc* 6–8: 184–202. https://doi.org/10.6084/m9.figshare.12941816.v1.

Ennabili, A. and Gharnit, N. (2003). Effetsd'aménagements du litoral tétouanais (Nord-ouest du Maroc) sur la végétation hygrophile (Spermatophyta). *Acta Botanica Barcinonensia* 48: 199–216.

Ennabili, A. and Radoux, M. (2020). Nitrogen and phosphorus removal efficiency of three helophytes in constructed surface flow wetlands for urban wastewater treatment. *IIUM Engineering Journal* 21 (2): 25–40. https://doi.org/10.31436/iiumej.v21i2.1207.

Ennabili, A. and Radoux, M. (2021). Nitrogen and phosphorus uptake and biomass production in four riparian plants grown in subsurface flow constructed wetlands for urban wastewater treatment. *Journal of Environmental Management* 280: https://doi.org/10.1016/j.jenvman.2020.111806.

Ennabili, A., Nabil, L., and Ater, M. (1996). Importance socio-économique des hygrophytes au Nord-ouest du Maroc. *Al Biruniya, Revue Marocaine de Pharmacognosie, d'Etudes Ethnomédicales et de Botanique Appliquée* 12 (2): 95–120. https://doi.org/10.6084/m9.figshare.12941951.v1.

Ennabili, A., Ezzahri, J., and Radoux, M. (2019). Performance of *Lemna gibba* bioreactor for nitrogen and phosphorus retention, and biomass production in Mediterranean climate. *Journal of Environmental Management* 252: https://doi.org/10.1016/j.jenvman.2019.109627.

Ennabili, A., Libiad, M., and Khabbach, A. (2021). Importance of wetlands in maintaining the richness of Morocco's vascular flora. *Wetlands* 41 (121): 1–10. https://doi.org/10.1007/s13157-021-01517-3.

Errochdi, S. (2015). Biodiversité, typologie et qualité des eaux de deux réseaux hydrographiques Nord Marocains Laou et Tahadart et taxonomie, Biogéographie et Atlas des Plécoptères du Maroc. PhD thesis. Abdelmalek Essaadi University of Tetouan.

Errochdi, S., Vinçon, G., and El Alami, M. (2014a). Contribution to the knowledge of the Rifan Stoneflies (Morocco). *Mitteilungen der Schweizerischen Entomologischen Gesellschaft* 87: 25–40.

Errochdi, S., Vinçon, G., El Alami, M. et al. (2014b). Contribution to the knowledge of Moroccan and Maghrebin stoneflies (Plecoptera). *Zootaxa* 3838 (1): 046–076. https://doi.org/10.11646/zootaxa.3838.1.2.

Ezzahri, J., Ennabili, A., and Radoux, M. (2006). Exemple de filières de traitement des eaux usées en zones agricoles méditerranéennes (M'Diq, Nord-ouest du Maroc). In: *Actes du Colloque international sur la gestion et la préservation des ressources en eau* (ed. Université Moulay Ismaïl). Rabat, Maroc: Imp. El Maarif Al Jadida.

Fadel, I., El Khaddari, A., Belahbib, N. et al. (2021). Les bryophytes du Maroc: état des lieux, importance et revue bibliographique. *Nowellia Bryologica* 61: 18–27.

Fadil, F. and Dakki, M. (2001). Remarques sur la présence du genre *Gammarus* (Crustacés, Amphipodes) au Maroc, avec description d'une nouvelle espèce. *Beaufortia* 51 (9): 153–161.

Fadil, F. and Dakki, M. (2003). A new species of *Echinogammarus* Stebbing, 1899 (Crustacea, Amphipoda) from Moroccan freshwaters. *Beaufortia* 53: 53–59.

Fadil, F. and Dakki, M. (2006). Deux espèces nouvelles du genre *Gammarus* (Crustacés, Amphipodes) du Maroc. *Bulletin de l'Institut Scientifique Maroc* 28: 13–24.

Fadil, F., Dakki, M., and Fadil, K. (2009). A new species of *Chaetogammarus* Martynov, 1925 (Amphipoda) from Moroccan fresh waters, including notes on the Moroccan epigean Gammaridae. *Crustaceana* 82 (8): 969–979. https://doi.org/10.1163/156854009x423184.

FAO (2018). The global status of seaweed production, trade and utilization. *Globefish Research Programme* 124: 120.

FAO (2021). Top 10 species groups in global aquaculture 2019. www.fao.org/3/cb5186en/cb5186en.pdf (accessed 1 October 2023).

Faraj, C. and Himmi, O. (2019). Liste actualisée des Phlebotominae (Diptera: Psychodidae) du Maroc updated list of Phlebotominae (Diptera: Psychodidae) from Morocco. *Bulletin de la Société de pathologie exotique* https://doi.org/10.3166/bspe-2020-0107.

Farid, Y., Etahiri, S., and Assobhei, O. (2009). Activité antimicrobienne des algues marines de la lagune d'oualidia (Maroc): criblage et optimisation de la période de la récolte. *Journal of Applied Biosciences* 24: 1543–1552.

Farid, Y., Chennaoui, M., Assobhei, O. et al. (2012). Evaluation de l'effet du lieu de récolte des algues marines des côtes atlantiques marocaines sur l'activité antibactérienne et anti-inflammatoire. *Revue de Microbiologie Industrielle, Sanitaire, et Environnementale* 6 (1): 54–66.

Ferreira, S., Velo-Anton, G., Brochard, C. et al. (2014). A critically endangered new dragonfly species from Morocco: *Onychogomphus boudoti* sp. nov. (Odonata: Gomphidae). *Zootaxa* 3856 (3): 349–365. https://doi.org/10.11646/zootaxa.3856.3.3.

Fishpool, L.D.C. and Evans, M.I. (ed.) (2001). *Important Bird Areas in Africa and Associated Islands: Priority Sites for Conservation*. Cambridge: Birdlife International, Pisces Publications.

Flower, R.J., Dobinson, S., Ramdani, M. et al. (2001). Recent environmental change in North African wetland lakes: diatom and other stratigraphic evidence from nine sites in the CASSARINA Project. *Aquatic Ecology* 35: 369–388.

Fochetti, R. and Tierno de Figueroa, J.M. (2008). Global diversity of stoneflies (Plecoptera; Insecta) in freshwater. *Hydrobiologia* 595: 365–377.

Ford, M., Brahimi, A., Baikeche, L. et al. (2020). Freshwater fish distribution in the Maghreb: a call to contribute. OSF Preprints. https://doi.org/10.31219/osf.io/kx4gc.

Fqih Berrada, D., Berrada, R., and Benzekri, A. (1999). Dynamique du phytoplancton en relation avec certains paramètres physico-chimiques dans la retenue El Kansera (Maroc). *Annales de Limnologie* 35 (3): 155–166.

Franchimont, J. and Saadaoui, E.M. (2001). Étude nationale sur la biodiversité. Rapport de synthèse, Programme des Nations Unies pour l'Environnement & Département de l'Environnement, Maroc.

Froufe, E., Gonçalves, D.V., Teixeira, A. et al. (2016). Who lives where? Molecular and morphometric analyses clarify which *Unio* species (Unionida, Mollusca) inhabit the southwestern Palearctic. *Organisms Diversity and Evolution* 16: 597–611. https://doi.org/10.1007/s13127-016-0262-x.

García, N., Cuttelod, A., and Abdul Malak, D. (2010). *The Status and Distribution of Freshwater Biodiversity in Northern Africa*. Gland, Cambridge, Malaga: IUCN.

Garnaud, B. and Rochette, J. (2012). Rôle et limites de l'approche projet dans l'aménagement du littoral à Nador (Maroc). *Revue Tiers Monde* 211: 169–188. https://doi.org/10.3917/rtm.211.0169.

Gerecke, R.F. (2004). Taxonomy and phylogeny in African water mites of the genus *Diplodontus* Dugés, 1834 (Acari, Hydrachnidia, Hydryphantidae). *Annales de Limnologie - International Journal of Limnology* 40 (1): 71–85. https://doi.org/10.1051/limn/2004007.

Ghamizi, M. (2020). New stygobiont genus and new species (Gastropoda, Hydrobiidae) from the Rif (Morocco). *Ecologica Montenegrina* 31: 50–56. https://doi.org/10.37828/em.2020.31.11.

Ghamizi, M. and Boulal, M. (2017). New stygobiont snail from groundwater of Morocco (Gastropoda: Moitessieriidae). *Ecologica Montenegrina* 10: 11–13.

Ghamizi, M., Bodon, M., Boulal, M. et al. (1999). *Atebbania bernasconii*, a new genus and species from subterranean waters of the Tiznit plain, Southern Morocco (Gatropoda: Hydrobiidea). *Journal of Molluscan Studies* 65: 89–98.

Glöer, P., Boeters, H.D., and Walther, F. (2015). Species of the genus *Mercuria* Boeters, 1971 (Caenogastropoda: Truncatelloidea: Hydrobiidae) from the European Mediterranean region, Morocco and Madeira, with descriptions of new species. *Folia Malacologica* 23 (4): 279–291.

Glöer, P., Mabrouki, Y., and Taybi, A.F. (2020a). Two new valvatoid genera (Gastropoda, Hydrobiidae) from Morocco. *Ecologica Montenegrina* 30: 124–128.

Glöer, P., Mabrouki, Y., and Taybi, A.F. (2020b). A new genus and two new species (Gastropoda, Hydrobiidae) from Morocco. *Ecologica Montenegrina* 28: 1–6.

Gomes-dos-Santosa, A., Froufea, E., Gonçalvesa, D.V. et al. (2019). Freshwater conservation assessments in (semi-) arid regions: testing river intermittence and buffer strategies using freshwater mussels (Bivalvia, Unionida) in Morocco. *Biological Conservation* 236: 420–434.

González García, J.A. and Conde Poyales, F. (1994). Catálogo del macrofitobentos del Mediterráneo de Marruecos. *Acta Botanica Malacitana* 19: 5–27.

Grichanov, I.Y., Nourti, M., and Kettani, K. (2020). A new species of *Chrysotus* Loew, 1857 (Diptera: Dolichopodidae) from Morocco. *Acta Zoologica Bulgarica*, 72 (3): 333–338.

Hajji, K. (2017). Les Trichoptères du Maroc: Atlas, Biogéographie et Degré de Vulnérabilité. PhD thesis. Abdelmalek Essaadi University of Tetouan.

Hajji, K., Zamora-Muñoz, C., Bonada, N. et al. (2012). Quelques notes sur l'écologie et distribution des Rhyacophilidae du Rif (Nord du Maroc). *Boletin Sociedad Entomológica Aragonesa* 50: 559–562.

Hajji, K., El Alami, M., Bonada, N. et al. (2013). Contribution à la connaissance des Tricoptères (Trichoptera) du Rif (Nord du Maroc). *Boletin de la Asociación española d'Entomologia* 37 (3–4): 181–216.

Hanif, N., Chair, M., and Chbani Idrissi, M. (2014a). Study of *Gelidium sesquipedale* overexploitation effects on *Sargassum muticum* proliferation of in El Jadida area, Morocco. *International Journal of Innovation and Scientific Research* 8: 283–290.

Hanif, N., Chair, M., Chbani, I.M. et al. (2014b). Contribution to the algal biodiversity study in the Moroccan Atlantic coast. *International Journal of Innovative Research in Science Engineering & Technology* 3: 12507–12524.

Hanif, N., Chair, M., Chbani Idrissi, M. et al. (2014c). L'exploitation des algues rouges *Gelidium* dans la région d'El-Jadida: aspects socio-économiques et perspectives. *Afrique Science* 10 (1): 103–126.

Hanif, N., Chair, M., Chbani Idrissi, M. et al. (2014d). Study of the red algae growth *Gracilaria multipartita* (Clemente) Harvey (Rhodophyta, Gracilaria Les) of the Moroccan Atlantic coast. *International Journal of Scientific and Research Publications* 4: 1–6.

Harrath, A.H., Sluys, R., Merzoug, D. et al. (2012). Freshwater planarians (Platyhelminthes, Tricladida) from the Palearctic section of the African continent: new records, with the description of a new species. *Zootaxa* 3182 (1): 1–15.

Hassi, M., Alouani, M., Ouaddi, O. et al. (2020). A review of Moroccan microalgae and their exploitation fields. *Journal of Environmental Science, Toxicology and Food Technology* 14 (7): 53–59. https://doi.org/https://doi.org/10.9790/2402-1407015359.

Hassoun, M., Salhi, G., Bouksir, H. et al. (2014). *Codium tomentosum* var. *mucronatum* et son epiphyte *Aglaothamnion pseudobyssoides*, deux nouvelles espèces d'algues benthiques pour la phycoflore du maroc. *Acta Botanica Malacitana* 39: 37–44.

Hassoun, M., Moussa, H., Salhi, G. et al. (2023). New records and observations of marine algae from Morocco (Eastern Atlantic Ocean). *Marine Biology Research* https://doi.org/10.1080/17451000.2023.2206139.

Hernandez, A. and Escoriza, D. (2019). A new subspecies of African fire Salamander *Salamandra algira* (Urodela, Salamandridae) from the Middle Atlas Mountains, Morocco. *ZooKeys* 893: 143–158. https://doi.org/10.3897/zookeys.893.46649.

Ichen, A., Messaoudi, C., El Malki,M. et al. (2021). Étude diachronique de l'occupation du sol de la zone humide de DayetAoua dans le Moyen Atlas au Maroc. *Ecologia Mediterranea* 47 (1): 107–115.

IUCN (2024). The IUCN red list of threatened species. Version 2023-1. https://www.iucnredlist.org (accessed November 2023).

Jäch, M.A. and Balke, M. (2008). Global diversity of water beetles (Coleoptera) in freshwater. *Hydrobiologia* 595: 419–442.

Jong, H., Adghir, A., Erik-Jan Bosch, E.J. et al. (2020). Taxonomy of *Nephrotoma guestfalica* (Westhoff, 1879) (Diptera, Tipulidae), with the description of a new subspecies from Moroccoin. *Tijdschrift voor Entomologie* 163 (1): 1–15. https://doi.org/10.1163/22119434-20192086.

Kaltenbach, T., Vuataz, L., Samraoui, B. et al. (2022). Two new species of *Centroptilum* Eaton, 1869 from North Africa (Ephemeroptera, Baetidae). *ZooKeys* 1131: 71–97. https://doi.org/10.3897/zookeys.1131.91017.

Kassila, J., Nhhala, H., Givernaud, T. et al. (2019). Opportunities for the development of seaweed farming as a supplementary income for small-scale fishermen in Nador lagoon: experimental cultivations of *Gracilaria gracilis* (Stackhouse). *Mediterranean Fisheries and Aquaculture Research* 2 (1): 12–26. https://doi.org/10.21608/maj.2022.5857.1003.

Kettani, K. and Moubayed-Breil, J. (2018). Communities of Chironomidae (Diptera) from four ecological zones delimited by the Mediterranean coastal ecosystems of Morocco (Moroccan Rif). Updated list and faunal data from the last two decades. *Journal of Limnology* 77 (s1): 141–144. https://doi.org/10.4081/jlimnol.2018.1727.

Kettani, K. and Negrobov, O. (2016). The updated check list of Dolichopodidae of Morocco (Diptera). *Cesa News* 128: 1–8.

Kettani, K., Ebejer, M.J., Ackland, D.M. et al. (2022). Catalogue of the Diptera (Insecta) of Morocco – an annotated checklist, with distributions and a bibliography. *ZooKeys* 1094: 1–466. https://doi.org/10.3897/zookeys.1094.62644.

Khabbach, A., Libiad, M., and Ennabili, A. (2019). Invasion increasing risk of Al Jawahir Wadi lentic hábitats by *Pistia stratiotes* L. (North-Central Morocco). *Botanica Complutensis* 43: 97–107. https://doi.org/10.5209/bocm.64280.

Khabbach, A., Libiad, M., and Ennabili, A. (2020). An updated checklist of the wetland vascular flora from Morocco. *Moroccan Journal of Biology* 17: 1–35. https://doi.org/10.6084/m9.figshare.12937823.v3.

Khadri, O., El Alami, M., El Bazi, R. et al. (2017). Ephemeroptera's diversity and ecology in streams of the ultramafic massif of Beni Bousera and in the adjacent non-ultramafic sites (NW, Morocco). *Journal of Materials and Environmental Science* 8 (10): 3508–3523.

Khaffou, M., Chillasse, L., and Abba, E.H. (2023). Nidification du Goéland railleur *Chroicocephalus genei* au barrage Ahmed El Hansali, Maroc. *Go-South Bulletin* 20: 59–62.

Kment, P., Jindra, Z., and Berchi, G.M. (2016). Review of West-Palaearctic Hebridae with description of a new species and redescription of *Hebrus fulvinervis* (Hemiptera: Heteroptera). *Zootaxa* 4147 (3): 201–239. https://doi.org/10.11646/zootaxa.4147.3.1.

Kraïem, M.M., Ben Hamza, C., Ramdani, M. et al. (2001). Some observations on the age and growth of thin-lipped grey mullet, *Liza ramada* Risso, 1826 (Pisces, Mugilidae) in three North African wetland lakes: Merja Zerga (Morocco), Garâat Ichkeul (Tunisia) and Edku Lake (Egypt). *Aquatic Ecology* 35: 335–345. https://doi.org/10.1023/A:1011900631096.

L'Mohdi, O. (2016). Les Hémiptères du Maroc. Atlas, Biogéographie et degré de vulnérabilité. PhD thesis. Abdelmalek Essaadi University of Tetouan.

L'Mohdi, O., Bennas, N., Himmi, O. et al. (2010). *Trichocorixa verticalis verticalis* (Fieber, 1851) (Hemiptera, Corixidae): une nouvelle espèce exotique au Maroc. *Boletín Sociedad Entomológica Aragonesa* 46: 395–400.

Lahrouz, S., Dakki, M., and Gmira, N. (2013). Cartographie des habitats naturels d'une zone humide périurbaine: la Merja de Fouwarate (Kenitra, Maroc). *Bulletin de l'Institut Scientifique, Rabat, Section Sciences de la Vie* 35: 27–34.

Laouina, A. (2010). Conservation des eaux et des sols au Maroc: prise en compte de la diversité géographique. *Norois* 214: https://doi.org/10.4000/norois.3156.

Leberger, R., Geijzendorffer, I.R., Gaget, E. et al. (2020). Mediterranean wetland conservation in the context of climate and land cover change. *Regional Environmental Change* 20: https://doi.org/10.1007/s10113-020-01655-0.

Libiad, M., Khabbach, A., and Ennabili, A. (2021). Checklist of the wetland facultative vascular flora from Morocco. *Ecology, Environment and Conservation* 27 (2): 860–875.

Maarouf, A. and Chemseddine, M. (1995). Surveillance de l'infestation des céréales par la mouche noire *Phorbia securis* Tiensu; Diptera, Anthomyiidae. *Ecologia Mediterranea* 21 (3–4): 93–99. https://doi.org/10.3406/ecmed.1995.1794.

Mabrouki, Y., Taybi, A.F., El Alami, M. et al. (2017). New and interesting data on distribution and ecology of Mayflies from Eastern Morocco (Ephemeroptera). *Journal of Materials and Environmental Science* 8 (8): 2839–2859.

Mabrouki, Y., Taybi, A.F., Skalli, A. et al. (2019a). Amphibians of the Oriental Region and the Moulouya river basin of Morocco: distribution and conservation notes. *Basic and Applied Herpetology* 33: 19–32. https://doi.org/10.11160/bah.134.

Mabrouki, Y., Ben Ahmed, R., Taybi, F.A. et al. (2019b). An annotated checklist of the leech (Annelida: Hirudinida) species of the Moulouya River basin, Morocco, with several new distribution records and a historical overview. *African Zoology* 54 (4): https://doi.org/10.1080/15627020.2019.1671218.

Mabrouki, Y., Taybi, A.F., El Alami, M. et al. (2020a). New data on fauna of caddisflies (Insecta: Trichoptera) from northeastern Morocco with notes on chorology. *Aquatic Insects* 41 (4): 356–390. https://doi.org/10.1080/01650424.2020.1797817.

Mabrouki, Y., Taybi, A.F., and Glöer, P. (2020b). New additions to gastropod fauna (Gastropoda: Hydrobiidae, Lymnaeidae) of Morocco. *Ecologica Montenegrina* 31: 40–44. https://doi.org/10.37828/em.2020.31.8.

Mabrouki, Y., Taybi, A.F., Bahhou, J. et al. (2020c). The first record of the swordtail *Xiphophorus helleri* Heckel, 1848 (Poeciliidae, Actinopterygii) established in the wild from Morocco. *Journal of Applied Ichthyology* 36 (6): 795–800. https://doi.org/10.1111/jai.14105.

Mabrouki, Y., Taybi, A.F., and Glöer, P. (2021a). Two new species of the genera *Islamia* and *Mercuria* (Gastropoda, Hydrobiidae) from Morocco. *Ecologica Montenegrina* 39: 76–80. https://doi.org/10.37828/em.2021.39.8.

Mabrouki, Y., Taybi, A.F., and Glöer, P. (2021b). Further records of freshwater Gastropods (Mollusca: HydrobiidaeLymnaeidae, Planorbidae) from Morocco. *Bonn Zoological Bulletin* 70 (2): 273–279. https://doi.org/10.20363/BZB-2021.70.2.27.

Mabrouki, Y., Taybi, A.F., and Glöer, P. (2022a). *Idrisiella bourkaizensis* gen. et sp.n. a new valvatoid snail (Gastropoda, Hydrobiidae) from Morocco. *Invertebrate Zoology* 9 (1): 18–23. https://doi.org/10.15298/invertzool.19.1.03.

Mabrouki, Y., Glöer, P., and Taybi, A.F. (2022b). *Gyraullus marocana* Sp. Nov., a new freshwater snail species (Mollusca, Gasteropoda, Planorbidae) from Morocco. *Nature Conservation Research* 7 (1): 96–100. https://doi.org/10.24189/ncr.2022.007.

Mabrouki, Y., Taybi, A.F., and Glöer, P. (2022c). *Mahrazia benlemlihi* gen. et sp. nov., a new subterranean snail (Gastropoda: Hydrobiidae) from Morocco. *Bonn Zoological Bulletin* 71 (2): 204–208. https://doi.org/10.20363/BZB-2022.71.2.204.

Mabrouki, Y., Taybi, A.F., and Ibrahimi, H. (2023a). *Hydropsyche nador* sp. n. (Trichoptera: Hydropsychidae), a new species of the *Hydropsyche guttata* species cluster from Morocco. *Aquatic Insects* 45 (1): 3–14. https://doi.org/10.1080/01650424.2023.2220311.

Mabrouki, Y., Taybi, A.F., and Glöer, P. (2023b). New data and records of the genus *Theodoxus*in Morocco (Gastropoda: Neritidae). *Natura Croatica* 32 (1): 143–148. https://doi.org/10.20302/NC.2023.32.9.

Mabrouki, Y., Glöer, P., and Taybi, A.F. (2023c). The first record of the North American freshwater limpet *Ferrissia californica* (Mollusca, Gasteropoda) in Morocco. *Nature Conservation Research* 8 (1): 108–112. https://doi.org/10.24189/ncr.2023.004.

Mabrouki, Y., Terec, A.B., Taybi, F.A. et al. (2023d). Taxonomic notes and key to the West Palearctic *Antocha (Antocha) Osten* Sacken, 1860 (Diptera, Limoniidae) with description of a new species from Morocco. *Biodiversity Data Journal* 11: https://doi.org/10.3897/BDJ.11.e103849.

Mabrouki, Y., Taybi, A.F., and VilaFarré, M. (2023e). First record of the globally invasive planarian *Girardia tigrina* (Girard, 1850) sensu lato in Morocco. *BioInvasions Records* 12 (1): 257–264. https://doi.org/10.3391/bir.2023.12.1.21.

Mansouri, I., Squalli, W., Agy, A. et al. (2021). Avifauna diversity in the gate between humid atlas and Saharan Desert: Midelt Province, Morocco. *Hindawi, International Journal of Zoology* 1311: 1–10. https://doi.org/10.1155/2021/5557921.

Mars, N., Haddad, H., Jaouani, M. et al. (2023a). A review of the avifauna visiting the wetland environments of the Martil plain (Tetouan region, northern Morocco): preliminary results. *E3S Web of Conferences*, 412. https://doi.org/10.1051/e3sconf/202341201068.

Mars, N., Maouni, A. and Saïdi, R. (2023b). Importance of urbanized wetlands environments in the reproduction of avifauna, city of Martil (Region of Tetouan) in Morocco. *E3S Web of Conferences*, 412 https://doi.org/10.1051/e3sconf/202341201029.

Martin, P. and Aït Boughrous, A. (2012). Guide taxonomique des oligochètes dulçaquicoles du Maghreb. *ABC Taxa* 12.

MedWet/Com (2016). Wetlands for Sustainable Development in the Mediterranean Region, A Framework for Action 2016–2030. *12th Meeting of the Mediterranean Wetlands Committee (MedWet/Com)*, Paris, France (7–11 February 2016).

Mhamdi, A.M., Assobhi, B., and Eljemli, M. (2018). *Microalgues d'eaux douces de la région Nord du Maroc, Série Botanique*, 41. Rabat: Travaux de l'institut scientifique.

Mhamdi, A.M., Bahhou, J., Bouhbouh, S. et al. (2020). Dynamique des populations et évolution métabolique du phytoplancon. In: *Le réservoir Idriss premier, biogéochimie, plancton et necton* (ed. Université Sidi Mohamed Ben Abdellah, Faculté des Sciences Dhar El Mahraz), 61–139. Morocco: Fez.

Millán, A., Sánchez-Fernández, D., Abellán, P. et al. (2014). *Atlas de los coleópteros acuáticos de España peninsular*. Madrid: Ministerio de Agricultura, Alimentación y Medio Ambiente.

Millán, A.O., L'Mohdi, J.A., Carbonell, F.A. et al. (2016). A new species of *Aphelocheirus* (Hemiptera: Heteroptera: Aphelocheiridae) from Morocco. *Zootaxa* 4173 (6): 577–582. https://doi.org/10.11646/zootaxa.4173.6.5.

Mouga, T. and Fernandes, I.B. (2022). The Red Seaweed Giant *Gelidium* (*Gelidium corneum*) for newbio-based materials in a circular economy framework. *Earth* 3 (3): 788–813. https://doi.org/10.3390/earth3030045.

Moussa, H., Hassoun, M., Salhi, G. et al. (2018). Checklist of seaweeds of Al-Hoceima National Park of Morocco (Mediterranean marine protected area). *Acta Botanica Malacitana* 43: 91–109. https://doi.org/10.24310/abm.v43i0.4966.

Muller, S.D., Rhazi, L., Soulie-Märsche, I. et al. (2017). Diversity and distribution of Characeae in the Maghreb (Algeria, Morocco, Tunisia). *Cryptogamie Algologie* 38 (3): 201–251. https://doi.org/10.7872/crya/v38.iss3.2017.201.

Mzibra, A., Aasfar, A., El Arroussi, H. et al. (2018). Polysaccharides extracted from Moroccan seaweed: a promising source of tomato plant growth promoters. *Journal of Applied Phycology* 30: 2953–2962. https://doi.org/10.1007/s10811-018-1421-6.

Nechad, I., El Harchli, E.H., and Fadil, F. (2014). Dynamique et cycle de reproduction de *Gammarus marmouchensis* Fadil (Crustacés, Amphipodes) de la source Tatawà Imouzzer Marmoucha (Maroc). *Afrique Science* 10 (3): 193–204.

Neveu, N., Allo, M.R., Nnon, J.P. et al. (1997). The fine structure of the egg shells of the cabbage maggot, *Delia radicum* L. (Diptera: Anthomyiidae), and its relation with developmental conditions and oviposition site. *Canadian Journal of Zoology* 75 (4): 535–541. https://doi.org/10.1139/z97-066.

Nogueira, J., Evangelista, H., Bouchaou, L. et al. (2022). Coastal wetland responses to a century of climate change in northern Sahara, Morocco. *Limnology and Oceanography* 67: 285–299. https://doi.org/10.1002/lno.11992.

Nourti, M., Grichanov, Y.I., and Kettani, K. (2019a). A new species of *Teuchophorus* (Dolichopodidae, Diptera) from Morocco. *Nature Conservation Research* 4 (4): 106–110. https://doi.org/10.24189/ncr.2019.064.

Nourti, M., Grichanov, Y.I., and Kettani, K. (2019b). New records of long-legged flies (Diptera, Dolichopodidae) from Morocco. *Acta BiologicaSibirica* 5 (3): 118–130. https://doi.org/10.14258/abs.v5.i3.6514.

Ouassou, A., Dakki, M., El Agbani, M.A. et al. (2021). Distribution and numbers of three globally threatened waterbird species wintering in Morocco: the Common pochard, Marbled teal, and White-headed duck. *International Journal of Zoology* 3: 1–17. https://doi.org/10.1155/2021/8846203.

Pârvu, C., Mirceni, R.P., and Zaharia, R. (2006). Faunistic data on some dipteran families (Insecta, Diptera) from Morocco (results of "Hamada" expedition 2005). *Travaux du Muséum National d'Histoire Naturelle «GrigoreAntipa»* 49: 271–281.

Pleguezuelos, J.M., Brito, J.C., Fahd, S. et al. (2010). Setting conservation priorities for the Moroccan herpetofauna: the utility of regional red lists. *Oryx* 44 (04): 501–508. https://doi.org/10.1017/S0030605310000992.

Qninba, A., Lieron, V., Dieuleveut, V. et al. (2011). Sur la présence de l'Anguille *Anguilla anguilla* (Linnaeus, 1758) dans l'Oued Tissint, un affluent de l'Oued Dr'a (Maroc). *Bulletin de l'Institut Scientifique, Rabat, section Sciences de la Vie* 33 (2): 65–66.

Qninba, A., Agbani, M.A., Radi, M. et al. (2012). Sur la présence de *Tilapia guineensis* (Teleostei, Cichlidae) dans les gueltas d'un affluent de l'Oued Chbeyka, l'Oued Aabar (Province de Tan Tan, Sud-ouest du Maroc). *Bulletin de l'Institut Scientifique, Rabat, section Sciences de la Vie* 34 (2): 125–126.

Rachiq, S. (2003). Structure et fonctionnement du peuplement phytoplanctonique et capacités phagotrophes des microalgues dans le réservoir Allal El Fassi (Maroc). PhD thesis. Sidi Mohammed Ben Abdellah University of Fez.

Rachiq, S., Raoui, M., Chadli, N. et al. (2002). Potentialités phagotrophes des phytoflagellés dans la retenue de barrage Allal El Fassi (Maroc). *Journal of Water Sciences* 15 (1): 87–99. https://doi.org/10.7202/705438ar.

Radi, M., Kimdil, L., Laghzaoui, M. et al. (2020). Reproduction de la Guifette moustac *Chlidonias hybrida* au barrage collinaire d'Ouled Abbas (Région de Marrakech - Safi). *Go-South Bulletin* 17: 55–60.

Radoux, M., Cadelli, D., Nemcova, M. et al. (2003). Optimisation of extensive wastewater treatment systems under Mediterranean conditions (Morocco): compared purification efficiency of artificial ecosystems. In: *Wetlands: Nutrients, Metals and Mass Cycling* (ed. J. Vymazal), 143–168. Leiden: Backhuys Publishers.

Ramdani, M.N., Elkhiati, R.J., Flower, H.H. et al. (2001a). Open water zooplankton communities in North African wetland lakes: the CASSARINA Project. *Aquatic Ecology* 35: 319–333. https://doi.org/10.1023/A:1011926310469.

Ramdani, M., Flower, R.J., Elkhiati, N. et al. (2001b). Zooplankton (Cladocera, Ostracoda), Chironomidae and other benthic faunal remains in sediment cores from nine North African wetland lakes: the CASSARINA Project. *Aquatic Ecology* 35: 389–403. https://doi.org/10.1023/A:1011965226399.

Ramdani, M., Flower, R.J., Elkhiati, N. et al. (2001c). North African wetland lakes: characterization of nine sites included in the CASSARINA Project. *Aquatic Ecology* 35: 281–302. https://doi.org/10.1023/A:1011957324901.

Ramdani, M., Elkhiati, N., Flower, R.J. et al. (2009). Environmental influences on the qualitative and quantitative composition of phytoplankton and zooplankton in North African coastal lagoons. *Hydrobiologia* 622: 113–131. https://doi.org/10.1007/s10750-008-9678-4.

Ramsar List (2023). The List of Wetlands of International Importance. Ramsar List 57 http://www.ramsar.org/sites/default/files/documents/library/strategic_framework_rsis_en.pdf (accessed 19 November 2023).

RCW (Ramsar Convention on Wetlands). (2016). *Introduction à la Convention sur les zones humides (Manuels Ramsar 5E Édition anciennement Le Manuel de la Convention de Ramsar)*. Gland: Secrétariat de la Convention de Ramsar.

RCW, Ramsar Convention on Wetlands (2018). *Perspectives mondiales des zones humides: état des zones humides à l'échelle mondiale et des services qu'elles fournissent à l'humanité*. Gland: Secrétariat de la Convention de Ramsar.

Rewicz, T., Rachalewski, M., and Grabowski, M. (2014). First record of *Echinogammarus pungens* (H. Milne Edwards, 1840) (Crustacea, Amphipoda) from Africa with the checklist of North African freshwater gammarids. *Mediterranean Marine Science* 15 (2): 443–448. https://doi.org/10.12681/mms.696.

Rhimou, B., Hassane, R., José, M. et al. (2010). The antibacterial potential of the seaweeds (Rhodophyceae) of the Strait of Gibraltar and the Mediterranean coast of Morocco. *African Journal of Biotechnology* 9 (38): 6365–6372. https://doi.org/10.5897/AJB09.1911.

Riadi, H. (1998). Etude nationale sur la biodiversité, Algues Marines. Observatoire National de l'environnement du Maroc, Direction de l'observation, des études et de la coordination, Secrétariat d'état chargé de l'environnement, Ministère de l'Aménagement du Territoire, de l'Environnement, de l'Urbanisme et de l'Habitat, Maroc. Programme des Nations unies pour l'environnement (PNUE). https://ma.chm-cbd.net/implementation/enb_ma/alg_mar_ma (accessed 10 November 2023).

Riadi, H. and Kazzaz, M. (1998). Inventaire bibliographique des algues benthiques du littoral Marocain. I. Chlorophyceaeet Phaeophyceae. *Acta Botanica Malacitana* 23: 23–41. https://doi.org/10.24310/abm.v23i0.8548.

Riadi, H., Kazzaz, M., and Ater, M. (2000). Contribution à la connaissance des algues marines benthiques du détroit de Gibraltar et de la méditerranée occidentale marocaine et. I. Chlorophyceae et Phaeophyceae. *Acta Botanica Malacitana* 25: 5–20. https://doi.org/10.24310/abm.v25i0.8469.

Rifai, N., Khattabi, A., Moukrim, S. et al. (2018). Évaluation de la dynamique de l'occupation du sol dans la zone humide Ramsar de Tahaddart (nord-ouest du Maroc). *Revue d'Ecologie (la Terre et la Vie)* 73 (2): 142–152. https://doi.org/10.3406/revec.2018.1922.

Rihane, A. and El Hamoumi, R. (2020). Reproduction of the Whiskered Tern *Chlidonias hybrida* near Tnine Al Gharbiya (Province of Sidi Bennour). *Arabian Journal of Chemical and Environmental Researches* 7 (1): 50–65.

Riservato, E., Boudot, J.P., Ferreira, S. et al. (2009). Statut de conservation et répartition géographique des Libellules du bassin Méditerranéen. In: *La Liste rouge de l'UICN des espèces menacéesTM. Évaluations régionales*. Gland, Malaga, Espagne: UICN.

Sabri, H., Cherifi, O., Maarouf, A. et al. (2021). The checklist and the ecological index of the brown seaweeds from Essaouira coastline (Morocco). *Journal of Applied Science and Environmental Studies* 4 (2): 406–417.

Sadki, M., Chahlaouil, A., Maliki, A. et al. (2022). Mapping the natural habitats of a continental wetland: Dayet Er Roumi lake (Khémisset, Morocco). *IOP Conference Series: Earth and Environmental Science* 1090: https://doi.org/10.1088/1755-1315/1090/1/012030.

Saguem, S.A., Sahli, E.H., Jdi, M. et al. (2020). The *Procambarus clarkia* (red swamp crayfish) in Merja Zerga wetland. *Journal of Biological Research* 93: 8778.

Sahib, S., Driauach, O., and Belqat, B. (2020). New data on the hoverflies of Morocco (Diptera, Syrphidae) with faunistic and bibliographical inventories. *ZooKeys* 971 (19): 59–103. https://doi.org/10.3897/zookeys.971.49416.

Sammoudi, R., Chahlaoui, A., Khaffou, M. et al. (2023). Biodiversity assessment and ecotypology of the Taanzoult plain running water (Aguelmam Sidi Ali, Morocco). *Scientific African* 20: https://doi.org/10.1016/j.sciaf.2023.e01631.

Samraoui, B., Boudot, J.P., Ferreira, S. et al. (2010). The status and distribution of dragonflies. In: *The Status and Distribution of Freshwater Biodiversity in Northern Africa* (ed. N. Garcia, A. Cuttelod, and D.A. Malak), 51–70. Gland, Cambridge, and Malaga: IUCN.

Sánchez-Fernández, D., Bilton, D., Abellán, P. et al. (2008). Are the endemic water beetles of the Iberian Peninsula and the Balearic Islands effectively protected? *Biological Conservation* 141 (6): 1612–1627. https://doi.org/10.1016/j.biocon.2008.04.005.

Sanchez-Jerez, P., Karakassis, I., Massa, F. et al. (2016). Aquaculture's struggle for space: the need for coastal spatial planning and the potential benefits of Allocated Zones for Aquaculture (AZAs) to avoid conflict and promote sustainability. *Aquaculture Environment Interactions* 8: 41–54. https://doi.org/10.3354/aei00161.

Schilling, J., Freier, K.P., Hertig, E. et al. (2012). Climate change, vulnerability and adaptation in North Africa with focus on Morocco. *Agriculture, Ecosystems & Environment* 156: 12–26. https://doi.org/10.1016/j.agee.2012.04.021.

Shine, C. and Klemm, C.D. (1999). *Wetlands, Water and the Law, Using Law to Advance Wetland Conservation and Wise Use*. Gland: IUCN.

Splendiani, A., Palmas, F., Sabatini, A. et al. (2019). The name of the trout: considerations on the taxonomic status of the *Salmo trutta* L., 1758 complex (Osteichthyes: Salmonidae) in Italy. *The European Zoological Journal* 86 (1): 432–442. https://doi.org/10.1080/24750263.2019.16865440.

Squalli, W., Mansouri, I., Dakki, M. et al. (2020). Nesting habitat and breeding success of *Fulica atra* in tree wetlands in Fez's region, central Morocco. *Journal of Animal Behaviour and Biometeorology* 8: 282–287. https://doi.org/10.31893/jabb.20037.

Squalli, W., Mansouri, I., Douini, I. et al. (2022). Diversity of avian species in peri-urban landscapes surrounding Fez in Morocco: species richness, breeding populations, and evaluation of menacing factors. *Diversity* 14 (11): 945. https://doi.org/10.3390/d14110945.

Taybi, A.F., Mabrouki, Y., Chavanon, G. et al. (2018). New data on the distribution of aquatic bugs (Hemiptera) from eastern Morocco with notes on chorology. *Zootaxa* 4459 (1): 139–163. https://doi.org/10.11646/zootaxa.4459.1.6.

Taybi, A.F., Mabrouki, Y., Berrahou, A. et al. (2019). New data on the dragonfly fauna (Odonata) of the Moulouya river basin and the oriental region, Morocco. *Arxius de Miscellània Zoològica* 17: 85–108. https://doi.org/10.32800/amz.2019.17.0085.

Taybi, A.F., Mabrouki, Y., and Doadrio, I. (2020). The occurrence, distribution and biology of invasive fish species in fresh and brackish water bodies of NE Morocco. *Arxius de Miscellània Zoològica* 18: 59–73. https://doi.org/10.32800/amz.2020.18.0059.

Taybi, A.F., Glöer, P., and Mabrouki, Y. (2021a). Description of a new valvatoid *Pikasia smenensis* n. gen. n. sp., (Gastropoda, Hydrobiidae) from Morocco. *Animal Biodiversity Conservation* 44 (2): 317–320. https://doi.org/10.32800/abc.2021.44.0317.

Taybi, A.F., Mabrouki, Y., and Keresztes, L. (2021b). First record of *Phyllolabis savtshenkoi* (Diptera, Limoniidae) in Morocco. *Travaux du Muséum National d'Histoire Naturelle "Grigore Antipa"* 64 (1): 115–118. https://doi.org/10.3897/travaux.64.e63917.

Taybi, A.F., Glöer, P., and Mabrouki, Y. (2022a). Four new species of the genus *Pseudamnicola* Paulucci, 1878 from Morocco (Gastropoda, Hydrobiidae). *Invertebrate Zoology* 19 (2): 191–198. https://doi.org/10.15298/invertzool.19.2.07.

Taybi, A.F., Glöer, P., and Mabrouki, Y. (2022b). *Ainiella zahredini* gen. et sp.n. a new valvatoid snail (Gastropoda: Hydrobiidae) from Morocco. *Invertebrate Zoology* 19 (3): 312–316. https://doi.org/10.15298/invertzool.19.3.04.

Taybi, A.F., Glöer, P., and Mabrouki, Y. (2022c). Two new species *Mercuria* Boeters, 1971 from Morocco (Gasteropoda, Hydrobiidae). *Natura Croatica* 31 (1): https://doi.org/10.20302/NC.2022.31.5.

Taybi, A.F., Glöer, P., and Mabrouki, Y. (2023a). *Gafaita gafaitensis* gen. et sp.n. a new hydrobiid snail (Gastropoda: Hydrobiidae) from Morocco. *Invertebrate Zoology* 20 (3): 280–286.

Taybi, A.F., Glöer, P. and Mabrouki, Y. (2023b). First record of the exotic seminole rams-horn Helisomaduryi (Wetherby, 1879), (Gastropoda, Planorbidae) in Morocco. *Graellsia* 79 (1): https://doi.org/10.3989/graellsia.2023.v79.362.

Taybi, A.F., Mabrouki, Y., and Piscart, C. (2023c). Distribution of freshwater alien animal species in Morocco: current knowledge and management issues. *Diversity* 15 (2): https://doi.org/10.3390/d15020169.

Trari, B., Harbach, R.E., Himmi, O. et al. (2004). An inventory of the mosquitoes of Morocco. I. Genus *Anopheles* (Diptera: Culicidae). *European Mosquito Bulletin* 18: 1–19.

Trari, B., Dakki, M., and Harbach, R.E. (2017). An updated checklist of the Culicidae (Diptera) of Morocco, with notes on species of historical and current medical importance. *Journal of Vector Ecology* 42 (1): 94–104. https://doi.org/10.1111/jvec.12243.

Tuzovskij, P. and Gerecke, R. (2003). A new diagnosis and status for Mixobates Thor, 1905 (Acari, Hydrachnidia, Hygrobatidae), with a revision of the palaearctic species. *Annales de Limnologie - International Journal of Limnology* 39 (2): 151–174. https://doi.org/10.1051/limn/20030130.

Van Damme, D., Ghamizi, M., Soliman, G. et al. (2010). The status and distribution of freshwater mollusks. In: *The IUCN Red List of Threatened Species™ – Regional Assessment* (ed. N. Garcia, A. Cuttelod, and D. Abdul Malak), 29–50. Gland, Cambridge, Malaga: IUCN.

Velo-Antón, G., El Marnisi, B., Fritz, U. et al. (2015). Distribution and conservation status of *Emys orbicularis* in Morocco. *Vertebrate Zoology* 65 (1): 131–134. https://doi.org/10.3897/vz.65.e31514.

Vikhrev, N. (2011). Review of the Palaearctic members of the *Lispe tentaculata* species-group (Diptera, Muscidae): revised key, synonymy and notes on ecology. *ZooKeys* 84: 59–70. https://doi.org/10.3897/zookeys.84.819.

Villastrigo, A., Hernando, C., Millán, A. et al. (2020). The neglected diversity of the Ochthebius fauna from Eastern Atlantic and Central and Western Mediterranean coastal rockpools (Coleoptera, Hydraenidae). *Organisms Diversity & Evolution* 20: 785–801. https://doi.org/10.1007/s13127-020-00463-y.

Wakass, S., Mounir, M., Squalli, W. et al. (2023). Diversity of avian species, their ecosystems and climate conditions in two zones of High Atlas (central Morocco) for ecotouristic purposes. *Journal of Animal Behaviour and Biometeorology* 11 (1): https://doi.org/10.31893/jabb.23004.

White, W.L. and Wilson, P. (2015). World seaweed utilization. In: *Seaweed Sustainability* (ed. B.K. Tiwari and D.J. Troy), 7–25. AcademicPress. https://doi.org/10.1016/B978-0-12-418697-2.00002-7.

Yahyaoui, A., Perea, S., Garzon, P. et al. (2020). *Atlas des poissons des eaux continentales du Maroc*. Madrid: Faculté des Sciences de Rabat, Museo Nacional Ciencias naturales.

Yimlahi, D., Üstüner, T., Zinebi, S. et al. (2017). New records of the soldier flies of Morocco with a bibliographical inventory of the North African fauna (Diptera, Stratiomyidae). *ZooKeys* 709: 87–125. https://doi.org/10.3897/zookeys.709.13364.

Zahidi, A., Akchour, A., Kreiter, S. et al. (2023). Phytoseiid mites (Acari: Mesostigmata) from Central West Morocco: new records and key to females of all recorded Moroccan species. *Acarologia* 63 (3): 691–724. https://doi.org/10.24349/v5of-5oe1.

Zerrouk, M., Dakki, M., El Agbani, M.A. et al. (2021). Evolution of the benthic communities in a north-African river, the upper Sebou (middle atlas-Morocco) between 1981 and 2017: Effects of global changes. *Biologia* 76 (10): 2973–2989. https://doi.org/10.1007/s11756-021-00787-w.

13

Does Malawi Need a Wetland Policy to Achieve the Wise Use Principle of the Ramsar Convention?

Rodgers Makwinja[1,2], Solomon G. Tesfamichael[1], and Christopher J. Curtis[1]

[1] *Department of Geography, Environmental Management and Energy Studies, University of Johannesburg, Johannesburg, South Africa*
[2] *Department of Fisheries, Ministry of Natural Resources & Climate Change, Lilongwe, Malawi*

13.1 Introduction

Although most Sub-Saharan African countries face challenges in managing wetlands, some have developed policies that respond to the 1971 Ramsar Convention. The Wise Use Principle of the Ramsar Convention highlights the significance of developing and implementing national wetland policies to address institutional, legislative, socioeconomic, and ecological challenges within a national context (Lindley et al. 2018). Uganda was the first African country to launch the National Wetland Policy for the Conservation and Management of Wetland Resources in 1995. The policy was framed to promote the protection of Uganda's wetlands to sustain their ecological and socio-economic functions (Mafabi 2018). Countries such as Ghana, Kenya, Namibia, Benin, Botswana, and Zambia also drafted wetland policies, though not yet launched (Dixon et al. 2021). Under the International Climate Initiative Project, Ethiopia, though not having ratified the Ramsar convention, developed a national wetlands directory depicting national wetlands status, threats, and opportunities (EWNHS 2018), a strategic document that could guide the development of national wetland policy. South Africa, though, has not developed a stand-alone wetland policy; it incorporates objectives relating to wetland conservation and wise use into relevant sectoral policies, including those covering environments, agriculture, biodiversity, and water (Dini and Everard 2016). Malawi joins countries such as Burkina Faso, Algeria, Chad, Congo, Côte d'Ivoire, Democratic Republic of Congo, Federal Republic of Comoros, Guinea, Mali, Morocco, Niger, Senegal, Tunisia, and Togo that do not have wetland policies and strategic documents (Simaika et al. 2022).

This chapter first discusses the fundamentals of wetland proposals to assist policymakers with a clear road map. It positions wetlands as the cradle of biological diversity and discusses their influence on human evolution. Considering the conflicting views from different societal groups, notably the contrast in perspectives between the use of wetlands for economic development and the need to support the conservation approaches, the chapter discusses the principles underpinning the balance between management and wise use of wetlands reflected in the Ramsar conservation guidelines. This is because so many competing Malawi sectoral policies compromise the effort to manage wetlands, and the country needs to establish well-defined priorities to help resolve these conflicts. Therefore, the chapter explores how Malawi's sectoral policies contradict each other and

Wetlands of Tropical and Subtropical Asia and Africa: Biodiversity, Livelihoods and Conservation,
First Edition. Edited by Thammineni Pullaiah.

how this contradiction affects wetland ecological restoration. It further discusses socioecological systems that link local wetland communities, vulnerability, resilience, and coping strategies. It lays down the right priorities for Malawi to adopt while demonstrating various approaches toward wetland ecological restoration.

13.2 Wetland Concepts and Definitions

13.2.1 Wetland Biological Diversity and Hominids Evolution Concepts

Significant gaps exist in finding a precise wetland concept, and there are no standardized criteria by which wetlands are defined to reflect national, regional, and global policies. Researchers have made a significant effort globally to find a universal definition of wetlands in recent years, and several concepts have been proposed. For example, the concept that describes wetlands as the cradle of biological diversity and human evolution (Anthwal et al. 2010; Marconi et al. 2022) was linked to the "aquatic ape" theory proposed by biologist Sir Alister Hardy in the 1960s. This theory explains how human evolution is linked to aquatic habitat. It argues that much more than other primates, man has several features seen more often in aquatic than in terrestrial mammals (Verhaegen 1985). The connection between aquatic ape theory and the concept of wetland is based on the precept that a few million years ago, our ancestors spent a considerable part of their days in aquatic habitats, and wetland fish's biological diversity played a significant role in shaping their evolution through brain capacity development (Morgan 1982). Indeed, there is indisputable evidence of the origin of the earliest microfossils between 3.46 and 3.85 million years ago and beyond, suggesting that wetlands were the site of the evolution of many highly specialized aquatic flora and fauna (Gopal 2013; Bassi et al. 2014) and is linked to the critical stages in human evolution.

There is also a theory claiming that the evolution of diverse fish species underscores the wetland concept. For example, it is believed that the evolution of fish biodiversity took place in wetland margins such as rivers, lakes, marshes, swamps, and coastal areas, evidenced by the fact that many aquatic flora and fauna survived in wetlands, and those outside wetland boundaries never survived (Maltby 2009). Lake Chilwa wetland fish biological diversity evolution supports this theory. For example, in 1865, Dr David Livingstone surveyed Lake Chilwa (Figure 13.1) and noticed the presence of cichlids and *Clarias* species.

In 1903, Prof. J.E.S. More claimed that Lake Chilwa cichlids were part of Lake Malawi species evolution and were later isolated when the lake disconnected itself from the Great African Rift Valley system. He argued that the only species that could survive was *Clarias mossambicus*, while the rest could not. Dixey supported this theory in 1926, suggesting that Lake Chilwa and Lake Malawi were once connected (Dixey 1926). However, a series of research studies conducted by Prof. Kirk and Denis Tweddle showed that almost 36 species in Lake Chilwa and Chiuta (lakes once connected, later separated by a sandbar, formed about 8000 years ago) (Njaya et al. 2011), except *Sarotherodon shriranus chlawae* (whose endemicity to Lake Chilwa was due to isolation), shared common traits with those found in Lugenda and Rovuma rivers in Mozambique (river systems that drain into the Indian Ocean) (Kirk 1967; Furse et al. 1979). Geologists also argued that no evidence either supports or contests Dixey's theory. However, what is clear is that Lake Chilwa was once part of an open lake through the Lugenda and Rovuma rivers that drain into the Indian Ocean. Linking the evolution of fish species to the "aquatic ape" theory, there is indisputable evidence that fish biological diversity shaped local societies' evolution and development in the Lake Chilwa wetland. Due to fishing and hunting, the wetland is currently ranked among Africa's most densely populated areas (162 people/km^2) (Njaya et al. 2011).

Figure 13.1 Lake Chilwa Wetland. *Source:* Photo by Mr. John Ray Kandapo – Deputy Director of Fisheries, Malawi, during observatory surveys.

13.2.2 Wetland and Psycho-spiritual Concept

Contrary to modern beliefs, early settlers linked wetlands to their spiritual identity (Fredengren 2011). It is thorough this connection that wetlands were reflected as sanctified places – a concept aligned with the principle of "the wisdom of wilderness and the infinite capacity of nature to uplift the human spirit." Within this scope, wetlands were considered a human heritage of sacred natural sites in which a spiritual conservation paradigm was entrapped. It should be noted that this concept does not preclude a scientific definition when perceived within the context of psycho-spiritual integration. Rivers, floodplains, and streams have been studied and found that they are considered areas of special spiritual significance upon which laws governing their management paradigm are framed. It is indisputable that these concepts still stand to date, though not explicitly recognized in a scientific domain. For example, Hook (2021) conceptualized a wetland as a place of spiritual purification and baptism, symbolic of Anglo-Saxon England-pre-Christians. On the other hand, wetlands were considered sites of liminality, perhaps a realm of gods and ancestors. In Ethiopia, several islands in freshwater wetlands carry a spiritual identity. For example, Lake Tana and Lake Ziway islands are designated as places of spiritual aspiration and governed by management paradigms within the spiritual context (Cheesman 1935; Berry and Smith 1972). At Chisi Island of Lake Chilwa wetland, Southern Malawi, the conservation of fish biodiversity was entrenched within the traditional practices and religious taboos or myths practiced to please ancestral spirits (Kalanda-Sabola et al. 2007). To date, several deep pools within Lake Chilwa wetlands are given special attention and identified as sacred places. The management of these pools follows the traditional belief that several species of snakes found in these sites are linked to ancestral spirits (Makwinja et al. 2014). Another exciting site is Khulubvi Hills in the Nsanje district, West of Ndinde Marsh. It is considered an important spiritual place among the Mang'anja tribe and is described as a sacred Mbona shrine – a legendary figure who lived in the area many years ago and is believed to have mystical powers to bring rain. It is the place where the local communities offer sacrifices to offset drought, diseases, heavy rains, and other calamities. The floodplains and chains of mountains rich with diverse aquatic and terrestrial biota are associated with spiritual aspiration in this part of Malawi – values also reflected among the

Maori tribe in New Zealand. The braided wetland ecosystems, such as the Rakaia River, have a solid genealogical connection with Maori communities and are intrinsic to Maori identities (Stewart-Harawira 2020). It is important to note that spiritual aspiration is linked to the perception of how the local population describes wetlands and how rules and regulations governing the use and management of wetlands within the local context are framed.

13.2.3 Wetlands and the Ramsar Concept

From ancient times to the present, new concepts of wetlands have emerged. Some researchers have described the wetland as the "kidney of the landscape" (Mitsch et al. 2015), a concept linked to the impact of human activities on wetland ecosystems and the wetland's capacity to respond to natural and human influence. On the other hand, it is also defined as an "ecological supermarket" (Phukan 2021), which depicts its extensive food chain and the rich biodiversity it supports. Its unique characteristics include standing water or waterlogged soils, anoxic conditions, and plant and animal adaptations. Wetlands further offer opportunities for testing ecological theories and principles involving succession and energy flow developed for aquatic or terrestrial ecosystems. Based on this precept, it is an excellent laboratory for studying principles related to transition zones, ecological interfaces, biodiversity, and human–nature interactions. In 1997, Malawi declared the Lake Chilwa Ramsar site a wetland of international ecological importance (Kalanda-Sabola et al. 2007). Later, in 2017, Elephant Marsh (Figure 13.2) in the Shire River basin was also granted a similar status, joining Lake Chilwa (Kosamu et al. 2017).

This development saw Malawi adopting 1971's universal conceptual definition of wetland – a definition proposed during the Convention on Wetlands of International Importance in Ramsar, Iran. The universal concept described wetlands as areas of marsh, fen, peatland, or water, whether natural or artificial, permanent or temporary, with water that is static or flowing, fresh, brackish or salt, including areas of marine water, low tides, and with depth that does not exceed 6 m (Scott and Jones 1995). Although Malawi adopted the Ramsar Wetland idea, conflicting views will still exist on how its concept is adopted in the proposed National Wetland policy framework.

Figure 13.2 Elephant Marsh wetland in Lower Shire Malawi. *Source:* Photo by Ms. Annie Magombo-Chikwawa district fisheries officer during the exploratory field surveys.

The Ramsar wetland concept is too complex and does not mention open coasts and coral reefs (Goodwin 2006; Christian and Mazzilli 2007; Rao et al. 2015; Sievers et al. 2019). Adopting this definition means excluding Lake Malawi coastal Wetlands such as Mpasanjoka in Salima, Central Malawi, Chia lagoon, Unaka lagoon, in Nkhotakota, Central Malawi, Limphasa and Majenene wetlands in Nkhata Bay District, and Lake Malombe' in Mangochi and Machinga, Southern Malawi, since they could be considered outside the scope of the Ramsar definition. Again, the Ramsar wetland concept does not mention hydrological systems, biological diversity, human evolution, and psycho-spiritual connections in its definition, a significant weakness as they are critical to wetland ecosystems such as lakes, rivers, floodplains, and water meadows and form part of the significant wetland features. For example, Elephant Marsh wetland in Malawi and the Ndinde marsh are depicted by psycho-spiritual importance such as the Mbona shrine and Lake Chilwa wetland is defined by its unique biological evolution of fish species. The concept also excludes man-made wetlands such as canals, reservoirs, rice paddy, and aquaculture ponds. In Malawi, wetlands, for example, in the shoreline of Lake Malawi districts, such as Salima, Nkhotakota, Nkhata Bay, Karonga, Mangochi, and Bwanje in Dedza; Lower Shire Valley such as Chikwawa and Nsanje; and Lake Chilwa basin in Zomba, Machinga, and Phalombe; and other districts such as Mchinji and Mzimba, used for rice production fall outside the definition (Yoon 2009; Hu et al. 2017). It implies that a total of 1154.625 ha for wetland aquaculture and 129 853.48 ha for wetland paddy rice are excluded from the definition. The Ramsar Convention definition further seems to miss significant terminology of river basins, catchments, and watersheds (Karageorgis et al. 2004; Igbadun et al. 2006; Giakoumis and Voulvoulis 2018), which are critical in framing Malawi's policy instruments related to the environment, water, fisheries, forestry, irrigation, land, agriculture, and other policies. It does not reflect the interlinkages between coastal waters, intertidal areas, and landward areas necessary to maintain critical ecological processes. This suggests that framing the National Wetland policy requires modifying the Ramsar concept in line with Malawi's biogeographic characteristics and detailed indigenous, scientific, and legal classification. Countries such as Spain, Costa Rica, and Uganda overlooked these shortfalls and incorporated the Ramsar concept into their national legislation without modification, resulting in conflicts with other critical national policies. It has also been challenging for them to balance conservation and economic policies, particularly in wetland farming and aquaculture. However, some countries, such as Canada and the United States, have retained their long-standing wetland definitions, enshrined in legislation and wetland policy, after noting these shortfalls.

13.3 Wetland's Benefits and Its Contribution to UN SDGs

Malawi wetlands' contribution is of economic, ecological, social, and cultural importance (Figure 13.3) (IPBES 2019; Makwinja et al. 2021a) and represents a strategic asset reflected in the United Nations Sustainable Development Goals (UN SDGs) (Nerini et al. 2018). They underpin the well-being and livelihood of rural populations through income generation, energy, medicine, water, and food supply – contributions aligned to pro-economic development, such as zero hunger (SDG2), clean water and sanitation (SDG6), no poverty (SDG1), affordable and clean energy (SDG7), decent work and economic growth (SDG8), and industry, innovation and infrastructure (SDG9) (UN 2019). They further play a vital role in providing good health and well-being (SDG3), quality education (SDG4), and climate action (SDG13) contributions linked to pro-conservation SDGs (Seifollahi-Aghmiuni et al. 2019; Dickens et al. 2020; Guo et al. 2021; Yin et al. 2021).

(a) (b) (c) (d)

Figure 13.3 (a–d) Wetland socioeconomic and ecological importance. *Source:* (a–c) Photo by Jimy during the Elephant Marsh video documentary shooting (accessed from https://www.youtube.com/watch?v=PTBzs9cd3yk). (d) Photo by Dr Rodgers Makwinja (chapter author) during Lake Malombe exploratory surveys.

Costanza et al. (2014), using a meta-analytical approach, estimated global wetland contributions of US$4.8 trillion, with US$1.82 billion contributed by Asia, US$677 million by North America, Europe US$300 million, Africa US$257 million, Latin America US$123 million, and other regions contributing the rest (WWF 2020) In Malawi, few studies on Lake Chilwa, Lake Malombe, Lake Chiuta, and Elephant Marsh underscore the significance of the four wetlands to the local populations. These studies indicate an overall contribution of US$24.36 million/year for Lake Chilwa (Schuyt 2005), US$20.35 million for Lake Malombe (Makwinja et al. 2021a), US$268 million for Elephant Marsh (Forsythe and Turpie, 2016), and US$17.2 million for Lake Chiuta (Mzuza 2013).

13.4 Threats to Wetland Sustainability

Malawi wetlands are facing multiple threats. However, among the threats is inadequate knowledge of wetland conditions and their contribution to local and national economies. The Intergovernmental Science-Policy Platform on Biodiversity and Ecosystem Services (IPBES) report pointed out that about 20% of Africa's land surface (6.6 million km^2) is degraded by soil erosion, pollution, vegetation loss, and salinization (IPBES 2019). In Southern Africa, where Malawi is located, wetlands are severely degraded and lost due to inappropriate policies and sociopolitical and cultural pressures compromising wetland management efforts (Coe and Foley 2001; Anthwal et al. 2010). Lake Malombe in Malawi has lost an average of US$45.58 million in wetland values and US$8.65 million from fisheries from 1989 to 2019, with the highest (US$79.83 million) net loss recorded from 1999 to 2019 (Makwinja et al. 2021d). The wetland economic loss is also reflected in many countries across the globe where inappropriate policies have significantly impacted the wetland and their ecological state (de Groot et al. 2010; Geist 2011; Cen et al. 2015; Yahdjian et al. 2015; Wangai et al. 2016). In Malawi, wetland exploitation has become a political priority, bringing trade-offs among stakeholders. These trade-offs include increased wetland exploitation for short-term economic benefits (Xu et al. 2014; Makwinja et al. 2022) at the expense of structural and functional biodiversity, habitat integrity, carbon sequestration from natural wetlands and forests, and hydrological systems (Xu et al. 2014). Lake Chilwa basin, Elephant Marsh, Lake Malombe, and other wetlands, where a general shortage of arable land instigated by rapid population growth triggers the competition over the wetland's resources (Peters and Kambewa 2007), provide the best examples. Unless wetlands are recognized and prioritized in political debate and national development policy, there will always be poor investments in restoration programs, and unprecedented degradation trends will continue.

Another significant threat is an unprecedented wetland landscape transformation driven by the need to increase food production and achieve SDG2 (zero hunger). Rapid population growth threatens food, water, energy, and health security, negatively impacting local communities' livelihoods (Hanjra and Qureshi 2010; Hernández-Delgado 2015). Besada and Werner (2015) pointed out that most African countries rely heavily on agriculture, as reflected in their national development agenda. Malawi streamlined its development agenda in line with SDGs 1, 2, and 8 by encouraging agricultural production to achieve food security and economic growth (Malawi Government 2004). It further states that almost 85% of Malawi's population are subsistence farmers – a scenario also depicted in Sub-Saharan African countries (Makoka 2008). In most wetlands, landscape transformation has been driven primarily by agriculture, energy demand, and settlements (Njaya et al. 2011). This significant threat has increased biodiversity loss, forest degradation, doubling nitrogen fixation, tripling phosphorous concentration, increasing nutrient loading, widespread eutrophication, and creating hypoxic conditions (Gil et al. 2019; Makwinja et al. 2021b). Jogo and Hassan (2010) acknowledged that global wetland areas had declined by 90% in the past years. They attributed to the high rate of conversion as more households shift into the wetlands for agricultural activities initiated by moisture-fertile soils during the water recession period – a scenario depicted in Lake Chilwa basin, Chia lagoon, Lake Chiuta, and Elephant Marsh in Malawi (Kosamu 2014; Nkhoma and Kayira 2016; Makwinja et al. 2019), Lake Tana basin in Ethiopia (Abera 2017; Minale 2019), Lake Chad (Kolawole 1988), and the Zambezi floodplain (Chimweta et al. 2021). It should be noted that Goal 2 of the UN SDGs suggested that agriculture is critical to achieving food security and human nutrition (UN 2019). However, agricultural production must be done to support food and nutrition security without compromising wetland sustainability.

Climate change (Figure 13.4) is another significant threat to wetland sustainability in Malawi (Kosamu et al. 2022). Severe impacts such as prolonged drought, floods, high temperatures, erratic rain, and high poverty rates have forced countries to overdepend on wetland ecosystems, resulting in overexploitation. For example, Malawi has one rainfall season, making food production vulnerable to climate change. The severe impacts of climate change, such as prolonged drought, floods, temperature, and rainfall fluctuations, have forced the country to overexploit wetlands for dry season farming and settlement. Wetlands, for example, in the shoreline of Lake Malawi districts such as Salima, Nkhotakota, Karonga, Mangochi, and Bwanje in Dedza; Lower Shire Valley such

Figure 13.4 (a–b) Lake Chilingali wetland (a) dried out in 2015 and Lake Chilwa wetland in 2023 (b). *Source:* (a) Photo by Mr Arab Pendame Msume during an observatory study in 2015. (b) Photo by Dr Rodgers Makwinja (chapter author) during an exploratory survey.

as Chikwawa and Nsanje; and Lake Chilwa basin in Zomba, Machinga, and Phalombe and other districts such as Mchinji and Mzimba, have been converted into rice production in line with National Food Security Policy. Coe and Foley (2001) and Schuyt (2005) had similar observations in the Lake Chad basin in Central Africa and Yala Swamp in Kenya, where local communities converted natural wetlands into agricultural production to offset drought impact.

Literature has isolated several scenarios to establish the critical evidence of climate change's impact on wetlands. Talling (1986), for example, observed that some internally closed wetlands in Africa, such as Lake Chilwa, are more vulnerable to evaporation and surface area decline than those of open or semiopen nature. According to Eissa and Zaki (2011), the dynamic changes in the hydrological cycle of water trigger a reliable number of aquatic species to migrate to alternative water bodies in case of open and semiopen types while leaving those in closed wetlands to live in danger of being threatened or even extinction. Jellison et al. (2008) demonstrated that high-temperature variation can cause closed wetlands to shrink, making them more saline. In arid Central Asia and the Middle East, broad swaths of Africa, and the Altiplano region of South America, a decrease in runoff has caused salt lakes to become more saline, resulting in biodiversity loss (Jellison et al. 2008). In Australia, Lake Corangamite shrank in 2002, and salinity exceeded 100 g per liter, with a remarkable reduction in aquatic biota (Jellison et al. 2008). Lake Chilwa in Malawi – an official Ramsar Site – has been experiencing severe water recessions due to prolonged dry spells and frequent droughts caused by climate change. Official documentation of Lake Chilwa indicates that the lake has undergone 12 recession periods between 1900 and 2012, making it more saline, consequently leading to aquatic biodiversity loss. According to the Government of Malawi report, the lake dried first in 1876, 1900, 1914–1915, 1922, 1931–1932, 1934, 1954, 1960–1961, 1967, 1973, 1995, and 2012, resulting in the disruption of the entire aquatic ecosystem (Njaya et al. 2011). In Ethiopia, the evidence of climate change's impact on the aquatic ecosystem has been demonstrated by the loss of Lake Hiromasa from 1989 to 2005 (Yilma 2010). Lake Naivasha-the official, Ramsar site (30 000 ha), turned into a shallow mud pool during the 2009 drought, resulting in a decline in the aquatic ecosystem (Ogola et al. 2012). Lake Chad, once the most significant freshwater lake in Africa, shrank dramatically in the last 40 years (Pham-Duc et al. 2020).

Figure 13.5 Elephant Marsh algal bloom (a) and Lake Chilwa (b) nutrient enrichment. *Source:* Photo by Dr Rodgers Makwinja (chapter author) during the field surveys.

Another scenario is that an increased precipitation regime could result in a heavy nutrient load (Figure 13.5a, b) entering the wetland, which may lead to seasonal plankton blooms (Whitehead et al. 2009). The accumulation of more organic materials than the capacity consumed at the bottom may lead to biogeochemical reactions that change the sulfates to sulphides (Zhao et al. 2023).

H_2S is more toxic to aquatic organisms such as fish and may result in fish kill. The possible signs of H_2S in water include an odor (rotten egg) and black, decaying organic matter on the windward shore. It should be noted that the future projection in Southern Africa, where Malawi is located, indicates that climate change's impact will be worse than in other regions (Nkhoma et al. 2021). The critical basin wetlands in southern Africa, such as the Zambezi, Shire River, and Lake Chilwa, have already been experiencing increased air temperatures, and it is predicted that potential evaporation will increase (Kambombe et al. 2023). At the same time, rainfall decreases, leading to a decline in river flows and increased reservoir losses (Hamududu and Killingtveit 2016).

Wetland resource overexploitation is another threat driven by rapid population growth and climate change impact. In Malawi, most of the riparian population depends on the fisheries – a typical situation in the least developed regions such as Africa, Asia, and Latin America (Cilliers et al. 2013; Cumming et al. 2017; IPBES 2019). The high contribution of fisheries in countries such as Malawi (Makwinja et al. 2021c) indicates (i) essential role in ensuring food security for the local population and (ii) the existential threat of overexploitation (Youn et al. 2014). Beeton (2002) acknowledged that many endemic species in wetlands had been overexploited and further predicted that by 2025, the species associations will have changed. On the other hand, the high contribution of agriculture production in wetlands could indicate the shifting of local communities' livelihood strategies as the wetland's fisheries collapse (Makwinja et al. 2021f). Jogo and Hassan (2010) reported that extensive modification of wetland landscapes in Africa is linked to increased agricultural activities instigated by the collapse of the fisheries. In Lake Chilwa, Lake Malawi wetlands, Lake Malombe, Lake Chiuta, and Elephant Marsh, the decline in the fisheries (Figure 13.6) has led to intense hunting and trapping pressure on water birds, threatening their population (Njaya et al. 2011; Makwinja et al. 2021e). In the Zambezi floodplain, the Chelonia population has become under severe threat from exploitation, particularly for international trade, and this further extends to the Malawi population of the Zambezi soft-shelled turtle (Forsythe and Turpie 2016).

(a) (b)

Figure 13.6 Lake Chilwa (a) and Lake Malawi wetland fishing activities (b). *Source:* (a) Photo by Dr Rodgers Makwinja (chapter author). (b) Photo by Idrissa Mkwanda.

In Malawi, the evidence of wetland riparian vegetation overexploitation is scarcely available. However, it could be suggested that it follows a similar global trend, suggesting that about 64–71% of wetland riparian vegetation has been lost due to multiple threats ranging from climate change to overexploitation (Davidson 2014; Gule et al. 2021).

13.5 Wetland Management Policy Dilemma

The Republic of Malawi (Constitution) Act, 1966 did not recognize the need for wetland protection and management (GoM 1966). The growing interest was in how to promote irrigation farming to achieve food security in the face of recurring drought, rapid population growth, and climate change. The construction of irrigation schemes in Malawi began between 1946 and 1960, with most schemes established between 1967 and 1982 along the Lake Malawi shoreline wetlands (Karonga, Nkhata Bay, and Nkhotakota), Lake Chilwa basin and Lower Shire – within Elephant and Ndinde Marsh catchments (Gwiyani-Nkhoma 2011). The schemes were designed to promote cash cropping, increase rice exports, and expand rural livelihoods. It was in 1994 that Malawi's constitution was revised, and Article 13, Chapter III recognized the need to implement policies and legislation to safeguard the environment, including wetlands (GoM 1994). Article 13(d) (iii) supported Ramsar Resolution X I.9, which stressed the need to establish the legal framework and policy to manage the wetland sustainably. Despite this effort, Malawi has never had a stand-alone policy or strategic guiding document for wetland management and exploitation. This signaled conflicting policies emanating from the different interests in the use and management of wetlands.

In 1996, the Malawi government drafted the Environmental Management Act. The act had a provision that underscored the contextual wise use Ramsar convention principle. For example, Chapter 60, section 02 explicitly highlighted the need to protect and manage environment which include wetlands, conserve and sustainable use of their natural endowment. Policies such as environment, fisheries, forestry, water, national parks and wildlife and other sectoral legislations framed from this act set a provision for wetland biodiversity conservation . . . However, few investments have been made in environmental conservation, evidenced by the country's lack of wetland restoration programs. National biodiversity targets reflected in the National Biodiversity Strategy and Action Plan (NBSAP) of 2015–2025 focus much on forestry due to its economic contribution through timber, fuelwood, and energy sources, while other wetland biodiversity gained little attention. Again, several national strategic documents such as the Malawi Growth and Development Strategy (MGDS) of the period between 2012 and 2016, Poverty Alleviation Programs (PAPs), and Malawi Vision 2063 strategic document, support the policies focusing on wetland conservation less than those that promote exploitation for economic benefits. Also, much of public and political discussion focuses on agriculture, mainly transforming wetlands to achieve sustainable food production and economic development, and little is mentioned about conservation. Consequently, public discussion, multiparty politics, new land, agriculture, irrigation, national food and nutritional policies have intensified interest in wetlands conversion into food production. It is within the same scope that government, donors, and nongovernment organizations have promoted small, medium, and large schemes in line with the country's development agenda. For example, in line with Vision 2063, the Malawi government embarked on the Shire Valley Transformation Program in 2016 – a program aimed at irrigating 43 370 ha of land by abstracting water from the Shire River at Kapichira and conveying it by gravity to the Elephant Marsh catchment (irrigated area) in Chikwawa and Nsanje districts through canals.

From the socioeconomic viewpoint, increasing farming activities will substantially increase food production, reaching the primary goal of Malawi's Vision 2063. However, in wetlands such as Lake Chilwa, Lake Malombe, Elephant Marsh, and Lake Malawi shoreline, small-scale, medium-scale, and large-scale irrigation have already intensified competition over stream banks and wetland areas for agricultural activities, compromising essential wetland ecological functions such as regulatory and supporting services. Peters and Kambewa (2007) acknowledged that the increased rate of landscape conversion in wetlands is linked to interacting processes among climatic, ecological, historical, political, and policy directions. It is essential to note that increased farming activities in the catchments are done at the expense of water bodies and biodiversity – a scenario depicted in the Lake Chilwa basin and Elephant Marsh, where increased cultivation and wetland conversion for rice production and settlements have amplified wood and wetland vegetation loss in the catchment (Peters and Kambewa 2007). An increase in wetland exploitation for economic benefits diminishes other ecological functions, such as regulation and supporting services – leading to a net loss of wetland value and fisheries (Makwinja et al. 2021b). Similar scenarios have been reported by Gordon et al. (2010), Jenkins et al. (2010), Barral et al. (2015), and Ricaurte et al. (2017) across the globe.

Lake Chilwa's declining fish biodiversity is linked to habitat degradation and overexploitation instigated by rapid population growth and unsustainable catchment farming systems (Nagoli and Chiwona-Karltun 2017). These changes have negative feedback on the wetland socioecological system itself through landscape degradation, hydrological alteration, habitat degradation, biodiversity loss, and loss of soil fertility- impacting on the livelihood of the local population (Gordon et al. 2010). Dybas (2005) pointed out that some trade-offs may be too expensive to reverse without urgent attention. Therefore, this chapter highlights the need to make hard choices between pursuing long-term benefits and fulfilling the local population's demands when reflecting on development policies. Pittock et al. (2012) pointed out that maximizing landscape transformation to boost agricultural production has historically resulted in ecological degradation – suggesting that it might be challenging to achieve a "win-win" situation in Malawi where wetlands landscape transformation, exploitation of natural resources such as fisheries and others form the main livelihood component of most of the local population.

Palacio et al. (2021) emphasized building a fact-based solution to demonstrate the relationship between ecosystem services and SDGs in wetland conservation and sustainable development. The researchers noted that SDG target 14.7 promotes economic benefits, focusing on exploiting wetlands resources such as fisheries, while SDGs 3.4 and 8.9 underpin conservation policies to promote culture, aesthetics, health, and well-being. However, they also expressed that achieving many SDGs depends on a well-functioning ecosystem and the benefits derived from ecosystem services (Palacio et al. 2021). Geijzendorffer et al. (2017) pointed out that global sustainability policies such as SDGs and Aichi Targets are framed to ensure sustainability, including well-being. They noted that 17 SDGs, 20 Aichi targets, 12 goals, and 13 targets are linked to ecosystem services, with ecosystem provisioning services mentioned 29 times, regulatory 33 times, and cultural services 23 times. Although Malawi has taken a significant step in formulating policies and conservation strategies to meet international obligations such as Agenda 2030 for Sustainable Development, Aichi Targets, and Convention on Biological Diversity, the conflict between the policies focusing on economic development and proconservation complicates the effort and underpins the need to have a wetland policy instrument in Malawi. A critical question that needs addressing in this regard is "How do we resolve these sectoral policy conflicts?" Another critical research question is: "How can different policies and SDGs complement each other to achieve sustainable wetland ecosystem functions?"

13.6 Wetland Management Options

Raworth (2017) suggested that negative interactions of sectoral policies cannot be addressed independently. Mooney et al. (2013) explained that this is because socioeconomic and environmental systems interact dynamically across time and space – suggesting that there is a need to consider such interactions in formulating national multisectoral policies while embracing integrated and complex socioecological system approaches (Johnson et al. 2019). This chapter proposes some wetland management options that should be considered when answering the question "Does Malawi Need a Wetland Policy to Achieve the Wise Use Principle of the Ramsar Convention". The first option is that there is a need to balance the opportunity costs among the complex wetland ecosystem services dynamics and landscape transformation to achieve management policy efficiency while avoiding adverse outcomes from human activities (de Groot et al. 2010; Jenkins et al. 2010; Wu 2013; Smiraglia et al. 2015). The second option points out the need to integrate the agricultural benefits into managing land and wetland ecosystem services to strike a balance. This can be achieved by bringing together the wetland and upland communities to negotiate unavoidable trade-offs between the upland and wetland livelihood activities to create synergies and achieve the sustainable flow of wetland ecosystem services. Summers et al. (2012), Green et al. (2015), Bai et al. (2018), and Sharafatmandrad and Mashizi (2021) supported this idea and reasoned that when the wetlands are degraded, their capacity to support the local and global communities decreases. Therefore, balancing people's demands and conservation in decision-making could offer a practical policy response. The third option is that there is a need to integrate various stakeholders' interests into the wetland policy framework and balance economic and ecological models to sustain the wetland ecosystems and promote multisectoral policy synergies. Liu et al. (2008) described this option as bridging the gap between science and decision-making to produce science-based data to support policy formulation. The first Dublin Principle for freshwater management recognized in the 1992 International Conference on Water and Environment in Dublin supports this option. It suggests that wetland challenges should be considered in land-use planning, socioeconomic development, and conservation (GWP 2000). Several models and approaches have been recommended based on this notion. For example, Lemos and Morehouse (2005) suggested an iterative approach to link science to policy through the interaction of policymakers, stakeholders, and the general public. The appropriateness framework was recommended by Xu et al. (2007) to help develop a decision-making supporting tool with an appropriate level of complexity to facilitate effective communication between policymakers, stakeholders, and the general public. The integrated framework based on Bayesian Networks was recommended by Ticehurts et al. (2007) to facilitate the active participation of relevant stakeholders in decision-making processes to restore the degraded wetlands.

The fourth option is based on the idea that fishing activities from wetlands such as Lake Chilwa, Elephant Marsh, Lake Chiuta, and Lake Malombe bind the livelihood sustenance of the local population. Therefore, when framing wetland policy, the fisheries sector is critical. As pointed out by Makwinja et al. (2021d), fisheries managers need to provide critical information on the socioeconomic complex of different groups involved in the fishery and how the fishery dynamics affect their livelihood, vulnerability, and adaptation strategies. This information is essential in aligning the proposed wetland policy with the fisheries policy to strengthen and diversify the local population's livelihood asset base to respond effectively to the wetland ecosystem's dynamics. By looking at social and economic benefits in terms of livelihood sustainability and general values, it is imperative to assess the implication of fish biomass dynamics on the ecosystem, local communities' livelihood, vulnerability, and adaptation strategies. The collapse of wetland fisheries could have a devastating impact on socioeconomic sectors and different subcomponents of the wetland ecosystem services (Makwinja et al. 2021d). Harrod (2016) noted that fishing communities face threats and prospects to which they must adapt. However, in Malawi, the socioeconomic and environmental drivers instigate wetlands fish biomass fluctuations, and the top-down management approach cannot solve the complexities underlying human-induced fluctuations (Makwinja et al. 2021d). It further demonstrates that fishing underpins the livelihood of most of the local population, and declining catches and other factors expose the fisherfolks to vulnerability (Makwinja et al. 2021d). Harrod (2016) suggested that understanding the response of fisherfolks to the wetland fishery dynamics can provide important information to policymakers on existing adaptation strategies that minimize vulnerability, maintain a fisheries-associated livelihood base, and manage the wetland ecosystem functions sustainably.

13.7 Conclusion

This chapter addressed a critical issue that guides the formulation of the Malawi Wetland Policy and how it can be aligned with other sectoral policies to achieve sustainable management without compromising the local population's livelihood sustenance. We provided a set of wetlands concepts and explored challenges reflected in the Ramsar definition. We provided important debate on how the wetland concept should be reflected in Malawi's strategic documents while providing a clear road map for framing wetland policy. This chapter underscores the significance of wetland while demonstrating how it aligns with UN SDGs. It further outlined the threats and conflicting policies that complicate the effort to manage wetlands while advocating the need to negotiate the ecosystem services trade-offs and landscape management options to achieve policy efficiency. We have cited practical examples that demonstrate the economic value losses from wetlands such as Lake Malombe emanating from the trade-offs among ecosystem services and the negative impact on the wetland ecosystem functions to provide strong evidence to help researchers and policymakers appreciate the gravity of wetland management challenges in Malawi and formulate appropriate policy responses to address the crises.

References

Abera, M. (2017). Agriculture in the Lake Tana Sub-basin of Ethiopia. In: *Social and Ecological System Dynamics.AESS Interdisciplinary Environmental Studies and Sciences Series* (ed. I.K. Stave, G. Goshu, and S. Aynalem). Cham: Springer https://doi.org/10.1007/978-3-319-45755-0_23.

Anthwal, A., Gupta, N., Sharma, A. et al. (2010). Conserving biodiversity through traditional beliefs in sacred groves in Uttarakhand Himalaya, India. *Resources, Conservation and Recycling* 54: 962–971.

Bai, Y., Wong, C., Jiang, B. et al. (2018). Developing China's Ecological Redline Policy using ecosystem services assessments for land use planning. *Nature Communications* 9: 3034. https://doi.org/10.1038/s41467-018-05306-1.

Barral, M., Benayas, J., Meli, P., and Maceira, N. (2015). Quantifying the impacts of ecological restoration on biodiversity and ecosystem services in agroecosystems: a global meta-analysis. *Agriculture, Ecosystems and Environment* 202: 223–231. https://doi.org/10.1016/j.agee.2015.01.009.

Bassi, N., Kumar, M., Sharma, A., and Pardha-Saradhi, P. (2014). Status of wetlands in India: a review of extent, ecosystem benefits, threats and management strategies. *Journal of HydrologyRegional Studies* 2: 1–19. https://doi.org/10.1016/j.ejrh.2014.07.001.

Beeton, A. (2002). Large freshwater lakes: present state, trends, and future. *Environmental Conservation* 29 (1): 21–38. https://doi.org/10.1017/S0376892902000036.

Berry, L. and Smith, R. (1972). Churches and monasteries of Lake Tana, Ethiopia. In: *Africa: Rivista Trimestrale Di Studi e Documentazione Dell'Istituto Italiano per* (ed. L. Berry and R. Smith). Istituto Italiano per l'Africa e l'Oriente, pp. 1/2: 1–34.

Besada, H. and Werner, K. (2015). An assessment of the effects of Africa's water crisis on food security and management. *International Journal of Water Resources Development* 31 (1): 120–133. https://doi.org/10.1080/07900627.2014.905124.

Cen, X., Wu, C., Xing, X. et al. (2015). Coupling intensive land use and landscape ecological security for urban sustainability: an integrated socio-economic data and spatial metrics analysis in Hangzhou City. *Sustainability* 7: 1459–1482. https://doi.org/10.3390/su7021459.

Cheesman, R. (1935). Lake Tana and its islands. *The Geographical Journal* 85 (6): 489–502. https://doi.org/10.2307/1785868.

Chimweta, M., Nyakudya, I., Jimu, L. et al. (2021). Flood-recession cropping in the mid-Zambezi Valley: a neglected farming system with. Potential to improve household food security and income. *The Geographical Journal* 188: 1–19. https://doi.org/10.1111/geoj.12418.

Christian, R. and Mazzilli, S. (2007). Defining the coast and sentinel ecosystems for coastal observations of global change. *Hydrobiologia* 577: 55–70. https://doi.org/10.1007/s10750-006-0417-4.

Cilliers, S., Cilliers, J., Lubbe, R., and Siebert, S. (2013). Ecosystem services of urban green spaces in African countries – perspectives and challenges. *Urban Ecosystem* 6 (2013): 681–702. https://doi.org/10.1007/s11252-012-0254-3.

Coe, M. and Foley, J. (2001). Human and natural impacts on the water resources of the Lake Chad basin. *Journal of Geographical Research* 106 (D4): 349–3356.

Costanza, R., de Groot, R., Sutton, P. et al. (2014). Changes in the global value of ecosystem services. *Global Environmental Change* 26: 152–158. https://doi.org/10.1016/j.gloenvcha.2014.04.002.

Cumming, T., Shackleton, R., Förster, J. et al. (2017). Achieving the national development agenda and the Sustainable Development Goals (SDGs) through investment in ecological infrastructure: a case study of South Africa. *Ecosystem Services* 27: 253–260. https://doi.org/10.1016/j.ecoser.2017.05.005.

Davidson, N. (2014). How much wetland has the world lost? Long-term and recent trends in global wetland area. *Marine and Freshwater Research* 65 (10): 934–941.

Dickens, C., McCartney, M., Tickner, D. et al. (2020). Evaluating the global state of ecosystems and natural resources: within and beyond the SDGs. *Sustainability* 12: 7381. https://doi.org/10.3390/su12187381.

Dini, J. and Everard, E. (2016). National wetland policy: South Africa. In: *The Wetland Book* (ed. C.M. Finlayson, M. Everard, K. Irvine, et al.), 795–800. Dordrecht: Springer Science https://doi.org/10.1007/978-90-481-9659-3_157.

Dixey, F. (1926). The Nyasaland section of the great rift valley. *The Geographical Journal* 68 (2): 117–137.

Dixon, A., Wood, A., and Hailu, A. (2021). Wetlands in Ethiopia: lessons from 20 years of research, policy and practice. *Wetlands* 41: 20. https://doi.org/10.1007/s13157-021-01420-x.

Dybas, C. (2005). Dead zones spreading in world oceans. *BioScience* 55 (7): 552–557.

Eissa, A. and Zaki, M. (2011). The impact of global climatic changes on the aquatic environment. *Procedia Environmental Sciences* 4: 251–259.

EWNHS (2018). *A First Directory of Ethiopian Wetlands: Descriptions, Ecosystem Services, Cause of Degradation & Recommendation for Restoration and Sustainability*. Ethiopian Wildlife and Natural History Society (EWNHS) in Collaboration with Wetlands International: Addis Ababa.

Forsythe, K. and Turpie, J. (2016). Climate resilient livelihoods and sustainable natural resources management in the Elephant Marsh, Malawi: sub-Study 3: ecosystem, services of the Elephant Marsh. ANCHOR environmental.

Fredengren, C. (2011). Where wandering water gushes – the depositional landscape of the Mälaren Valley in the Late Bronze Age and Earliest Iron Age of Scandinavia. *Journal of Wetland Archaeology* 10 (1): 109–135. https://doi.org/10.1179/jwa.2011.10.1.109.

Furse, M., Kirk, R., Morgan, P., and Tweddle, D. (1979). Fishes: distribution and biology in relation to changes. In: *Lake Chilwa: Monographiae Biologicae* (ed. M. Kalk, A. McLachlan, and C. Howard-Williams), 175–208. Dordrecht: Springer.

Geijzendorffer, I., Cohen-Shacham, E., Cord, A. et al. (2017). Ecosystem services in global sustainability policies. *Environmental Science and Policy* 74: 40–48.

Geist, J. (2011). Integrative freshwater ecology and biodiversity conservation. *Ecological Indicators* 11 (6): 1507–1516. https://doi.org/10.1016/j.ecolind.2011.04.002.

Giakoumis, T. and Voulvoulis, N. (2018). The transition of EU water policy towards the water framework directive's integrated river basin management paradigm. *Environmental Management* 62: 819–831. https://doi.org/10.1007/s00267-018-1080-z.

Gil, J., Reidsma, P., Giller, K. et al. (2019). Sustainable development goal 2: improved targets and indicators for agriculture and food security. *Ambio* 48: 685–698. https://doi.org/10.1007/s13280-018-1101-4.

GoM. (1966). *Republic of Malawi Constitution*. Zomba: Malawi Government Press.

GoM. (1994). *The Republic of Malawi Constitution*. Lilongwe: Malawi Government Press.

Goodwin, E. (2006). Conservation of coral reefs under the Ramsar Convention on wetlands. *Journal of International Wildlife Law & Policy* 9 (1): 1–31. https://doi.org/10.1080/13880290500343657.

Gopal, B. (2013). Future of wetlands in tropical and subtropical Asia, especially in the face of climate change. *Aquatic Sciences* 7 (5): 39–61. https://doi.org/10.1007/s00027-011-0247-y.

Gordon, L., Finlayson, C., and Falkenmark, M. (2010). Managing water in agriculture for food production and other ecosystem services. *Agricultural Water Management* 97 (4): 512–519.

Green, P., Vörösmarty, C., Harrison, I. et al. (2015). Freshwater ecosystem services supporting humans: pivoting from water crisis to water solutions. *Global Environmental Change* 34: 108–118. https://doi.org/10.1016/j.gloenvcha.2015.06.007.

de Groot, R., Alkemade, R., Braat, L., and Hein, L. (2010). Challenges in integrating the concept of ecosystem services and values in landscape planning, management, and decision-making. *Ecological Complexity* 7 (2010): 260–272.

Gule, T., Tsheboeng, G., Diamini, W. et al. (2021). Change detection of riparian woodland vegetation in the Okavango Delta, Botswana. *Remote Sensing Applications: Society and Environment* 22: 100497. https://doi.org/10.1016/j.rsase.2021.100497.

Guo, Z., Boeing, W., Borgomeo, E. et al. (2021). Linking reservoir ecosystems research to the sustainable development goals. *Science of the Total Environment* 781: 146769. https://doi.org/10.1016/j.scitotenv.2021.146769.

Gwiyani-Nkhoma, B. (2011). Irrigation development and its socio-economic impact on rural communities in Malawi. *Development Southern Africa* 28 (2): 210–223.

GWP (2000). *Integrated Water Resources Management*. Stockholm: Global Water Partnership.

Hamududu, B. and Killingtveit, A. (2016). *Energies* 9 (7): 502. https://doi.org/10.3390/en9070502.

Hanjra, M. and Qureshi, M. (2010). Global water crisis and future food security in an era of climate change. *Food Policy* 35 (5): 365–377.

Harrod, C. (2016). Climate change and freshwater fisheries. In: *Freshwater Fisheries Ecology* (ed. J. Craig), 641–694. Wiley.

Hernández-Delgado, E. (2015). The emerging threats of climate change on tropical coastal ecosystem services, public health, local economies and livelihood sustainability of small islands: Cumulative impacts and synergies. *Marine Pollution Bulletin* 101 (1): 5–28. https://doi.org/10.1016/j.marpolbul.2015.09.018.

Hook, D. (2021). The sacred nature of rivers, wells, springs, and other wetlands in Anglo-Saxon England. In: *Meanings of Water in Early Medieval England*, 33–57. Brepols Publishers.

Hu, S., Niu, Z., Chen, Y. et al. (2017). Global wetlands: potential distribution, wetland loss, and status. *Science of the Total Environment* 586: 319–327. https://doi.org/10.1016/j.scitotenv.2017.02.001.

Igbadun, H., Mahoo, H., Tarimo, A., and Salim, B. (2006). Crop water productivity of an irrigated maize crop in Mkoji sub-catchment of the Great Ruaha River Basin, Tanzania. *Agricultural Water Management* 85 (1–2): 141–150. https://doi.org/10.1016/j.agwat.2006.04.003.

IPBES (2019). *Summary for Policymakers of the Global Assessment Report on Biodiversity and Ecosystem Services of the Intergovernmental Science-Policy Platform on Biodiversity and Ecosystem Services*. Bonn: IPBES.

Jellison, R., Williams, D., Timms, B., and Alcocer, J. (2008). *Salt Lakes: Values, Threats and Future*. Cambridge: Cambridge University Press.

Jenkins, W., Murry, B., Kramer, R., and Faulkner, S. (2010). Valuing ecosystem services from wetlands restoration in the Mississippi Alluvial Valley. *Ecological Economics* 69 (5): 1051–1061. https://doi.org/10.1016/j.ecolecon.2009.11.022.

Jogo, W. and Hassan, R. (2010). Balancing the use of wetlands for economic well-being and ecological security: the case of the Limpopo wetland in southern Africa. *Ecological Economics* 69 (7): 1569–1579. https://doi.org/10.1016/j.ecolecon.2010.02.021.

Johnson, J., Jones, S., Wood, S. et al. (2019). Mapping ecosystem services to human well-being: a toolkit to support integrated landscape management for the SDGs. *Ecological Applications* 29 (8): e01985.

Kalanda-Sabola, M., Henry, E., Kayambazinthu, E., and Wilson, J. (2007). Use of indigenous knowledge and traditional practices in fisheries management: a case of Chisi Island, Lake Chilwa, Zomba. *Malawi Journal of Science and Technology* 8: 9–29.

Kambombe, O., Ngongondo, C., Monjerezi, M., and Eneya, L. (2023). Investigating human-induced threat to hydrological regime of Lake Chilwa Basin, Malawi. *Applied Water Science* 13 (2023): 161. https://doi.org/10.1007/s13201-023-01965-8.

Karageorgis, A., Skourtos, M., Kapsimalis, V. et al. (2004). An integrated approach to watershed management within the DPSIR framework: Axios River catchment and Thermaikos Gulf. *Regional Environmental Change* 5: 138–160. https://doi.org/10.1007/s10113-004-0078-7.

Kirk, R. (1967). The fishes of Lake Chilwa. *The Society of Malawi Journal* 20 (1): 35–48.

Kolawole, A. (1988). Cultivation of the floor of Lake Chad: a response to environmental hazard in Eastern Borno, Nigeria. *The Geographical Journal* 154 (2): 243–250.

Kosamu, I. (2014). Conditions for sustainability of the elephant marsh fishery in Malawi. *Sustainability* 6: 4010–4027. https://doi.org/10.3390/su6074010.

Kosamu, I., de Groot, W., and Kambewa, P. (2017). Actor-based design of a management system for the elephant marsh fishery in Malaw. *Society & Natural Resources* 30 (3): 299–314. https://doi.org/10.1080/08941920.2016.1209604.

Kosamu, I., Makwinja, R., Kaonga, C. et al. (2022). Application of DPSIR and Tobit models in assessing freshwater ecosystems: the case of Lake Malombe, Malawi. *Water* 14: 619. https://doi.org/10.3390/w140406.

Lemos, M. and Morehouse, B. (2005). The co-production of science and policy in integrated climate assessments. *Global Environmental Change* 15: 57–68.

Lindley, S., Pauleit, S., Yeshitela, K. et al. (2018). Rethinking urban green infrastructure and ecosystem services from the perspective of sub-Saharan African cities. *Landscape and Urban Planning* 180: 328–338. https://doi.org/10.1016/j.landurbplan.2018.08.016.

Liu, Y., Gupta, H., Springer, E., and Wagener, T. (2008). Linking science with environmental decision making: Experiences from an integrated modeling approach to supporting sustainable water resources management. *Environmental Modelling & Software* 23: 846–858. https://doi.org/10.1016/j.envsoft.2007.10.007.

Mafabi, P. (2018). National Wetland Policy: Uganda. In: *The Wetland Book* (ed. C.M. Finlayson et al.). Dordrecht: Springer https://doi.org/10.1007/978-90-481-9659-3_154.

Makoka, D. (2008). *The Impact of Drought on Household Vulnerability: The Case of Rural Malawi.* Bonn: Centre for Development Research, University of Bonn.

Makwinja, R., Chapotera, M., Likongwe, P. et al. (2014). Location and roles of deep pools in Likangala River during the 2012 recession period of Lake Chilwa Basin. *International Journal of Ecology* 2014: 294683. 4 pages. https://doi.org/10.1155/2014/294683.

Makwinja, R., Kosamu, I., and Kaonga, C. (2019). Determinants and values of willingness to pay for water quality improvement: insights from Chia Lagoon, Malawi. *Sustainability* 11: 4690. https://doi.org/10.3390/su11174690.

Makwinja, R., Mengistou, S., Kaunda, E., and Alamirew, T. (2021a). Economic value of tropical inland freshwater shallow lakes: Lesson from Lake Malombe, Malawi. *African Journal of Ecology* 39 (4): 1–15. https://doi.org/10.1111/aje.12935.

Makwinja, R., Mengistou, S., Kaunda, E., and Alamirew, T. (2021b). Land use/land cover dynamics, trade-offs and implications on tropical inland shallow. *Sustainable Environment* 7 (1): 1969139. https://doi.org/10.1080/27658511.2021.1969139.

Makwinja, R., Mengistou, S., Kaunda, E. et al. (2021c). Modeling of Lake Malombe annual fish landings and catch per unit effort (CPUE). *Forecasting* 3: 39–55. https://doi.org/10.3390/forecast3010004.

Makwinja, R., Kaunda, E., Mengistou, S., and Alamirew, T. (2021d). Impact of land use/land cover dynamics on ecosystem service value—a case from Lake Malombe, Southern Malawi. *Environmental Monitoring and Assessment* https://doi.org/10.1007/s10661-021-09241-5.

Makwinja, R., Mengistou, S., Kaunda, E., and Alamirew, T. (2021e). Lake Malombe fish stock fluctuation: ecosystem and fisherfolks. *Egyptian Journal of Aquatic Research* 47 (3): 321–327. https://doi.org/10.1016/j.ejar.2021.07.001.

Makwinja, R., Kaunda, E., Mengistou, S. et al. (2021f). Lake Malombe fishing livelihood, vulnerability, and adaptation strategies. *Current Research in Environmental Sustainability* 3: 10055. https://doi.org/10.1016/j.crsust.2021.100055.

Makwinja, R., Mengistou, S., Kaunda, E., and Alamirew, T. (2022). Managing ecosystem services demand under a changing catchment: a case study of Lake Malombe Catchment, Malawi. *GeoJournal* https://doi.org/10.1007/s10708-022-10575-x.

Malawi Government (2004). *Malawi Economic Growth Strategy Volume II: Main Report*. Lilongwe: Ministry of Economic Planning and Development.

Maltby, E. (2009). The Changing Wetland Paradigm. In: *The Wetlands Handbook* (ed. I.E. Maltby and T. Barker), 4–42. Blackwell Publishing Ltd.

Marconi, P., Arengo, F., and Clark, A. (2022). The arid Andean plateau waterscapes and the lithium triangle: flamingos as flagships for conservation of high-altitude wetlands under pressure from mining development. *Wetlands Ecology and Management* 30: 827–852. https://doi.org/10.1007/s11273-022-09872-6.

Minale, A. (2019). Water level fluctuations of Lake Tana and its implication on local communities livelihood, northwestern Ethiopia. *International Journal of River Basin Management* 18 (4): 503–510. https://doi.org/10.1080/15715124.2019.1700512.

Mitsch, W., Bernal, B., and Hernandez, M. (2015). Ecosystem services of wetlands. *International Journal of Biodiversity Science, Ecosystem Services & Management* 11 (1): 1–4. https://doi.org/10.1080/21513732.2015.1006250.

Mooney, H., Duraiappah, A., and Larigauderie, A. (2013). Evolution of natural and social science interactions in global change research programs. *Proceedings of the National Academy of Sciences USA* 110 (Supplement 1): 3665–3672.

Morgan, E. (1982). *The Aquatic Ape*. London: Souvenir Press.

Mzuza, S. (2013). Measuring the economic value of wetland ecosystem services in Malawi: a case study of Lake Chiuta wetland. MSc thesis. University of Zambia, Lusaka.

Nagoli, J. and Chiwona-Karltun, L. (2017). Uncovering human social networks in coping with Lake Chilwa recessions in Malawi. *Journal of Environmental Management* 192: 134–141.

Nerini, F., Tomei, J., To, L. et al. (2018). Mapping synergies and trade-offs between energy and the sustainable development goals. *Nature Energy* 3 (2018): 10–15. https://doi.org/10.1038/s41560-017-0036-5.

Njaya, F., Snyder, K., Jamu, D. et al. (2011). The natural history and fisheries ecology of Lake Chilwa, southern Malawi. *Journal of Great Lakes Research* 37: 15–25.

Nkhoma, B. and Kayira, G. (2016). Gender and power contestations over water use in irrigation schemes: lessons from the Lake Chilwa. *Physics and Chemistry of the Earth* 92: 79–84. https://doi.org/10.1016/j.pce.2015.10.007.

Nkhoma, L., Ngongondo, C., Dulanya, Z., and Monjerezi, M. (2021). Evaluation of integrated impacts of climate and land use change on the river flow regime in Wamkurumadzi River, Shire Basin in Malawi. *Journal of Water and Climate Change* 12 (5): 1674–1693.

Ogola, P., Davidsdottir, B., and Fridleifsson, I. (2012). Potential contribution of geothermal energy to climate change adaptation: a case study of the arid and semi-arid eastern Baringo lowlands, Kenya. *Renewable and Sustainable Energy Reviews* 16 (6): 4222–4246.

Palacio, E., van Beukerin, P., van Zante, B. et al. (2021). Linking ecosystem services and the sustainable development goals in small island developing states: the case of Aruba. *One Ecosystem* 6: e71033. https://doi.org/10.3897/oneeco.6.e71033.

Peters, P. and Kambewa, D. (2007). Whose security? Deepening social conflict over 'customary' land in the shadow of land tenure reform in Malawi. *The Journal of Modern African Studies* 45 (3): 447–472. https://doi.org/10.1017/S0022278X07002704.

Pham-Duc, B., Sylvestre, F., Papa, F. et al. (2020). The Lake Chad hydrology under current climate change. *Scientific Reports* 10: 5498. https://doi.org/10.1038/s41598-020-62417-w.

Phukan, M. (2021). Ecosystem services of wetland. *International Journal of Aquatic Science* 12 (3): 1–9.

Pittock, J., Cork, S., and Maynard, S. (2012). The state of the application of ecosystem services in Australia. *Ecosystem Services* 1: 111–120.

Rao, N., Ghermandi, A., Portela, R., and Wang, X. (2015). Global values of coastal ecosystem services: A spatial economic analysis of shoreline protection values. *Ecosystem Services* 11: 95–105. https://doi.org/10.1016/j.ecoser.2014.11.011.

Raworth, K. (2017). A doughnut for the anthropocene: humanity's compass in the 21st century. *Lancet Planetary Health* 1: e48–e49.

Ricaurte, L., Olaya-Rodríguez, M., Cepeda-Valencia, J. et al. (2017). Future impacts of drivers of change on wetland ecosystem services in Colombia. *Global Environmental Change* 44: 158–169. https://doi.org/10.1016/j.gloenvcha.2017.04.001.

Schuyt, K. (2005). Economic consequences of wetland degradation for local populations in Africa. *Ecological Economics* 53 (15): 177–190. https://doi.org/10.1016/j.ecolecon.2004.08.003.

Scott, D. and Jones, T. (1995). Classification and inventory of wetlands: a global overview. *Vegetatio* 118: 3–16. https://doi.org/10.1007/BF00045186.

Seifollahi-Aghmiuni, S., Nockrach, M., and Kalantari, Z. (2019). The potential of wetlands in achieving the sustainable development goals of the 2030 agenda. *Water* 11: 609. https://doi.org/10.3390/w11030609.

Sharafatmandrad, M. and Mashizi, A. (2021). Linking ecosystem services to social well-being: an approach to assess land degradation. *Frontiers in Ecology and Evolution* https://doi.org/10.3389/fevo.2021.654560.

Sievers, M., Brown, C., Tulloch, V. et al. (2019). The role of vegetated coastal wetlands for marine megafauna conservation. *Trends in Ecology & Evolution* 34 (9): 807–817.

Simaika, J., Chakona, A., and van Dam, A. (2022). Editorial: towards the sustainable use of African wetlands. *Sec. Freshwater Science* https://doi.org/10.3389/fenvs.2021.658871.

Smiraglia, D., Ceccarelli, T., Bajocco, S. et al. (2015). Linking trajectories of land change, land degradation processes and ecosystem service. *Environmental Research* https://doi.org/10.1016/j.envres.2015.11.03.

Stewart-Harawira, M. (2020). Troubled waters: Maori values and ethics for freshwater management and New Zealand's freshwater crisis. *WIREs Water* 7 (5): e1464. https://doi.org/10.1002/wat2.1464.

Summers, J., Smith, L., Case, J., and Linthurst, R. (2012). A review of the elements of human well-being with an emphasis on the contribution of ecosystem services. *Ambio* 41: 327–340. https://doi.org/10.1007/s13280-012-0256-7.

Talling, J. (1986). The seasonality of phytoplankton in African lakes. *Hydrobiologia* 138: 139–160 (1986). https://doi.org/10.1007/BF00027237.

Ticehurts, J., Newham, L., Rissik, D. et al. (2007). A Bayesian network approach for assessing the sustainability of coastal lakes in New South Wales. *Australia Environmental Modelling & Software* 22 (8): 1129–1139.

UN. (2019). *The Sustainable Development Goals Report 2019*. New York: United Nations.

Verhaegen, M. (1985). The aquatic ape theory: evidence and a possible scenario. *Medical Hypotheses* 16: 17–32.

Wangai, P., Burkhard, B., and Muller, F. (2016). A review of studies on ecosystem services in Africa. *International Journal of Sustainable Built Environment* 5: 225–245.

Whitehead, P., Wilby, R., Battarbee, R. et al. (2009). A review of the potential impacts of climate change on surface water quality. *Hydrological Sciences Journal* 54 (1): 101–123. https://doi.org/10.1623/hysj.54.1.101.

Wu, J. (2013). Landscape sustainability science: ecosystem services and human well-being in changing landscapes. *Landscape Ecology* 28: 999–1023. https://doi.org/10.1007/s10980-013-9894-9.

WWF (2020). Bending the curve of biodiversity loss. In: *Living Planet Report 2020* (ed. R. Almond, M. Grooten, and T. Petersen), 1–83. Switzerland: WWF.

Xu, X., Booij, M., and Mynett, A. (2007). An appropriateness framework for the Dutch Meuse decision support system. *Environmental Modelling & Software* 22 (11): 1667–1678.

Xu, J., Grumbine, R., and Beckschäfer, P. (2014). Landscape transformation through the use of ecological and socio-economic indicators in Xishuangbanna, Southwest China, Mekong Region. *Ecological Indicators* 6: 749–756. https://doi.org/10.1016/j.ecolind.2012.08.023.

Yahdjian, L., Sala, O. and Havstad, K. (2015). Rangeland ecosystem services: shifting focus from supply to reconciling supply and demand. *Frontiers in Ecology and the Environment*, 13 (1): 4–51. https://doi.org/10.1890/140156.

Yilma, A. (2010). *Climate Change & Adaptation in Africa//Ethiopia*. Addis Ababa: International Water Management Institute (IWI).

Yin, C., Zhao, W., Cherubini, F., and Pereira, P. (2021). Integrate ecosystem services into socio-economic development to enhance achievement of sustainable development goals in the post-pandemic era. *Geography and Sustainability* 2 (1): 68–73. https://doi.org/10.1016/j.geosus.2021.03.002.

Yoon, C. (2009). Wise use of paddy rice fields to partially compensate for the loss of natural wetlands. *Paddy and Water Environment* 7: 357–366. https://doi.org/10.1007/s10333-009-0178-6.

Youn, S., Tylor, W., Lynch, A. et al. (2014). Inland capture fishery contributions to global food security and threats to their futur. *Global Food Security* 3 (3–4): 142–148.

Zhao, Y., Dong, Y., Chen, X. et al. (2023). Using sulfide as nitrite-oxidizing bacteria inhibitor for the successful coupling of partial nitrification-anammox and sulfur autotrophic denitrification in one reactor. *Chemical Engineering Journal* https://doi.org/10.1016/j.cej.2023.146286.

14

Wetland Ecosystems in Nigeria

Ecological Significance, Carbon Sequestration Potential, and Conservation Challenges

Isa Olalekan Elegbede[1,3,4], *Amira M. Hotaiba*[2], *Adebola Rashidat Adewale*[1], *Abdulwakil Olawale Saba*[1], *Shehu Akintola*[1], *Emily Osa Iduseri*[3], *Abdul Rahamon Olodo*[4], *and Kehinde Moyosola Ositimehin*[5]

[1] *Department of Fisheries, Faculty of Science, Lagos State University, Lagos, Nigeria*
[2] *Department of Environmental Sciences, Faculty of Science, Alexandria University, Alexandria, Egypt*
[3] *Department of Environmental Science, National Open University of Nigeria, Abuja, Nigeria*
[4] *Department of Environmental Planning, Brandenburg University of Technology, Cottbus, Germany*
[5] *Department of Environmental Studies, Ohio University, Athens, OH, USA*

14.1 Introduction

Wetlands, as delineated by Matthews (1993), constitute an array of environments including marshes, fens, peatlands, or bodies of water, whether naturally occurring or human-made, permanent or ephemeral. Despite their remarkable productivity, wetlands stand among the most imperiled ecosystems globally (Mahapatra et al. 2024).

These ecosystems, characterized by distinctive hydrological regimes, soil compositions, and vegetative assemblages, hold indispensable roles within the Earth's biosphere. Their contributions to climate regulation, water quality preservation, and biodiversity maintenance are profound (Mitsch and Gosselink 2000). In the context of Nigeria, a country graced with a rich tapestry of wetland habitats spanning from inland floodplains to coastal mangrove forests, their significance is both ecological and economic, offering benefits to local communities and harboring invaluable biological diversity (Pandit et al. 2018). This introduction serves to lay the foundation for an exploration of Nigeria's wetland ecosystems, delving into their biodiversity, sustainable management practices, and conservation imperatives within the backdrop of this vibrant and biologically diverse nation.

Stretching across the varied landscapes of Nigeria, wetland ecosystems encompass an extensive array of habitats, including riverine expanses, lakeshores, swamps, marshlands, and mangrove swaths. These habitats serve as sanctuaries for numerous plant and animal species, many of which are endemic or endangered, while concurrently providing critical ecological services such as flood mitigation, water purification, carbon sequestration, and support for fisheries (Cunliffe 2023). However, the intrinsic value of these ecosystems is under threat from an array of human activities, which inflict habitat degradation, pollution, biodiversity depletion, and alteration of fundamental ecological processes.

In the face of burgeoning human pressures, Nigeria's wetlands are increasingly imperiled, with activities such as resource extraction, industrial expansion, agricultural intensification, and

Wetlands of Tropical and Subtropical Asia and Africa: Biodiversity, Livelihoods and Conservation, First Edition. Edited by Thammineni Pullaiah.

urban sprawl exerting mounting stressors on these delicate ecosystems (Adekola and Mitchell 2011). These anthropogenic actions precipitate habitat degradation, water contamination, loss of biodiversity, and disruption of ecosystem functions, rendering wetland ecosystems highly vulnerable to environmental perturbations, including those stemming from climate change. Furthermore, the efficacy of conservation efforts is hampered by inadequacies in legislative frameworks, governance structures, and public awareness regarding the ecological significance of wetlands.

Spanning from the expansive wetlands of the Niger Delta to the elevated lakeshores of the Jos Plateau, Nigeria's wetland ecosystems embody unparalleled biodiversity and furnish a myriad of ecosystem services (Okonkwo et al. 2015). Each ecosystem enclave boasts a unique mosaic of flora and fauna and contributes distinct services that accrue benefits to both human and nonhuman populations alike. By focusing on Nigeria's wetlands, this study endeavors to offer a holistic appraisal of their current status, the challenges confronting them, and the prospects for their sustainable preservation and management. The ensuing discourse aims to underscore the exigency of concerted action to safeguard Nigeria's wetland ecosystems and secure their enduring vitality amid escalating anthropogenic and climatic pressures.

14.2 Wetlands in Nigeria

Wetlands constitute integral components of Nigeria's hydrological landscape. Projections suggest that a sea-level rise of up to 59 cm could inundate several coastal states in Nigeria, exacerbating flood risks. Nigeria harbors a wealth of diverse wetland ecosystems, predominantly situated within the Niger, Benue, and Chad basins (Wali 2015). Despite their ecological significance, wetlands in Africa, including those in Nigeria, suffer from a dearth of comprehensive data and research (Adekola and Mitchell 2011).

Approximately 14 major wetlands (Figure 14.1) have been identified across Nigeria, encompassing regions such as Sokoto–Rima, Komadugu Yobe, Makurdi, Cross River, Lower Niger, Upper Niger, Kainji Lake, Middle Niger–Lokoja–Jebba–Lower Kaduna, Niger Delta, Benin–Owena, Lake Chad, Badagry, Yewa creeks, and Lagos lagoon (Orie 2019). Among these, the Lagos lagoon emerges as a prominent wetland in the southern region of Nigeria. Spanning an extensive area, both Lagos and Lekki lagoons collectively cover 646 km^2. The Commodore Channel serves as a crucial conduit linking the Bight of Benin and the Lagos lagoon, serving as the primary western outlet for these two lagoons. Moreover, the Lagos lagoon traverses 10 out of the 20 Local Government Areas (LGA) in the State, including Epe, Etiosa, Ibeju/Lekki, Ikorodu, part of Amuwo–Odofin, Apapa, Kosofe, Lagos Island, Mainland, and Shomolu LGAs (Orie 2019; Adebayo et al. 2021).

14.2.1 Classification of Wetlands in Nigeria

Wetlands encompass a diverse array of ecosystems, each characterized by unique water distribution, soil composition, vegetation, and aquatic life. These ecosystems serve as crucial interfaces between aquatic and terrestrial systems (Rafferty 2010; Elegbede et al. 2014). The classification of wetlands in Nigeria is based on several factors, including their hydrology, vegetation, and water chemistry. Below are the main types of wetlands found in Nigeria, along with their characteristics and distribution, which are Swamps, Marches, Fens, and Bogs (Dey and Roy 2022; Garba et al. 2023)

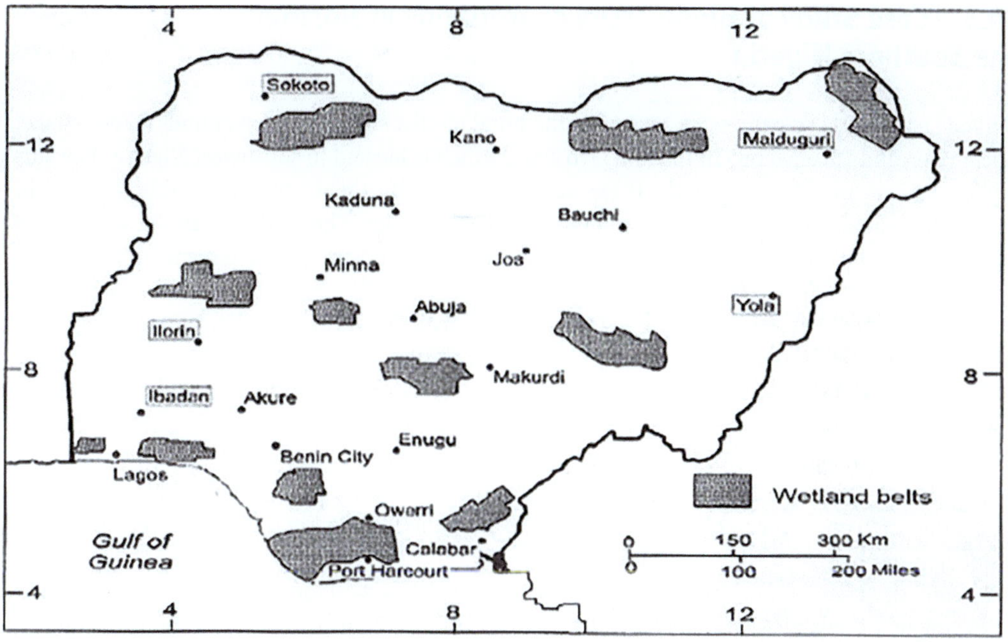

Figure 14.1 Major wetland belts in Nigeria. *Source:* Nwankwoala (2012)/European Center of Sustainable Development.

14.2.1.1 Swamps

Swamps are wetlands dominated by woody vegetation and typically experience seasonal inundation. They are commonly situated in low-lying floodplains along rivers or slow-moving streams (Idris 2008).

14.2.1.2 Marshes

Marshes are wetlands characterized by their frequent flooding, particularly during periods of high water flow in rivers, streams, and lakes/ponds. They support various vegetation, including submerged floating leaves, pond weeds, water lilies, sedges, rushes, grasses, and forbs. Marshes can be further classified into two subtypes: emergent marshes, which are found along the shoreline, and hemi marshes, which occur in deeper water.

14.2.1.3 Fens

Fens are wetlands fed by groundwater seepage, with water slowly emerging from the soil surface. The vegetation composition in fens is influenced by water chemistry and pH levels (Idris 2008).

14.2.1.4 Bogs

Bogs are wetlands where water primarily comes from precipitation, resulting in waterlogged conditions near the surface. These wetlands are characterized by the abundant growth of *Sphagnum* mosses.

Each type of wetland in Nigeria supports unique flora and fauna adapted to its specific hydrological and environmental conditions. Understanding the classification and distribution of wetlands is essential for effective management and conservation efforts aimed at preserving these valuable ecosystems and their associated biodiversity.

14.3 Case Study of Biodiversity of Wetlands in Northern and Southern Nigeria

Olubode et al. (2011) conducted a survey on the floral biodiversity of three wetlands (Eleyele lake, Apete river, and Obada dam) in Ibadan, Oyo state, which is located in southwest Nigeria. The survey results indicated that despite the three wetlands' proximity, there was a noticeable variation in the species composition and population density of each species. The wetlands are located within 5 km of each other and share the same ecozone. These variations might have resulted from variations in edaphic variables (Olubode et al. 2011).

Similarly, Adeniyi et al. (2016)'s survey on the floral diversity of three wetland sites in the Ibeju-Lekki area of Lagos State, which is also located in the southwest of Nigeria, within a 5-km radius of the Lekki port, revealed that, despite their proximity, there was variation in the species composition and population density of each species among the three sites. According to Adeniyi et al. (2016) and Olubode et al. (2011), these indices imply that human or edaphic influences may be responsible for the floral dissimilarities in the locations. Ogunyemi (1977) found that edaphic factors, rather than crop type or climate, had a greater influence on the weed flora of several cropping systems in southwestern Nigeria (Table 14.1).

Wetland sizes have a substantial impact on bird variety but not on the abundance of bird species, according to a study by Sulaiman et al. (2015) on the Hadejia-Nguru wetlands, which are situated in northeastern Nigeria. The findings demonstrated that bird diversity decreases as wetlands size increases; that is, for every 2 hectares that wetlands are expanded, bird diversity decreases by 0.02 (Table 14.2).

Furthermore, the diversity indices for all sectors showed greater values at the fringing zones, followed by the submerged zones and, finally, the intermittently flooded zones, according to Ali and Bichi (2021) on their study of the Hadejia-Nguru wetlands. This research revealed that most birds' basic needs – nesting sites, food for omnivorous birds, and cover – may be met by the fringing zones. This suggests that there are thriving bird populations in the research region.

The local ecology and biodiversity are greatly impacted by wetlands (Mitsch et al. 1994). Therefore, in order to preserve or safeguard these locations as well as other undisturbed wetlands in Nigeria, suitable laws and policies on biodiversity conservation must be created and implemented.

Table 14.1 Species diversity indices for the studied areas at Ibeju-Lekki.

	Site A	Site B	Site C
Taxa (species)	35	15	26
Individuals	670	560	714
Dominance	0.0615	0.1113	0.07943
Simpson index	0.9385	0.8887	0.9206
Shannon index	3.057	2.335	2.738
Evenness	0.6077	0.6887	0.5945
Margalef	5.225	2.212	3.805

Source: Adapted from Adeniyi et al. (2016).

Table 14.2 The species list of Avifauna resources in the Hadejia-Nguru wetlands.

Family	Species	Nguru	Dagona	Baturiya
Ardeidae	*Ardea malanocephala*	+	+	+
	Ardea pupurea	+	+	+
	Mesophyx intermedia	+	+	+
	Casmerodius albus	–	+	+
	Egretta gularies	+	+	+
	Egretta garzetta	+	+	+
	Bubulcus ibis	+	+	+
	Scopus umbretta	–	+	+
	Ardeola ralloides	+	+	+
	Nycticorax nycticorax	+	+	+
Ciconiidae	*Ciconia abdinii*	+	+	+
	Anastomus lamelligerus	+	+	–
Rallidae	*Microparra capensis*	+	+	+
	Actophilornis africana	+	+	+
	Gallinula chloropus	+	+	+
	Porphyrio alleni	+	+	+
	Caracias abyssinicus	+	+	+
Viduidae	*Vidua macroura*	+	+	+
Estrildidae	*Lagonosticta senegala*	+	+	+
	Estrilda melpoda	+	+	+
	Scopus umbretta	+	–	–
Ploceidae	*Euplectes franciscanus*	+	+	+
Burhinidae	*Burhinus capensis*	+	+	+
Phasianidae	*Francolinus bicalcaratus*	+	+	+
Accipitridae	*Elanus caeruleus*	+	+	+
	Circus aeruginosus	+	+	+
	Circus pygargus	+	+	+
	Milvus mirans	+	+	+
Sagittariidae	*Sagittaius serpentarius*	+	+	+
	Polyboroides typus	+	+	+
	Haliaeetus vocifer	+	+	+
Ploceidae	*Placeus luteolus*	+	+	+
	Placeus nigrimentus	+	+	+
	Quelea quelea	+	+	+
Nectariniidae	*Chalcomitra rubescens*	+	+	
Caprimugidae	*Caprimulgus inornatus*	+	+	+
Cuculidae	*Centropus senrgalensis*	+	+	+
	Clamator glandarius	+	+	+

(*Continued*)

Table 14.2 (Continued)

Family	Species	Nguru	Dagona	Baturiya
Columbidae	*Streptopelia decipiens*	+	+	+
	Columba guinea	+	+	+
	Streptopelia senegalensis	+	+	+
Apodidae	*Apus apus*	+	+	+
Muscicapidae	*Muscicapa pammelaina*	–	–	+
	Muscicapa gambage	+	+	+
	Muscicapa pallidus	+	+	–
Corvidae	*Carvus albus*	+	+	+
Charadriidae	*Venellus tectus*	+	+	+
	Vannellus spinosus	+	+	+
	Charadrius alexandrines	+	+	+
	Charadrius dublus	+	+	+
	Charadrius pecuarius	+	+	+
	Limosa limosa	+	+	+
Laridae	*Larus cirocephalus*	+	+	+
Scolopacidae	*Calidris alba*	+	+	+
	Tringa erythropus	+	+	+
	Philomachus pugnax	+	+	–
	Tringa tetanus	+	+	+
	Tringa nebularia	+	+	+
	Tringa stagnatilis	+	+	+
	Tringa hypoleocus	+	+	–
	Tringa glareola	+	+	+
Sternidae	*Chlidonias hybrida*	+	+	+
	Gellochenidon nilotica	+	+	+
	Sterna albifrons	+	+	+
Pandionidae	*Pandion haliactus*	+	+	+
Ardeidae	*Butorides striata*	+	+	+
Threskiornithidae	*Platalea leocorodia*	+	+	+
	Bostrychia hagedash	+	+	+
	Plagades falcinellus	+	+	+
Anatidae	*Dendrocygna bicolor*	+	+	+
	Plectopterus gambensis	+	+	+
	Sarkidiornis melanotus	+	+	+
	Nettapus auritus	+	+	+
	Anas acuta	+	+	+
	Anas querquedula	+	+	+
	Dendrocydna viduata	+	+	+

Table 14.2 (Continued)

Family	Species	Nguru	Dagona	Baturiya
	Anas clypeata	+	+	+
	Aythya nyroca	+	+	+
Rallidae	*Amaurornis flavirosta*	+	+	+
	Porphyrio porphyrio	+	+	+
Haematopodidae	*Himantopus himantopus*	+	+	+
	Recurvirostra avosetta	+	+	+
Alcedinidae	*Cirycle rudis*	+	+	+

+, present; – absent.
Source: Ali and Bichi (2021)/The University of Maiduguri.

14.4 Wetland of Nigeria in Relation to Blue Carbon and Sequestration

Nigeria's coastal and inland environments encompass various ecosystems, including mangroves, wetlands, seagrasses, lagoons, estuaries, coastal wetlands, and sandy beaches (Akanni et al. 2018). Despite their ecological significance, these habitats have not received adequate attention in terms of their role as carbon sinks in Nigeria (Filani 2021; Elegbede et al. 2023b). However, Wetlands serve as crucial carbon sinks, actively sequestering carbon dioxide from the atmosphere through the process of photosynthesis. Among these ecosystems, coastal wetlands stand out as highly effective in this capacity and are identified as blue carbon ecosystems. Through their growth and development, these coastal wetlands absorb carbon dioxide and deposit a substantial portion of it into the nutrient-rich organic soils encapsulated by their intricate root systems (Herr and Landis 2016; Owoyemi et al. 2016).

14.4.1 Carbon Sequestration in Wetlands

Carbon sequestration in wetlands involves the transfer and accumulation of carbon dioxide (CO_2) from the atmosphere into the soil carbon pool as soil organic matter. This accumulation occurs over time because of the balance between carbon inputs and outputs (Elegbede et al. 2017). Inputs primarily consist of carbon from decomposed organic matter from vegetation within and surrounding the wetland (autochthonous) as well as carbon dissolved and suspended in incoming water and runoff (allochthonous). Outputs can include dissolved and suspended organic carbon in outflowing water, but the main sources are inorganic forms emitted as CO_2 and methane (CH_4) during the decomposition of organic matter (Villa and Bernal 2018; Valach et al. 2021).

While high carbon inputs can lead to increased carbon accumulation, they can also result in elevated gas emissions. For instance, studies have shown a correlation between CH_4 emissions and aboveground biomass productivity (ANPP), indicating that higher carbon inputs to the soil can potentially lead to greater carbon sequestration in soil organic matter (SOM). However, this also means that there is more substrate available for degradation and export to the atmosphere and downstream waters (Lorenz et al., 2018; Meena et al., 2020). Therefore, understanding the factors affecting carbon decomposition and mineralization is crucial for comprehending carbon sequestration in wetlands (Elegbede et al. 2023c).

Wetlands act as significant carbon sinks, capable of storing up to three times more carbon than terrestrial forests. Through photosynthesis, wetland vegetation captures atmospheric carbon dioxide, converting it into organic carbon. A considerable portion of this carbon is then buried in the anaerobic sediments of wetlands, where it is protected from decomposition and remains sequestered. Sea level rise contributes to the expansion of soil volume and creates additional space in the sediments, enhancing the burial and preservation of organic matter (Were et al. 2019).

In recent years, carbon sequestration in wetlands has garnered significant attention within climate science. Wetlands have been recognized for their pivotal role in regulating the Earth's atmospheric radiative balance, prompting their integration into climate change adaptation strategies and ecosystem conservation initiatives. However, the effectiveness of wetlands as carbon sinks varies widely depending on their hydrogeomorphic characteristics and landscape position. While wetland ecosystems can sequester carbon in their soils, their overall capacity as carbon sinks is influenced by various factors, including methane emissions. Many freshwater wetlands struggle to function as net carbon sinks due to methane emissions, whereas coastal wetlands like salt marshes and mangroves can accumulate substantial soil carbon pools with minimal methane emissions (Adhikari et al. 2009).

An overview of global biome carbon pools underscores the significant contribution of wetlands, despite their relatively small surface area. Boreal peatlands and coastal wetlands are identified as significant carbon sinks, although precise global estimates for other wetland ecosystems remain challenging (Villa and Bernal 2018).

14.4.2 Role of Wetlands in Climate Change Mitigation

Wetland ecosystems play a vital role in mitigating climate change by sequestering carbon dioxide and other greenhouse gases. However, human activities such as industrialization, urbanization, and agriculture exert pressure on these ecosystems, leading to contamination, oil spillage, and environmental degradation (Anule and Ujoh 2017). Despite their importance, wetlands are often overlooked in climate change mitigation efforts, highlighting the need for enhanced conservation measures.

14.4.3 Soil Organic Carbon Dynamics

Soil organic carbon, a key component of soil health, influences carbon sequestration in wetland ecosystems. Soil formation processes involve the breakdown of rocks, migration of soil biota, and accumulation of organic matter containing carbon (Kadiri et al. 2022). The balance of organic carbon in the soil is influenced by factors such as plant residues, decomposition rates, soil erosion, and soil type. Soils rich in clay and organic matter tend to store more carbon, while sandy soils have lower organic matter content.

14.4.4 Impact of Industrialization and Human Activities

Industrialization and human activities, including power generation, petrochemical production, motor vehicle emissions, and agricultural practices, contribute to atmospheric carbon dioxide levels (Yahaya-Shiru et al. 2022). Nigeria, in particular, faces challenges related to gas flaring, making it one of the largest contributors to greenhouse gas emissions. Despite efforts to promote renewable energy sources, fossil fuels remain the primary source of energy, exacerbating the greenhouse effect and climate change.

Understanding the dynamics of carbon sequestration in wetland ecosystems is crucial for developing effective strategies for climate change mitigation and conservation. By recognizing the importance of wetlands as carbon sinks and implementing sustainable practices, Nigeria can contribute to global efforts to combat climate change while preserving its valuable ecosystems.

14.5 Wetlands and Their Ecosystem Services

The evaluation of wetlands' ecological services represents a relatively recent endeavor within the scientific community. Traditionally, wetlands were perceived as wastelands ripe for development, leading to their drainage and degradation (Mitsch et al. 2015). However, contemporary perspectives acknowledge the invaluable services rendered by wetlands, sparking debates on their economic worth and the allocation of resources for their conservation and restoration. Consequently, there arises a pressing need to quantify the value of wetland services through rigorous assessment methodologies.

Wetlands offer a myriad of essential ecosystem services, including habitat provision for diverse species, flood mitigation, water purification, and recreational opportunities. Despite their profound societal benefits, these services often lack a market price, necessitating the utilization of nonmarket valuation techniques to ascertain their economic value. Numerous wetland valuation studies have been conducted, yielding a wide range of estimates that underscore the multifaceted nature of wetland ecosystems and the complexity of valuing their services.

14.5.1 Ecosystem Services Provided by Wetlands

Ecosystem services, as outlined by the Millennium Ecosystem Assessment (Leemans and De Groot 2003), encompass the benefits derived from ecosystems. The MEA delineates four categories of ecosystem services: provisioning (e.g. food), regulating, supporting, and cultural services. While these classifications are widely adopted, modifications have been proposed due to challenges in distinguishing between supporting and regulatory services (MEA 2003; Saba et al. 2024). The MEA framework identifies ecosystem services delivered by agricultural grasslands, focusing on agricultural production, water quality regulation, hydrological regulation, and biodiversity conservation. These services are particularly relevant in the context of the small wetlands under consideration, given the limited literature on other ecosystem services. They include provisioning services, regulating services, supporting services, and cultural services (Figure 14.2).

14.5.1.1 Provisioning Services

Provisioning services refer to the products obtained from ecosystems that directly benefit human well-being. Wetlands offer various provisioning services, including the provision of food resources such as fish, shellfish, and aquatic plants. Additionally, wetlands support timber and nontimber forest products, medicinal plants, and raw materials for traditional crafts, contributing to local livelihoods and economies.

14.5.1.2 Regulating Services

Regulating services involve the regulation of essential ecosystem processes that support environmental quality and stability. Wetlands play a crucial role in regulating water quality by filtering pollutants, sediment retention, and nutrient cycling. These processes help improve water clarity, reduce nutrient runoff, and mitigate the impacts of pollutants on aquatic ecosystems. Moreover, wetlands regulate hydrological processes by attenuating floods, recharging groundwater, and maintaining base flows in rivers, thereby reducing the risk of flooding and enhancing water availability during dry periods.

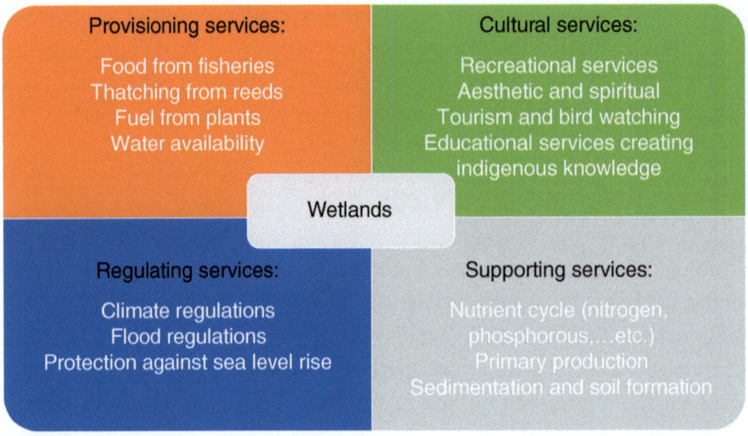

Figure 14.2 Ecological services provided by wetlands according to the Millennium Ecosystem Assessment (2005).

14.5.1.3 Supporting Services

Supporting services are fundamental to the provision of other ecosystem services, sustaining ecosystem structure and function. Wetlands serve as vital habitats for a diverse range of flora and fauna, supporting biodiversity and ecosystem resilience. They provide breeding grounds, nesting sites, and foraging areas for numerous plant and animal species, including migratory birds, amphibians, and aquatic organisms. The structural complexity and habitat heterogeneity of wetlands create niche environments that support a rich array of species, contributing to overall ecosystem health and functioning.

14.5.1.4 Cultural Services

Cultural services encompass the nonmaterial benefits that ecosystems provide to humans, including aesthetic, spiritual, recreational, and educational values. Wetlands hold significant cultural importance as sites of cultural heritage, spiritual significance, and recreational activities. They offer opportunities for outdoor recreation, such as bird watching, fishing, boating, and nature photography, promoting physical and mental well-being. Wetlands also serve as educational resources, facilitating environmental education and ecological awareness among local communities and visitors.

The primary mechanism through which small wetlands influence climate is via the production and emission of greenhouse gases such as methane (CH_4) and nitrous oxide (N_2O). The quantity of greenhouse gases emitted by small wetlands can be disproportionately large and is influenced by factors such as interface length with other systems and nutrient supply. While wetlands can affect air quality by promoting the emission of gases like ammonia (NH_3), emissions from natural small wetlands are generally low. However, constructed treatment wetlands receiving high ammoniacal loadings may exhibit elevated emissions.

Wetlands rank among the world's most productive and valuable ecosystems, as emphasized by the Ramsar Convention. They provide diverse benefits, contributing significantly to provisioning, regulating, and cultural services. Despite their high value, wetland ecosystems have been adversely impacted by land use changes, urban expansion, and development, leading to the loss of nearly 50% of the world's wetland ecosystems (Figure 14.2). The Ramsar Convention advocates for the wise use of wetlands to ensure the preservation and restoration of ecosystem services for societal benefit.

14.5.2 Ecological and Economic Valuation for a Holistic Understanding

Understanding the intricate relationship between ecosystems and human well-being requires a comprehensive assessment of ecosystem services and their economic valuation. The seminal work by Costanza et al. (1997) marked a watershed moment in this endeavor, providing a pioneering estimate of the global value of ecosystem services. Their study, published in Nature, unveiled a wide-ranging valuation ranging from $16 to $54 trillion annually, exceeding the collective global GDP at the time. This groundbreaking revelation illuminated the indispensable role of ecosystems in supporting human welfare and highlighted the urgent need for further investigation.

Building upon this foundational research, Costanza et al. (2014) revisited their methodologies with updated data, yielding an even more staggering estimate of $125 trillion per year for global ecosystem services in 2011. This updated valuation underscored the immense economic significance of natural capital and underscored the imperative for continued efforts in ecosystem assessment and valuation.

Concurrently, the Millennium Ecosystem Assessment (MA) of 2005, a collaborative effort involving over 1360 experts worldwide, provided a comprehensive evaluation of ecosystem change and its implications for human well-being. The MA's findings revealed a distressing trend of declining ecosystem services, with two-thirds of the planet's services in decline. Moreover, it warned of the increasing likelihood of nonlinear changes in ecosystems, posing significant risks to human welfare.

In response to the MA's call to action, the "The Economics of Ecosystems and Biodiversity" (TEEB) initiative was launched in 2010. TEEB aimed to raise awareness about the economic benefits of biodiversity and the escalating costs of biodiversity loss. Through a series of reports and assessments, TEEB highlighted the crucial role of ecosystem service valuation in informing policy decisions and promoting sustainable resource management.

As part of the TEEB initiative, de Groot et al. (2012) developed the "Ecosystem Service Valuation Database" (ESVD), a comprehensive repository comprising over 300 valuation studies and 1350 valuation estimates worldwide. This database serves as a vital resource for policymakers and researchers, facilitating the synthesis of findings from multiple studies and enabling informed decision-making. By quantifying the economic value of ecosystem services, the ESVD empowers policymakers to assess the trade-offs associated with different land use options and develop strategies for sustainable development and conservation.

14.6 Resilience of Wetland Ecosystems

Sustainability, as a fundamental concept in conservation biology, bridges the gap between human needs and ecosystem services (Morelli 2011). It encapsulates the ability to meet present human needs while safeguarding the capacity of future generations to meet their own needs, all while preserving ecosystem health (Munier 2005). Assessing the ecological resilience of ecosystems is crucial for evaluating their potential sustainability. It serves as an indicator of system health and the ability to rebound from disturbances, thereby contributing to environmental sustainability (Wu and Wu 2013).

Early conceptualizations of ecological resilience, such as Holling's (1973), defined it as the persistence of ecosystems in the face of change and disturbance while maintaining their structure and functions. A more contemporary definition by Walker et al. (2004) characterizes resilience as the capability of an ecosystem to endure disturbances, adapt, and maintain its function and structure. In essence, ecological resilience denotes an ecosystem's capacity to withstand and recover from disturbances, with resilient ecosystems capable of assuming multiple stable states despite induced perturbations.

Disturbances to ecosystems can arise from both natural and human-induced sources. Natural disturbances encompass events like forest fires, floods, and infectious diseases, while human-induced disturbances result from activities such as land use changes, habitat transformation (e.g. agriculture, aquaculture, and urbanization), and pollution. Human interventions in natural processes can disrupt ecosystem balance, hindering the ecosystem's ability to return to its predisturbance state. These disturbances may lead to adverse changes in ecosystem dynamics, with unpredictable consequences (Baho et al. 2017).

Given the ongoing modification and degradation of ecosystems, there is concern that the ability of natural ecosystems to cope with environmental disturbances may diminish over time. Hence, understanding how human interactions with ecosystems influence ecological resilience and the provisioning of ecosystem services is paramount for effective management (Gunderson 2000). Evaluating ecosystems' resilience aids natural resource managers and decision-makers in determining the necessary interventions to restore ecosystems to their predisturbance state (Standish et al. 2014).

14.6.1 Integrated Approaches for Assessing Ecosystem Resilience

Assessing ecosystem resilience necessitates integrated approaches that combine various metrics and methodologies. For instance, De Keersmaecker et al. (2015) employed vegetation stability metrics to measure vegetation resilience and resistance to climate change. Their study utilized remote sensing data, including NDVI bimonthly time-series, and climate data, such as monthly temperature, to develop a model establishing the relationship between vegetation metrics and global temperature changes. This model provides insights into vegetation resilience and resistance to drought and temperature changes, applicable across ecosystems.

High spatial resolution and remotely sensed data facilitate change detection studies (Hotaiba et al. 2023), enabling the evaluation of past and present changes in wetlands. By integrating climatic data as an indicator of system well-being against climate change, these studies contribute to assessing wetlands' resilience and adaptive capacity.

Another approach to assessing ecosystem resilience is exemplified by Comer et al. (2018), who measured the resilience of different vegetation ecosystems based on sensitivity and adaptive capacity. Sensitivity indicators encompass changes in land use/land cover, landscape fragmentation, invasive species, and other disturbances, while adaptive capacity factors in ecosystem characteristics like species diversity and keystone species. This integrated approach provides insights into ecosystems' ability to cope with environmental changes and informs management strategies for enhancing resilience.

14.7 Geospatial Assessment of Nigeria's Wetlands

Wetlands play a pivotal role globally, contributing significantly to climate change mitigation, hydrological and biogeochemical cycle management, and sustaining livelihoods (Wilen and Bates 1995; Junk et al. 2013). In Nigeria's southern region, wetlands, predominantly comprising marshes, mangroves, and freshwater swamps, are abundant (Nwankwoala and Okujagu 2014). However, these ecosystems face considerable environmental pressures, including land reclamation for agriculture and industrialization in regions like the Niger delta (Chidumeje et al. 2015), urbanization around areas such as the Lekki lagoon of Lagos (Obiefuna et al. 2013), and pollution from sources like oil spills (Igu and Marchant 2017).

Despite their significance, existing wetland maps, such as the Global Land Cover GLC250-2010 and the Global Lakes and Wetlands Database (GLWD-3), offer only coarse resolutions and exhibit discrepancies due to methodological variations (Gumbricht et al. 2017; Xu et al. 2018). Therefore, improving wetland mapping using diverse techniques is imperative to gain a comprehensive understanding of wetland distribution, types, and fragmentation. Detailed wetland maps are essential for economic assessments of ecosystem services and informed decision-making by regional and international authorities (McInnes et al. 2017).

In addition to delineating wetlands, understanding their classification is crucial. Wetlands encompass marshes, swamps, bogs, and fens, each characterized by specific hydrology, soil type, and vegetation (Mitsch and Gosselink 2000). However, tropical wetlands, particularly peatlands, present classification challenges. While some areas mapped as peatlands may resemble swamps due to limited peat cover, defining clear criteria helps in accurate classification.

14.7.1 Geographic Information System (GIS) and Remote Sensing

A Geographic Information System (GIS) serves as a comprehensive tool for generating, organizing, analyzing, and visualizing various types of data, integrating location data with descriptive information to create detailed maps. On the other hand, remote sensing involves observing the Earth's surface or atmosphere using satellites in space or aircraft. This method utilizes different parts of the electromagnetic spectrum to capture radiation emitted or reflected by the Earth's surface. While human eyes detect solar light, remote sensing extends beyond the visible spectrum, incorporating near-infrared, middle-infrared, thermal-infrared, and microwave wavelengths.

14.7.2 Applications of Remote Sensing in Wetland Mapping

For over three decades, remote sensing techniques have proven effective for mapping and monitoring wetlands, utilized by both academic researchers and government agencies (Tiner 1996; Dahl 2006; Klemas 2011). Satellite remote sensing offers numerous advantages for wetland inventorying and monitoring, including repeat coverage enabling seasonal or yearly assessments and providing insights into surrounding land use changes over time (Klemas 2011). Various remote sensing platforms, such as aerial photography, S, and Landsat MSS, have been reviewed for mapping coastal wetlands and estimating biomass (Hardisky et al. 1986).

Spatial analysis techniques for wetland mapping rely on remote sensing data, GIS, and various classification algorithms to delineate wetland boundaries and characterize their attributes (Figure. 14.3).

Satellite Imagery Classification: Multispectral or hyperspectral satellite imagery, combined with classification algorithms such as maximum likelihood, support vector machines (SVMs), random forests, and deep learning techniques like convolutional neural networks (CNNs), is employed to identify and classify wetland areas (Li et al. 2018). These algorithms analyze spectral signatures to differentiate between wetland and non-wetland land cover types.

LiDAR Data Processing: Light Detection and Ranging (LiDAR) data provide high-resolution elevation information, enabling the identification of topographic signatures associated with wetland ecosystems. LiDAR data processing techniques assist in wetland delineation by capturing terrain features indicative of wetland presence (Rapinel et al. 2019).

Object-Based Image Analysis (OBIA): OBIA involves segmenting satellite imagery into meaningful objects based on spectral, spatial, and contextual information. This approach enables more accurate wetland classification compared to pixel-based methods by considering the spatial arrangement and relationships between image objects (Dronova 2015).

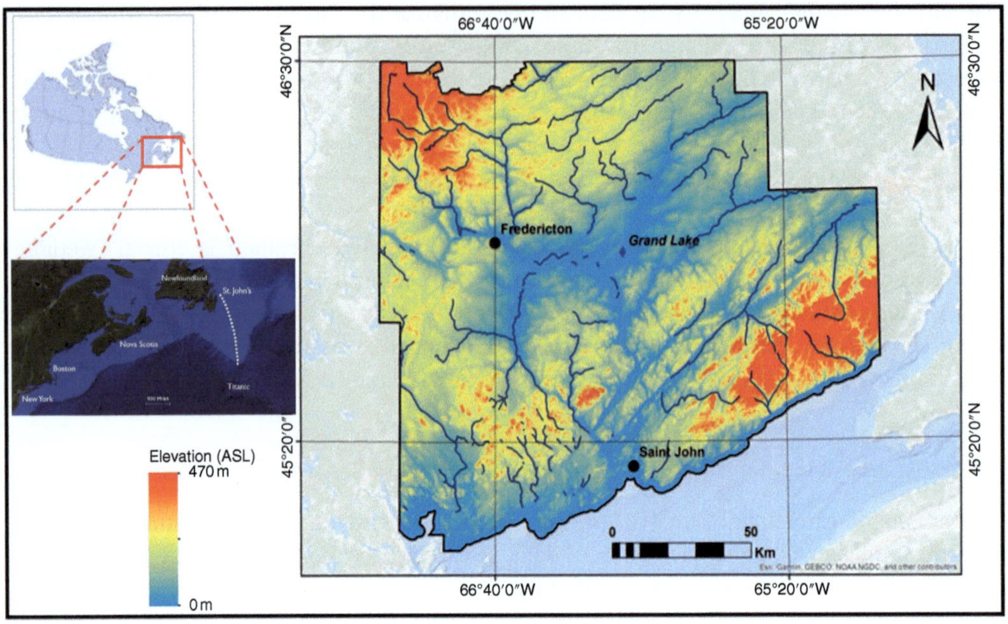

Figure 14.3 Example of wetland mapping using remote sensing. *Source:* LaRocque et al. (2020)/MDPI/CC BY 4.0.

Hydrological Modeling: Analysis of hydrological parameters such as water flow patterns, groundwater levels, and soil moisture content contributes valuable information for wetland mapping and delineation. Hydrological modeling techniques help assess the hydrological connectivity and functioning of wetland ecosystems (Ameli and Creed 2017).

Integration of Field Data: Ground truth data collected through field surveys, vegetation sampling, and wetland inventories are integrated into mapping models to validate remote sensing-derived classifications. Field data verification enhances the accuracy and reliability of wetland mapping outputs by providing empirical evidence of wetland characteristics (Wu et al. 2019).

Geospatial Representation and Comprehensive Cartographic Assessment of Nigeria's Wetlands: Mapping and evaluating Nigeria's wetlands requires a comprehensive approach that integrates remote sensing, GIS analysis, field surveys, and data integration.

Acquisition of Remote Sensing Data: Utilizing satellite imagery with suitable spatial and spectral resolutions, such as Landsat, Sentinel, and MODIS data, to delineate wetland extent and characteristics across Nigeria. GIS Analysis and Spatial Modeling: Employing GIS techniques for integrating data, conducting spatial analysis, and modeling to identify wetland distribution, types, and temporal changes. This includes overlay analysis, land cover classification, and hydrological modeling to understand wetland dynamics (Singh et al. 2020).

Wetland Classification and Typology: Developing classification systems and typologies based on ecological characteristics, hydrology, and vegetation cover to categorize wetlands across different regions of Nigeria. This aids in understanding wetland diversity and functions (Olaniyi et al. 2017).

Socioeconomic and Ecological Assessment: Integrating socioeconomic data with ecological parameters to assess the importance of wetlands for local communities, biodiversity preservation, and provision of ecosystem services. Understanding the socioecological dynamics of wetlands informs conservation and management strategies (Ohl et al. 2010).

Threat Mapping and Conservation Priorities: Identifying and mapping anthropogenic threats such as urbanization, agricultural expansion, and pollution to prioritize conservation efforts and implement sustainable wetland management strategies. Threat mapping helps identify areas requiring immediate conservation action (Tulloch et al. 2020).

14.7.3 Case Study of Wetland Mapping in Southern Nigeria

The study evaluated the mapping of wetlands in southern Nigeria, revealing that they cover a vast area of 29 924 km^2, representing over one-fifth of the region's total landmass (Figure 14.4). These wetlands primarily comprise swamps (44%), mangroves (31%), marshes (20%), and shallow lakes (5%), with the majority clustered along the coastal areas, notably in the Niger delta and Lagos (Garba et al. 2023).

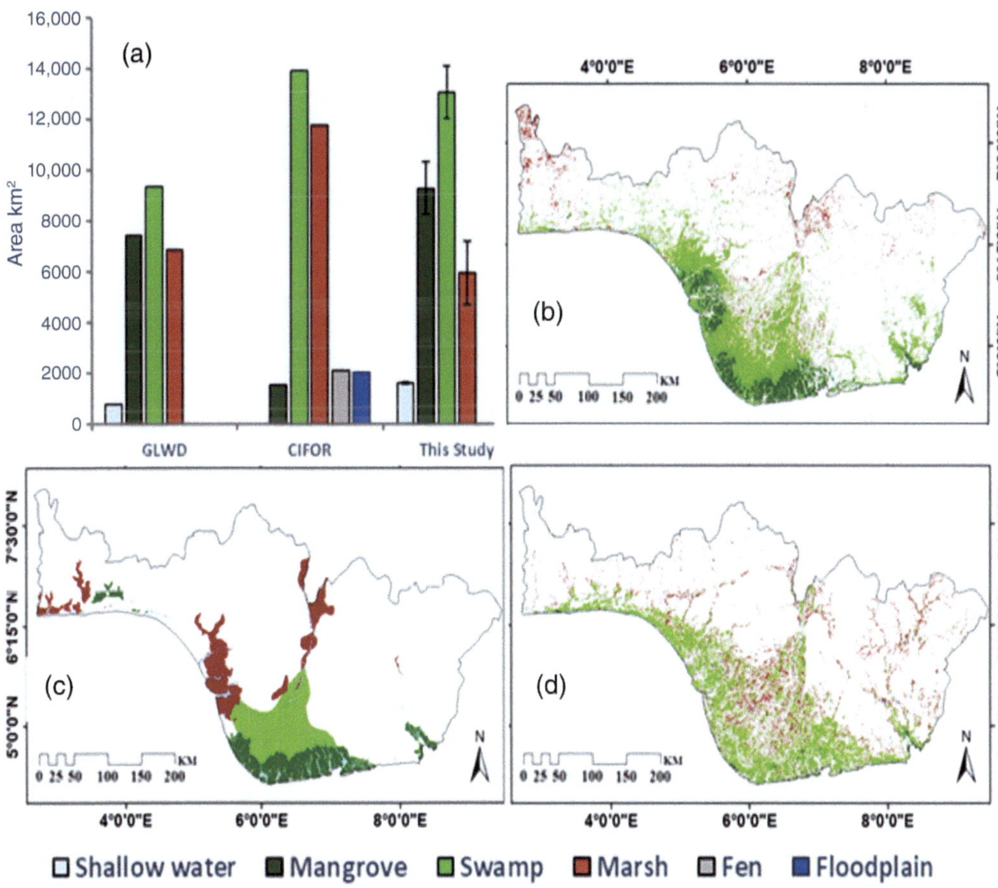

Figure 14.4 Comparison of wetland mapping images for the southern part of Nigeria: (a) Areas of different wetland classes in southern Nigeria are depicted, with error bars indicating misclassification levels based on the accuracy achieved for each wetland type in our study. (b) A map showing only the wetland classes identified in our study across southern Nigeria. (c) The Global Lakes and Wetlands Database (GLWD). (d) The global wetland database by the Center for International Forestry Research (CIFOR). *Source:* Garba et al. 2023/ Springer Nature/CC BY 4.0.

The mapping efforts provided insights into the intricate details of these wetland ecosystems, highlighting discrepancies in previous estimates. While our total wetland coverage estimates fall below that of the Center for International Forestry Research (CIFOR), it surpasses the GLWD. This variance is primarily attributed to our extensive mapping of mangroves and the identification of fewer marsh wetlands. Although maps may appear similar briefly, closer inspection reveals significant differences in detail.

The study meticulously classified various wetland types, including swamps, marshes, shallow water bodies, and coastal mangroves, and also addressed the challenge of distinguishing between peatlands and swamps, given the lack of a standardized classification system for tropical peatlands. By strictly defining swamps as tree-dominated mineral soil wetlands, we aimed to minimize confusion in our mapping efforts.

The study utilized composite images derived from optical and radar imagery, considering seasonal variations in wetland signatures. By employing advanced techniques and integrating optical indices and SAR polarimetric features, we successfully mapped four wetland types at a remarkable 10-m resolution. Our mapping accuracy reached an impressive 88%, providing valuable insights into wetland distribution and fragmentation across southern Nigeria.

The study underscores the urgent need for enhanced wetland conservation efforts, particularly in the face of rapid population growth in Nigeria. It is imperative that wetland protection organizations leverage high-resolution data for change detection and take proactive measures to safeguard these invaluable ecosystems. Additionally, our data holds immense potential for informing future modeling efforts aimed at promoting sustainable land use practices and preserving wetland biodiversity for generations to come (Garba et al. 2023).

14.8 Policy and Institutional Framework for Wetlands Protection in Nigeria

14.8.1 Wetlands of International Importance (Ramsar Sites)

Wetlands are undergoing massive transformations and degradation due to severe anthropogenic interventions. Such losses and degradation of wetlands' services are increasingly expressed in global initiatives as the Ramsar Convention on Wetlands of International Importance Especially Waterfowl Habitat, also known as the Convention on Wetlands, and the Ramsar Convention. Ramsar Convention is an international treaty adopted in 1971 and named after the Ramsar town in Iran in which it was conducted (Bowman 2009). The convention aims to promote the international cooperation for conservation of wetlands worldwide and the sustainable uses of its resources. The convention addresses three main pillars: the wise use of all wetland resources, the sustainable and effective management of wetlands sites, and promotion of international cooperation in management of shared resources and sharing of knowledge and information (Gardner and Davidson 2011).

The criteria for identifying a wetland site as of international importance depend on nine characteristics: The wetland represents rare or unique wetland type, it supports threatened and critically endangered species, it supports population of important species that maintain the biodiversity of a biogeographic region, it provides refuge to critical species during adverse and harsh conditions, it supports more than 20 000 of waterbirds, it represents an important area for nursery or migration pathway or a spawning area, and it supports a substantial proportion of native fish species and subspecies (Ramsar Handbooks 2016). According to the Ramsar Convention, the number of wetlands recognized under the Ramsar Convention is 2439 with a total surface area of 254 691 993 ha (https://www.ramsar.org/).

The conservation of wetlands is integral to the broader objectives of environmental sustainability and biodiversity conservation in Nigeria. This section elucidates the national policies, legal frameworks, and institutional structures aimed at protecting wetlands, with a focus on the scientific and regulatory dimensions (Matemilola and Elegbede 2017).

The 1999 National Policy on the Environment underscores the imperative of securing a high-quality environment conducive to human health and well-being, while also emphasizing the sustainable utilization of natural resources for present and future generations. Complementing this policy is the commitment to biodiversity conservation, as outlined in the Convention on Biological Diversity (CBD). Furthermore, specific policies such as the National Forest Policy and the National Policy on Erosion, Flood Control, and Coastal Zone Management contribute to the overarching goal of wetlands protection by promoting sustainable land management practices.

Constitutionally, Nigeria mandates the protection and improvement of the environment, including water bodies, air quality, forests, and wildlife. The establishment of the Federal Environmental Protection Agency (FEPA) in 1988 initially spearheaded these efforts, albeit it was later replaced by a more comprehensive regulatory framework in 2007. This new legislation emphasizes the enforcement of environmental regulations and the sustainable management of natural resources, aligning with international standards and best practices.

To operationalize these policies, the Federal Government has developed 24 Environmental Regulations, covering various aspects of environmental management. Additionally, several laws complement federal statutes, including the Natural Resources Conservation Act and the Endangered Species (Control of International Trade and Traffic) Act, providing a comprehensive legal framework for biodiversity conservation.

Internationally, Nigeria is a signatory to numerous treaties and conventions aimed at promoting biodiversity conservation and sustainable development. Active participation in the negotiation and ratification of the CBD underscores Nigeria's commitment to global environmental stewardship. The preparation of the Biodiversity Strategy and Action Plan further demonstrates the country's proactive approach to conservation efforts.

The primary institution responsible for wetlands protection in Nigeria is the Federal Ministry of Environment, which operates under the framework of the National Environmental Standards and Regulations Enforcement Agency (NESREA) Act. This regulatory body is tasked with enforcing environmental laws and regulations to ensure compliance and mitigate environmental degradation.

In addition to governmental efforts, Nigeria benefits from international collaborations and donor support. Organizations such as the Canadian International Development Agency (CIDA) and the United Kingdom's Department for International Development (DFID) provide financial and technical assistance for capacity building and policy implementation. Furthermore, multilateral agencies like the World Bank and United Nations Environment Programme (UNEP) play crucial roles in supporting environmental management projects within the country.

Nongovernmental Organizations (NGOs) also contribute significantly to wetlands protection through community engagement and advocacy initiatives. Entities such as the Nigerian Conservation Foundation International and Wetlands International collaborate with government agencies to address challenges and promote sustainable wetlands management practices.

Despite these concerted efforts, several challenges impede effective wetlands protection in Nigeria. These include inadequate funding, insufficient capacity-building initiatives, and limited public awareness. Addressing these challenges requires enhanced collaboration between government agencies, NGOs, and international partners, coupled with robust legislative frameworks and enforcement mechanisms.

The protection of wetlands in Nigeria necessitates a multidimensional approach encompassing policy formulation, legal frameworks, institutional capacities, and international collaborations. By addressing existing challenges and leveraging available resources, Nigeria can enhance its efforts toward sustainable wetlands management and biodiversity conservation for the benefit of current and future generations.

14.8.2 Overcoming Hurdles in Wetlands Conservation

14.8.2.1 Legal Vacuum: The Case of Wetlands Legislation

Nigeria's commitment to the CBD remains unfulfilled until integrated into domestic law. Despite having related legislation, the absence of specific wetlands laws poses a significant challenge. For instance, while there are regulations governing environmental protection, the lack of targeted wetlands legislation leaves enforcement agencies with limited legal tools. To illustrate, imagine a scenario where a wetland ecosystem is threatened by unplanned development. Without specific laws addressing wetland protection, authorities may struggle to take effective legal action to prevent environmental degradation.

14.8.2.2 Weak Enforcement: Bridging the Gap Between Policy and Practice

Nigeria's enforcement framework for biodiversity and wetlands conservation faces inherent weaknesses. The Ministry of Environment primarily relies on the NESREA Act and other regulations for wetlands management. However, the inconsistent adoption of these regulations at the state level undermines enforcement efforts. Consider a situation where a wetland area is illegally encroached upon for agricultural purposes. In the absence of consistent regulations and enforcement mechanisms, unauthorized activities may continue unabated, leading to irreversible damage to the wetland ecosystem.

14.8.2.3 Capacity-Building and Awareness Deficiency: Empowering Stakeholders for Effective Conservation

A critical challenge in wetlands conservation is the lack of understanding and capacity among enforcement agencies and local communities. Limited awareness of existing laws and insufficient training hinder effective compliance and enforcement. For example, imagine a community residing near a wetland area where unsustainable harvesting practices threaten biodiversity. Without proper training and awareness initiatives, community members may inadvertently contribute to the degradation of the wetland ecosystem due to a lack of understanding of its ecological importance.

14.8.2.4 Top-Down Approach: Integrating Local Perspectives for Sustainable Management

The top-down approach to biodiversity management in Nigeria often fails to consider local perspectives and interests. Conflicts arise between government policies, landowners, and conservationists, leading to challenges in achieving consensus on conservation strategies. For instance, consider a wetland area designated for conservation by the government, but also traditionally used by local communities for fishing and gathering resources. Without meaningful engagement and compensation for affected communities, conservation efforts may face resistance, undermining their effectiveness.

14.8.2.5 Threat of Non-native Species to Wetland Ecosystems

Invasive species pose a significant threat to wetland ecosystems in Nigeria. Species like the Nipa palm and water hyacinth disrupt native habitats and alter ecosystem dynamics. Increased ponding and fluctuating water levels exacerbate their proliferation, leading to ecological imbalances. Picture a scenario where water hyacinths rapidly spread across a wetland, choking waterways and outcompeting native vegetation. Without effective management strategies, invasive species can lead to the loss of biodiversity and ecosystem services vital for local communities.

Addressing these challenges requires a holistic approach that integrates legal reforms, capacity-building initiatives, community engagement, and proactive measures to combat invasive species. By bridging the gap between policy and practice and fostering collaboration among stakeholders, Nigeria can work toward sustainable wetlands conservation for the benefit of present and future generations.

14.8.3 Approach to Wetland Management and Sustainability

A comprehensive strategy that incorporates ecological, social, economic, and governance factors is required to achieve the protection and sustainable management of Nigeria's wetlands (Adebayo and Ugwu 2020). The conservation and restoration of wetland habitats, the encouragement of sustainable land use practices, and the inclusion of local people in decision-making procedures need to be the top priorities of such an approach (Mitsch and Gosselink 2000). Furthermore, in order to guarantee coordinated action and optimize the effectiveness of conservation efforts, productive collaboration between government agencies, civil society groups, academia, and the commercial sector is crucial (Adekola et al. 2022; Elegbede et al. 2023a). Nigeria can protect these priceless ecosystems for future generations while fostering sustainable development and resilience to environmental change by using an all-encompassing and inclusive strategy to wetland management.

14.9 Conclusion

Wetlands in Nigeria represent critical components of the country's natural landscape, playing essential roles in water management, biodiversity conservation, and climate regulation. With approximately 11 major wetlands identified across various regions, including the Niger, Benue, and Chad basins, Nigeria boasts rich biodiversity and diverse ecosystems. However, these wetlands face numerous threats, including sea-level rise, urbanization, industrialization, and agricultural expansion, which compromise their ecological integrity and resilience. The concept of blue carbon and carbon sequestration underscores the significant role of wetlands in mitigating climate change by capturing and storing atmospheric carbon dioxide. Despite their potential as nature-based solutions to climate change, wetlands in Nigeria and globally continue to face anthropogenic pressures, leading to habitat degradation and loss of ecosystem services. An analysis of wetlands and their ecosystem services reveals the need for a holistic understanding of their ecological and economic value.

References

Adebayo, A.O. and Ugwu, F.O. (2020). Integrating wetland conservation and sustainable livelihoods: Insights from Nigeria's Niger Delta region. *Wetlands Ecology and Management* 28 (1): 85–98.

Adebayo, A.T., Adewole, H.A., Akindele, E.O., and Olaleye, V.F. (2021). Planktonic flora and fauna of Opa Reservoir wetlands, Obafemi Awolowo University, Ile-Ife, Nigeria. *The Journal of Basic and Applied Zoology* 82: 1–10.

Adekola, O. and Mitchell, G. (2011). The Niger Delta wetlands: threats to ecosystem services, their importance to dependent communities and possible management measures. *International Journal of Biodiversity Science, Ecosystem Services & Management* 7 (1): 50–68.

Adekola, O., Oguntoke, O., and Adedokun, S. (2022). Assessing the ecological status and conservation implications of selected wetlands in Nigeria. *Journal of Environmental Management* 300: 113667.

Adeniyi, T.A., Oyebanji, O.O., and Adeonipekun, P.A. (2016). Floral diversity in the wetlands of Ibeju-Lekki area, Lagos, Nigeria. *IFE Journal of Science* 18 (3): 729–737.

Adhikari, S., Bajracharaya, R.M., and Sitaula, B.K. (2009). A review of carbon dynamics and sequestration in wetlands. *Journal of Wetlands Ecology* 2 (1): 42–46.

Akanni, A., Onwuteaka, J., Uwagbae, M. et al. (2018). The values of mangrove ecosystem services in the Niger Delta Region of Nigeria. In: *The Political Ecology of Oil and Gas Activities in the Nigerian Aquatic Ecosystem*, 387–437. Academic Press.

Ali, A. and Bichi, H. (2021). Species list and diversity of avifauna resources of Hadejia-Nguru wetlands. *Journal of Forest Science Environment* 6: 157–162.

Ameli, A.A. and Creed, I.F. (2017). Quantifying hydrologic connectivity of wetlands to surface water systems. *Hydrology and Earth System Sciences* 21 (3): 1791–1808.

Anule, P. and Ujoh, F. (2017). Geospatial analysis of wetlands degradation in Makurdi, Nigeria. *The International Archives of the Photogrammetry, Remote Sensing and Spatial Information Sciences* 42: 1075–1081.

Baho, D.L., Allen, C.R., Garmestani, A. et al. (2017). A quantitative framework for assessing ecological resilience. *Ecology and Society* 22 (3): https://doi.org/10.5751/ES-09427-220317.

Bowman, M.J. (2009). The Ramsar convention comes of age. *Netherlands International Law Review* 42 (1): 1–52. https://doi.org/10.1017/S0165070X00003363.

Comer, P., Reid, M., Schulz, K. et al. (2018). *Assessing Vulnerability and Resilience of Major Vegetation Types of the Western Interior U.S – Forests and Woodlands*, Prepared for the U.S. Department of the Interior, Bureau of Land Management. Arlington, VA: NatureServe 288 pp.

Costanza, R., d'Arge, R., De Groot, R. et al. (1997). The value of the world's ecosystem services and natural capital. *Nature* 387 (6630): 253–260.

Costanza, R., De Groot, R., Sutton, P. et al. (2014). Changes in the global value of ecosystem services. *Global Environmental Change* 26: 152–158.

Cunliffe, B. (2023). *Facing the Sea of Sand: The Sahara and the Peoples of Northern Africa*. Oxford University Press.

Dahl, D.B. (2006). Model-based clustering for expression data via a Dirichlet process mixture model. *Bayesian Inference for Gene Expression and Proteomics* 4: 201–218.

De Groot, R., Brander, L., Van Der Ploeg, S. et al. (2012). Global estimates of the value of ecosystems and their services in monetary units. *Ecosystem Services* 1 (1): 50–61.

De Keersmaecker, W., Lhermitte, S., Tits, L. et al. (2015). A model quantifying global vegetation resistance and resilience to short-term climate anomalies and their relationship with vegetation cover. *Global Ecology and Biogeography* 24 (5): 539–548.

Dey, D. and Roy, J. (ed.) (2022). *Biorights: The Neo-economic Conservation Paradigm for Global South*, vol. 7, 51–79. Springer Nature.

Dronova, I. (2015). Object-based image analysis in wetland research: a review. *Remote Sensing* 7 (5): 6380–6413.

Elegbede, I.O., Abimbola, L.H.M., Matemilola, S., and Damilola, O.A. (2014). Wetland resources of Nigeria: case study of the Hadejia-Nguru wetlands. *Poultry, Fisheries & Wildlife Sciences* 2: 123.

Elegbede, I.O., Matemilola, S., Kies, F. et al. (2017). Risk analysis and development of algae biofuel from aquatic and terrestrial systems. *Energy Procedia* 128: 324–331.

Elegbede, I.O., Akintola, S.L., Jimoh, A.A.A. et al. (2023a). *Blue Economy (Sustainability). Encyclopedia of Sustainable Management.* Cham: Cham https://doi.org/10.1007/978-3-030-02006-4_401-1.

Elegbede, I.O., Fakoya, K.A., Adewolu, M.A. et al. (2023b). Understanding the social–ecological systems of non-state seafood sustainability scheme in the blue economy. *Environment, Development and Sustainability* 24 (11): 1–32.

Elegbede, I.O., Kaullysing, D., Aderonke, L.A. et al. (2023c). Ocean sustainability. In: *Encyclopedia of Sustainable Management*, 2482–2488. Cham: Springer International Publishing.

Filani, O. T. (2021). Blue carbon marine habitats in Nigeria: ecosystem services, threats and implication for ecosystem based management. World Maritime University dissertations 1694. https://commons.wmu.se/all_dissertations/1694 (accessed 12 June 2024).

Garba, S.I., Ebmeier, S.K., Bastin, J.F. et al. (2023). Wetland mapping at 10 m resolution reveals fragmentation in southern Nigeria. *Wetlands Ecology and Management* 31 (3): 329–345.

Gardner, R.C. and Davidson, N.C. (2011). The Ramsar convention. In: *Wetlands: Integrating Multidisciplinary Concepts* (ed. B.A. LePage), 189–203. Dordrecht: Springer.

Gumbricht, T., Román-Cuesta, R.M., Verchot, L.V. et al. (2017). Tropical and subtropical wetlands distribution version 2.

Gunderson, L.H. (2000). Ecological resilience – in theory and application. *Annual Review of Ecology and Systematics* 31: 425–439.

Hardisky, M.A., Gross, M.F., and Klemas, V. (1986). Remote sensing of coastal wetlands. *Bioscience* 36 (7): 453–460.

Herr, D. and Landis, E. (2016). *Coastal Blue Carbon Ecosystems. Opportunities for Nationally Determined Contributions. Policy Brief.* Gland, Washington, DC: IUCN, TNC.

Holling, C.S. (1973). Resilience and stability of ecological systems. *Annual Review of Ecology and Systematics* 4 (1): 1–23.

Hotaiba, A.M., Salem, B.B., and Halmy, M.W.A. (2023). Assessment of wetland ecosystem's health using remote sensing – case study: Burullus wetland – Ramsar site. *Estuaries and Coasts* https://doi.org/10.1007/s12237-023-01274-y.

Idris, M. (2008). *Types of Wetlands.* U.S. Forest and Fish Wild Life Service. 99 pp.

Igu, N.I. and Marchant, R. (2017). Freshwater swamp forest use in the Niger Delta: perception and insights. *Journal of Forest Research* 22 (1): 44–52.

Junk, W.J., An, S., Finlayson, C.M. et al. (2013). Current state of knowledge regarding the world's wetlands and their future under global climate change: a synthesis. *Aquatic Sciences* 75: 151–167.

Kadiri, W.O.J., Ogunleye, K.S., Fasina, A.S., and Babalola, T.S. (2022). Soil properties affecting soil organic carbon stock of different land use types in two agro-ecological zones of Nigeria. *Agro-Science* 21 (3): 37–48.

Klemas, V. (2011). Remote sensing of wetlands: case studies comparing practical techniques. *Journal of Coastal Research* 27 (3): 418–427.

LaRocque, A., Phiri, C., Leblon, B. et al. (2020). Wetland mapping with Landsat 8 OLI, sentinel-1, ALOS-1 PALSAR, and LiDAR data in Southern New Brunswick, Canada. *Remote Sensing* 12 (13): 2095.

Leemans, R., & De Groot, R. S. (2003). Millennium Ecosystem Assessment: Ecosystems and human well-being: a framework for assessment.

Li, N., Lu, D., Wu, M. et al. (2018). Coastal wetland classification with multiseasonal high-spatial resolution satellite imagery. *International Journal of Remote Sensing* 39 (23): 8963–8983.

Lorenz, K., Lal, R., Lorenz, K., and Lal, R. (2018). Carbon sequestration in grassland soils. *Carbon Sequestration in Agricultural Ecosystems* 175–209.

Mahapatra, A., Hore, U., Singh, A., and Kumari, M. (2024). The effect of urbanization on the shrinkage of wetlands in the Noida-Greater Noida region and its surrounding sub-urban areas. *Ecological Frontiers* 44 (1): 96–104.

Matemilola, S. and Elegbede, I. (2017). Transparency: a tool for effective environmental governance in Nigeria. *Open Access Library Journal* 4 (12): 1–18.

Matthews, G.V.T. (1993). *The Ramsar Convention on Wetlands: Its History and Development*. Gland: Ramsar Convention Bureau.

McInnes, R.J., Simpson, M., Lopez, B. et al. (2017). Wetland ecosystem services and the Ramsar Convention: an assessment of needs. *Wetlands* 37 (1): 123–134.

Meena, R.S., Kumar, S., and Yadav, G.S. (2020). Soil carbon sequestration in crop production. *Nutrient Dynamics for Sustainable Crop Production* 1–39.

Millennium Ecosystem Assessment. (2005). https://www.millenniumassessment.org/en/About.html.

Mitsch, W.J. and Gosselink, J.G. (2000). The value of wetlands: importance of scale and landscape setting. *Ecological Economics* 35 (1): 25–33.

Mitsch, W.J.R., Mitsch, R.H., and Turner, R.E. (1994). Wetlands of the old and new worlds: ecology and management. In: *Global Wetlands: Old World and New*, 80. Amsterdam: Elsevier Science BV, Research.

Mitsch, W.J., Bernal, B., and Hernandez, M.E. (2015). Ecosystem services of wetlands. *International Journal of Biodiversity Science, Ecosystem Services & Management* 11 (1): 1–4. https://doi.org/10.1080/21513732.2015.1006250.

Morelli, J. (2011). Environmental sustainability: a definition for environmental professionals. *Journal of Environmental Sustainability* 1 (1): 2.

Munier, N. (2005). *Introduction to Sustainability*, 3558–3556. Amsterdam: Springer.

Nwankwoala, H.O. (2012). Case studies on coastal wetlands and water resources in Nigeria. *European Journal of Sustainable Development* 1 (2): 113–113.

Nwankwoala, H.O. and Okujagu, D.C. (2014). A review of wetlands and coastal resources of the Niger Delta: potentials, challenges and prospects. *Environment & Ecosystem Science* 5: 37–46.

Obiefuna, J.N., Nwilo, P.C., Atagbaza, A.O., and Okolie, C.J. (2013). Spatial changes in the wetlands of Lagos/Lekki Lagoons of Lagos, Nigeria. *Journal of Sustainable Development* 6 (7): 123.

Ogunyemi, S. (1977). The distribution of some weeds of arable crops in Southwestern Nigeria. *Proceedings of 7th Annual Conference of The Weed Science Society of Nigeria*, pp. 29–45.

Ohl, C., Johst, K., Meyerhoff, J. et al. (2010). Long-term socio-ecological research (LTSER) for biodiversity protection – a complex systems approach for the study of dynamic human–nature interactions. *Ecological Complexity* 7 (2): 170–178.

Okonkwo, C.N.P., Kumar, L., and Taylor, S. (2015). The Niger Delta wetland ecosystem: what threatens it and why should we protect it? *African Journal of Environmental Science and Technology* 9: 451–463.

Olaniyi, O.E., Esan, D.B., and So, A. (2017). Forest dynamism of a green space: a perspective from Ta Afolayan Wildlife Park, Akure, Ondo state, Nigeria. *Nigerian Journal of Technological Research* 12 (1): 90–95.

Olubode, O.S., Awodoyin, R.O., and Ogunyemi, S. (2011). Floral diversity in the wetlands of Apete River, Eleyele Lake and Oba Dam in Ibadan, Nigeria: its implication for biodiversity erosion. *West Africa Journal of Applied Ecology* 18: 109–119.

Orie, E.G. (2019). Ecosystem services: legal issues on Nigeria's wetlands. In: *Law| Environment | Africa*, 315–334. Nomos Verlagsgesellschaft mbH & Co. KG.

Owoyemi, J.M., Zakariya, H.O., and Elegbede, I.O. (2016). Sustainable wood waste management in Nigeria. *Environmental & Socio-Economic Studies* 4 (3): 1–9.

Pandit, R., Parrota, J., Anker, Y. et al. (2018). *Responses to Halt Land Degradation and to Restore Degraded Land*. IPBES.

Rafferty, J.P. (ed.) (2010). *Lakes and Wetlands*. Britannica Educational Publishing.

Ramsar Handbooks (2016). *An Introduction to the Ramsar Convention on Wetlands*, 7e. Gland: Ramsar Convention Secretariat.

Rapinel, S., Fabre, E., Dufour, S. et al. (2019). Mapping potential, existing and efficient wetlands using free remote sensing data. *Journal of Environmental Management* 247: 829–839.

Saba, A.O., Eyo, V.O., Elegbede, I.O. et al. (2024). Sustaining the blue bounty: fish food and nutrition security in Nigeria's evolving blue economy. *AIMS Agriculture and Food* 9 (2): 500–530.

Singh, S., Bhardwaj, A., and Verma, V.K. (2020). Remote sensing and GIS based analysis of temporal land use/land cover and water quality changes in Harike wetland ecosystem, Punjab, India. *Journal of Environmental Management* 262: 110355.

Standish, R.J., Hobbs, R.J., Mayfield, M.M. et al. (2014). Resilience in ecology: abstraction, distraction, or where the action is? *Biological Conservation* 177: 43–51. https://doi.org/10.1016/j.biocon.2014.06.008.

Sulaiman, I.M., Abubakar, M.M., Ringim, A.S. et al. (2015). Effects of wetlands type and size on bird diversity and abundance at the Hadejia-Nguru wetlands, Nigeria. *International Journal of Research Studies in Zoology* 1 (1): 15–21.

Tiner, R.W. (1996). Technical Aspects of Wetlands: Wetland Definitions and Classifications in the United States. US Geological Survey Water-Supply Paper, Report: W, 2425.

Tulloch, V.J., Turschwell, M.P., Giffin, A.L. et al. (2020). Linking threat maps with management to guide conservation investment. *Biological Conservation* 245: 108527.

Valach, A.C., Kasak, K., Hemes, K.S. et al. (2021). Productive wetlands restored for carbon sequestration quickly become net CO_2 sinks with site-level factors driving uptake variability. *PLoS ONE* 16 (3): e0248398.

Villa, J.A. and Bernal, B. (2018). Carbon sequestration in wetlands, from science to practice: an overview of the biogeochemical process, measurement methods, and policy framework. *Ecological Engineering* 114: 115–128.

Wali, E. (2015). Urbanization and Loss of Wetlands in Port-Harcourt Metropolis, Nigeria. An unpublished M.Sc. thesis.

Walker, B., Holling, C.S., Carpenter, S.R., and Kinzig, A. (2004). Resilience, adaptability and transformability in social–ecological systems. *Ecology and Society* 9 (2): 1–9.

Were, D., Kansiime, F., Fetahi, T. et al. (2019). Carbon sequestration by wetlands: a critical review of enhancement measures for climate change mitigation. *Earth Systems and Environment* 3: 327–340.

Wilen, B.O. and Bates, M.K. (1995). The US fish and wildlife service's national wetlands inventory project. *Classification and Inventory of the World's Wetlands* 118 (1–2): 153–169.

Wu, J. and Wu, T. (2013). *Ecological Resilience as a Foundation for Urban Design and Sustainability. Resilience in Ecology and Urban Design*, 211–230. Dordrecht: Springer.

Wu, Q., Lane, C.R., Li, X. et al. (2019). Integrating LiDAR data and multi-temporal aerial imagery to map wetland inundation dynamics using Google Earth Engine. *Remote Sensing of Environment* 228: 1–13.

Xu, L.D., Xu, E.L., and Li, L. (2018). Industry 4.0: state of the art and future trends. *International Journal of Production Research* 56 (8): 2941–2962.

Yahaya-Shiru, M., Igwe, O., and Obafemi, S. (2022). 3D structural and stratigraphic characterization of X field Niger Delta: implications for CO_2 sequestration. *Journal of Petroleum Exploration and Production Technology* 12 (4): 959–977.

15

Wetlands of Senegal: Biodiversity, Livelihoods, and Conservation

Abdoul Aziz Diouf[1], Paterne Arnaud Bernard Mingou[2], Taïbou Ba[3], and Mathieu Gueye[4]

[1] *Ecological Monitoring Center, Research & Development Program, Dakar, Senegal*
[2] *Department of Botany and Geology, Laboratory of Botany, IFAN Ch. A. Diop, Dakar, Senegal*
[3] *Department of Education, Science, Technology & Innovation, African Union Commission, Addis Ababa, Ethiopia*
[4] *Department of Botany and Geology, Laboratory of Botany, IRL 3189, Environnement, Santé et Société, IFAN Ch. A. Diop, Dakar, Senegal*

15.1 Introduction

Wetlands are of major ecological and socioeconomic importance. Indeed, many species live and reproduce in these areas, which are also the source of subsistence for local populations, without forgetting ecotourism, landscape aesthetics, and their cultural function (Ambus and Lowrance 1991; Sène et al. 2006). They provide a range of ecosystem services essential to human life and well-being. They are the natural repository of biological diversity but have long been regarded by rural development specialists as sterile and unhealthy environments, breeding grounds for disease, and a haven for mosquitoes and crocodiles. With the evolution of technology, many efforts have been made to develop them and the space they occupy has been devoted to uses that are supposed to be more productive, such as irrigated agriculture. Thus, some wetlands have even been completely dried up.

Despite the recognition of the importance of wetlands in production systems and the conservation of biological diversity, these ecosystems remain threatened by multiple human activities and global change. Climate change, notably an increase in temperatures of 0.87 °C (IPCC 2019) and the extreme variability of rainfall patterns, especially in the Sahelian strip, seriously threaten ecosystem services and undermine the economies and development efforts of the African continent (Janicot et al. 2015). Thus, the combined effect of all factors constitutes a major threat to the maintenance and sustainable use of their biodiversity.

Africa's wetlands cover an area of over 131 million hectares and range from saline coastal lagoons to fresh and brackish water lakes (Wetlands International 2023). In West Africa, and more particularly in the Sahelian regions that are often qualified as arid or semiarid, including Senegal, wetlands fulfill important functions for the environment (Sène et al. 2006). They provide a wide range of ecosystem services for the well-being of the planet (Wetlands International 2023).

Wetlands of Tropical and Subtropical Asia and Africa: Biodiversity, Livelihoods and Conservation, First Edition. Edited by Thammineni Pullaiah.

Despite their great usefulness, wetlands continue to decline worldwide, in both quantity and quality, and their global extent has fallen by between 64 and 71% since the twentieth century (Gardner et al. 2015). Senegal has not been left behind, but efforts are being made by the authorities to safeguard and restore them on a national scale with, in particular, the development and implementation of the National Wetland Management Policy since 2015 (MEDD 2015). This study describes the current situation of Senegal's wetlands, including their location, biodiversity, income livelihoods, and the conservation measures taken to maintain them.

15.2 Wetland Types and Biodiversity

In Senegal, wetlands are fairly diversified and located mainly in the basins of the Senegal, Gambia, Sine Saloum, and Casamance rivers and in catchment areas with nonperennial runoff, on the Grande and Petite Côte, and at the Ferlo (Figure 15.1). They are mainly of three types (coastal, continental, and artificial), distributed across the six ecogeographic zones (Figure 15.1) defined by Senegal's scientific community (CSE 2015b). Senegal's wetlands are among the richest and most extensive in West Africa (Diop 2012). Overall, they are characterized by the almost permanent presence of water, migratory birds, and particularly rich biodiversity. They therefore offer numerous opportunities in terms of meeting safety and economic growth objectives (MEDD 2015). These wetlands are presented according to their specific characteristics (Figure 15.2).

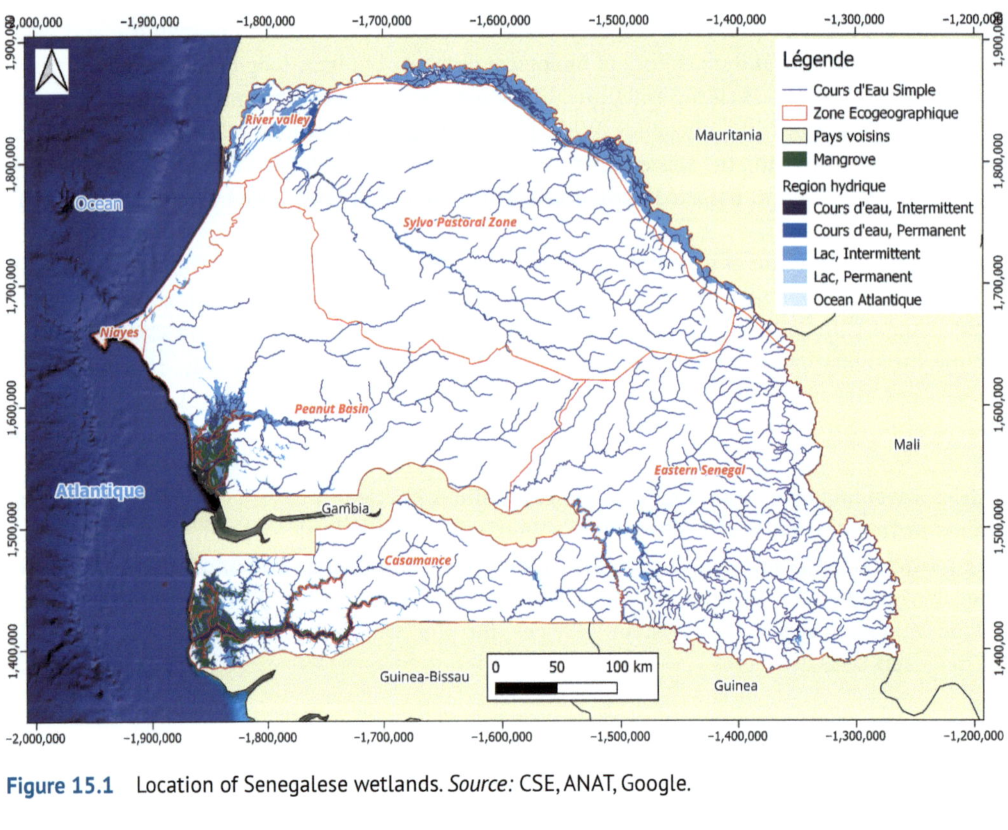

Figure 15.1 Location of Senegalese wetlands. *Source:* CSE, ANAT, Google.

Figure 15.2 Wetland examples in each Senegalese eco-geographical zone: (a) Colony of pelicans at PNOD, (b) aerial view of a pond at Ferlo, a place where livestock converge, (c) dry Niayes stream with a stand of palm trees, (d) Saloum delta, (e) landscape characteristic of Casamance (rice field and palm trees in the background), and (f) Dindéfélo waterfall in eastern Senegal.

15.2.1 Wetlands of the Senegal River Valley

15.2.1.1 Location and Type Diversity

The Senegal River Valley is home to 20 types of wetlands, grouped into 3 categories (Peeters 1998) according to the classification system defined by the Ramsar Convention (*Annexe 2B, recommendation C.4.7*) : (i) coastal wetlands comprising beaches and sandy islets, permanent estuarine waters, intertidal marshes, mudflats, intertidal woodlands, and brackish lagoons; (ii) inland wetlands including seasonal streams, temporary pools, seasonally flooded forests, permanent rivers and streams, settling basins, and natural depressions; and (iii) artificial wetlands such as fish ponds, irrigated land, irrigation canals, and water storage basins.

River valley wetlands showed an overall regression of 9% between 2000 and 2020. This is mainly due to a decrease in the area of the cropland irrigated or postflooding (−7.0%), tree cover-flooded saline water (−6%), and shrub or herbaceous cover-flooded water (−12.0%) classes, whereas an increase was observed for the tree cover-flooded fresh or brackish water (+4.9%) and water bodies (+2.4%) classes (Figure 15.3).

15.2.1.2 Diversity of Plant and Animal Species

The Senegal delta is one of the most important areas in Africa for migratory waterbirds, particularly those from the Western Palearctic. It is also an important refuge for African birds. The Djoudj National Bird Park (PNOD) welcomes around 300,000 waterbirds in January, with a minimum of 133,000 and a maximum of 568,000 individuals (Triplet et al. 2014), for a total of a hundred species between 1989 and 2017. Over the past 40 years, 365 species of waterbirds have been seen in the site. The latest and 35th consecutive count of Delta birds, carried out in 2023, once again confirms the prepoderal role of the PNOD, which hosts the bulk of Senegal's waterbirds. Figure 15.3 presents the waterbird species recently counted with more than 1000 individuals in the PNOD. Only two species *Dendrocygna viduata*, and *Anas acuta* exceed 100,000 individuals, while for *Anas querquedula* (Sarcelle d'été) a decline in numbers has been noted over the past 15 years, with the lowest number (13,648) observed in the year 2023 (Figure 15.3) (SWR 2023). The PNOD is a living but fragile sanctuary for around 1.5 million birds of 365 species, including over 120 species of Palaearctic migrants (UNESCO 2023). The PNOD is also a nesting sanctuary for many birds, including nearly 5000 pairs of *Pelecanus onocrotalus, Ardea purpurea, Platalea alba, Nycticorax nycticorax, Casmerodius albus*, and *Phalacrocorax carbo* (Treca 1989; Triplet et al. 2018). There is also an undetermined number of other birds, including *Riparia riparia, Pelecanus rufescens, Mycteria ibis*, and *Platalea alba* (Triplet et al. 2018). The Ndiaël Wildlife Reserve is also home to a large number of birds, in particular, *Pelicanus* spp., *Scopus* sp., *Egretta* spp., *Microcarbo africanus*, etc. (Diouf et al. 2016).

The river delta is rich in 111 fish species while ichthyofauna of the PNOD is rich in 92 species belonging to 26 families, covering an aquatic surface area of 380 km^2. The most abundant families in the PNOD are, in descending order, the *Cichlidae, Characidae, Bagridae, Cyprinidae, Clariidae, Mochokidae*, and *Cyprinodontidae* (DPN 2017).

The warthog (*Phacochoerus africanus*) is the most frequently observed mammal in the PNOD. However, the red-fronted gazelle (*Eudorcas rufifrons*) and the dorcas (*Gazella dorcas*) are only seen intermittently. Red monkeys or patas (*Erythrocebus patas*) and the golden jackal (*Canis aureus*) are also very abundant. The African wolf (*Canis lupus lupaster*), the Caracal (*Caracal caracal*), the genet (*Genetta genetta*), the civet (*Civettictis civetta*), the ichneumon mongoose (*Herpestes ichneumon*), and the porcupine (*Hystrix cristata*) are present without specification on their abundance (DPN 2017). Ba et al. (2000) identified nine species of rodents belonging to three subfamilies

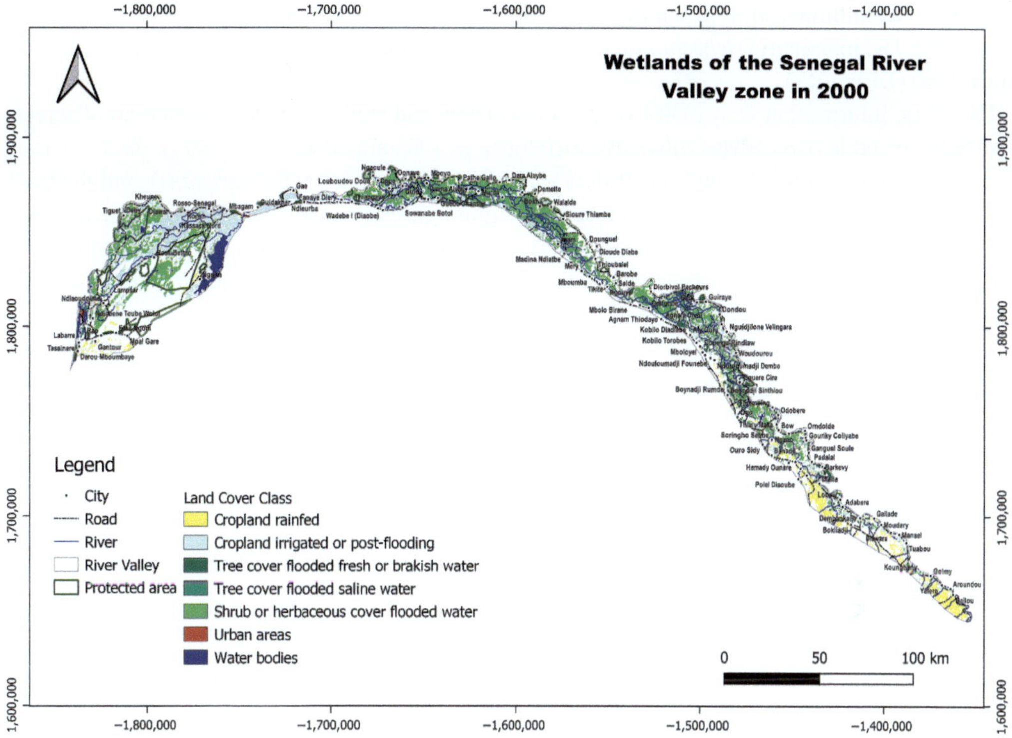

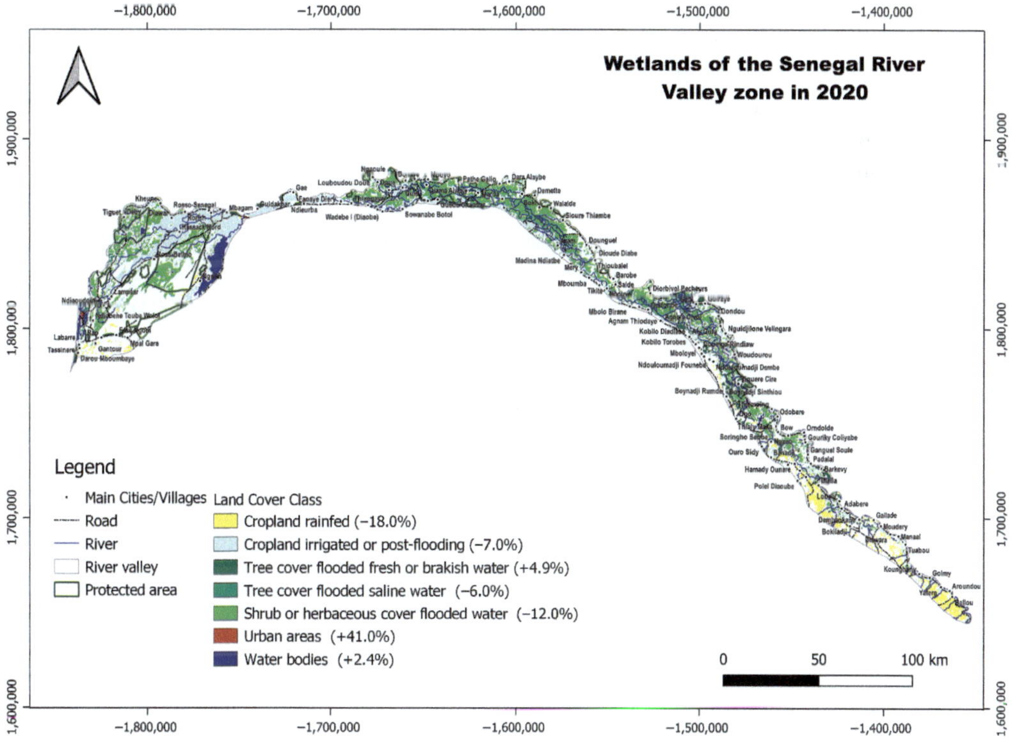

Figure 15.3 Wetlands of the Senegal River Delta and Valley in 2000 and 2020 (numbers in brackets correspond to the percentage change (positive or negetive) in class area between the two years). *Source:* CSE, ESA CCI and EC C3S Land cover (C3S & CDS, 2019; ESA, 2017).

(*Murinae*, Gerbillinae, and *Dipodinae*) and four species of shrews belonging to the genera *Crocidura*. The manatee (*Trichechus africans*) is present, but none has been observed in the PNOD since 1990 (DPN 2017).

Very little information is available on reptiles (turtles and snakes), most of which are discreet. The Nile crocodile *(Crocodylus niloticus)* introduced in 1980 and 1982 (DPN 2017), the Nil Varan (*Varanus niloticus*), the seba python (*Python sebae*), the striking viper (*Bitis arietans*), and the land tortoise are all present in the PNOD, but no population figures are available. However, the number of *C. niloticus* is on the rise, as *V. niloticus* and *P. sebae* are frequently observed in water bodies. The land tortoise is commonly observed in the rainy season or when temporary pools appear (DPN 2017).

The flora of the PNOD, rich in 127 species, distributed in 97 genera themselves distributed in 46 families, is characterized by the diversity of its ecosystems (aquatic, semiaquatic, and terrestrial). The best-represented families are Poaceae, Cyperaceae, Fabaceae, Amaranthaceae, Asclepiadaceae, Convolvulaceae, and Chenopodiaceae, which account for over 50% of the total (DPN 2017). This herbaceous-dominated flora is distributed according to soil salinity and flooding conditions (Triplet et al. 2018).

15.2.2 Sine Saloum and Petite Côte Wetlands

15.2.2.1 Location and Type Diversity

The Sine Saloum basin is part of the large Senegal-Mauritania sedimentary basin and covers an area of 29,720 km^2, of which 2800 km^2 is permanently submerged by tides (Dacosta and Gomez 1998). This area of the Sine Saloum and Petite Côte is characterized by the presence of three types of wetlands: (i) coastal wetlands such as sandy islets and permanent estuary waters in the Saloum delta, and permanent rivers and streams such as the Sine and Saloum; (ii) continental wetlands consisting of numerous floodplains and freshwater swamps and several temporary ponds; and (iii) artificial wetlands represented by water storage areas (retention basins and water retention dykes). The delta owes its name to the Saloum River, which flows into the Atlantic through a mouth that receives 3 main branches: the Saloum to the north (110 km), the Diombos (30 km), and the Bandiala to the south (18 km), which interlock through small channels filled with brackish water, known locally as "bolons," creating some 20 islands bordered by dense mangrove (Camara et al. 2017) and the Saloum Delta Biosphere Reserve (RBDS) (Figure 15.4).

The RBDS is a vast complex wetland made up of three main areas (Thomas et al. 2003): a maritime complex covering the entire western fringe, from Djifère to Djinack; an amphibious ensemble comprising the islands, separated by interconnected inlets, and a continental ensemble, commonly known as the edge region, characterized by a few mangrove and tanne formations (a vast expanse of salty land); and an upper part made up of the low plateaus of the Continental Terminal.

This area of the Groundnut Basin showed an overall decrease of 5.7% in wetland area between 2000 and 2020. This is mainly due to a decline in the area of the cropland irrigated or postflooding (−15.8%), shrub or herbaceous cover-flooded water (−11.7%), and water bodies (−12.3%) classes. However, an increase in area was noted for the tree cover-flooded saline water class (+9.9%), whereas the tree cover-flooded fresh or brackish water class was absent from the area for both years of the study (Figure 15.5).

15.2.2.2 Diversity of Plant and Animal Species

The Mangrove forests, mudflats, sandbanks, and sandy islets form a complex of ecosystems, which enable the delta to support large populations of waterbirds, including numerous palearctic migrants (Peeters 1998). So it is no surprise that the Saloum Delta area, with 100,000–120,000 individuals for

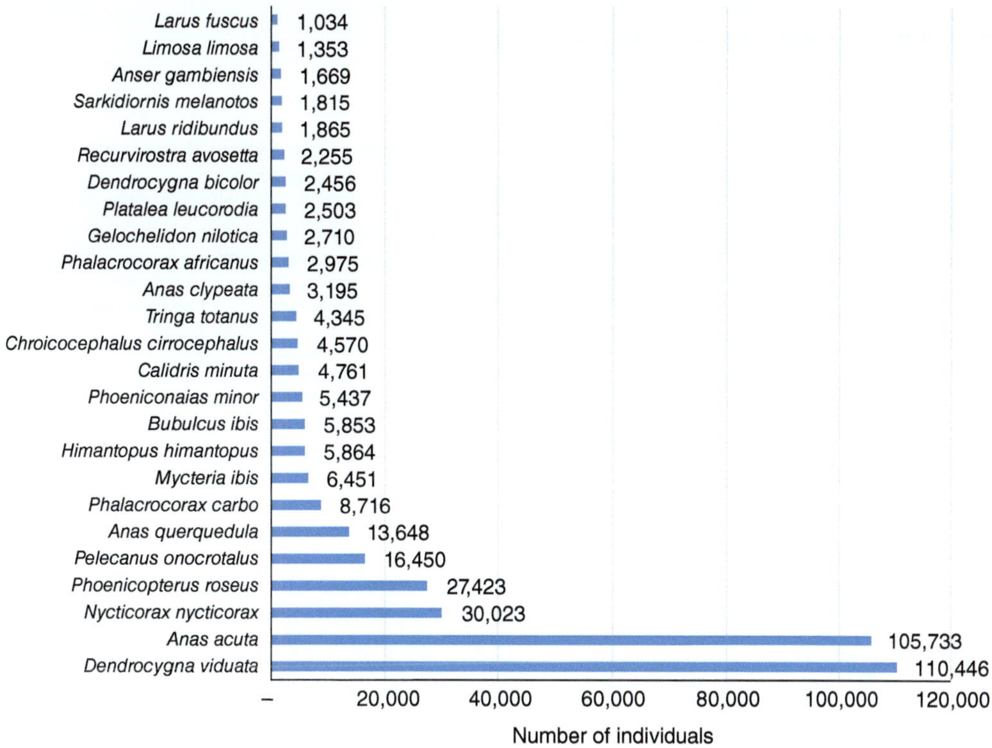

Figure 15.4 Waterbird species with more than 1000 individuals recorded in the Senegal River Delta in 2023 (DPN 2023).

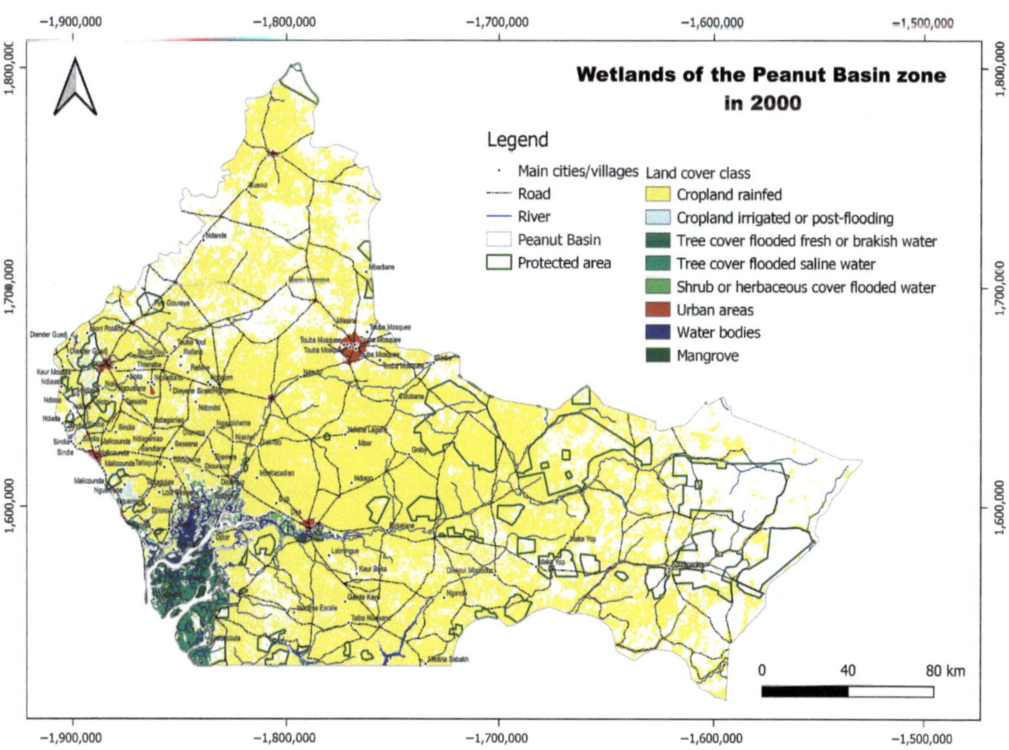

Figure 15.5 (Continued)

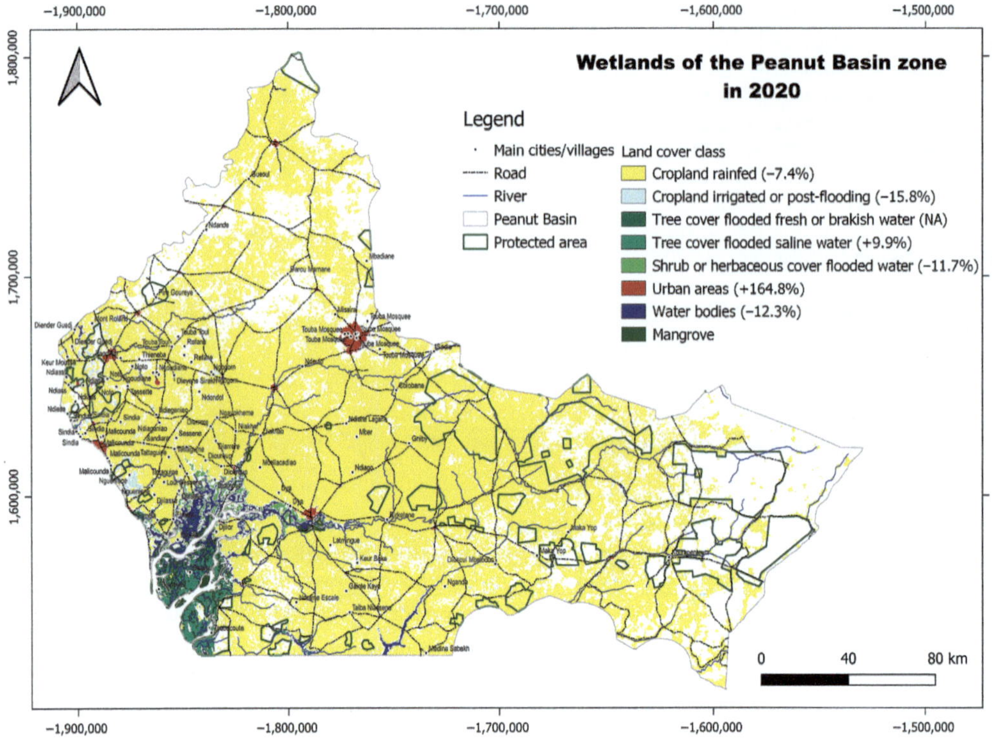

Figure 15.5 Wetlands in the Groundnut Basin in 2000 and 2020 (numbers in brackets correspond to the percentage change in class area between the two years). *Source:* CSE, ESA CCI and EC C3S Land cover (C3S & CDS, 2019; ESA, 2017).

80–95 bird species, is ranked as the third most important ornithological site in West Africa, behind the Banc d'Arguin and the PNOD. The Delta's avifauna consists mainly of limicolous birds, including sandpipers (*Calidris* sp.), the knights (*Tringa* sp.), the collared turnstone (*Arenaria interpes*), the white stilt (*Himantopus himantopus*), the Spotted thick-knee (*Burhinus* sp.), the Common Pied Oystercatcher (*Haematopus ostralegus*), the Plovers (*Charadriinae*), little ringed plover (*Charadrius* sp.) and the lapwing (*Vanellus* sp.) (Ngo Samnick 2021). The Delta is also a nesting ground for many birds such as *Phœniconaias minor*, *Pelecanus rufescens*, *Chroicocephalus genei*, *Chroicocephalus cirrocephalus*, *Thalasseus maximus*, *Hydroprogne caspia*, *Egretta dimorpha*, and *Limosa limosa*. It is also home to other types of birds such as the *Dendroperdix sephaen*, *Numida meleagris*, *Streptopelia* spp., and even nocturnal and diurnal raptors (Ngo Samnick 2021).

The latest inventory of waterbirds on the Petite Côte reveals a predominance of *Dendrocygna viduata*, followed by *Anas acuta* and *Nycticorax nycticorax* in the same proportions, after *Phœnicopterus roseus* (Figure 15.6).

Waterbird species with more than 1000 individuals recorded in the Petite côte in 2023 (Diouf et al. 2021) including *Tragelaphus spekei*, *Crocuta crocuta*, *Phacochaerus porcus*, *Redunca arundinum*, *Chlorocebus sabaeus*, *Erythrocebus patas*, and *Piliocolobus badius* (Sène et al. 2006). The latter is locally threatened (Diouf et al. 2021). In the opinion of Ngo Samnick (2021), *Tragelaphus scriptus* and *Cephalophus* sp. are rare in this area. The waters of the Delta are home to four species of marine mammals: *Sousa teuszii*, *Delphinus delphus*, *Delphinus capensis*, and *Trichechus senegalensis* (Ngo Samnick 2021).

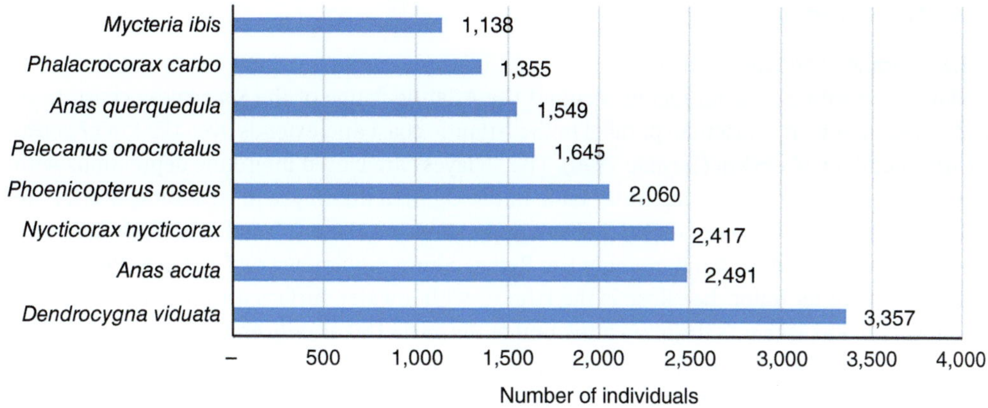

Figure 15.6 Waterbird species with more than 1000 individuals recorded in the Petite côte in 2023.

The Delta estuary is a spawning and feeding area for ichthyofauna, mainly marine due to its hypersalinity (Baran 2000). It is rich in fish, with 114 species from 42 families. A comparison with 60 estuarine and lagoon environments around the world places the Saloum Delta in sixth place. For an average biomass of 27 kg/ha, it ranks second in comparison with five estuaries and lagoons sampled by Diouf (1996). It seems that a species of carp (*Lisa bandialensis*) which is endemic to the area is threatened with extinction. At least 50 species reproduce there or begin their sexual maturation. This area is also a habitat for several crustaceans and molluscs (e.g. shrimp, oysters, and many shellfish). The main molluscs species collected are *Murex* sp., *Anadara senilis, Crassostrea gasar,* and *Cymbium sp.*

The herpetofauna of the Delta is little studied, and the rare documents have focused on turtles. Six species of marine turtle frequent the Delta, four of which reproduce there: the olive ridley turtle (*Lepidochelys olivacea*) vulnerable, the green turtle (*Chelonia mydas*), the caret turtle (*Caretta caretta*), and the leatherback turtle (*Dermochelys coriacea*) threatened with extinction. Two others, the hawksbill turtle (*Eretmochelys 'imbricata*) and Kemp's ridley (*Lepidochelys kempii*), threatened with extinction, have also been reported (Diouf et al. 2021). Snakes such as the garter snake (*Bitis arietans*), the Nile varan (*Varanus niloticus*), and the sand varan (*Varanus panoptes*), lizards, and the crocodile (*Crocodylus* sp.) are present in the Saloum Delta.

There are two main types of plant formation in the Delta: submersible and unsubmersible zones. The mangrove occupies the submersible zones and, like all mangroves, that of the RBDS is not very diversified, with five plant species zoned according to their tolerance of salt and the soil's low oxygen content. Starting from the inside to the outside (the "*tannes*" zone), the observed species are as follows *Rhizophora mangle*, *Rhizophora racemosa*, and *Avicennia germinans* are predominant, followed by sparse formations of *A. germinans*, *Conocarpus erectus*, and *Laguncularia racemosa*. The only mangrove fern (*Acrostichum aureum*) can be found here, surrounded by *Avicennia germinans, Phragmites* sp., and *Typha* sp. at Koular and in the Patako forest (Mingou 2023). In the nonsubmersible zones, we have gallery forests, savannahs (wooded), and open forests. The woody flora of the Saloum Delta Reserve is made up of 188 species (Dia 2003) representing around 5–6% of the national flora, in 50 families. The most species-rich sites are the Fathala (125 species), Sangako (81 species), Keur Sambel (58 species), Falia (54 species), and Sipo (51 species) forests (Dia 2003). Some species, such as *Khaya senegalensis*, *Prosopis africana*, *Detarium senegalense*, and *Neocarya macrophylla*, are abundant there and widely exploited for various purposes.

15.2.3 Niayes Wetlands

15.2.3.1 Location and Type Diversity

The Niayes or northern coastal region occupies the Atlantic fringe of the Senegalese coast, from Dakar to Saint Louis. It is oriented parallel to the Atlantic coast and extends over 180 km of coastline with a depth of 30–35 km (Dryade 1990). The "Niayes" are closed interdune depressions with an outcropping or sub-outcropping water table in the Ogolian dune system (red dunes). They are characterized by a succession of dunes and wet depressions on the one hand, and freshwater and saltwater lakes on the other hand, notably Lac Retba, commonly known as Lac Rose (Pink Lake) due to the color of its water. Lac Rose is the largest, with a water surface area of 5 km by 1.7 km wide at low tide. The lake was the main attraction of the Paris-Dakar rally and is also a popular site for tourists visiting Dakar. The Niayes wetlands are of three types: (i) coastal wetlands with sandy beaches occupied by the filaos strip, which plays an important role in protection against wind erosion, forestry production, and tourism; (ii) inland wetlands with seasonal streams and ponds, brackish lakes, permanent rivers and streams, seasonally flooded forests and valleys; and (iii) artificial wetlands such as the water storage areas in the flooded suburbs of Dakar (MEDD 2015).

This ecogeographical zone is characterized by a sharp decline in wetland ecosystems between 2000 and 2020, with a variation of −19.9% in the surface area of the classes concerned. This decrease in wetlands is mainly due to the reduction in the cropland irrigated or postflooding (−19.5%), tree cover-flooded saline water (−100%), shrub or herbaceous cover-flooded water (−20.7%), and water bodies (−18.9%) classes, bearing in mind that the tree cover-flooded fresh or brackish water class does not exist in this zone (Figure 15.7).

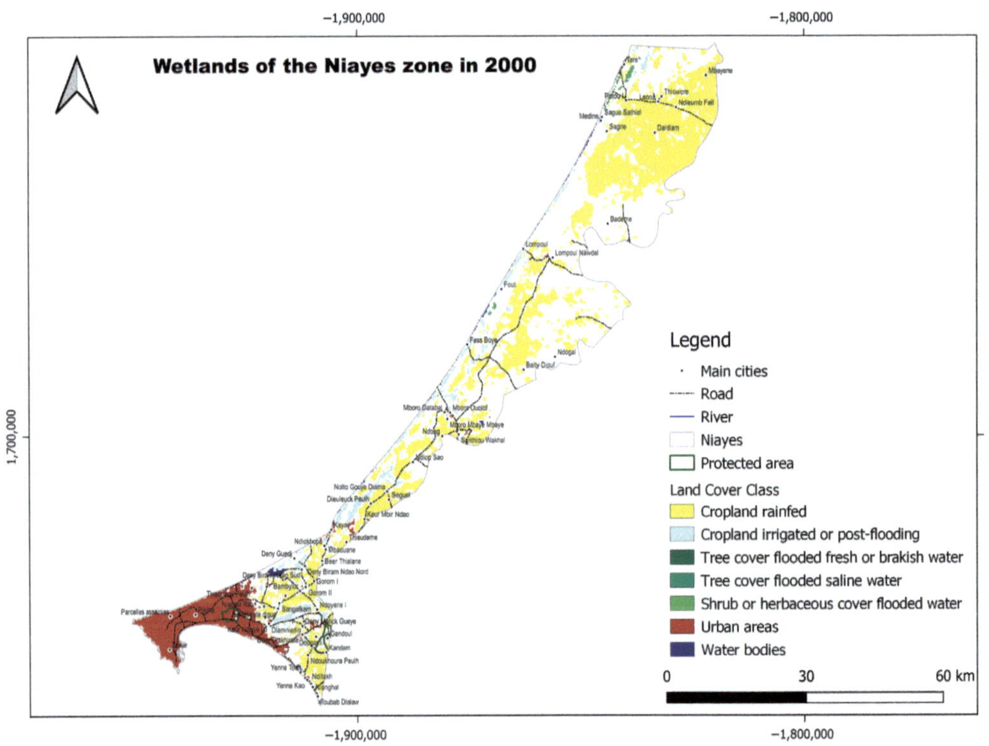

Figure 15.7 Niayes wetlands in 2000 and 2020. Numbers between brackets correspond to the percentage change in class area between the two years. *Source:* CSE, ESA CCI and EC C3S Land cover (C3S & CDS, 2019; ESA, 2017).

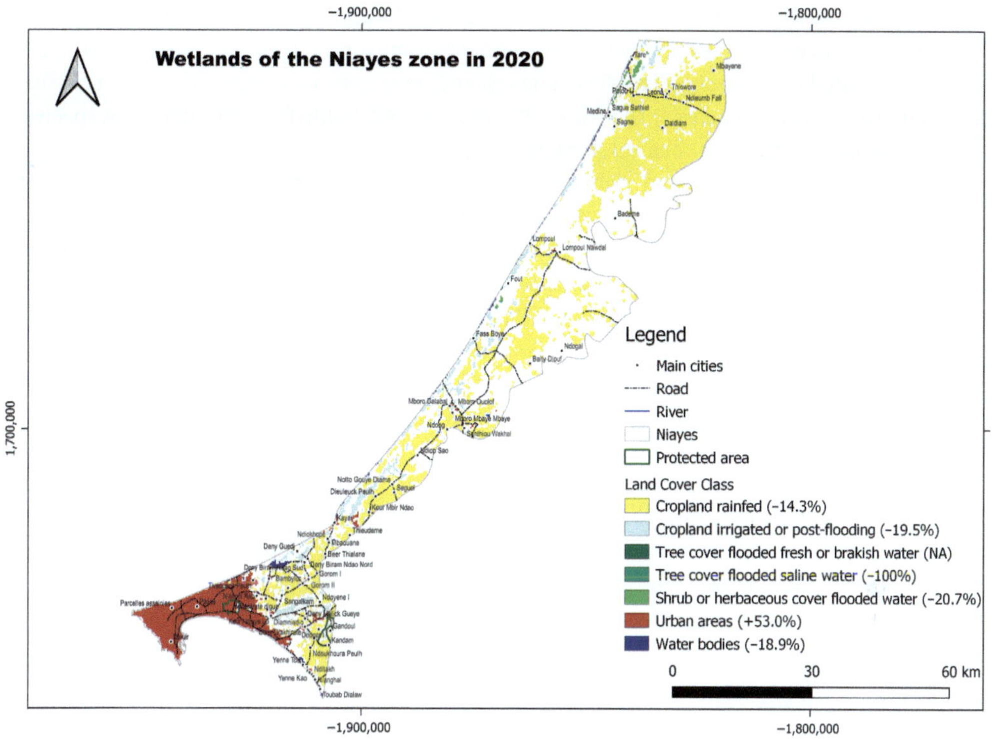

Figure 15.7 (Continued)

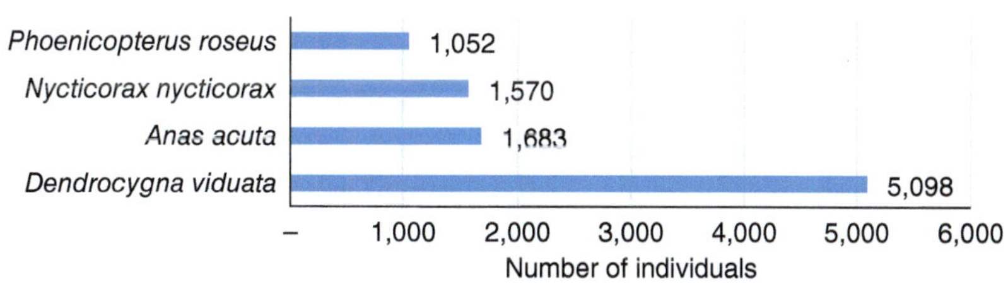

Figure 15.8 Waterbird species with more than 1000 individuals recorded in the Niayes in 2023.

15.2.3.2 Diversity of Plant and Animal Species

In terms of animal diversity, the Niayes are rich in ornithological species, largely dominated by *Dendrocygna viduata* followed by *Anas acuta, Nycticorax nycticorax*, and *Phœnicopterus roseus* (Figure 15.8). IUCN (2002) states that 133 bird species have been counted in the Grande Niaye de Pikine, including 40 endemic, 25 migratory, and 51 nesting. Between October 1996 and March 2001, 182,000 birds distributed in 142 species were observed in the Niaye (Reynaud 2001). Among these species, 25 are migratory and 11 were only reported in less than 5% of observations. The presence of *Necrosyrtes monachus*, *Numida meleagris*, and *Columba palumbus* is also noted, even if they are rare.

With regard to the fauna of the Niayes, the diagnostic report on biodiversity in Senegal, drawn up in 2013 by the DPN, mentions the existence of species such as *Accipiter badius*, *Becorvus lead beater*, *Canus anthus*, *Genetta maculata*, *Hystrix cristata*, *Lepus timidus*, *Rattus rattus*, *Tragelaphus scriptus*, and *Xerus rutilus*. Others, such as *Varanus griseus*, *Varanus niloticus*, *Grayia tholloni*,

Dasypeltis gansi, and *Geochelone sulcata*, are also present. Animal species including *Crocodylus niloticus* and *Crocuta crocuta* and many fish species have disappeared from the Niayes. In the past, the area was dotted with crocodile-infested lakes and marigots teeming with fish. The fish population in the ponds and lakes of the Niayes has plummeted. Some formerly abundant species, including those of the genus *Protopterus*, have become very rare.

The Niayes had a complexity and richness of flora found nowhere else in Senegal north of latitude 13°. The Niayes includes some 419 plant species (MEPN 1997) representing 12% of Senegal's flora. The vegetation appears as a string of oases surrounded by dunes, of sub-Guinean type, whose characteristic species remains *Elaeis guineensis*. Sudanian species such as *Prosopis africana, Ficus sur, Detarium senegalense*, and *Dialium guineense* are always found there, while others such as *Vitex doniana* and *Afzelia africana* are rare. The herbaceous layer comprises hydrophytes such as *Typha domingensis, Typha elephantina, Phragmites australis, Sporobolus spicatus*, and *Nymphaea lotus*. Filaos plantations (around 10,000 ha) constitute the artificial vegetation along the entire ocean coast. Today's Niayes vegetation is far from the sub-Guinean type described by Trochain (1940). The Niayes zone had a pteridoflora with 18 species among which *Adiantum philippense, Ampelopteris prolifera, Ceratopteris thalictroides, Christella dentata, Cyclosorus striatus, Isoëtes melanotheca*, and *Lygodium microphyllum* (Mingou 2023). Unfortunately, 17 of the 18 species are now extinct, as only the *Marsilea minuta* var. *minuta* was recently recorded there (Mingou 2023).

15.2.4 Wetlands of the Ferlo Sylvopastoral Zone

15.2.4.1 Location and Type Diversity

Located between latitudes 16°15 and 14°30 North and longitudes 12°50 and 16° West (Ndiaye 2013), Ferlo is located in the Sahelian zone. It extends over more than 60,000 km^2 from the Senegal River valley in the north and east to the Niayes in the west and the Groundnut Basin and eastern Senegal in the south (Wane et al. 2006; Sy 2009). Although located south of the Sahara in an area with an average annual rainfall of around 300 mm/year, the Ferlo is home to a wide variety of wetlands. In addition to the vast alluvial plain of the Senegal River (Walo), there is the large Lac de Guiers (reservoir of 400–600 million m^3 of fresh water, depending on its elevation) located in the upper delta of the river and fed both by the discharge of part of the Senegal River and by rainy-season runoff from the Ferlo hydrographic network (Faye 2021). The Silvopastoral Zone contains only continental wetlands, the most important of which is the southern part of Lac de Guiers. There are also numerous pools in the beds of the ancient Sine and Bas-Ferlo valleys (Figure 15.9) as well as artificial wetlands such as former quarries, which play an important role in collecting runoff water. Ponds are the main wetlands in the Ferlo and the only watering point for these populations and their livestock during the rainy season and most of the dry season (Diop et al. 2004). In this way, dwellings are often organized around ponds, whose names reflect their diversity and value. Each pond is given a name to distinguish and characterize it according to its importance, the vegetation that surrounds it and the animals that frequent it (Diop et al. 2004). Unfortunately, with rainfall deficits, some ponds dry up very early or do not even have water.

The area of wetland ecosystems in the sylvopastoral zone declined by 12.8% over the analysis period from 2000 to 2020. This reduction is linked to the regression of the cropland irrigated or postflooding (−10.7%), shrub or herbaceous cover-flooded water (−21.3%), and water bodies (−13.5%) classes. However, the tree cover-flooded fresh or brackish water (+8.3%) and tree cover-flooded saline water (+31.6%) classes were extended, probably due to the revitalization of fossil valleys (Figure 15.9).

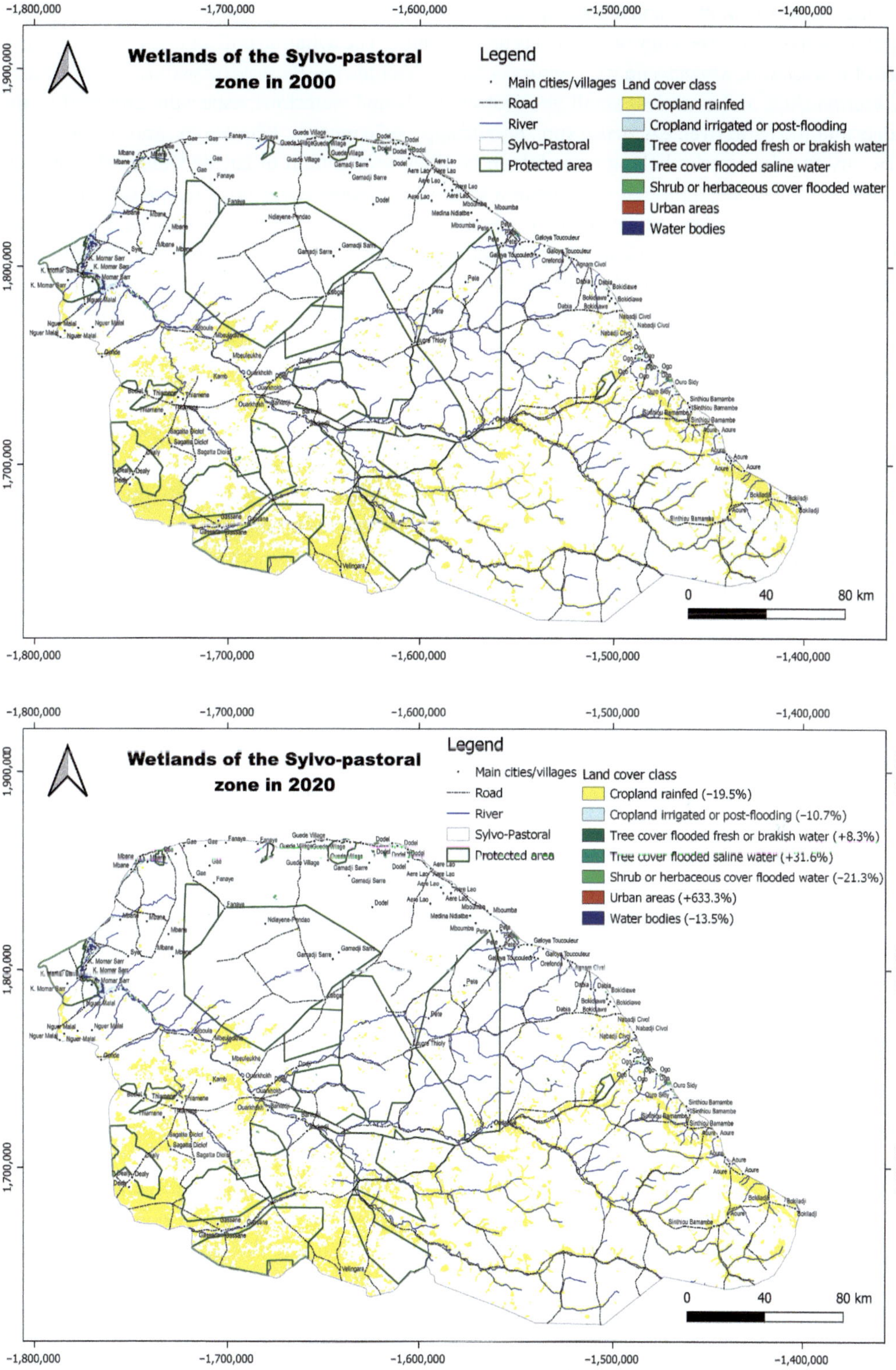

Figure 15.9 Wetlands in the Ferlo sylvopastoral zone in 2000 and 2020 (numbers in brackets correspond to the percentage change in class area between the two years). *Source:* CSE, ESA CCI and EC C3S Land cover (C3S & CDS, 2019; ESA, 2017).

15.2.4.2 Diversity of Plant and Animal Species

Wildlife adapted to the ecology of the area is present there. The avifauna is diversified with the presence of *Pterocles* sp., *Streptopelia* sp., *Dendroperdix* sp., *Numida* sp., *Bucorvus abyssinicus*, and *Tockus erythrorhynchus*, and countless small granivorous birds and waterfowl, especially around Lac de Guiers and the Ferlo river, e.g. *Ardea* spp., *Egretta* spp., *Pelicanus* sp., *Platalea* sp., and *Anser* sp.

Twenty-five fish species belonging to fifteen families were caught in Lac de Guiers. Freshwater species still dominate, although some genera such as the *Heterobranchus*, *Protopterus*, and *Lates* are less and less abundant. Since the dams were built, brackish water species have been unable to migrate to Lake Guiers. *Chrysichthys nigrodigitatus* and *Tilapia dageti* have not been caught, but their presence in the lake should not be surprising (Diouf et al. 2016).

Mammals are represented by different groups: insectivores (hedgehogs), chiropterans (bats), lagomorphs (hares), sciuromorphs (squirrels), carnivores such as ratels (*Millivora*), wildcats (*Felis libyca*), common jackals (*Canis aureus*), and myomorphs or rodents (Diallo 1995). This area is home to the red-fronted gazelle (*Eudorcas rufifrons*), ruddy-necked ostriches (*Struthio camelus*), and great bustards, many of which have disappeared due to poaching (Sow and Akpo 2011). This area is characterized by the existence of amodiated zones to the west of Lake Guiers and sylvopastoral reserves (wildlife reserves). Many emblematic species (*Gazella dorcas, Oryx dammah, Hippotragus* sp., *Panthnera leo, Panthera pardus*, and *Syncerus* sp.) would have lived here until recently. Today, the most visible species are *Canis* sp., *Erythrocebus patas, Vulpes zerda*, and several other small carnivores.

Reptiles include the land tortoise and the Nile and desert varan (*Varanus niloticus* and *Varanus axanthematicus*).

At the end of the rainy season, at its peak, Ferlo vegetation takes the form of a more or less continuous herbaceous carpet, up to 50 cm–1 m in height, composed mainly of annual species. This carpet is interspersed with mainly thorny trees and shrubs, never forming a continuous stratum (Akpo 1992). The woody flora of the Ferlo is estimated at around 50 species, dominated by Fabaceae, Combretaceae, Apocynaceae, Asclepiadaceae, Rubiaceae, and Capparaceae. In the Ferlo, however, there is combretinization and changes in vegetation. Indeed, *Combretum glutinosum* and *Guiera senegalensis* are very frequent and occupy more and more space. Alongside these ligneous species are *Balanites aegyptiaca, Acacia senegal, Boscia senegalensis, Pterocarpus lucens, Ziziphus mauritiana*, and *Acacia nilotica* var. *raddiana* (Ngom et al. 2013). *Mitragyna inermis* is the pond indicator plant par excellence, accompanied by *Anogeissus leiocarpa, Acacia nilotica* subsp. *adstringens, Acacia nilotica* subsp. *tomentosa*, and *Ziziphus mauritiana* (Ba et al. 2017). Occasionally, *Balanites aegyptiaca* grows more or less densely around the edges of ponds.

15.2.5 Wetlands of Eastern Senegal

15.2.5.1 Location and Type Diversity

The zone is characterized by continental wetlands such as rivers and permanent streams, including the Dindéfélo, Ségou, and Fongolimbi waterfalls. The area is also crisscrossed by two permanent watercourses, the Falémé and the Gambia rivers, and numerous seasonal watercourses and temporary pools (e.g. Niokolo, Touroukoto, and Niériko). It is home to several flood-prone areas and seasonally flooded forests (Ndiambour classified forest, Sédou and Niokolo Koba National Park).

A decrease of around 8.4% in wetland areas was observed in the Senegal eastern between 2000 and 2020. The decrease mainly concerns the cropland irrigated or postflooding (−13.0%), shrub or herbaceous cover-flooded water (−29.3%), and water bodies (−26.1%) classes. As in the Ferlo zone, the tree cover-flooded fresh or brackish water (+110%) and tree cover-flooded saline water (+300%) classes have expanded significantly, probably due to the same revitalization of fossil valleys and gallery forests that are characteristic of this part of Senegal (Figure 15.10).

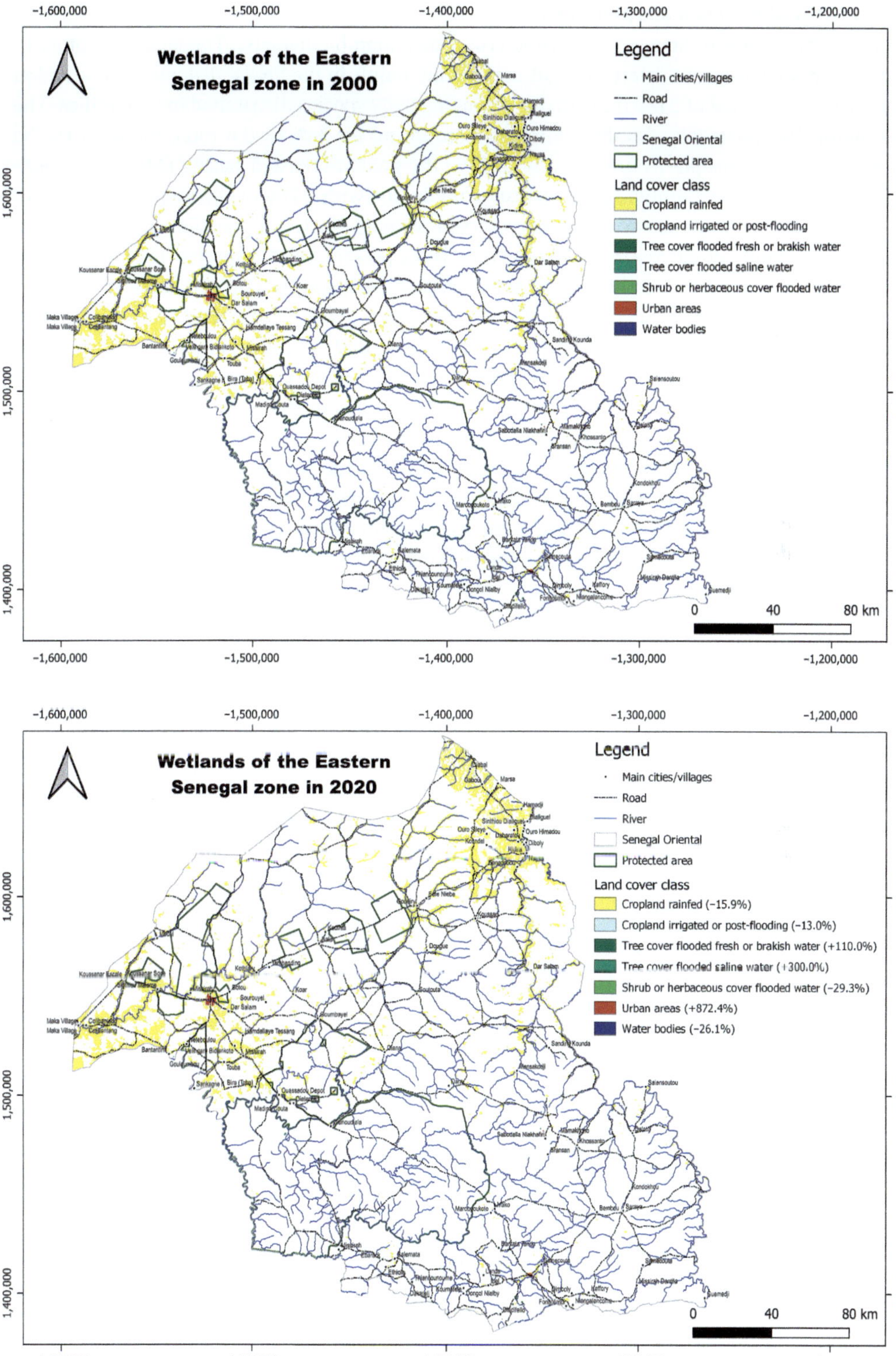

Figure 15.10 Wetlands of Eastern Senegal in 2000 and 2020 (numbers between brackets correspond to the percentage change in class area between the two years). *Source:* CSE, ESA CCI and EC C3S Land cover (C3S & CDS, 2019; ESA, 2017).

15.2.5.2 Diversity of Plant and Animal Species

The wetlands of Eastern Senegal are characterized by a high biodiversity. The latest inventory of the waterbirds avifauna of Eastern Senegal, carried out mainly in the PNKK, Falémé, and Boundou reserves, yielded a total of 2180 individuals, belonging to 77 species distributed in 22 families. The cattle heron (*Bubulcus ibis*) family dominates, accounting for 36% of the total, followed by the spurred lapwing (*Hoplopterus armlatus*) with 17%, the golden-breasted Jacana (*Metopidius indicus*) with 13%, and the Gambian goose (*Anser gambiensis*) with 137% of the total (Figure 15.10) (DPN 2023). The avifauna of the Dindéfélo reserve is unique due to the specific nature of its ecosystem. An inventory of the reserve's avifauna reveals the presence of 220 species, 16 of which have never before been recorded in this part of south-eastern Senegal (Fernández-García et al. 2013). Prior to this inventory, *Important Bird and Biodiversity Areas (IBAs)* recognized 17 sites (e.g. PNOD, Delta du Saloum, PNKK, and Basse Casamance) today, Dindéfélo has joined this select group thanks to the presence of mountain birds and rocky areas, which are very rare in the rest of Senegal. The Dindéfélo reserve is currently the only place where you can find *Falco alopex*, *Apaloderma narina*, *Ptyonoprogne fuligula*, *Thamnolaea cinnamomeiventris*, *Onychognathus neumanni*, and *Euschistospiza dybowskii* (Fernández-García et al. 2013). Six species deserve special conservation attention: one is classified as endangered (*Necrosyrtes monachus*), two are vulnerable (*Torgos tracheliotus* and *Circaetus beaudouini*), and three as near-threatened (*Gyps africanus*, *Terathopius ecaudatus*, and *Polemaetus bellicosus*) (Fernández-García et al. 2013) (Figure 15.11).

The site's gallery forests are home to a number of birds typical of Guinean forests, which are very rare or nonexistent in Senegal's semiarid zones (*Apalis flavida*, *Cyanomitra verticalis*, *Turtur afer*,

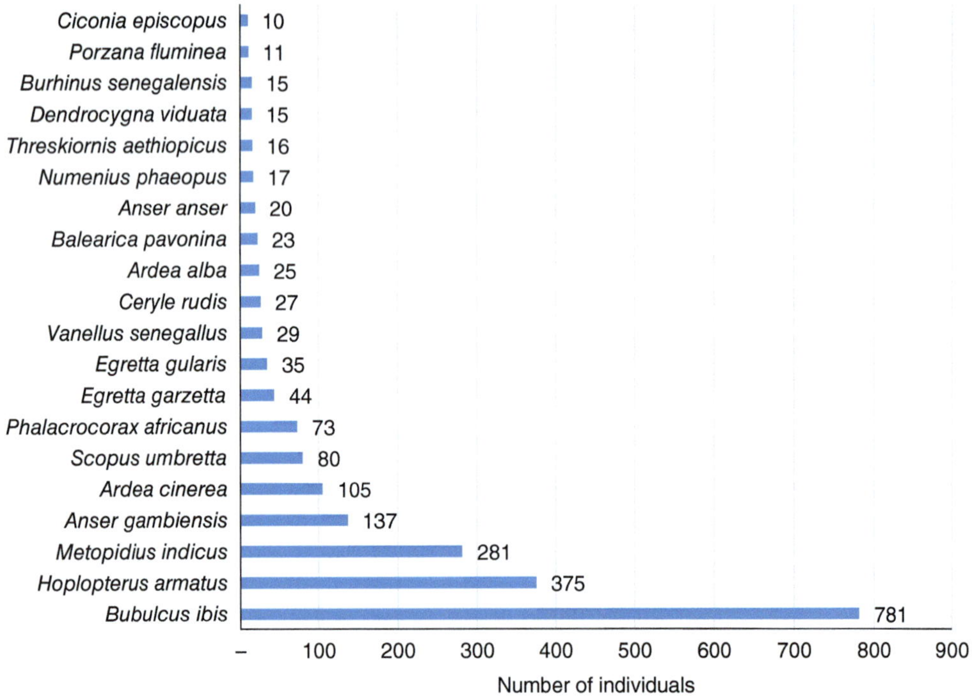

Figure 15.11 Bird species with more than 10 individuals recorded in eastern Senegal in 2023.

Tauraco persa, *Cynniris cupreus*, and *C. venustus*). Dindéfélo is also the habitat of a small bird (*Lagonosticta virata*) or Kulikoro pigweed endemic to western Senegal and eastern Mali (Fernández-García et al. 2013). The specificity of the avifauna of this eastern part is due to its cliffs and gallery forests, which are unique in Senegal.

The PNKK is Senegal's mammal sanctuary, with 80 species recorded. Some of them are endangered, e.g. *Lycaon pictus*, *Pan troglodytes*, *Loxodonta africana*, and *Taurotragus derbianus*. The only Senegalese populations of *Taurotragus derbianus* and *Loxodonta africana* are observed in this region. Chimpanzees (*Pan troglodytes*) live in the gallery forest and on the slopes of Mount Assirik, toward the northwest limit of their distribution (Coulthard 2001). Chimpanzees are also present in the Dindéfélo Reserve and are its emblematic species. *Pan troglodytes* has been classified as critically endangered by the IUCN since 2016.

Ecosystems are characterized by a great diversity of plant formations ranging from dense gallery forests, shrub savannahs, and open forests to wooded savannahs and tree savannahs (Goudiaby 1996; CSE 2015a, 2020) according to geomorphological units (valleys, ponds, river banks, waterfalls, etc.). In eastern Senegal, there are various forest galleries. The Falémé gallery is characterized by discontinuous vegetation dominated mainly by *Anogeisus leiocarpa*, *Diospyros mespilifoprmis*, *Borassus akeasii*, *Mitragyna inermis*, and *Sarcocephalus latifolius* (Gomis 2016). Along the Gambia River, vegetation is continuous, sometimes interrupted in shallow areas. It is characterized by species such as *Elaeis guineensis*, *Borassus akeasii*, *Alchornea cordifolia*, *Pandanus senegalensis*, *Pterocarpus santaloides*, *Cola laurifolia*, and *Dialium guieneense* (Traoré 1997). In the galleries and waterfalls, the characteristic species are mainly *Alchornea cordifolia*, *Carapa procera*, *Tetracera alnifolia*, *Pentaclethra macrophylla*, and *Pseudospondia microcarpa* (Goudiaby 1996). It should also be noted the existence of several pure stands of *Oxytenanthera abyssinica* (bamboo), *Raphia* sp. (raffia), and *B. akeasii* (palmyra palm) in marshy meadows. The flora of secondary ponds and watercourses is essentially characterized by *Mitragyna inermis*, *Raphia sudanica*, *Alchornea cordifolia*, and *Sarcocephalus latifolius*. Alongside this woody flora, we note the presence of a more or less neglected flora: ferns (Pteridophytes). Indeed, seven (7) species of Pteridophytes were recently discovered for the first time in Senegal in this eastern part of the country, at the level of waterfalls (Mingou and Gueye 2017; Mingou 2023; Mingou et al. 2021, 2023).

15.2.6 Casamance Wetlands

15.2.6.1 Location and Type Diversity

Casamance area has different types of wetlands: (i) coastal wetlands consisting of sandy beaches and islets, permanent estuary waters, mudflats and intertidal woodlands, permanent rivers and streams, freshwater lakes, and flood-prone areas such as the numerous valleys; (ii) inland wetlands with numerous floodplains and seasonally flooded forests; and (iii) artificial wetlands such as water storage areas like dams and antisalt dykes, fish-farming ponds, and vast tracts of irrigated land and irrigation canals.

Like all the ecogeographical zones of Senegal, the Casamance zone showed a decrease of −8.8% in water-related ecosystems over the period 2000–2020. With the exception of the tree cover-flooded saline water class, which showed a slight increase of 0.6%, all wetland classes showed a regressive trend: Cropland irrigated or postflooding (−15.1%), tree cover-flooded fresh or brackish water (−14.9%), shrub or herbaceous cover-flooded water (−3.9%), and water bodies (−11.3%) (Figure 15.12).

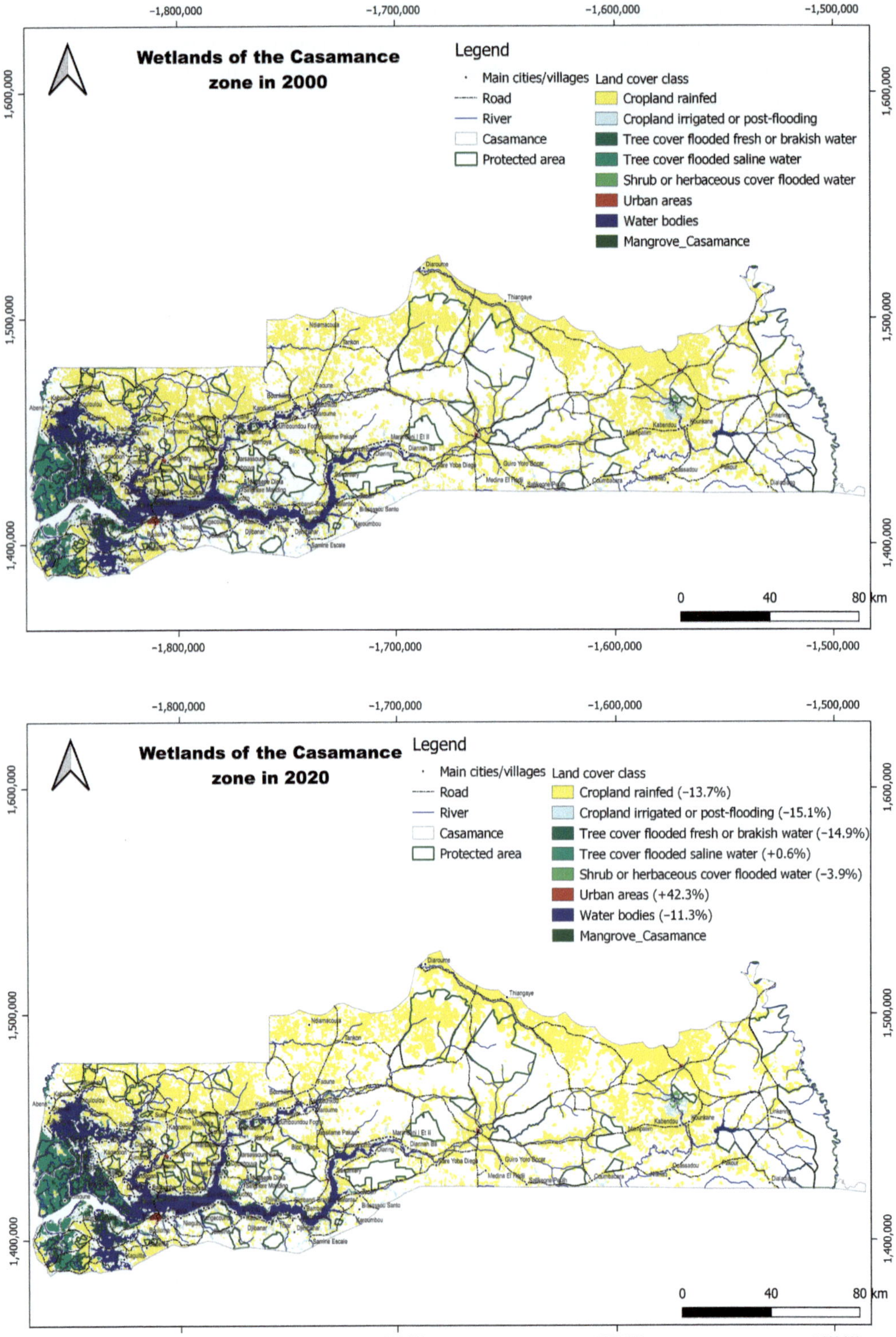

Figure 15.12 Casamance wetlands in 2000 and 2020 (numbers between brackets correspond to the percentage change in class area between the two years). *Source:* CSE, ESA CCI and EC C3S Land cover (C3S & CDS, 2019; ESA, 2017).

15.2.6.2 Diversity of Plant and Animal Species

Casamance's avifauna is rich and diverse with the presence of pelicans (*Pelucanus onocratus* and *Pelucanus rufescens*), kingfishers (*Halcyon senegalensis*), storks (*Iconia cigonia*), flamingos (*Phoenicopterus ruber*), crowned eagles (*Stephanoactus coronutus*), African spoonbills (*Platalea alba*), tiger herons (*Tigriormis leucoopha*) (Sané and Mbaye 2007; Bassène 2016). The latest inventory is largely dominated by *Dendrocygna viduata* followed at a distance by *Anas acuta*, *Nycticorax nycticorax*, *Phoenicoptgerus roseus*, and *Pelecanus onocrotalus* in similar proportions (Figure 15.13). Doves (*Streptopelia*), francolins (*Dendroperdix*), hornbills (*Tokus* spp.), blackbirds (*Turdus* spp., *Lamprotornis* sp.), and crows (*Corvus* spp.) are present. Mangrove birds include marabou (*Leptoptilos* sp.), herons (*Ardea* spp., *Bubulcus* spp.), egrets (*Egretta* spp.), kingfishers (*Alcedo atthis*), and pelicans (*Pelecanus* spp.). The Kalissaye ornithological reserve is home to around 10,000 pairs of Caspian terns (*Hydroprogne caspia*), numerous king terns (*Thalasseus maximus*), a large colony of white pelicans (*Pelecanus onocrotalus*), and a heronry where more than a dozen species breed (Ndiaye 2005; CSE 2020). Endangered birds include the least sandpiper (*Calidris minuta*), red knot (*Calidris canutus*), sanderling (*Calidris alba*), black-tailed godwit (*Limosa limosa*), red-breasted godwit (*Limosa lapponica*), curlew (*Numenius arquata*), whimbrel (*Numenius phaeopus*), and great egret (*Ardea alba*). Kalissaye is also a feeding ground for Atlantic humpback dolphins and the African manatee (Ramsar 2023). The Kalissaye reserve is also a refuge for marine turtles found in West Africa, notably the leatherback turtle (*Dermochelys coriacea*), olive ridley turtle (*Lepidochelys olivacea*), hawksbill turtle (*Eretmochelys imbricata*), and green turtle (*Chelonia mydas*) (Ramsar 2023).

There are few big cats in the region, apart from *Crocuta crocuta* and *Panthera pardus*. On the other hand, numerous antelopes and several species of monkeys are easily encountered in the forests and sometimes on the roads. The fauna includes over 50 mammal species, including the buffalo (*Syncerus caffernanus*), panther (*Panthera pardus*), demidoff's galago (*Galagoides demidoff*) and giant pangolin (*Manis gigantea*), the mangrove monkey (*Cercopithecus campbelle*), reptiles such as the crocodile (*Crocodylus miloticus*), the Seba python (*Python sebae*), the royal python (*Python regius*), and the spitting naja (*Naja nigricollis*) (Sané and Mbaye 2007; CSE 2015a, 2020; Bassène 2016).

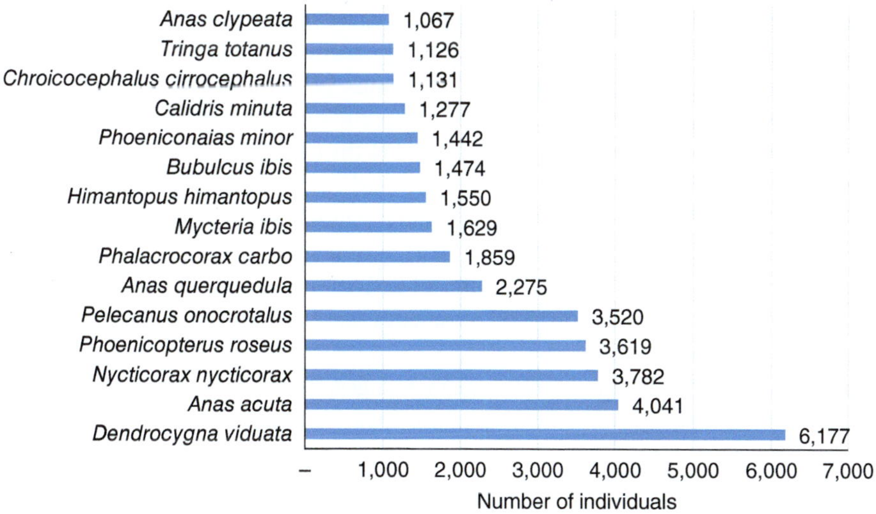

Figure 15.13 Waterbird species with more than 1000 individuals recorded in Casamance in 2023.

The Casamance region has Sudano-Guinean vegetation. It is the last forest reserve in Senegal. In the lower Casamance, the wettest part, the flora is made up of species with Guinean affinities, including *Elaeis guineensis, Parinari excelsa, Dialium guineense, Erythrophleum guineense*, and *Treculia africana*. Dense dry relict forests are dominated by *Erythrophleum suaveolens, Detarium senegalense, Malacantha alnifolia, Parinari excelsa, Treculia africana, Pentaclethra macrophylla*, and *Raphia sudanica* (Sané and Mbaye 2007; CSE 2015a, b, 2020). The middle and upper Casamance remain par excellence the domain of the savannah under different physiognomies with species of Sudanian affinity such as *Bombax costatum, Pterocarpus erinaceus, Khaya senegalensis, Parkia biglobosa*, and *Prosopis africana* (Sané and Mbaye 2007; CSE 2015b). Gallery forests occupy the valleys and are characterized by species such as *Elaeis guineensis, Erythrophleum suaveolens, Ceiba pentandra, Alchornea cordifolia*, and *Sarcocephalus latifolius* (CSE 2015b). The mangrove, like all the others, is made up of a low diversity of species. The six species *Rhizophora racemosa, Rhizophora mangle, Rhizophora harisonnii*, and *Avicennia germinans* are on mudflats, while *Laguncularia racemosa* and *Conocarpus erectus* are set back.

Like the Niayes and Eastern Senegal, Casamance has a rich pteridoflora with 21 species (Mingou 2023). In fact, most of these species deserve special attention, such as *Acrostichum aureum*, which has only been observed in the mangrove south-west of Boukitengho in the Oussouye region (Mingou 2023), while the species *Bolbitis gemmifera, Lycopodium cernuum, Lygodium microphyllum, Pityrogramma calomelanos, Pteris similis*, and *Selaginella tenerrima* may be considered disappeared from the region due to habitat fragmentation (Mingou 2023).

15.3 Livelihoods

Various international summits on the environment and biodiversity and their reports refer to ecodevelopment or sustainable development as development that meets the needs of today's populations without compromising the ability of future generations to meet their own needs (CMED 1987). This development must enable: social progress in line with everyone's needs, effective protection of the environment, rational use of natural resources and economic growth, and job creation. However, we need to change the model of relations between administrations and populations, which often does not encourage consultation and participative management (Sène et al. 2006). Local populations must benefit fully from the enjoyment of resources, but they must also be the first conservers of their environment. All this must be done in harmony between administrative agents and local populations, without forgetting the specificity of each wetland.

The promotion of ecotourism in Senegal's wetlands has led to the establishment of a number of accommodation facilities and village camps, as well as local tourist circuits, thanks to the support of administrative conservationists, scientists, and public and private development partners. Local services (dugout canoe rides, cart rides, etc.) are sold as part of ecological tours of specific ecosystems and cultural tours related to the many sociocultural activities of Senegal's various ethnic groups. Local products of all kinds (cultural, food, cosmetics, etc.) are also promoted through stores organized and run by local people (Figure 15.14).

Fishing is a very important activity, and even in coastal villages, it is the predominant activity and the main source of income. Men go out to sea, while women are responsible for processing and selling fish products.

A total of 111 fish species have been identified along the Senegal River (Peeters 1998) forming the basis of a fishing industry that is important not only for human consumption but also for birdlife. In fact, supporting the promotion of this activity in the area would make a major

Figure 15.14 Different livelihoods in some wetlands of Senegal: (a) mangrove wood exploitation, (b) salt exploitation in Saint-Louis, (c) arch harvesting, (d) hawk fishing on the Gambia river, (e) Hotel complex in the Delta, and (f) restaurant on stilts on the Casamance river.

socioeconomic contribution, especially for riverside populations, who could at the same time be better integrated into ecosystem management and conservation. *Typha domingensis* is used locally by local craftsmen to make mats, which are sold locally and at weekly markets. The same invasive plant (*T. domigensis*) is also targeted for use as a source of energy (e.g. biochar) and as a building material due to its insulating properties.

The Saloum Delta and Petite Côte region has great ecotourism potential, with its numerous waterways, strings of islands, islets, and meanders. What's more, the Saloum Delta is part of the World's Most Beautiful Bays Association, with five marine protected areas (MPAs) in the Delta (Delta du Saloum, Bamboung, Gandoul, Sangomar, and the Palmarin Reserve). By way of example, the Sangomar Air Marine Protegée (AMP) with its 87,437 ha surface area, a large part of which lies within the RBDS, offers great ecotourism potential thanks to its ecological and biological diversity and its historical, archaeological, and cultural wealth as well as their economic potential. The diversity of local products (fishery, horticultural, forestry, etc.) has encouraged many of the inhabitants to become itinerant traders (DAMCP 2018). The mudflats and sandbanks offer a wealth of shellfish: *Anadara senilis.*, *Semifisus morio, Murex* sp., *Cymbium* spp. *Crassostea gasar, Typanotonus* sp., and *Sepia officinalis* harvested for self-consumption and sale. Salt is mined industrially in the Kaolack and Fatick salt works and artisanally in several villages, the basis of a business that provides women with a substantial income (Sène et al. 2006). In the lowlands, food crops such as rice are grown in the central valleys during the rainy season. Tree plantations of *N. macrophylla* and *D. senegalense* are an asset for the Saloum Delta. Their exploitation and development could make a significant contribution to the economic autonomy of local populations and to the preservation of the resource by promoting the use of these species in reforestation (Sène et al. 2006).

The Niayes area represents an undeniable and rapidly developing market-gardening potential (CSE 2015b). A farmer can achieve an average annual gain of 5,000,000 FCFA (7634 euros) at Technopôle, 2,000,000 FCFA (3053 euros) at Lake Wouye, and 800,000 FCFA (1221 euros) at Lake Mbeubeuss (Badiane and Mbaye 2018). The valorization of what remains of Dakar's wetlands can be envisaged as part of urban planning. Alongsidé market garden production, it is worth noting the growing importance of fruit arboriculture in the area. Orchards have been set up in many villages. Rain-fed agriculture is practiced in the area on "Dior" soils (CSE 2015b). A coherent approach to development would help safeguard these natural environments as well as the various compatible socioeconomic activities that develop there (Badiane and Mbaye 2018). In economic terms, the Pikine Niaye plays an extremely important role in supplying the Greater Dakar area with vegetables, ornamental plants, fruit, and poultry (Ndao 2012). The uncontrolled growth of recent decades has led to massive urbanization of this area, which is becoming increasingly fragile and exposed.

The sylvopastoral zone has enormous natural resource potential. The revitalization of the fossil valleys has enabled a number of important breeding sites to be rewatered. In fact, the Ferlo is essentially a pastoral zone. The existence of boreholes means that the area can be used for pastoral purposes all year round. Dairy production has expanded significantly with the support of the state, private structures and development aid programs. Cereal production (maize, millet) has also seen a strong revival, alongside cash crops such as cowpeas and groundnuts. In addition, local people have mastered the marketing of gum arabic and harvested forest products, e.g. *Balanites aegyptiaca* and *Ziziphus mauritiana* (CSE 2015b).

The eastern region is a veritable melting pot of natural, cultural, and artistic riches. These include (i) a flora and fauna reserve, which is unique in Senegal: the Niokolo Koba National Park (PNNK) and the area of hunting interest (ZIC), (ii) several waterfalls: the Cascades de Dindéfélo, de Ségou, Toumania, etc., (iii) ethnic and cultural diversity: Bassari, Malinké, Peul, Dialonké, Bedik, Bambara, Diakhanké countries; and (iv) UNESCO World Heritage status for Bassari country (SRSDK 2021). The rich and varied cultural heritage of the region is an asset for the development of its magnificent natural sites. The different types of tourism practiced in the region are as follows: (i) sport hunting in areas allocated to hotels and campsites; (ii) discovery tourism at waterfalls and specific ecosystems; (iii) cultural tourism (cultural events in Bedik, Iwol, Ethiowar, and Dandé countries and in Bassari Ethiolo, Eganga, and Koté countries) (SRSDK 2021). Fishing activity is marginal and only practiced in the Falémé, Gambia, and Niokolo rivers. However, it should be noted that the authorities have launched a program to develop fish farming in this region, the furthest from the ocean at the national level.

Fishing plays a key role in Casamance, as the region's ecosystem boasts significant aquaculture potential, illustrated by a diversified maritime and continental hydrographic network. Casamance boasts 85 km of coastline and a major hydrographic network, including a 300 km-long axial river and numerous bolongs, giving it a wealth of fishery resources and enormous potential for maritime, lagoon, and river fishing (fish, crustaceans, and molluscs). In 2019, the Ziguinchor region recorded 74,828,541 kg of fishery resources, compared with 67,490,972 kg in 2018, an increase of 7,337,569 kg in 2019 (SRSDZ 2021). As a result, the Ziguinchor region boasts considerable exploitable fishery resources. In addition, the diversity and abundance of nontimber forest products (fruit, bark, oil, roots, etc.) from these plant formations contribute to satisfying local and national needs and improving living conditions in rural areas (SRSDZ 2021). Casamance's forest resources include many valuable species such as *Pterocarpus erinaceus*, *Melicia regia*, *Khaya senegalensis*, *Ceiba pentandra*, *Prosopis africana*, *Afzelia africana*, and *Cordyla pinnata*. The vast natural resources of the region offer enormous potential in a wide variety of sectors.

15.4 Wetland Conservation

Almost all of the country's development sectors rely on biodiversity. For this reason, the loss of biodiversity could inevitably affect people's livelihoods and jeopardize the sustainable development to which Senegal aspires. Wetlands are biodiversity hotspots and therefore play a key role in preserving and enhancing biodiversity.

Fortunately, biodiversity conservation is still a national concern, as demonstrated by the creation of a number of protected areas, almost all of which contain wetlands that are central to their preservation. The ratification of numerous international conventions, the development of a concerted management policy at the national level and with neighboring countries, and the existence of national texts, documents, and policies in favor of biodiversity and wetland conservation attest to a firm and ongoing commitment to biodiversity conservation (Gueye et al. 2018).

The Ministry of the Environment, Sustainable Development and Ecological Transition defines guidelines through its sectoral policy letter, but does so in close collaboration with several ministries, given the cross-cutting nature of biodiversity. Coordination between these different entities is facilitated by the National Committee on Biodiversity, whose main tool is the Clearing House and its synergy with other structures in the field (Gueye et al. 2018).

Major efforts to improve the conservation status of wetlands and their biodiversity, not to mention their sustainable and harmonious use, are being made in all the country's agroecological zones. Wetland conservation in Senegal began in the 1970s with the creation of protected areas including the PNOD, the Saloum Delta, Madeleine Island, and Langue de Barbarie national parks. This commitment to conserving the country's natural ecosystems was confirmed in 1977 with the ratification of the Ramsar Convention, making Senegal one of the first African countries to become a party to the convention and the first West African country to join (Smart 1988). The convention is a global treaty on the conservation and sustainable use of wetlands. The Parc national des Oiseaux du Djoudj (PNOD), the Réserve spéciale de Faune de Gueumbeul (RSFG), and the Parc national de la Langue de Barbarie (PNLB) are protected areas in the Senegal delta. The first two

(PNOD and RSFG) and the Ndiaël Basin Wildlife Reserve are protected under the Convention on Wetlands. The Ndiaël Reserve was removed from the Montreux Record of Ramsar sites in difficulty in 2018 following numerous efforts by the State of Senegal to restore its water level (OLAC et al. 2018). In November 2004, the State of Senegal set up a MPA covering almost 496 km^2 (49,600 ha) in Saint Louis. It is located in an area where geophysical factors and hydrodynamic conditions have contributed to the development of a multitude of refuges for numerous marine and estuarine species. These refuges constitute fishing grounds known as "Xer" for rocky bottoms or "joxoor" for sandy-muddy bottoms, which are characterized by the presence of shellfish. The proliferation of invasive aquatic plants (e.g. *Pistia stratiotes, Typha domingensis, Salvinia molesta*, and *Mimosa pigra*) is a nuisance for wetlands and human activities alike. Fortunately, efforts have been made to contain them, particularly in the PNOD and PNKK, where their harmful effects are no longer felt to any great extent. The PNOD regularly carries out an inventory, locates and estimates the spread of invasive plants.

The Niayes area has benefited from several protection initiatives in recent years. Indeed, there have been the Niayes de Dakar wetlands conservation and enhancement project, the Conservation Migratory Birds (CMB) project, and the rainwater management and climate change adaptation project. All this should lead to a protected status for the Niayes for sustainable use (Badiane and Mbaye 2018). Indeed, the recent classification of part of the Niayes as "protected wetlands" will very soon lead to its official inclusion on the list of sites protected by the International Ramsar Convention.

In recent years, public authorities have opted to implement a program to improve the availability of surface water in the Ferlo region. Ponds are an important part of this program. Some of them have therefore been developed (by overdigging), extending their lifespan (Diop et al. 2004) and promoting a diversification of biodiversity and activities for local populations. The ostrich (*Struthio camelus*), whose range once covered a large part of the African continent, including the sylvopastoral zone of Senegal, is highly threatened. Its presence in the wild in small numbers was confirmed in the Louguignibi locality by observations made in 2013 (Gueye et al. 2018). Its restoration in this area is a priority for Senegal. Initiatives to raise ostriches in enclosures in Katané have been carried out. Today, these efforts are ongoing and the species is regularly monitored (Guèye et al. 2018).

The reintroduction of species into wildlife reserves in different agro-ecological zones is underway. For example, the algazelle oryx (*Oryx dammah*), the dama gazelle (*Gazella dama mhorr*), and the dorcas gazelle (*Gazella dorcas neglecta*) have been reintroduced into the Gueumbeul Special Wildlife Reserve (river valley zone) and the Ferlo Nord Wildlife Reserve (sylvopastoral zone).

Eastern Senegal remains one of the most ecosystem-rich regions in Senegal. In terms of natural resource management, despite the absence of classified forests in the Kédougou region, there is the PNKK, five Community Nature Reserves (Niéméniké, Thiabédji, Dindéfélo, Oubadji, and Bandafassi-Tomboronkoto), and zones of hunting interest (ZIC) (SRSDK 2021). The Tambacounda region, with most of the PNKK, boasts several classified forests (SRSDT 2021). A number of fish farms have been set up in the eastern part of Senegal, the part furthest from the ocean and least endowed with icthyofauna, to develop this activity.

The Lower Casamance Park is a biodiversity sanctuary par excellence, thanks to its particular diversity of flora and fauna. Unfortunately, access to this jewel of Senegalese biodiversity has been restricted for over 40 years by the rebellion. Among the region's protected areas is the Kalissaye ornithological reserve, located at Pointe de Sankoye in the Bignona Department. It is home to several breeding colonies of seabirds and is also a breeding site for sea turtles.

15.5 Conclusion

Senegal's wetlands are highly diversified in nature and spread across the country's six ecogeographical zones. They are among the richest and most extensive in West Africa and play a major role in maintaining production systems and conserving biological diversity. They offer a wide range of ecosystem services and are home to numerous species of birds, reptiles, mammals, and fish. Given their generally favorable hydric and hydrological conditions, many income-generating activities such as agriculture and fishing are commonly practiced by local populations. Other ecodevelopment activities are also practiced, notably in connection with ecotourism through sport hunting, discovery tourism, and cultural tourism.

However, the combined effect of climate change and human activities of all kinds poses a major threat to the sustainability of these areas. Hence, the conservation actions undertaken by the State of Senegal with the creation of protected areas since the 1970s to enable their sustainable and harmonious use in all the country's ecogeographical zones. Overall, Senegal has a network of 6 national parks, 4 wildlife reserves and 3 special reserves, 213 classified forests, 22 community nature reserves, and 27 pastoral units and several community forests. In each of these areas, there are several wetlands that concentrate most of the site's biological diversity and income-generating activities. In addition to national parks with marine bangs, a national network of MPAs exists. Because of their globally recognized importance, these protected areas are listed as biosphere reserves (5), Ramsar sites (9), or natural world heritage sites (2). The drive to protect and designate Ramsar sites continues, as the Ministry for the Environment, Sustainable Development and Ecological Transition continues to encourage and assist the creation of community nature reserves, in line with the recommendations of the CBD.

In addition, prospects for regular monitoring of these wetlands are emerging thanks to geospatial applications currently being developed by the Ecological Monitoring Centre as part of the African Union Commission's GMES and Africa program (Global Monitoring for Environment and Security in Africa).

References

Akpo, L.E. (1992). Influence du couvert ligneux sur la structure et le fonctionnement de la strate herbacée en milieu sahélien : Les déterminants écologiques. Thèse de Doctorat de 3ème cycle. Université Cheikh Anta Diop, Dakar-Sénégal

Ambus, P.L. and Lowrance, R. (1991). Comparaison of denitrification in two Riparian soils. *Soil Science Society of America Journal* 55 (4): 994–997. https://doi.org/10.2136/sssaj1991.03615995005500040017x

Ba, K., Granjon, L., Hutterer, R., and Duplantier, J.-M. (2000). Les micromammifères du Djoudj (Delta du Sénégal) par l'analyse du régime alimentaire de la chouette effraie, Tyto alba. *Bonn. Zool. Beitr* 49 (1–4): 31–38.

Ba, T., Akpo, L.E., and Diouf, A.A. (2017). Dynamique spatio-temporelle des écosystèmes du bassin versant du Ferlo (Nord-Sénégal). *Journal of Animal & Plant Sciences* 33 (1): 5255–5273.

Badiane, S. and Mbaye, E. (2018). Zones humides urbaines à double visage à Dakar: opportunité ou menace ? *Sciences, Eaux & Territoires* https://doi.org/10.14758/set-revue.2018.hs.08.

Baran, E. (2000). Biodiversity of estuarine fish faunas in West Africa. *Naga, The ICLARM Quarterly* 23 (4): 4–9.

Bassène, O.A. (2016). L'évolution des mangroves de la Basse Casamance au sud du Sénégal au cours des 60 dernières années: surexploitation des ressources, pression urbaine et tentatives de mise en place d'une gestion durable. Thèse de Doctorat en cotutelle. Université de Lyon.

Camara, A., Hardy, K., Dioh, E. et al. (2017). Amas et sites coquilliers du delta du Saloum (Sénégal): passé et présent. *L'Anthropologie* https://doi.org/10.1016/j.anthro.2017.03.018.

CMED (1987). *Notre avenir à tous*. Rapport, Brundtland: Commission Mondiale sur l'Environnement et le Développement 349 p.

Coulthard, N.D. (2001). Senegal. In: *Important Bird Areas in Africa and Associated Islands: Priority Sites for Conservation* (ed. L.D.C. Fishpool and M.I. Evans), 733–750. Newbury & Cambridge: Pisces Publications and BirdLife International.

CSE (2015a). *Rapport annuel sur l'état de l'environnement au Sénégal*. Ministère de l'Environnement et du Développement Durable 194 p.

CSE (2015b). *Actualisation du découpage et de la caractérisation des zones éco-géographiques du Sénégal. Rapport élaboration de la 3ème Communication nationale*. Ministère de l'Environnement et du Développement Durable 75 p.

CSE (2020). *Rapport annuel sur l'état de l'environnement au Sénégal*. Ministère de l'Environnement et du Développement Durable 267 p.

C3S & CDS (2019). Land cover classification gridded maps from 1992 to present derived from satellite observation. In: *Copernicus Climate Change Service (C3S) Climate Data Store (CDS)*. https://doi.org/10.24381/cds.006f2c9a.

Dacosta, H. and Gomez, R. (1998). *Inventaire des zones humides de la Casamance*. Commission Niayes. Réseau national Zones humides, Rapport UICN, 33 p.

DAMCP (2018). *Plan d'Aménagement et de Gestion de l'Aire Marine Protégée de Sangomar. Rapport Final*. Ministère de l'Environnement et du Développement Durable, 61 p.

Dia, I.M.M. (2003). Élaboration et mise en œuvre d'un plan de gestion intégrée : La Réserve de biosphère du delta du Saloum, Sénégal. Rapport, UICN, Gland, Suisse et Cambridge, Royaume-Uni. 130 p.

Diallo, M. (1995). *Dynamique comparée des populations de Culicidae à Kédougou (zone soudano-guinéenne) et à Barkédji (zone de savane sahélienne): conséquences dans la transmission des arbovirus. Mémoire de DEA*. Dakar-Sénégal: Université Cheikh Anta Diop.

Diop, S.E. (2012). Les écosystèmes marins et côtiers de l'Afrique de l'Ouest : enjeux de leur gestion à long terme et de leur développement durable - Exemple du Sénégal. Rapport du Programme des Nations Unies pour l'Environnement, 15 p.

Diop, T., Diaw, O.T., Diémè, I. et al. (2004). Mares de la zone sylvopastorale du Sénégal : tendances évolutives et rôle dans les stratégies de production des populations pastorales. *Revue Élev. Méd. vét. Pays Trop.* 57 (1–2): 77–85.

Diouf, P.S. (1996). Les peuplements en poissons des milieux estuariens de l'Afrique de l'Ouest : l'exemple de l'estuaire hyperhalin du Sine-Saloum. Thèse de Doctorat. Université Montpellier II, France.

Diouf, P.S., Ngom, M., and Fall, M. (2016). Ichtyofaune et pêche dans le lac de Guiers et la réserve du Ndiaël. Rapport UICN et OLAG, 58 p.

Diouf, A.C., Ba, A.I.S.W., Ba, M.D., and Diallo, M.S. (2021). Biogéographie des zones humides face aux contraintes multiples : Cas de la réserve de Biosphère du Delta du Saloum au Sénégal. *DELLA/Afrique* 3 (8): 325–341.

DPN (2017). Plan de gestion du Parc National des oiseaux du Djoudj 2017–2021.République du Sénégal, 112 p.

DPN (2023). Rapport technique sur le dénombrement international des oiseaux d'eau (Sénégal). République du Sénégal, 25 p.

Dryade (1990). Inventaire Biophysique de la région des Niayes. In: *Synthèse, document de recherche dans le cadre du Project Conservation des Terroirs du Littoral*. République du Sénégal: Ministère du Développement Rural et de l'Hydraulique 262 p.

ESA (2017). Land Cover CCI Product User Guide Version 2. Tech. Rep. maps.elie.ucl.ac.be/CCI/viewer/download/ESACCI-LC-Ph2-PUGv2_2.0.pdf

Faye, N. (2021). Végétation des mares temporaires du Ferlo, Nord Sénégal : Impact des facteurs anthropiques et topographiques cas du village de Widou. Thèse de Doctorat. Université Cheikh Anta Diop, Dakar-Sénégal.

Fernández-García, J.M., de Azua, N.R., and Pacheco, L. (2013). Birds of Dindéfello Nature Reserve, Senegal. *Bulletin of African Bird Club* 20 (1): 49–59.

Gardner, R.C., Barchiesi, S., Beltrame, C. et al. (2015). *State of the World's Wetlands and Their Services to People: A Compilation of Recent Analyses. Ramsar Briefing Note No. 7*, 20. Gland: Ramsar Convention Secretariat.

Gomis, F. (2016). Composition floristique, structure et usages de la végétation ligneuse de la Réserve Naturelle Communautaire (RNC) Boundou (Sénégal/Tambacounda) : éléments d'analyse pour un aménagement. Mémoire master. Université Cheikh Anta Diop, Dakar-Sénégal.

Goudiaby, A. (1996). Étude de la flore et de la végétation du site de Dindéfelo (Sud-est du Sénégal): éléments pour un aménagement. Thèse de Doctorat. Université Cheikh Anta Diop de Dakar-Sénégal.

Gueye, M., Ba, T., and Diop Sall, A. (2018). Biodiversity in Senegal: terrestrial ecosystems flora and great fauna. In: *Global Biodiversity Volume 3 Selected Countries in Africa* (ed. T. Pullaiah), 175–212. Oakville: Apple Academic Press.

IPCC (2019). Summary for policymakers. In: *Climate Change and Land: an IPCC Special Report on Climate Change, Desertification, Land Degradation, Sustainable Land Management, Food Security, and Greenhouse Gas Fluxes in Terrestrial Ecosystems* (ed. P.R. Shukla, J. Skea, E. Calvo Buendia, et al.), 7–190. IPCC.

IUCN/Réseau zones humides (2002). La Grande Niaye de Dakar, problématique urbaine et enjeux environnementaux. Rapport IUCN, 22 p.

Janicot, S., Aubertin, C., Bernoux, M. et al. (2015). Zones semi-arides : le Sahel sensible aux variations de pluies. In: *Changement climatique : Quels défis pour le Sud?* (ed. IRD), 115–127. Marseille: IRD https://doi.org/10.4000/books.irdeditions.294363.

MEDD (2015). *Politique nationale de gestion des zones humides*. Rapport République du Sénégal, 124 p.

MEPN (1997). *Rapport national sur la Biodiversité*. Projet SEN/96/G31/A/1G/99.

Mingou, P.A.B. (2023). Diversité taxonomique, écologie et distribution des Ptéridophytes du Sénégal. Thèse de doctorat. Université Cheikh Anta Diop, Dakar-Sénégal.

Mingou, P. and Gueye, M. (2017). La flore ptéridologique de quelques endroits humides de la région de Kédougou (Sénégal). *European Scientific Journal* 13 (12): 127–144. https://doi.org/10.19044/esj.2017.v13n12p127.

Mingou, P.A.B., Gueye, M., Abotsi, K.E. et al. (2021). Three new records of fern species (Polypodiopsida) in Senegal, from Dindefelo Falls, Kedougou region. *Check List* 17 (3): 923–930. 10.15560/17.3.923.

Mingou, P.A.B., Bayet, T., Cambier, C., and Gueye, M. (2023). New fern records for the Senegal: *Adiantum capillus-veneris* and *Pteris togoensis* (Pteridaceae). *Fern Gazette* 22 (2): 45–60.

Ndao, M. (2012). Dynamiques et gestion environnementales de 1970 à 2010 des zones humides au Sénégal: étude de l'occupation du sol par télédétection des Niayes avec Djiddah Thiaroye Kao (à Dakar), Mboro (à Thiès et Saint-Louis). HAL Id: tel-00718050, https://theses.hal.science/tel-00718050. Thèse Doctorat. Université Toulouse 2 Le Mirail (UT2 Le Mirail).

Ndiaye, O. (2013). Caractéristiques des sols, de la flore et de la végétation du Ferlo, Sénégal. Thèse de Doctorat. Université Cheikh Anta Diop, Dakar-Sénégal.

Ndiaye, S. (2005). Stratégie nationale de gestions des aires protégées au Sénégal : (cas des aires protégées par la Direction des Parcs Nationaux). Rapport, 42 p.

Ngo Samnick, E.L. (2021). Le Delta du Saloum au Sénégal, un écosystème d'exception. *Liaison Energie Francophone* 101: 20–24.

Ngom, D., Fall, T., Sarr, O. et al. (2013). Caractéristiques écologiques du peuplement ligneux de la réserve de biosphère du Ferlo (Nord Sénégal). *Journal of Applied Biosciences* 66: 5008–5023. https://doi.org/10.4314/jab.v65i0.89644.

OLAC, IUCN, DEFCCS (2018). Request for the withdrawal of the Réserve Spéciale de Faune du Ndiaël, Ramsar Site from the Montreux Record. PREFELAG project, 34 p.

Peeters, J. (1998). État des lieux de la conservation des zones humides au Sénégal. In: *Deuxième conférence internationale sur les zones humides et le développement, Dakar, Sénégal, (Novembre 1998).* Direction des Parc Nationaux.

Programme de Gestion Durable de la Faune Sauvage, SWR (2023). Site des zones humides sahéliennes. https://www.swm-programme.info/fr/ressource-project (accessed 23 November 2023).

Ramsar (2023). Service d'information sur les Sites Ramsar. https://rsis.ramsar.org/fr/ris/2326?language=fr (accessed 15 November 2023).

Reynaud (2001). *L'avifaune des Niayes*. IRD Sénégal. 47 p.

Sané, T. and Mbaye, I. (2007). Etat des lieux et étude diagnostique de l'environnement de la Casamance. *Annales de la Faculté des Lettres et Sciences Humaines* 37/B: 183–201.

Sène, E.H.M., Thiaw, I., and Lamizana-Diallo, B. (ed.) (2006). *Gestion des zones humides en milieux arides : leçons d'expériences*. Gland: IUCN 86 p.

Smart, M. (1988). *Réserve spéciale de faune du N'diaël, Sénégal*, Procédure de surveillance continue des zones humides d'importance Internationale. Rapport n° 9. Djoudj, Sénégal: Mission Consultative Ramsar 5 p.

Sow, A.A. and Akpo, L.E. (2011). *Plan de coopération de la réserve de biosphère du Ferlo*. Rapport de consultance, 51 p.

SRSDK (2021). *Rapport annuel Situation économique et sociale régionale 2019*, 145. Agence Nationale de la Statistique et de la Démographie.

SRSDT (2021). *Rapport annuel Situation économique et sociale régionale 2019*, 89. Agence Nationale de la Statistique et de la Démographie.

SRSDZ (2021). *Rapport annuel Situation économique et sociale régionale 2019*, 132. Agence Nationale de la Statistique et de la Démographie.

Sy, O. (2009). Rôle de la mare dans la gestion des systèmes pastoraux sahéliens du Ferlo (Sénégal). *Cybergeo : European Journal of Geography* 440: 1–11. https://doi.org/10.4000/cybergeo.22057.

Thomas, Y.-F., Diaw, A.-T., and Cesaraccio, M. (ed.) (2003). *Problème d'érosion côtière : la Réserve de Biosphère du delta du Saloum (Sénégal). Patrimoines et développement dans les pays tropicaux, IXe Journées de Géographie Tropicale, La Rochelle (13–14 septembre 2001)*, 57–66. Bordeaux: Presses Universitaires de Bordeaux (*Espaces tropicaux*, 18); https://www.persee.fr/doc/etrop_1147-3991_2003_act_18_9_1107.

Traoré, S.A. (1997). Analyse de la flore ligneuse et de la végétation de la zone de Simenti (Parc national du Niokolo Koba), Sénégal oriental. Thèse de 3e cycle. Université Cheikh Anta Diop, Dakar-Sénégal.

Treca, B. (1989). *Les dégâts aux semis de riz causés par les oiseaux d'eau dans le Delta du fleuve Sénégal*, 1–13. Dakar: Orstom.

Triplet, P., Diop, I., Sylla, S.I., and Schricke, V. (2014). *Les oiseaux d'eau dans le delta du fleuve Sénégal (rive gauche). Bilan de 25 années de dénombrements hivernaux (mi-janvier 1989-2014).* Rapport, OMPO, ONCFS, DPN, SMBS 125 p.

Triplet, P., Gueye, I., and Guissé, S.S. (2018). Sénégal : Le Parc national des oiseaux du Djoudj, un monument naturel menacé. *Connaissances & Gestion des Habitats* 319: 43–48.

Trochain, J. (1940). *Contribution à l'étude de la végétation du Sénégal. Mémoires de l'Institut Français d'Afrique Noire* 2: 433 p.

UNESCO (2023). Mesures de protection contre les impacts des projets agricoles et agro-industriels dans la Réserve nationale d'oiseaux du Djoudj. https://whc.unesco.org/fr/activites/981/ (accessed 5 February 2023).

Wane, A., Ancey, V., and Grosdidier, B. (2006). Les unités pastorales du Sahel sénégalais, outils de gestion de l'élevage et des espaces pastoraux. *Développement durable et territoires* [En ligne], Dossier 8 | 2006, mis en ligne le 04 janvier 2013, consulté le 23 novembre 2023. http://journals.openedition.org/developpementdurable/3292. https://doi.org/10.4000/developpementdurable.3292.

Wetlands International (2023). Quels sont les zones humides? https://africa.wetlands.org/zones-humides/quels-sont-les-zones-humides/ (accessed 15 November 2023).

16

Wetlands of Zimbabwe: Biodiversity, Livelihoods, and Conservation

Tatenda Musasa[1,2], *Oshneck Mupepi*[2], *and Thomas Marambanyika*[2]

[1] *Institute of Water Studies, Faculty of Natural Sciences, University of the Western Cape, Cape Town, South Africa*
[2] *Department of Geography Environmental Sustainability and Resilience Building, Midlands State University, Gweru, Zimbabwe*

16.1 Introduction

Wetlands are biologically diverse and productive ecosystems in the world, which are essential for human livelihood. Global research estimates indicate that coastal and inland wetland areas occupy more than 12.1 million km^2 (Davidson et al. 2018; Ramsar Convention Secretariat 2018). This includes areas that are permanently inundated (54%) and seasonally inundated (46%) (Ramsar Convention Secretariat 2018). Given the role of wetland ecosystems for biodiversity conservation that enhances various cultural, provisioning, and regulatory services, there is a great need for their conservation globally. However, their loss due to climate change and human activities is increasing owing to changing hydro systems and agricultural activities. This threatens the biodiversity wealth within these ecosystems, hence the need for conservation initiatives so that humanity and biodiversity (flora, fauna, and avifauna) continue to benefit. Many livelihoods especially in the developing world are sustained by wetland ecosystems' species diversity. Therefore, conservation measures need to be understood such that biodiversity conservation and derivation of wetland livelihood benefits are balanced.

Wetland degradation in Europe has been associated with loss of gray heron (*Area camera*), common goldeneye (*Bucephala clangula*), smew, and great cormorant (*Phalacrocorax carbo*) bird species, and other many animals and plant species (Soultan et al. 2022). However, wetland protection has increased and some positive changes in species diversity have been noted. For example, in Northern Italy, wetlands with wide water body perimeters with diverse vegetation habitats were found key for biodiversity (birds, mammals, amphibians, reptiles, fish, invertebrates, and plants) conservation (Guareschi et al. 2020). However, increased exploitation of some wetlands in Northern Italy to derive benefits like water and crop production resulted in decline in wetland areas and loss of some plant, animal, and bird species (Guareschi et al. 2020; Soultan et al. 2022). This indicates that wetlands of the developed world are also facing threats from human activities and climate change, which in turn affects biodiversity within these habitats.

In the United States, reptile, bird, and amphibian diversity were found high in wetland areas, which include southeastern Plains and Piedmont, the Mojave and Sonoran Desert Basins, and the

Wetlands of Tropical and Subtropical Asia and Africa: Biodiversity, Livelihoods and Conservation,
First Edition. Edited by Thammineni Pullaiah.

Great Plains where species richness and diversity positively correlate with wetland land covers (Dertien et al. 2020). These were found hotspot areas where conservation is recommended especially grasslands, which inhabits wading birds and waterfowls (Steen et al. 2016). However, threats from human activities like construction and agriculture were confirmed threating biodiversity in these regions (Dertien et al. 2020). Conservation efforts by federal conservation laws and executive orders that regulate wetland depletion and incentives for the creation of some wetland habitat were confirmed but the challenge is that isolated wetlands were being ignored (Cimon-Morin et al. 2016). This calls for attention to isolated inland wetlands as they accommodate important biodiversity, which benefits humanity. The same was observed in Patagonia's unprotected wetlands, which were found to be oases of water in an arid and semiarid steppe, containing not only a large proportion of the aquatic biodiversity but also acting as a refuge for terrestrial flora and fauna (Epele et al. 2021). Different management actions have been proposed, for instance, fencing and creation of artificial ponds (Epele et al. 2021). These conservation measures are important to maintain wetland health as well as the diversity of habitats and species within these ecosystems.

Studies in the Asian continent indicate that wetlands are rich in a wide variety of fish which they use for food, animals which are hunted for meat, and various edible plants and grasses which are important for local people (Bhatta et al. 2016). This is some of biodiversity that shapes livelihoods of humanity while at the same time important for a healthy wetland ecosystem. In Bangladesh, wetlands are rich in floral and faunal biodiversity (300 plant species, 400 vertebrate species, and 260 freshwater fish species) with some of the major wetlands of importance being Tanguar haor Hakaluki haor Sundarbans and St. Martin's Coral Island) (Yousuf Haroon and Kirkia 2017). It is reported that two-thirds of Bangladesh's population depend on these wetlands for medicinal plants, thatch grass harvesting, and cattle grazing among others. Utilization of these wetlands is now more than conservation efforts being made which is threatening the diversity of animals and plants available (Yousuf Haroon and Kirkia 2017).

In sub-Saharan Africa, wetlands cover about 7%, with freshwater marshes and floodplains being the most extensive types (Rebelo et al. 2010; Gxokwe et al. 2022). These inhabit a number of bird, mammal, reptile, and plant species, hence important biodiversity reserves. In the valley bottom wetlands of Tanzania, for instance, the Ruaha River Basin's valley bottom wetlands, it was found that faunal biodiversity is very high because of consistently wet conditions that support perennial growth of various plant species including the endangered *Prunus africana,* which also have medicinal value for the treatment of prostate cancer (Munishi et al. 2011). In Ethiopia, several bird species including marabou stork (*Leptoptilos crumenifer*), great white pelican (*Pelecanus onocrotalus*), striated heron (*Butorides striata*), gray heron (*Ardea cinerea*), African darter (*Anhinga rufa*), and reed cormorant (*Microcarbo africanus*) are inhabited in wetlands in areas like Baro River, Dabus Wetland, Lake Zeway, Lake Langano, and Lake Chamo among others. As a strategy of conserving this biodiversity, integration of wetland resources with ecotourism was found to be one of the noble options, which is being implemented in Ethiopia (Israel and Timar 2018). Despite these benefits, wetland degradation and loss continues in Africa and Asia due to increasing food demand by growing populations and climate change impacts that are adversely affecting rain-fed agriculture which is the mainstay of rural economy.

Management and conservation of wetland ecosystems and their biodiversity in Africa remain an important element as these precious ecosystems are exposed to exploitation mainly by human beings for survival (Rivers-Moore and Cowden 2012; Matamanda et al. 2018; Mandishona and Knight 2022). Wetland management and conservation is extremely difficult and frustrating due to the complex social, economic, and biophysical processes underpinning their existence (Turner et al. 2000; Marambanyika and Beckedahl 2016a). Moreover, local communities generally lack an understanding of biophysical processes peculiar to wetlands in their vicinity, which results in unsustainable use (Rebelo et al. 2010; Matamanda et al. 2018). Many research that have been conducted tend to be focusing on wetland use dynamics, their extent and distribution, understanding the biophysical components of wetlands using different methods such as remote sensing (Dube 2005; Rivers-Moore and Cowden 2012; Lu and Chang 2023; Moiss et al. 2023). However, a holistic understanding of conservation and governance issues related to wetland ecosystems, their biodiversity, and associated human livelihoods is lacking. Therefore, it is important to understand wetland use, biodiversity conservation, and livelihoods at local, national, and regional levels as a key factor for reversing the trend of wetland loss and degradation (Chuma et al. 2008).

It is against this background that the chapter seeks to enhance understanding of the complex nexus between wetland types, biodiversity, livelihoods, and conservation approaches, with a specific focus on Zimbabwe. The chapter examines wetland types and distribution in Zimbabwe, biodiversity characteristics, and categories of livelihood benefits derived from wetlands by the communities. It also explores the conservation initiatives to promote wetland conservation and human livelihoods, such as legislation, policies, institutional arrangements, programs, and projects.

16.2 Wetland Type and Distribution in Zimbabwe

Zimbabwe despite being a dry country in global terms with a mean annual precipitation of 700 mm is endowed with various wetland types (Mazvimavi 2008, 2010) (Figure 16.1). Wetlands make up 34.96% of Zimbabwe's total surface area, which is approximately 13,659,579 ha. Of the existing

Figure 16.1 The Intunjambili wetland, Matobo district highlighting the wetland in full flow with the various grass and trees (i.e. channel valley bottom).

wetlands, 17.63% are in a pristine condition, 55.65% moderately degraded, and 26.72% severely degraded. The country has seven Ramsar sites designated as Wetlands of International Importance (Ramsar sites), with a surface area of 453,828 ha.

Although Zimbabwe adapted the Ramsar Convention definition of wetlands, to include natural and artificial wetlands in its conservation scope, this study focuses on natural wetlands which despite constituting the largest spatial extent in the country and largely unprotected, but are a key cornerstone of people's livelihoods. These include dambos, marshes or swamps, pans, and floodplains (Musasa and Marambanyika 2020). Dambos are the most widespread wetland type, forming most of the 75% of those wetlands that fall under the headwaters (Whitlow 1990; EMA 2021), which consist of very sensitive wetlands. In the country, this special type of wetlands (dambos) is identified using local terms (Bullock and McCartney 1995; Marambanyika and Beckedahl 2016b), whereby local people define them based on the seasonal or permanent accumulation of water, using various terms such as "*matoro*" *or "dekete*" (Shona) and "*inuta*" (Ndebele) (Maseko and Bussink 1995; Mbereko et al. 2015; Mandishona and Knight 2019; Figure 16.1).

The majority that is about 60% of Zimbabwe's wetlands are located in communal and resettlement areas (Musasa and Marambanyika 2020). In Zimbabwe, communal areas cover 42% while resettlement areas occupy 28% of the total country area (Chimbindi 2018). This implies that 60% of the wetlands are found on 70% of the country's total land area.

Most of the wetlands in Zimbabwe are found in the northern parts of the country on high veld plateaus where rainfall is considerably high. Zimbabwe is divided into five natural farming regions, with region I receiving an annual average rainfall of above 1000 mm while region V b gets rainfall below 600 mm (Manatsa et al. 2020). Approximately 84% of the dambos are located on the high veld plateau, specifically in natural farming regions one and two (Dube 2005; Figure 16.2). The fact that large wetlands of over 20,000 ha (Manapools and Driefontein) and small wetlands of

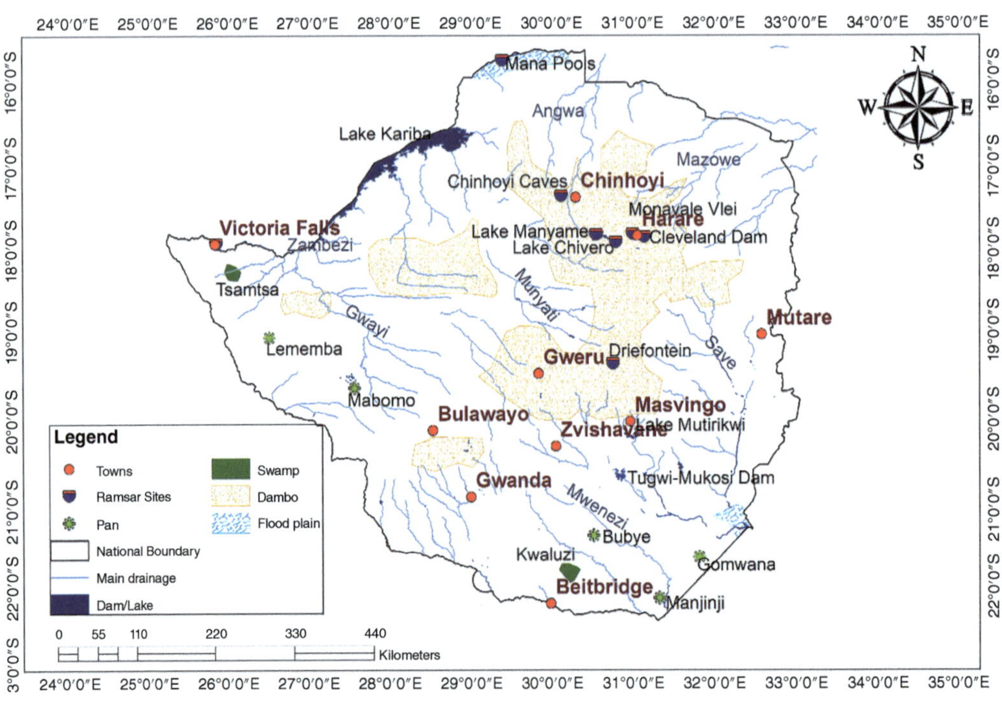

Figure 16.2 Distribution of wetlands in Zimbabwe. *Source:* Adapted from Musasa and Marambanyika (2020).

20 ha and below exist in the country (Timberlake 2000; EMA 2015; Fakarayi et al. 2015) indicates that Zimbabwe is largely endowed with wetlands of various sizes, which give them potential to contribute to human livelihood owing to the biodiversity within which they are associated with (Sithole and Goredema 2013).

Due to its position on the Great Central Plateau of Southern Africa (Matiza 1992; Maseko and Bussink 1995; Blessing 2020), Zimbabwe does not have many swamps (Whitlow 1990). A swamp is a type of wetland that is dominated by woody vegetation with also traces of water, which implies that it is permanently inundated. The existing few are found in natural farming regions IV and V, which are low rainfall areas (Matiza and Crafter 1994; Figure 16.2). Notable examples include Kwaluzi Swamp in the Lowveld and Tsamtsa in Hwange (Figure 16.2). Pans, which are depressions that collect and retain water from the surrounding uplands, are found in the southern and western districts of the country (Musasa and Marambanyika 2020). Some examples of pans in southern Zimbabwe are Bubye and Mwenezi, whereas the western part of the country, around Hwange, has the Mabomo-Joweni and Lememba complexes (Figure 16.2). Generally, the literature on wetlands extent and distribution in Zimbabwe is rather outdated as the last comprehensive assessments were done by Chabwela in the 1990s (Frenken and Mharapara 2002). More research is therefore required to expand the knowledge base. Such data is essential in providing baseline information for planning sustainable use and management of these ecosystems (Sithole and Goredema 2013).

16.3 Wetland Biodiversity

Wetlands are unique water-dominated ecosystems that provide the habitat for various aquatic organisms, wildlife, and plants, including rare, threatened, endangered, and endemic species.

16.3.1 Wetland Flora

Wetlands are home to several plant species which in turn provide shelter and food for animals and human beings (Madebwe and Madebwe 2005). Dambos are largely interspersed by Miombo woodlands dominated by *Brachystegia* and *Julbernadia* with a well-developed grass layer. The Miombo woodlands therefore dominate on wetland fringes and catchments. However, *Terminalia* vegetation species (*Mususu*) are invading other areas once dominated by *Julbernardia globiflora* and *Brachystegia spiciformis,* including wetlands. Wetland ecosystems constitute diverse flora (Table 16.1), important for the habitat of faunal and avifaunal species.

16.3.2 Wetland Fauna

Some of the threatened bird species such as the Wattled Crane (*Bugeranus carunculatus*), Gray Crowned Crane (*Balearica regulorum*), and Secretary bird (*Sagittarius serpentarius*) have their habitats in wetland area (Fakarayi et al. 2015; Musasa and Marambanyika 2022). One of Zimbabwe's Ramsar sites that is the Driefontein Grasslands is rich in birds and other biodiversity. Driefontein Grasslands Important Bird Area is the key breeding and foraging area for cranes, supporting more than half of the total cranes population found in the country. Many other specially protected bird species that includes Kori Bustard, Black-bellied Korhan, Saddle Billed Tork, White-backed and Lappet-faced Vulture, and Batelaur Eagle are found in this area.

Floodplains and dambos in the Zambezi basin support about 95% of the world's bird species (Timberlake 2000; Figure 16.3). Approximately 700 species of birds are found in the Zambezi basin

Table 16.1 Wetland type and their associated common flora.

Wetland type	Grass	Tree
Dambo	*Andropogon eucomus* *Helictotrichon turgidulum* *Ctenium concinnum*, *Setaria verticillata, Kyllinga erecta*, Love grass, *Fimbristylis dichotoma*, Fleabane *(Conyza albida)* Rapoko grass *(Eleusine indica)* Sawgrass *(Cladium maricus)* Yellow nutsedge *(Cyperus esculentus)* Black-footed water grass *(Eriochloa meyeriana)*	Terminalia vegetation species (*Mususu*) are dominant as they are invading other areas once dominated by *Julbernardia globiflora* and *Brachystegia spiciformis, Cyrtanthus contractus* Fire Lily *Eulophia cucullata*, Muonde (*Ficus capensis*)
Floodplains	Black jack *(Bidens pilosa)*, Chickweed *(Galinsoga parviflora)*, Salt weed *(Oxalis latifolia)* Mexican marigold *(Tagetes minuta)* Pigweed (*Amaranthus hybridus*), Thatching reed *(Phragmites mauritianus)*, Couch grass *(Cynodon dactylon)* Mexican clover *(Richardia scabra)* Common corkwood *(Commiphora pyracanthoides)* Goat apple (*Solanum aculeastrum)* Giant reed *(Arundo donax)* Giant sedge *(Cyperus dives)*	Zambezi teak (*Baikiaea plurijuga*) and Mukwa (*Pterocarpus angolensis*), *Colophospermum mopane*, Monkey ball *(Strychnos spinosa)*, Mobola plum *(Parinari curatellifolia* Planch. ex Benth.), Cutch *(Acacia polyacantha* Willd.), Balsam *(Colophospermum mopane)* Zebrawood *(Brachystegia spiciformis)* Weeping fig *(Ficus)* Large-fruited sycamore fig *(Ficus gnaphalocarpa)* Large-leaved guarri *(Euclea natalensis* A. DC.) The Weeping wattle *(Peltophorum africanum* Sond) Camel-foot *(Bauhinia thonningii* Schumach.*)*, Prickly pear *(Opuntia ficus-indica* (L.) Mill), Bbuffalo thorn *(Ziziphus mucronata)*, Poplar-leaved fig *(Ficus fischeri* Mildbr. & Burret) Common bushweed *(Flueggea virosa)* Snot-apple *(Azanza garckeana* (F. Hoffm. Exell & Hillc) African sicklebush *(Acacia)*
Pans	Yellow water lily *(Nymphaea mexicana)* Blue water lily *(Nymphaea nouchali)* Common reed *(Phragmites australis)* Water reed *(Schoenoplectus brachyceras)*	*Ficus capensis, Ficus natalensis, Syzigium cordatum*

Table 16.1 (Continued)

Wetland type	Grass	Tree
Dams/lakes	Sedge/rush *(Cyperus sexangularis)* Sedge *(Eleocharis acutangula)* Leafy juncos *(Juncus lomatophylus)* Yellow water lily *(Nymphaea mexicana)* Blue water lily *(Nymphaea nouchali)* Common reed *(Phragmites australis)* Water reed *(Schoenoplectus brachyceras)*	

Figure 16.3 The Driefontein wetland, highlighting the wetland in full flow with the various vegetation species and grass (i.e. flood plain wetland).

of which 167 are closely related to wetlands in the basin (Timberlake 2000). Therefore, wetlands in the basin are important sites for Afrotropical and Palearctic migrants that come to breed in Southern Africa (Timberlake 2000). Some of the species found within the Zambezi River include Carmine bee-eater (*Merops nubicoides*). The Zambezi basin is well known for supporting an assemblage of large mammals such as lions, giraffe, black and white rhinos, elephants, and buffaloes that can lure regional tourists (Timberlake 2000).

Urban wetlands, for example, Monavale vlei and Budiriro Harare, are rich in birds, insects (grasshoppers, locusts, and butterflies), and other biodiversity. The most common bird species include red bishops, black necked weaver, black headed/gray heron, long-tailed paradise whydah, and pied crow (Table 16.2). Efforts are being made by organizations such as Birdlife to establish a comprehensive bird's population database (Fakarayi et al. 2015). Musasa and Marambanyika (2022) also reported that catfishes (*Mirambas*) and breams are dominant aquatic fish species in wetlands such as Intunjambili, resulting in fishing being a significant economic activity in those areas. Four endemic fish species found in wetlands of the Eastern Highlands include the *Labeo baldasseronii, Amarginops hildae, Varicorhinus pungweensis,* and *Parakneria mossambica* (Freshwater Ecosystems of the World 2019). These are important resources required for human nutritional balance through protein addition, thus making dambos and pans important ecosystems in Zimbabwe. Wetlands in the Limpopo basin such as Matolokosi and Inyakuni also provide habitats for mammals, which include monkeys, baboons, squirrels, and jackals hence resulting in ecosystems diversification (Finlayson and Pollard 2009).

16.4 Nature of Livelihood Benefits Derived from Wetlands by Surrounding Communities

Healthy wetlands provide direct and indirect benefits to communities through goods and services (Mandishona and Knight 2019). Their contribution to human survival has been confirmed by Millennium Ecosystem Assessment which demonstrated the intricate relationships between wetland ecosystems and human livelihoods (Chuma et al. 2008). Hence, rural communities consider wetlands to be important ecosystems that should be used wisely to sustain essential services (Chifamba 2013).

Wetland use in Zimbabwe dates back to the precolonial period (Whitlow 1990; Madebwe and Madebwe 2005). Crop cultivation in dambos has enabled poor people in communal areas, to grow crops throughout the year (Whitlow 1990; Table 16.3). This implies that wetlands enable rural communities to cushion themselves from food insecurity especially where horticultural activities are done (Madebwe and Madebwe 2005; Chifamba 2013; Marambanyika et al. 2021; Figure 16.4). In Zimbabwe's communal areas, where poverty levels are high, the wetland area under cultivation has increased by almost 50% between 1980 and 2015 (Marambanyika and Beckedahl 2016a; Chikodzi and Mufori 2018). Agricultural activities provide employment and income for 60–70% of the population, supplies 60% of the raw materials required by the industrial sector, and contributes 40% of total export earnings (Government of Zimbabwe 2015). This clearly demonstrates that agriculture is an important aspect supported by various wetland types in Zimbabwe.

Wetlands such as floodplains also provide water and pastures for livestock grazing due to the perennial availability of water especially in resource-constrained communities (Mbereko et al. 2007; Chuma et al. 2008; Svotwa et al. 2008; Table 16.3). Given the importance of livestock in augmenting livelihood options through income generation, grazing will remain persistent in the communal areas of Zimbabwe. It is for this reason that livestock grazing is the second most popular land use in Zimbabwe's wetlands (Musasa and Marambanyika 2022). Magwenzi wetland in Chivi District is overgrazed as evidenced by a decline in wetland vegetation species diversity and density accompanied by increased infestation with alien non-wetland invasive species such as

Table 16.2 Wetland type and their associated fauna.

Wetland type	Large mammals	Small mammals and insects	Bird	Reptiles and fish
Dambo	Waterbuck, zebra, vervet warthog, spring, impalas, jackals monkeys, baboons	Squirrels, hares, and jackals Grasshoppers, locusts, and butterflies	Wattled Crane (*Bugeranus carunculatus*), gray crowned crane (*Balearica regulorum*) and secretary bird (*Sagittarius serpentarius*) red bishops, black-necked weaver, black-headed/gray heron, long-tailed paradise whydah, pied crow, the secretary bird, *Sagittarius serpentarius*, saddle-billed stork *Ephippiorhynchus senegalensis*, African marsh harrier, *Circus ranivorus,* and the duck species, *Circaetus pectoralis* Black-chested Snake Eagle	Banded Yellow Cobra and Puff adder, Black Mamba, catfishes (*Mirambas*) and breams *Labeo baldasseronii, Amarginops hildae, Varicorhinus pungweensis* and *Parakneria mossambica*
Floodplains	Lions, giraffe, black and white rhinos, elephants, and buffaloes. Duiker *(Cephalophus)*	African savanna hare *(Lepus victoriae)* The Tiger gecko *(Pachydactylus tigrinus),*	Kori Bustard, Black-bellied Korhan, Saddle Billed Tork, White-backed and Lappet-faced Vulture, and Batelaur Eagle, Carmine bee-eater (*Merops nubicoides*), Monotonous lark *(Mirafra passerine)* Ring-necked dove *(Streptopelia capicola)* Mourning collared-dove *(Streptopelia decipiens),* Namaqua dove *(Oena capensis),* Crowned lapwing *(Vanellus coronatus),* White-headed vulture *(Trigonoceps occipitalis),* Red-faced mousebird *(Urocolius indicus),* African hobby *Falco (cuvierii),* Crested Guinea Fowl (*Guttera pucherani),* Helmeted guinea fowl *(Numida meleagris),* African pied wagtail, *(Motacilla aguimp)*	Slender blind snakes or thread snakes (*Leptotyphlopida)* Giant Bullfrog *(Pyxicephalus adspersus)*
Pans			African Fish Eagle, and Secretary bird. Greater flamingo *(Phoenicopterus roseus),* Montane blue swallow *(Hirundo atrocaerulea)* Red-backed shrike *(Lanius collurio)*	Snakes (Python, green mamba), Crabs, frogs

(*Continued*)

Table 16.2 (Continued)

Wetland type	Large mammals	Small mammals and insects	Bird	Reptiles and fish
Riverine			Namaqua dove *(Oena capensis)*, crowned lapwing *(Vanellus coronatus)*, white-headed vulture *(Trigonoceps occipitalis)*, red-faced mousebird *(Urocolius indicus)*, African hobby *Falco (cuvierii)*	Snakes (Pythons, Green Mamba)
Dams/lakes	Hippopotamus,		African marsh harrier, *Circus ranivorus*, African Fish Eagle, Secretary bird	Crocodiles, snakes, fish (the introduced Lake Tanganyika sardine (*Limnothrissa miodon*), Largemouth Bass (*Micropterus salmoides*), the Rainbow Trout (*Oncorhynchus mykiss*), the Nile Tilapia (*Oreochromis niloticus niloticus*), the native Greenhead Tilapia (*Oreochromis macrochir*), Brown Squeaker (*Synodontis zambezensis*), Redbreast Tilapia (*Tilapia rendalli*), Tiger Fish (*Hydrocynus vittatus*) and Sharptooth Catfish (*Clarias gariepinus*)).

Lantana camara (Chimhiti 2019). Analysis of information clearly shows that wetland ecosystems are vital since they augment human livelihood through cattle grazing hence income generation for sustenance.

In communal areas of Zimbabwe, wetlands are commonly used for both commercial and subsistence crop production (Svotwa et al. 2008; Fakarayi et al. 2015). Different varieties of vegetables grown in communal gardens include beans and tomatoes (Madebwe and Madebwe 2005). These wetlands have a prominent role in diet diversification since they support several wetland crops that cannot be found in dryland areas (Timberlake 2000). This therefore shows that wetlands are essential ecosystems since they contribute to diet diversification and provide a cushion against food insecurity, thus contributing to human livelihood sustenance.

Wetland ecosystems in Zimbabwe specifically floodplains and marshes are essential for economic activities like fishing and tourism (Timberlake 2000). The most important wetlands used for commercial fishing and tourism in the country include the Lake Kariba, the Kaluzi swamps, and Driefontein floodplains. Fishing in Lake Kariba and Zambezi River is mainly done by the young and economically active groups as a livelihood option (Timberlake 2000). Rural communities of Binga and Kariba districts are earning significant income and food from fishing as indicated by the growth and development of fishing villages along the Zambezi River as is the case with Mbila,

Table 16.3 Wetland type and their associated livelihood option.

Wetland type	Livelihood option	Literature sources
Dambo	Crop production, livestock grazing domestic water supply, grass for thatching, wild plant collection for relish and medicinal use, tourism	Whitlow (1990); Bullock (1992); Maseko and Bussink (1995) Madebwe and Madebwe (2005); Mbereko et al. (2007); Finlayson and Pollard, (2009) Hove and Chapungu (2013); Fakarayi et al. (2015); Mbereko et al. (2015); Nyamadzawo et al. (2015); Marambanyika and Beckedahl (2016a);Mandishona and Knight (2019)
Floodplain	Tourism, livestock grazing, brick molding, wild fruits and material collection, different varieties of vegetables grown in communal gardens include beans and tomatoes	Svotwa et al. (2008); Dahwa et al. (2014); Fakarayi et al. (2015), Marambanyika and Beckedahl (2016a); Chikodzi and Mufori (2018); Chimhiti (2019); Marambanyika et al. (2012)
Marshes	Provisioning of clean water for drinking, food and fiber production, regulating services such as water balance, groundwater recharge, flood mitigation and storm protection, cultural and social functions such as sacred and religious importance, providing recreation, economic activities like fishing and tourism, groundwater recharge	Whitlow (1990); Timberlake (2000), Madebwe and Madebwe (2005); Finlayson and Pollard, (2009); Chifamba (2013); Musasa and Marambanyika (2020)
Pans	Water for domestic supply, provide important habitat for a wide variety of wildlife, recreation activities	Whitlow (1990); Madebwe and Madebwe (2005); Musasa and Marambanyika (2020)
Rivers	Provide water for irrigation, domestic supply, provide an important habitat for a wide variety of wildlife, natural water quality improvement, conserves reservoirs of wildlife and biodiversity, bolsters the recreational economy, helps limit and control flooding, transfers nutrients	Timberlake (2000); Finlayson and Pollard (2009); Mbereko et al. (2015); Nyamadzawo et al. (2015); Marambanyika and Beckedahl (2016a); Mandishona and Knight (2019)
Swamp	Natural water quality improvement, food for wildlife including nesting areas, shelter, and water, opportunities for recreation and aesthetic appreciation, absorb, and manage rainwater so that it does not become a flood disaster	Timberlake (2000); Finlayson and Pollard (2009); Chikodzi and Mufori (2018); Chimhiti (2019); Mandishona and Knight (2019)

Mujele, Lokola, Simatelele, and Musuna fishing camps in Binga District. This clearly demonstrates that wetland nests key benefits that can be tapped for economic and social well-being hence aiding human sustenance. Moreover, taking into account predicted future changes in hydrological patterns caused by global climate change, wetlands can increase resilience to drought, floods, and damage caused by extreme weather scenarios (Junk et al. 2013; Table 16.4). This therefore clearly

Figure 16.4 Horticultural activities being practiced by local communities in rural areas of Zimbabwe.

Table 16.4 Initiatives to promote wetland conservation and human livelihoods.

Institutions	Wetland type	Conservation approach	Literature sources
Environmental Management Agency (EMA)	Dambo, floodplains pans, rivers, marshes	They play a role in governing the type of activities carried out on the wetland. The government of Zimbabwe through EMA is mainly involved in the development of policies that help govern the use, promoting community-based clubs that teach on the sustainable use of wetlands as well as the importance of conserving this natural resource. Proper mapping of wetlands to help them determine wetland boundaries so as to create a database that contains all necessary information needed for developing and implementing conservation strategies.	Sithole and Goredema (2013), Matamanda et al. (2018), Mandishona and Knight (2019)
Monavale Wetland Trust	Dambo, floodplains, vleis	Awareness campaigns have been carried out in collaboration with EMA in an attempt to raise awareness on the sustainable use and conservation of the Monavale vlei enhance sound environmental practices	Murungweni (2013), Sithole and Goredema (2013), Mutisi and Nhamo (2017)
Harare Wetland Trust	Dambo, floodplains, rivers	Fostering the protection and preservation of wetlands. Advocating for safe, sustainable, and ecoaware methods of waste disposal, which will go a long way in the protection of wetlands. Wetland conservation and environmental education in the heart of Harare. Strengthen the public notification and consultation process for any planned interference with wetlands. To establish nature reserves on urban wetlands and programs for local management and restoration of wetland ecosystems for purposes of water provisioning and biodiversity conservation.	Murungweni (2013), Sithole and Goredema (2013), Sharai et al. (2020)

Table 16.4 (Continued)

Institutions	Wetland type	Conservation approach	Literature sources
Conservation Society of Monavale (COSMO)	Dambo, floodplains, vleis	COSMO is also responsible for governing access and activities undertaken on the wetland. Community-based initiative hence they use the wetland to carry out meetings and sometimes hold public lectures on the dangers of destroying wetlands. COSMO is also responsible for governing access and activities undertaken on the wetland. This is due to the fact that they are in a position to educate on how different activities like urban agriculture and improper waste dumping affect the wetland. COSMO is also an active participant in the development of conservation policies.	Sharai et al. (2020), Musasa et al. (2023)
Local Community Groups	Dambo, floodplains, vleis, rivers	Wetland committees at the local level strengthen wetland governance efforts, thereby enhancing sustainable wetland utilization by local communities. Involved in formulation of wetland management plans.	Sithole and Goredema (2013), Nyamadzawo et al. (2015), Marambanyika and Beckedahl (2016a, b)
Zimbabwe National Water Authority (ZINWA)	Dambo, floodplains, vleis, rivers	ZINWA manages water resources on a catchment basis with involvement of stakeholders in each catchment area. Management of the water permit system, the pricing of water, operating and maintaining existing infrastructure, and executing development projects. ZINWA, specifically the Upper Manyame Catchment Council, is interested in the quality of water that goes through the Monavale vlei, for example, hence they tend to carry out regular tests on the wetland. It is also involved in governing wetland access and activities and development of policies.	Government of Zimbabwe (2015), Musasa et al. (2023)
Environment Africa	Dambo, floodplains, rivers	NGOs are allowed to frequently assist local communities as they come with incentives aimed at conservation. Training and awareness on sustainable wetland use activities.	Murungweni (2013), Musasa et al. (2023)

shows that wetlands provide a broad range of goods and essential services that are invaluable to human livelihood (Figure 16.5).

Swamps ensure water quality improvement, food for wildlife including nesting areas, shelter, and water, opportunities for recreation and aesthetic appreciation (Chikodzi and Mufori 2018; Chimhiti 2019; Mandishona and Knight 2019; Table 16.3). These precious ecosystems also absorb and manage rainwater so that it does not become a flood disaster (Timberlake 2000; Finlayson and Pollard 2009). This therefore clearly shows that wetlands provide a broad spectrum of goods essential for human livelihood.

Figure 16.5 Vegetable and banana plantation in wetland in semiarid areas.

16.5 Initiatives to Promote Wetland Conservation and Human Livelihoods

In a bid to ensure sustainable use and conservation, Zimbabwe has made great strides, which have seen the development of a National Wetland Policy and Wetlands Management Guidelines, both launched in February 2022. These instruments are expected to play a prominent role as far as the governance framework for wetland use and conservation is concerned. Musasa et al. (2023) observed that Zimbabwe's wetland policy has brought a paradigm shift as far as wetland use, protection, and management are concerned. This is because it has clearly spelt guidelines that are informative and clearly identified key threatened areas, which inform conservation initiatives thereby enhancing the likelihood of improved environmental conservation. Wetland degradation and loss have been a cause of concern over the past decades (Sithole and Goredema 2013; Mutisi and Nhamo 2017). The Environmental Management Act (Chapter 20:27) prohibits unlicensed wetland use for agriculture and the enunciation of the policy and guidelines enhance enforcement of the law. Community sustainable wetland agriculture projects are also being implemented with the support of development agencies in different wetlands, though their degree of success in wise use tends to vary (Marambanyika et al. 2017).

Several studies such as those by Mandishona and Knight (2022) and Matamanda et al. (2018) confirmed that one of the strategies that have been put in place to promote wetland conservation is proper wetland mapping. Wetland mapping is the first step to monitoring wetlands and it helps to develop wetland inventories that can be used for informed decision-making regarding wise use and conservation (Murungweni 2013; Matamanda et al. 2018). This enables the detection of key threatened areas that will assist decision-makers and policy implementers to come up with meaningful wetland management and restoration strategies (Murungweni 2013; Musasa et al. 2023). The same situation was also observed by Matamanda et al. (2018) in a study conducted in Harare as they observed that wetlands continue to be degraded despite having sound policies and strategies and wetland mapping seems beneficial to counter this challenge. Adequate wetland mapping will ensure that required resources will be deployed when and where they are critically required (Environmental Management Agency 2015; Musasa et al. 2023). This reduces setbacks that may compromise wetland conservation efforts through improper application of efforts.

Literature also shows that awareness campaigns that have been carried out by state and nonstate institutions significantly contribute to wetland conservation (Murungweni 2013; Sithole and Goredema 2013; Mutisi and Nhamo 2017). For instance, Musasa et al. (2023) observed that in Harare's Monavale vlei, organizations such as Monavale Wetland Trust have launched a series of trainings that are aimed at enhancing awareness levels on wetland conservation. Awareness is important since it is a key capacity measure for the development and success of any particular initiative. Wetlands are better conserved if there is increased awareness among the local users as they will be aware of any potential move and its associated implications, thus preventing further degradation (Murungweni 2013; Mandishona and Knight 2019). It can be noted that raising awareness on the sustainable use and conservation of the wetlands will contribute to meaningful wetland conservation, which results in sustainable provisioning of essential goods and services for human livelihood.

Another important element is the fact that the government through the Environmental Management Agency (EMA) has fostered and strengthened private partnerships that benefit wetland conservation (Sithole and Goredema 2013; Matamanda et al. 2018; Mandishona and Knight 2019). Of notable importance is the Conservation Leadership Programme (CLP) Team in Zimbabwe, which is currently conducting research on testing environmental friendly models that could be applied to reduce maize crop damage on wetland health. Historically, Bird Life Zimbabwe engaged in livelihoods improvement initiatives like community gardens to reduce pressure on wetlands as habitats for floral, faunal, and avifaunal biodiversity (Fakarayi et al. 2015). The Conservation Society of Monavale (COSMO), a branch of the Monavale Residents' Environmental Action Group that intends to prevent the degradation of the Monavale wetland has also contributed to wetland conservation in Harare (Musasa et al. 2023). This society has joined hands with the EMA to assist local communities with resources and the required skills, and knowledge transfer in line with conservation of wetland ecosystems (Sharai et al. 2020; Musasa et al. 2023). As a result, most of the residents in the areas surrounding the Monavale wetland are involved in policymaking procedures (Sharai et al. 2020). The inclusion of the general public in policymaking helps foster a sense of ownership in the locals and encourages transparency, which is key toward wetland conservation (Zinhiva et al. 2014). This ensures sustainable implementation of set policies, hence sound wetland protection and management.

Previous studies (Marambanyika et al. 2012; Clare and Creed 2014) have shown that wetland management plans and strategies work well where there is local community participation which has resulted in establishment of wetland committees. Involving local people in wetland use planning and management brings a sense of ownership among themselves, which results in balancing the human needs and the natural resource base, hence sustainable utilization of wetland resources (Mandishona and Knight 2019). Marambanyika and Beckedahl (2016a, b) observed that establishment of wetland committees at the local level strengthened wetland governance efforts, thereby enhancing sustainable wetland utilization by local communities. Wetland committees consist of a team of members that are selected by the community to guide wetland use as well as come up with strategies that ensures that wetland ecological conditions are preserved, which contributes to sound conservation (Sharai et al. 2020; Marambanyika and Musasa 2023). Based on this information, it can be observed that local-level initiatives that involve communities and establishment of local wetland committees bring a sense of ownership, hence contributing to wetland protection.

Various studies have been undertaken on wetlands and show that the success of wetland conservation is anchored on people's perceptions, norms, values, attitudes, sustainable use of resources and services, and development of pro-poor policies and initiatives (Matamanda et al. 2018). Hence, there is also a need to adopt the same approach for Zimbabwe where emphasis will be on examining the values and uses of wetlands so as to identify the socioeconomic factors relating to wetlands. This will help and contribute in the formulation and development of policies, which relate to wetland use and management.

16.6 Shortcomings and Enforcement Challenges of Wetland Protection and Conservation Approaches

Wetlands are fragile ecosystems that need to be protected from unsustainable use activities. However, they are seldom explicitly covered at the national level in other existing natural resource management policies, which bring into question efforts aimed at conserving them (Bethune et al. 2005; Mwale 2018). It is therefore key to note that wetland management policies and regulations put in place are considerate of the impacts of surrounding communities that share the same river basin (Matamanda et al. 2018). Overall, in Zimbabwe, implementation and enforcement of wetland policies and legislation has been hindered by; inadequate stakeholder participation, fragmented sectorial approaches (Frenken and Mharapara 2002; Mwale 2018), different government policies and agendas as well as inadequate financing toward wetland conservation (Sithole and Goredema 2013; Matamanda et al. 2018).

Inconsistent dissemination of provisions of the law and policies presents challenges as far as wetland conservation is concerned in Zimbabwe (Marambanyika and Beckedahl 2016a; Matamanda et al. 2018). In most cases, there is contradictory dissemination of wetland law provisions by government agencies. For example, officers from the same agency (for instance, EMA) may either encourage or forbid specific wetland activities, in which local communities are actively involved despite local people being deemed custodians of the resources (Marambanyika et al. 2012), hence confusing local people's conservation initiatives. The same sentiments were echoed by Chabwela and Haller (2010) in a study carried out in Kafue Flats, Zambia. This means that law enforcement by statutory bodies should be taken into consideration through an approach that encourages citizens' participation in decision-making and implementation so much so that rather than being a degrading force they have a sense of ownership thus ensuring wetland protection.

To add on, there are issues related to inadequate financing to spearhead wetland management efforts as well as incapacitation through human resource skills and other valuable resources, which hinders effective wetland policy and legislation implementation (Marambanyika and Beckedahl 2016b; Mandishona and Knight 2022).

In several instances, wetland laws are not respected due to the lack of monitoring and weak enforcement by the responsible authorities as they are financially incapacitated to conduct their duties (Sithole and Goredema 2013; Marambanyika and Beckedahl 2016b). Therefore, encroachment into fragile and sensitive wetland areas continues despite having a raft of pieces of legislation. Therefore, in order to improve policy implementation and enforcement mechanisms, well-resourced institutions should be established so as to provide commensurate resources and support.

Political interference has also been cited as a major stumbling block, which has resulted in the belittling of wetland law. In some cases, political leaders pursued populist policies to win the hearts of the electorate thereby condoning illegal use of the resource (Marambanyika and Beckedahl 2016b; Mandishona and Knight 2019). Politicians sometimes persuade regulatory officers to ignore illegal wetland activities with impunity, thereby threatening wetland ecosystems. The prime effect of such a scenario is wetland use without permits, which has weakened the capacity of the responsible institutions to enforce legal stipulations where political ambitions were under threat. More similarly, in the Kafue Flats of Zambia, many farmers joined local leaders in unsustainable exploitation as they feared losing farming space in the wetlands (Chabwela and Haller 2010). Politics should therefore not be allowed to interfere with the implementation of measures aimed at conserving wetlands in Zimbabwe if wetlands are to be conserved.

Lack of awareness on the rationale and existence of laws governing wetland conservation is a major constraint toward the implementation of policies and legislation (Marambanyika and Beckedahl 2016a). This is specifically true in areas where there is a disjuncture between the legal provisions and practical implementation. This brings the question; does legislation remain only in theory yet sound wetland management lies within practical applicability? To this end, for as long as there is a limited understanding of the law, enforcement will remain a challenge. In communal areas of Zimbabwe, it was observed that there was a missing link between awareness and policy implementation a scenario, which frustrated efforts aimed at enhancing sustainable wetland use and conservation efforts (Marambanyika and Beckedahl 2016a).

In Zimbabwe, environmental legislation and its administration are plagued by poor organization and rampant noncompliance. It has been noted that, despite persistent noncompliance, little if not nothing has been done to enforce the environmental laws (Marambanyika and Beckedahl 2016a). One of the most heralded reasons cited for this is the fact that authorities with the mandate to enforce legislation are more concerned with promoting national and local development initiatives. In developing countries, history has shown that when faced with a choice between resource conservation and development, enforcing authorities tend to choose the latter which contributes to wetland degradation and loss (Phethi and Gumbo 2019).

16.7 Conclusion and Recommendations

The chapter examined wetland types and distribution in Zimbabwe, biodiversity characteristics, and categories of livelihood benefits derived from wetlands by the local communities. Results demonstrate that wetlands nest a number of key livelihood benefits with the major being opportunities for crop cultivation, providing pasture and water for livestock drinking especially in rural communities in semiarid lands of Zimbabwe. Urban wetlands in Zimbabwe provide opportunities for tourism and act as regulation of flow accumulation and buffer to floods. However, increasing population has resulted in encroachment of these precious systems, which puts their sustainability into question if there is no balance between conservation and use. In order to protect these ecosystems, there is a need to come up with a holistic approach that encourages wise use as enshrined in the Ramsar wise use guidelines. Wetlands have been providing local communities by supporting crop cultivation and several horticultural activities, which assist toward the attainment of Sustainable Development Goals (SDGs) 1 on ending poverty and 2 on zero hunger, respectively. To promote the sustainability of wetland use, several conservation initiatives have been implemented to promote wetland conservation and human livelihoods, such as legislation, policies, institutional arrangements, programs, and projects with varying degrees of success. It can be noted that raising awareness on the sustainable use and conservation of the wetlands will contribute to meaningful wetland conservation, which results in sustainable provisioning of essential goods and services for human livelihood. The chapter proposes sustained longitudinal research to build a database of essential information that informs continual improvement of wetland management decision-making.

References

Bethune, S., Amakali, M., and Roberts, K. (2005). Review of Namibian legislation and policies pertinent to environmental flows. *Physics and Chemistry of the Earth* 30 (11–16): 894–902.

Bhatta, L.D., Chaudhary, S., Pandit, A. et al. (2016). Ecosystem service changes and livelihood impacts in the Maguri-Motapung Wetlands of Assam, India. *Land* 5 (2): 15.

Blessing, N. (2020). Construction in the Republic of Zimbabwe: rural and urban buildings. In: *V Международный студенческий строительный форум-2020*, 30–39.

Bullock, A. (1992). The role of dambos in determining river flow regimes in Zimbabwe. *Journal of Hydrology* 134: 349–372.

Bullock, A. and McCartney, M.P. (1995). Wetland and river interactions in Zimbabwe. *Tropical Hydrology* 238: 305–321.

Chabwela, H. and Haller, T. (2010). Governance issues, potentials and failures of participative collective action in the Kafue Flats, Zambia. *International Journal of the Commons* 4 (2): 621–642. https://doi.org/10.18352/ijc.189.

Chifamba, E. (2013). Livelihood resilience through utilization of wetland patches in dry areas of Buhera North, Zimbabwe. *Wudpecker Journal of Agricultural Research* 2 (11): 303–314.

Chikodzi, D. and Mufori, R.C. (2018). Wetland fragmentation and key drivers: a case of Murewa District of Zimbabwe. *Journal of Environmental Science Toxicology Food Technology* 12 (9): 49–61.

Chimbindi, F. (2018). *Land Ownership in Zimbabwe: Factsheet*. Zimfact. https://zimfact.org/factsheet-land-ownership-in-zimbabwe/.

Chimhiti, N. (2019). The effects of grazing encroachment on plant species density and invasive species infestation in Magwenzi wetland, Chivi district. MSc thesis. Midlands State University.

Chuma, E., Masiyandima, M., Finlayson, M. et al. (2008). *Guideline for Sustainable Wetland Management and Utilisation: Key Cornerstones. CPWF Project Report*. Colombo: CGIAR Challenge Programme on Water and Food https://hdl.handle.net/10568/21605.

Cimon-Morin, J., Darveau, M., and Poulin, M. (2016). Site complementarity between biodiversity and ecosystem services in conservation planning of sparsely-populated regions. *Environmental Conservation* 43 (1): 56–68.

Clare, S. and Creed, I.F. (2014). Tracking wetland loss to improve evidence-based wetland policy learning and decision making. *Wetlands Ecology and Management* 22 (3): 235–245.

Dahwa, E., Kativu, S., Hungwe, T. et al. (2014). Influence of grazing intensity on plant community diversity, structure and composition in Dambo wetlands of communal Zimbabwe. *Global Journal of Biology, Agriculture & Health Sciences* 3 (1): 259–266.

Davidson, N.C., Fluet-Chouinard, E., and Finlayson, C.M. (2018). Global extent and distribution of wetlands: trends and issues. *Marine and Freshwater Resources* https://doi.org/10.1071/MF17019.

Dertien, J.S., Self, S., Ross, B.E. et al. (2020). The relationship between biodiversity and wetland cover varies across regions of the conterminous United States. *PLoS ONE* 15 (5): e0232052.

Dube, T. (2005). The impact of communal land use on dambos in Lower Gweru, Zimbabwe. Dissertation. University of Zimbabwe.

Environmental Management Agency (2015). *Annual Report*. Harare: Government of Zimbabwe.

Environmental Management Agency (2021). *Zimbabwe Wetland Policy*. Harare: Government of Zimbabwe.

Epele, L.B., Grech, M.G., Manzo, L.M. et al. (2021). Identifying high priority conservation areas for Patagonian wetlands biodiversity. *Biodiversity and Conservation* 30: 1359–1374.

Fakarayi, T., Mashapa, C., Gandiwa, E., and Kativu, S. (2015). Pattern of land-use and land cover changes in Driefontein grassland important bird area, Zimbabwe. *Journal of Tropical Conservation Science* 8 (1): 274–283.

Finlayson, C.M. and Pollard, S. (2009). *A Framework for Undertaking Wetland Inventory Assessment and Monitoring in the Limpopo Basin Southern Africa. CPWF Project Report*. Colombo: CGIAR Challenge Program on Water and Food.

Frenken, K. and Mharapara, I. (ed.) (2002). *Wetland Development and Management in SADC Countries*. Harare: FAO Sub-Regional Office for East and Southern Africa.

Freshwater Ecosystems of the World (2019). FEOW. https://www.feow.org/ (accessed 26 September 2023).

Government of Zimbabwe (GoZ) (2015). *Fifth National Report to the Convention on Biological Diversity*. Harare: Ministry of Environment Water and Climate.

Guareschi, S., Laini, A., Viaroli, P., and Bolpagni, R. (2020). Integrating habitat-and species-based perspectives for wetland conservation in lowland agricultural landscapes. *Biodiversity and Conservation* 29: 153–171.

Gxokwe, S., Dube, T., and Mazvimavi, D. (2022). Leveraging Google Earth Engine platform to characterize and map small seasonal wetlands in the semi-arid environments of South Africa. *Science of the Total Environment* 803: 150139.

Hove, C. and Chapungu, L. (2013). Human perceptions on degradation of wetland ecosystems: the case of Magwenzi Wetland in Chivi District, Zimbabwe. *Journal of Geography, Environment and Earth Science* 1 (1): 13–22.

Israel, P.M. and Timar, P.M. (2018). Wetland ecosystems in Ethiopia and their implications in ecotourism and biodiversity conservation. *Journal of Ecology and the Natural Environment* 10 (6): 80–96.

Junk, W.J., Shuqing, A.N., Finlayson, C.M. et al. (2013). Current state of knowledge regarding the world's wetlands and their future under global climate change: a synthesis. *Journal of Aquatic Sciences* 75 (1): 151–167.

Lu, D. and Chang, J. (2023). Examining human disturbances and inundation dynamics in China's marsh wetlands by using time series remote sensing data. *Science of the Total Environment* 863: 160961.

Madebwe, V. and Madebwe, C. (2005). An exploratory analysis of the social, economic and environmental impacts on wetlands: the case of Shurugwi District, Midlands Province, Zimbabwe. *Journal of Applied Sciences Research* 1 (2): 228–233.

Manatsa, D., Mushore, T.D., Gwitira, I. et al. (2020). *Revision of Zimbabwe's Agro-Ecological Zones. Produced by the Government of Zimbabwe under Zimbabwe National Geospatial and Space Agency (ZINGSA) for the Ministry of Higher and Tertiary Education, Innovation*. Science and Technology Development, Zimbabwe Government Printers.

Mandishona, E. and Knight, J. (2019). Users' perceptions and understanding of two urban wetlands in Harare, Zimbabwe. *South African Geographical Journal* 101 (3): 326–348.

Mandishona, E. and Knight, J. (2022). Historical changes in wetland management legislation and practices in Zimbabwe, and their implications. *Wetlands Ecology and Management* 30 (6): 1197–1211.

Marambanyika, T. and Beckedahl, H. (2016a). Wetland utilisation patterns in semi-arid communal areas of Zimbabwe between 1985 and 2013 and the associated benefits to livelihoods of the surrounding communities. *Transactions of the Royal Society of South Africa* 71 (2): 175–186.

Marambanyika, T. and Beckedahl, H. (2016b). The missing link between awareness and the implementation of wetland policy and legislation in communal areas of Zimbabwe. *Wetland Ecology and Management* 24 (5): 545–563.

Marambanyika, T. and Musasa, T. (2023). Contestations and struggles in the use and protection of wetland ecosystems in communal areas of Zimbabwe: a case study of Intunjambili wetland in Matobo district. *Wetlands Ecology and Management*, 1 (3) pp.1–17.

Marambanyika, T., Mutsiwegota, C., and Muringaniza, K.C.R. (2012). Importance of community participation in sustainable utilization of wetlands: case of Chebvute in Zvishavane District of Zimbabwe. *Journal of Environmental Science and Engineering. B* 1 (7B): 832.

Marambanyika, T., Beckedahl, H., Ngetar, N.S. et al. (2017). Assessing the environmental sustainability of cultivation systems in wetlands using the WET-health framework in Zimbabwe. *Physical Geography* 38 (1): 62–82.

Marambanyika, T., Musasa, T., Mupfiga U. and Ngwenya, K (2021) Local perceptions on the impact of drought on wetland ecosystem services and associated household livelihood benefits: the case of Driefontein Ramsar site in Zimbabwe. *Land, MDPI*, 10(6): p.587. https://doi.org/10.3390/land10060587

Maseko, P.Z. and Bussink, M. (1995). Wetlands and water harvesting as resources in communal farmers' food security. In: *Proceedings for a Workshop on Dambo Farming in Zimbabwe: Water Management, Cropping and Soil Potentials for Smallholder Farming in the Wetlands* (ed. R. Owen), 54–67. Harare: University of Zimbabwe Publications.

Matamanda, A.R., Chirisa, I., Mukamuri, B.B. et al. (2018). Harare's deteriorating wetlands: why sound policies and legislations are not enough. *Case Studies in the Environment* 2 (1): 5–11.

Matiza, T. (1992). Wetlands in Zimbabwe: an overview. In: *Wetlands Ecology and Priorities for Conservation in Zimbabwe: Proceedings of a Seminar on Wetlands of Zimbabwe* (ed. T. Matiza and S.A. Crafter), 3–10. Harare: IUCN.

Matiza, T. and Crafter, S.A. (ed.) (1994). *Wetlands Ecology and Priorities for Conservation in Zimbabwe: Proceedings of a Seminar on Wetlands of Zimbabwe*. Gland: IUCN.

Mazvimavi, D. (2008). Investigating possible changes of extreme annual rainfall in Zimbabwe. *Journal of Hydrology and Earth Systems Science, Discuss* 5 (4): 1765–1785.

Mazvimavi, D. (2010). Investigating changes over time of annual rainfall in Zimbabwe. *Journal of Hydrology and Earth Systems Science* 14 (12): 2671–2679.

Mbereko, A., Chimbari, M.J. and Mukamuri, B.B. (2007). An analysis of institutions associated with wetlands use, access and management in communal areas of Zimbabwe: a case study of Zungwi vlei, Zvishavane. *Journal of Physics and Chemistry of the Earth Parts A/B/C.* 32 (15–18):1291–1299. https://doi.org/10.1016/j.pce.2007.07.038

Mbereko, A., Mukamuri, B.B., and Chimbari, M.J. (2015). Exclusion and contests over wetlands used for farming in Zimbabwe: a case study of broad-ridge and broad-furrow tillage system on Zungwi Vlei. *Journal of Political Ecology* 21: 322–338.

Moisa, M.B., Bulto, T.W., Werku, B.C. et al. (2023). Analyzing wetland dynamics using geospatial techniques: a case of Abay Choman and Jimma Geneti Watershed, Horo Guduru Wollega Zone, Western Ethiopia. *Air, Soil and Water Research* 16: 11786221221150183.

Munishi, P.K., Wilfred, N.N., Nshare, J.S. et al. (2011). Valley bottom wetlands can serve for both biodiversity conservation and local livelihoods improvements. *Ecosystems Biodiversity* 2 (1): 310–338.

Murungweni, F.M. (2013). Effect of land use change on quality of urban wetlands: a case of Monavale wetland in Harare. *Geoinfor Geostat: An Overview* https://doi.org/10.4172/2327-4581.S1-015.

Musasa, T. and Marambanyika, T. (2020). Threats to sustainable utilization of wetland resources in Zimbabwe: a review. *Wetlands Ecology and Management* https://doi.org/10.1007/s11273-020-09732-1.

Musasa, T. and Marambanyika, T., (2022). Assessing the sustainability of land uses in Driefontein and Intunjambili wetlands, Zimbabwe. *Scientific African*, 2 (3) p. e01195., Elsevier.

Musasa, T., Muringaniza, K.C., and Manyati, M. (2023). The role of stakeholder participation in wetland conservation in urban areas: a case of Monavale Vlei, Harare. *Scientific African* 19: e01574.

Mutisi, L. and Nhamo, G. (2017). An analysis of water quality in belvedere north and the epworth wetlands of Harare, Zimbabwe. books.google.com. https://books.google.com/books?hl=en&lr=&id=pl1DDgAAQ (accessed 20 October 2021).

Mwale, C. (2018). Exploring the implications of institutional fragmentation and the role of community participation in sustainable natural resource management in Zambia. A case study of Lochinvar National Park. Doctoral dissertation. University of Zambia.

Nyamadzawo, G., Wuta, M., Nyamangara, J. et al. (2015). Optimizing dambo (seasonal wetland) cultivation for climate change adaptation and sustainable crop production in the smallholder farming areas of Zimbabwe. *International Journal of Agricultural Sustainability* 13 (1): 23–39.

Phethi, M.D. and Gumbo, J.R. (2019). Assessment of impact of land use change on the wetland in Makhitha village, Limpopo province, South Africa. *Jàmbá Journal of Disaster Risk Studies* 11 (2): 1–6. https://doi.org/10.4102/jamba.v11i2.693.

Ramsar Convention Secretariat (2018). *Global Wetland Outlook: State of the World's Wetlands and Their Services to People*. Gland: Ramsar Convention Secretariat.

Rebelo, L.M., McCartney, M.P., and Finlayson, C.M. (2010). Wetlands of Sub-Saharan Africa: distribution and contribution of agriculture to livelihoods. *Wetlands Ecology and Management* 18 (5): 557–572.

Rivers-Moore, N.A. and Cowden, C. (2012). Regional prediction of wetland degradation in South Africa. *Wetlands Ecology and Management* 20 (6): 491–502.

Sharai, S., Tawanda, T., and Gladman, C. (2020). Threats and conservation strategies on urban wetlands: a case of Monavale and surrounding areas in Harare, Zimbabwe. *American Journal of Environmental Protection* 9 (2): 36–43.

Sithole, A. and Goredema, B. (2013). Building in wetlands to meet the housing demand and urban growth in Harare. *International Journal of Humanities and Social Science* 3 (8): 1–13.

Soultan, A., Pavón-Jordán, D., Bradter, U. et al. (2022). The future distribution of wetland birds breeding in Europe validated against observed changes in distribution. *Environmental Research Letters* 17 (2): 024025.

Steen, V.A., Skagen, S.K., and Melcher, C.P. (2016). Implications of climate change for wetland-dependent birds in the Prairie Pothole Region. *Wetlands* 36 (Suppl 2): 445–459.

Svotwa, E., Manyanhaire, I.O., and Makombe, P. (2008). Sustainable gardening on wetlands in the communal lands of Zimbabwe. *Electronic Journal of Environmental, Agricultural and Food Chemistry* 7 (3): 2754–2760.

Timberlake, J. (2000). *Biodiversity of the Zambezi Basin*. Biodiversity Foundation for Africa Bulawayo.

Turner, R.K., van den Bergh, J.C.J.M., Soderqvist, T. et al. (2000). Ecological-economic analysis of wetlands: scientific integration for management and policy. *Journal of Ecological Economics* 35: 7–23. https://doi.org/10.1016/S0921-8009(00)00164-6.

Whitlow, R. (1990). Conservation status of wetlands in Zimbabwe: past and present. *GeoJournal* 20 (3): 191–202. Springer.

Yousef Haroon, A. K., and Kirkia, G. (2017). Wetlands: biodiversity and livelihood values and significance with special context to Bangladesh. Wetland Science: Perspectives from South Asia. pp.317–346.

Zinhiva, H., Chikodzi, D., Mutowo, G. et al. (2014). The implications for loss and degradation of wetland ecosystems on sustainable rural livelihoods. Case of Chingombe Community, Zimbabwe. *Greener Journal of Environmental Management and Public Safety* 3 (2): 43–52.

17

Wetlands of Sudan: Types, Conservation, and Socioeconomics

Mohamed Elmekki Ali Elbadawi Hussien[1] and Ibrahim Mohammed Hashim[2]

[1] *Department of Wildlife, University of Sinnar, Suki, Sudan*
[2] *Sudanese Wildlife Society, Khartoum, Sudan*

17.1 Introduction

Wetlands in Sudan are part of the diverse and ecologically significant wetlands found across Africa. These wetlands are crucial for biodiversity (Qiu et al. 2024), supporting various flora and fauna (Grasel et al. 2021), and they play a vital role in the livelihood of local communities (Raburu et al. 2012). The wetlands provide essential ecosystem services such as water filtration, flood control, and habitats for a wide range of species (Mitsch et al. 2015; Xu et al. 2020). They also have significant cultural and economic value for the people of Sudan and beyond (Reimold et al. 1980; Emerton 2018).

The wetlands of Sudan are primarily associated with river systems (riverine), Red Sea system (marine), irrigated systems (man-made), and montane system (saline). Based on their importance in these systems, they are either protected or inscribed as Biosphere Reserves or Ramsar sites.

In this chapter, we intend to overview aspects of Sudan's wetlands, policies and laws, classification, biodiversity, conservation, and socioecological importance.

17.2 Wetlands of Sudan

17.2.1 Wildlife Policies and Laws

17.2.1.1 Laws

The following laws and regulations are in place:

a) Wildlife regulations 2005; according to the Interim Constitution, Article 11, Para (b): The State shall not follow policies or do or allow any action that may have a negative effect on the presence of any animals' diversity or their natural habitat; Para (c) denotes: The State, by legislation, shall develop the best practices for natural resources utilization and management.
b) Wildlife and Protected Area Act 1986; focuses on organizing, utilization, and degree of protection to species, ownership, trade, transport of species, types of protracted areas, forbidden actions inside National Parks, and powers given to wildlife personnel and punishment and penalties on convection.

Wetlands of Tropical and Subtropical Asia and Africa: Biodiversity, Livelihoods and Conservation,
First Edition. Edited by Thammineni Pullaiah.

c) Wildlife and National Park regulations of 2000:
d) **Environmental Act of 2000**: It is an umbrella law to protect environment in general and natural resources in particular. It gives special attention to the protection of wildlife species and their habitat with emphasis on protected areas. The act authorizes any citizen to open a case if any article is violated.
e) The Presidential Declaration of 2000 in which the president affirmed and directed all concerned institutions and people to protect wildlife and the Protected Areas.

17.2.1.2 Policies

Sudan is a party to the following global and regional multilateral agreements:

a) The Convention on the International Trade in Endangered Species of Wild Fauna and Flora (CITES) in 1973.
b) The Ramsar Convention on Wetlands in 1971.
c) The African-Eurasian water bird agreement (AEWA) in 1999.
d) The Convention on Migratory Species (CMS) in 1979.
e) The United Nations Convention on the Law of Sea in 1982 and the Convention on International Marine Organization in 1958.
f) The Regional Convention for the Conservation of the Environment of Red Sea and the Gulf of Aden (PERSGA) in 1982.
g) Bilateral agreement in tourism between Sudan, Kenya, Zanzibar, Central African Republic, Tunisia, Saudi Arabia, Libya, Uganda, Turkey, Algiers, Lebanon, Ethiopia, Jordan, and Bahrain.
h) Bamako Convention on the Ban of the Import into Africa and the Control of Transboundary Movement and Management of Hazardous Wastes within Africa in 1993.
i) Convention Concerning the Protection of the World Cultural and Natural Heritage in 1974.
j) Convention on Biological Diversity (CBC) in 1995.
k) Convention on the Law of the Sea in 1985.
l) The Nile Basin Initiative in 1999.
m) The guidelines of International Union for Conservation of Nature (IUCN), World Wide Fund for Nature (WWF), and Birdlife Conservation International are adopted.
n) Four management plans for four protected areas are developed, which is considered a policy guide for the management of these protected areas.

17.2.1.3 Regulatory Bodies and Mechanisms

The Ministry of Tourism, Antiquities and Wildlife (MTAW) is technically in charge of opening, closing, or stopping the hunting periods. The Wildlife Conservation General Administration (WCGA), administered by the Ministry of Interior, oversees tackling all issues of hunting licenses, according to the policies of the MTAW. Privately owned hunting agencies whose clients come from the Middle East and Europe take the hunters to the hunting sites under the supervision of WCGA. The Wildlife Research Centre, under the Ministry of Animal Resources, Fisheries and Range, is the official governmental body in charge of conducting wildlife research in Sudan. However, there is no official link between the Wildlife Research Centre and the MTAW and The Ministry of Interior and WCGA.

17.2.1.4 Roles and Responsibilities

There seems to be no control over the national hunters although safari hunters are always accompanied by wildlife personnel. The role of the companion is to make sure that international hunters do not exceed their bag limits and possession limits. Usually, safari companies hire guides from the local community.

The bag and the possession limits issued by WCGA are not research-based. Hunting licenses are issued without knowing what is available in the field. On many occasions, hunting was banned for five years after which licenses were issued. This was based on no solid data to support the decision that animals' population numbers are decreasing. Literally, the role of the Wildlife Research Centre is to address such research problems. Literature abounds with reports, theses, and publications by the Wildlife Research Centre and Universities since the 1970s but WCGA may have no access to this volume of research. This is because there is no mechanism for transferring technological packages generated by the research institutions to WCGA to be adopted for wildlife conservation.

17.2.1.5 NGOs

The Sudanese Wildlife Society usually conducts wildlife conservation projects in collaboration with WCGA. The society participates in WCGA workshops, meetings, and sometimes vital decisions about conservation issues. The society is a member of the scientific authority that addresses CITES issues.

17.3 Wetland Types

There are four sources of water in Sudan that feed the wetlands with water: rainfall, River Nile system, perennial streams, flood water, and ground aquifer, in addition to wetland on the Red Sea. Thus, the Sudanese wetlands are Marine Coastal Wetlands, Inland Wetlands, and man-made wetlands (Table 17.1).

Table 17.1 Classification of Sudanese wetlands.

Inland wetlands				
Freshwater	Lotic	Permanent	M	Permanent rivers/streams/creeks; includes waterfalls.
		Seasonal	N	Seasonal/intermittent/irregular rivers/streams/creeks.
	Lakes, Basins	Permanent	O	Permanent freshwater lakes (over 8 ha); includes large oxbow lakes.
			Tp	Permanent freshwater marshes/pools; ponds (below 8 ha), marshes and swamps on inorganic soils; with emergent vegetation waterlogged for at least most of the growing season.
		Seasonal	P	Seasonal/intermittent freshwater lakes (over 8 ha); includes floodplain lakes.
			Ts	Seasonal/intermittent freshwater marshes/pools on inorganic soils; includes sloughs, potholes, seasonally flooded meadows, and sedge marshes.
			W	Shrub-dominated wetlands; shrub swamps, shrub-dominated freshwater marshes, shrub carr, alder thicket on inorganic soils.
			Xf	Freshwater, tree-dominated wetlands; includes freshwater swamp forests, seasonally flooded forests, wooded swamps on inorganic soils.

(*Continued*)

Table 17.1 (Continued)

Inland wetlands			
Saline	Permanent	Q	Permanent saline/brackish/alkaline lakes.
		Sp	Permanent saline/brackish/alkaline marshes/pools.
		R	Seasonal/intermittent saline/brackish/alkaline lakes and flats.
	Seasonal	Ss	Seasonal/intermittent saline/brackish/alkaline marshes/pools.
Non-forested peatlands		U	Non-forested peatlands; includes shrub or open bogs, swamps, fens.
Forested peatlands		Xp	Forested peatlands; peat swamp forests.
		Y	Freshwater springs; oases.
		Zg	Geothermal wetlands
		Zk(b)	Karst and other subterranean hydrological systems, inland
Marine wetlands			
A	Permanent shallow marine waters; in most cases, less than 6 m deep at low tide; including sea bays and straits		
B	Marines' tidal aquatic beds; including kelp beds, seagrass beds, and tropical marine meadows.		
C	Coral reefs.		
D	Rocky marine shores; including rocky offshore islands, sea cliffs.		
E	Sand, shingle, or pebble shores; including sand bars, spits, and sandy islets; dune systems and humid dune slacks.		
F	Estuarine waters; permanent water of estuaries and estuarine systems of deltas.		
G	Intertidal mud, sand, or salt flats.		
H	Intertidal marshes; including salt marshes, salt meadows, salting, raised salt marshes; tidal brackish and freshwater marshes.		
I	Intertidal forested wetlands; including mangrove swamps, Nipah swamps, and tidal freshwater swamp forests.		
J	Coastal brackish/saline lagoons; brackish to saline lagoons with at least one relatively narrow connection to the sea.		
K	Coastal freshwater lagoons; including freshwater delta lagoons.		
Karst/ZK(a)	Karst and other subterranean hydrological systems, marine/coastal		
Man-made wetlands			
1	Aquaculture (e.g. fish/shrimp) ponds		
2	Ponds; includes farm ponds, stock ponds, small tanks (generally below 8 ha).		
3	Irrigated land; including irrigation channels and rice fields.		
4	Seasonally flooded agricultural land (including intensively managed or grazed wet meadow or pasture).		
5	Salt exploitation sites; salt pans, saline, etc.		
6	Water storage areas; reservoirs/barrages/dams/impoundments (generally over 8 ha).		
7	Excavations; gravel/brick/clay pits; borrow pits, mining pools.		
8	Wastewater treatment areas; sewage farms, settling ponds, oxidation basins, etc.		
9	Canals and drainage channels, ditches.		
Zk(c)	Karst and other subterranean hydrological systems, human-made		

Among the wetlands in various ecosystems in Sudan (Table 17.2), only 35 of them in 6 ecosystems are prioritized by the national conservation authorities (Table 17.2); of these 2.7% were in saline (Figure 17.1), 11.1% in man-made (Figures 17.2 and 17.3), 13.8% in marine (Figure 17.4), 65.7% in riverine ecosystems, and 5.7 in desert (Figure 17.5) ecosystem. The conservation status of the wetlands is not satisfactory because all those in saline, man-made and desert ecosystems are not protected. The situation is different for the wetlands in the marine ecosystem. Among the five wetlands in this system, two are protected. The conservation status of the wetlands in the riverine ecosystem is even worse: Only 5 of 23 wetlands in this system are protected. Altogether, 17.1% of the wetlands across ecosystems in Sudan are protected. This deficiency in conservation can negatively impact biodiversity supported by the wetlands.

Table 17.2 Wetlands of Sudan.

#	Ecosystem	Wetland	Location	Area/ha	Status
1	Saline	Jebel Marra Volcanic lakes	13°04′N 24°21′E	150,000	Unprotected
2	Man-made Agricultural Schemes	Gezira	14°30′N 33°10′E	850,000	Unprotected
3	Man-made, River Nile	Wadi Halfa (Lake Nubia) Reservoir	21°55′N 31°19′E	10,000	Unprotected
4	Man-made, Blue Nile	Roseires Dam Reservoirs	11°47′53″N 34°23′15″E	29,000	Unprotected
5	Man-made, Blue Nile	Sinnar Dam reservoir	13°32′29.99″N 33°38′4.19″E	400,000	Unprotected
6	Marine	Khor Arba'at	19°48′N 37°03′E	200,000	Unprotected
7	Marine	Mukawwar island and Dunganab bay	20°50′N 37°17′E	12,000 ha	Protected
8	Marine	Suakin archipelago	18°50′N 38°00′E	150,000	Unprotected
9	Marine	Sanganeb Atoll	19°44′10″N 37°26′35″E	199,524	Protected
10	Marine	14 Mangroves	Along Red Sea coast	329–721	Unprotected
11	Riverine	Sunut forest and UmShugeira island	15°34′N 15°35′N 32°3′0E 32°29′E	202	Protected
12	Riverine	Um-Jar	13°32′N–12°52/′N 32°19.5′E–32°36.5E	25,000	Unprotected
13	Riverine	Khor Abu Habil Delta	13°03′N 32°06′E	946,409	Unprotected
14	Riverine	Gladeema	13°20′30″N 33°44′33″E	3,000	Unprotected

(*Continued*)

Table 17.2 (Continued)

#	Ecosystem	Wetland	Location	Area/ha	Status
15	Riverine	Rahad kobri 45	12°37′1.6″N 34°2′35.3″	Not estimated	Unprotected
16	Riverine	Lawni	12°49′22.1″N 34°01′21.7″	2,500	Unprotected
17	Riverine	Elban	13°4′51.96″N 33°35′9.06″E	2,000	Unprotected
18	Riverine	Homrani	13°18′14.461″N 33°52′28.56″E	3,000	Unprotected
19	Riverine	Lake Kundi	10°26′N 25°I0′E	2,000	Unprotected
20	Riverine	Lake Keilak	10°50′N 29°17′E	500 in dry season; 3,000 in wet season	Unprotected
21	Riverine	Turdat el Rahad	12°44′N 30°38′E	Not estimated	Unprotected
22	Riverine	Radom National Park	09°10′N 24°00′E	1,250,970	Protected
23	Riverine	Dinder National Park	12°26′N 35°13′E	1,029,100	Protected
24	Riverine	Taya-Basonda-Algalabat National Park	36°02′ 35°02′N 12°:46′12°:42′E	209,200	Protected
25	Riverine	Lake Um Bader	14°15′N 27°50′E	2,000	Unprotected
26	Riverine	Lake Abiad	10°17′N 29°59′E	500,000	Unprotected
27	Riverine	Tordat umdafog			Unprotected
28	Riverine	Nabagah			Unprotected
29	Riverine	Alsafya			Unprotected
30	Riverine	Dahal deleibba			Unprotected
31	Riverine	Hamidia pool	12°54′806″N 23°29°21″1E	64	Unprotected
32	Riverine	Zour lake	12°57′295N 23°31′785″E	100	Unprotected
33	Riverine	Gugoldonga pool	12°53′606″N 23°29′978″E	9	Unprotected
34	Desert	Nukhaila oasis	14°34′48″N 29°29′24″E		Unprotected
35	Originated from borehole in Nubian Aquifer in desert	El-Ga'ab oasis	19° 38′ N/30° 18.25′ E, 19° 39′ N /30° 7.5′ E, 18° 33.8′ N /29° 59.8′ E and 18° 32.25′ N /30° 22.8′ E	24	Unprotected

Figure 17.1 Jebel Marra/Deriba Caldera (crater lake). *Source:* J Williams/Wikimedia Commons/ https://commons.wikimedia.org/w/index.php?curid=1557097/CC BY 2.5.

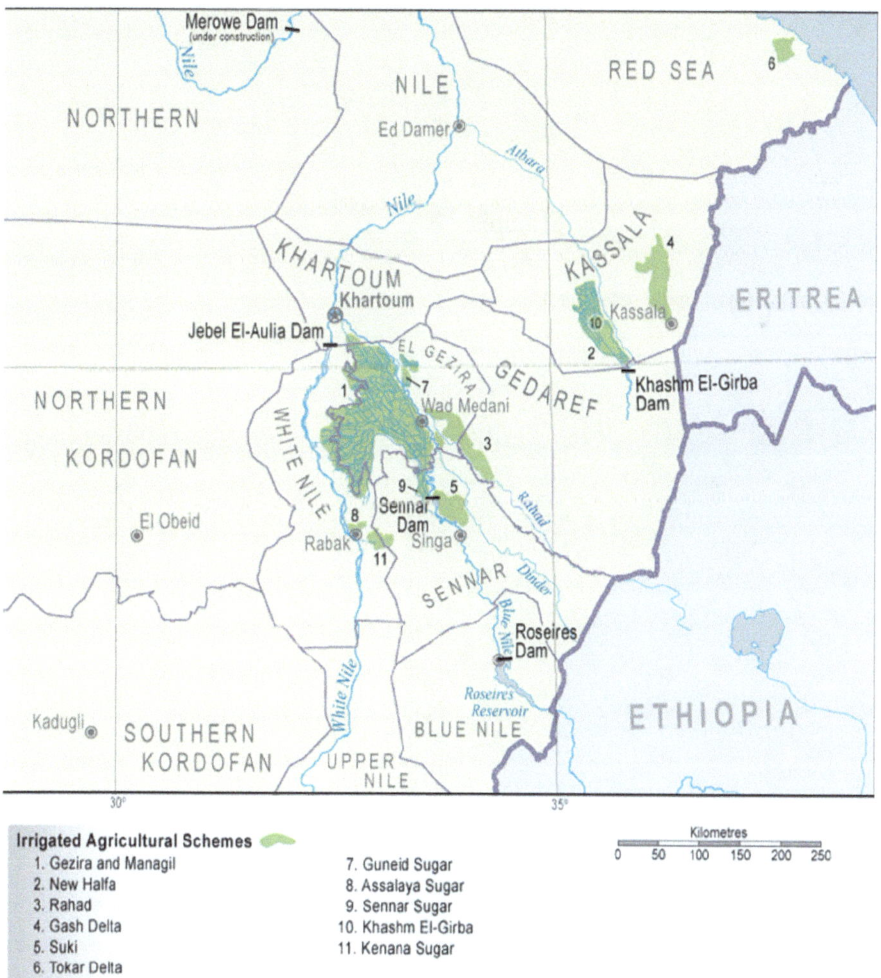

Figure 17.2 Irrigated agricultural schemes (man-made wetlands).

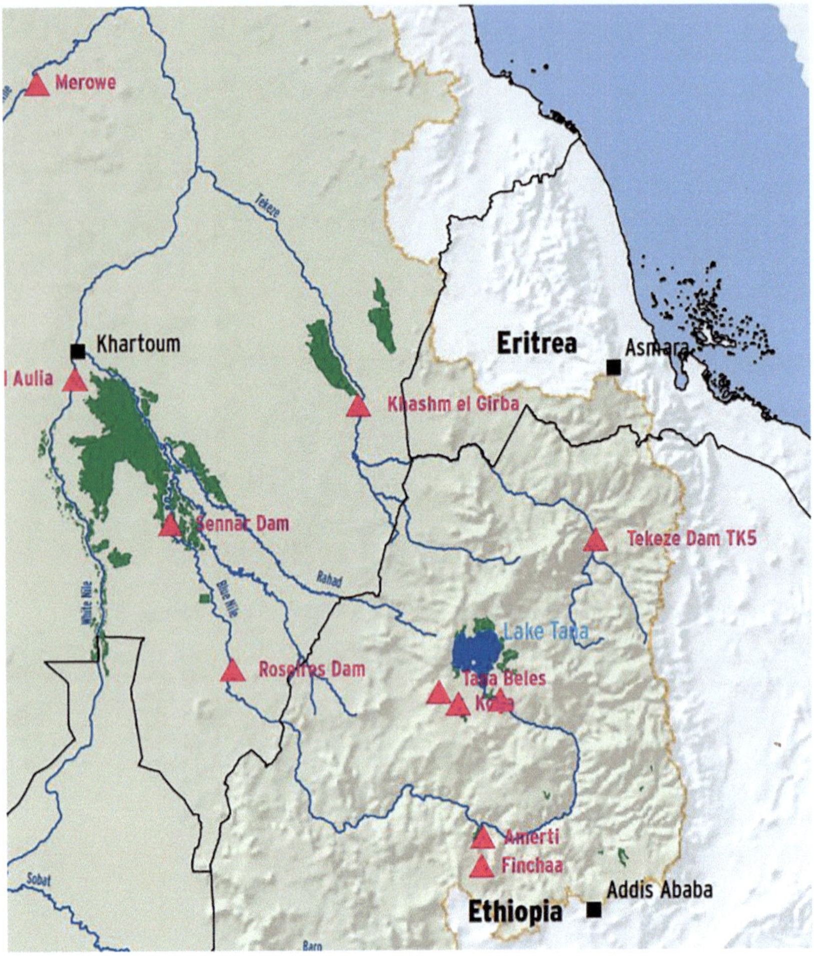

Figure 17.3 Dams of Sudan (man-made reservoirs).

Figure 17.4 Sanganeb Marine National Park and Dungonab Bay–Mukkawar. *Source:* Island Marine National Park/UNESCO/https://whc.unesco.org/en/list/262/CC-BY-SA IGO 3.0.

Figure 17.5 Nukhaila Oasis, desert ecosystem.

17.4 Biodiversity

Ecosystem services provided depend on the integrity of biodiversity in wetlands. Agriculture, urbanization, and industry are environmental factors that mostly affect mangroves and marine waters, leading to the loss of wetlands (Ballut-Dajud et al. 2022). Drought can change the role of wetlands from carbon sinks to carbon sources, whereas high temperature and more rain can maintain the role of wetlands as carbon sinks (Salimi et al. 2021). Vegetated coastal wetlands are among the most carbon-rich sinks, sequestering approximately as much carbon as do global forest ecosystems (Moomaw et al. 2018). Hence, it is important to protect wetlands from drying by disturbance agents to maintain ecosystem integrity.

Disturbance agents can be detected and ameliorated by continuous monitoring of the wetlands' biodiversity. In this context, Sudan wetlands largely are deficient. The biodiversity studies conducted averaged 1.2 study/wetland. The single wetland in saline ecosystem (Tables 17.2 and 17.3) received four studies. Studies of wetlands in man-made ecosystem averaged two studies per wetland, marine ecosystem 1.6. The riverine ecosystem is the least studied, being 1.2 study per wetland, although it comprises 65.7% of Sudan wetlands. This deficiency in biodiversity studies renders Sudan wetland vulnerable to environmental vicissitudes. It is worth mentioning that all wetlands having unpublished data need to be studied, in addition to wetlands on which only a few studies are conducted.

Table 17.3 Biodiversity studies on the important wetlands of Sudan.

#	Wetland	Biodiversity studies
1	Jebel Marra Volcanic lakes	Cloudsley-Thompson (1967), Hammerton (1968), Green et al. (2009), Robertson (2001)
2	Gezira	Sawicka and Egbuna (2020), Subedi et al. (2023), Robertson (2001)
3	Wadi Halfa (Lake Nubia)	Magdi and Ali (2006), Zuheir (2020), Robertson (2001)
4	Roseires dam reservoirs	Sinada and Abdelrahman (2019)
5	Sinnar dam reservoir	Hussien (2015)
6	Khor Arba'at	Nikolaus (1983), Robertson (2001)
7	Mukawwar island and Dunganab bay	Claudino-Sales (2019), Robertson (2001)
8	Suakin archipelago	BirdLife International (2024_c), Robertson (2001)
9	Sanganeb Atoll	Claudino-Sales (2019)
10	Mangroves	Mubarak et al. (2018)
11	Sunut forest and UmShugeira island	Eltayeb et al. (2012), Magboul et al. (2013)
12	Um-Jar	Deficient
13	Khor Abu Habil Delta	Deniau et al. (2022)
14	Gladeema	Twenty-two species of plants, nineteen birds, and seven species of fish purported.
15	Rahad kobri 45	Thirty species of birds and seventeen species of plants purported.
16	Lawni	Twenty-nine species of birds, twenty-two of plants purported.
17	Elban	Twenty-six species of plants and nineteen species of birds purported.
18	Homrani	Twenty-six species of plants and nineteen species of birds purported.
19	Lake Kundi	Green et al. (1984), Burgis and Symoens (1987), BirdLife International (2024_a).
20	Lake Keilak	Green et al. (1984), Burgis and Symoens (1987), Robertson (2001).
21	Turdat el Rahad	Deficient.
22	Radom National Park	El Khidir and Ali (2018), Robertson (2001).
23	Dinder National Park	Mohammed et al. (2023), Hashim and Mahgoub (2008), Ibrahim and Hashim (2011), Meina and Hashim (2012), Robertson (2001).
24	Taya-Basonda-Algalabat National Park	Abdelkarim et al. (2024).
25	Lake Um Bader	Robertson (2001).
26	Lake Abiad	BirdLife International (2024_b), Robertson (2001).
27	Tordat umdafog	Twenty-eight species of birds, seven species of reptiles, eight species of mammals, and eight species of insects purported.
28	Nabagah	Seventeen species of birds, eight species of reptiles, and five species of mammals purported.
29	Alsafya	Thirty species of birds, six species of insects, four species of reptiles, and two species of mammals purported.
30	Dahal deleibba	Thirty-five species of birds purported.
31	Hamidia pool	Korssi and Hamad (2019).

Table 17.3 (Continued)

#	Wetland	Biodiversity studies
32	Zour lake	Grmot (*Clarias lazera*) and Bulty (*Tilapia zillii*) occur in the lake in addition to 49 species of birds; the lake is the watering point for wildlife small mammals, birds, and cattle of nomadic herders (Korssi and Hamad 2019)
33	Gugoldonga pool	Fifty-one species of birds occur in the lake which is the watering point for wildlife small mammals, birds, and cattle of nomadic herders (Korssi and Hamad 2019)
34	Nukhaila Oasis	*Artemia artemia* recorded (el Moghraby 1995)
35	Ga'ab oasis	Species such as *Acacia ehrenbergiana, Hyphaene thebiaca, Boerhavia repens, Senna alexandarina, Portulaca quadrifida,* and *Corchorus depressus* recorded (Madani et al. 2015)

17.5 Conservation Aspects

As previously stated, only 23% of the wetlands of Sudan are protected. The protected areas are Biosphere Reserves, Ramsar sites, and mere protected areas.

17.5.1 Biosphere Reserves

These are as follows: (i) Dinder National Park, (ii) Radom National Park, (ii) Mukawwar island and Dunganab bay, and Sanganeb Atoll.

17.5.1.1 Dinder National Park

The park was proclaimed during the colonial rule in 1935 and inscribed Biosphere Reserve in 1979. Ecological characteristics that qualified the park to be biosphere reserve are that it harbors an extensive network of wetlands (called mayas), which are a refuge for many migratory birds in the savannah ecosystem. There are also mammals the most important species are Tiang (*Damaliscus lunatus tiang*), Bohor Reedbuck (*Redunca redunca*), Defassa Waterbuck (*Kobus ellipsiprymnus defasa*), Roan Antelope (*Hippotragus equinus*), Oribi (*Ourebia ourebi*), Common Warthog (*Phacochoerus africanus*), and African Buffalo (*Syncerus caffer*). Other animals such as Olive Baboon (*Papio anubis*), Grivet Monkey (*Chlorocebus aethiops*), and Patas Monkey (*Erythrocebus patas*) are frequent. Greater Kudu (*Tragelaphus strepsiceros*) and Heuglin's Gazelle (*Eudorcas tilonura*) are restricted to certain locations in the park.

About 7200 people live within the boundaries of Dinder National Park. To conserve biodiversity and use resources sustainably, a Management Plan is developed (Anonymous 2004) emphasizing the integration of local communities in the utilization and management of park's natural resources.

17.5.1.2 Radom National Park

Within the savannah woodland ecosystem, the park was declared as protected in 1980, a biosphere reserve in 1982. The park is drained by four rivers interspersed by numerous wetlands (called Dahals), in addition to a few water resources including boreholes and Birkas (Pools). The most important animals inhabiting the reserve include Bushbuck (*Tragelaphus scriptus*), Spotted Hyaena (*Crocuta crocuta*), Stripped Hyaena (*Haeana hyaena*), Olive Baboon (*Papio anubis*),

Common Warthog (*Phacochoerus africanus*), Patas Monkey (*Erythrocebus patas*), Tantalus Monkey (*Chlorocebus tantalus*), Honey Badger (*Mellivora capensis*), and considerable numbers of African wild birds including tufted Helmeted Guineafowl (*Numida meleagris*), Saddlebill (*Ephippiorhynchus senegalensis*), Abyssinian ground horn-bill (*bucorvus abussinian*), secretary bird (*Sagittarius serpentarius*), Savile's Bustard (*Lophotis savilei*) Marabou stork (*Leptoptilos crumentiferus*), and crowned crane (*Balaerica Pavonina*); reptiles are Central African Rock Python (*Python sebae*), African Spurred Tortoise (*Centrochelys sulcata*); crocodile (*Crocodilus niloticus*), Nile Monitors (*Varanus niloticus*), and lizards (*Agama* spp.). To conserve biodiversity, the inhabitants of the park and its surroundings are to be integrated within a management plan to be developed for the park.

17.5.1.3 Sanganeb Marine National Park and the Dungonab Bay–Mukkawar Island Marine National Parks

The parks (Figure 17.1) lie in the northern part of the Red Sea, with a buffer zone of 504,600 ha. These two marine protected areas were inscribed UNISCO Biosphere Reserve in 2016. The property consists of two separate areas: "Sanganeb is an isolated, coral reef structure in the central Red Sea and the only atoll, 25 km off the shoreline of Sudan. The second component of the property is made up of Dungonab Bay and Mukkawar Island, situated 125 km north of Port Sudan. It includes a highly diverse system of coral reefs, mangroves, seagrass beds, beaches, and islets. The site provides a habitat for populations of seabirds, marine mammals, fish, sharks, turtles, and manta rays. Dungonab Bay also has a globally significant population of Dugong (*Dugong dugon*)".

The two sites consist of both marine and terrestrial ecosystems. Fauna and flora of the marine ecosystem are from an Indian Ocean origin, developed unique and different ecosystems and species due to their semienclosed nature. The parks contain impressive natural phenomena, reef formations, and areas of great natural beauty that is relatively undisturbed. The two parks are connected by a coastal stretch extending 125 km including mersas, inlets, fringing reefs, and offshore reef formations, and the whole serial site is geologically and ecologically connected via the open flows that facilitate the exchange of biotic and abiotic elements within the marine ecosystems of the Red Sea. They encompass a large bay that contains islands, several small islets, and some of the most northerly coral reefs in the world associated with species (including seagrass and mangroves) at the limits of their global range and evolutionary expansion, which are therefore important from a scientific and conservation perspective. Sanganeb atoll is the only atoll-like feature in the Red Sea, and a submerged and overhanging predator-dominated coral reef ecosystem. It consists of thirteen different bio-physiographic reef zones, each providing typical coral reef assemblages, supporting a wealth of marine life and breathtaking underwater vistas, hosting over three hundred fish species with that include endemic and rare species. Besides providing important nurseries and spawning grounds for key species, it also hosts resident populations of dolphins, sharks, and marine turtles, which use the atoll as a resting, breeding, and feeding area.

Dungonab Bay, including Mukkawar Island and other islands, contains an array of habitat types, such as extensive coral reef complexes, mangroves, seagrasses, and intertidal and mudflat areas which all enable the survival (breeding, feeding, and resting) of endangered dugong, sharks, manta rays, dolphins, and migratory birds.

17.5.1.4 Jebel Marra/Deriba Caldera (Crater Lake)

It lies within a plain of semidesert with desert to the north. The natural vegetation is composed of herbaceous plants and trees, and it includes the gallery and riverain forests, savanna grassland with scattered trees, savanna woodlands and Boswellia mixed stands on hillside. Deriba Caldera (crater lake) is at the highest point of Jebel Mara located at latitude 12°95′N–longitude 24°27′E at an elevation of 3042 m (9980 ft), Darfur, western Sudan. The 5-km wide Deriba Caldera resulted from the explosive eruption of the Jebel Mara Volcano approximately 3500 years ago. The volcano is dormant, rather than extinct, as hot springs and fumaroles (gas and steam vents) are still present. While Jebel Mara is high enough (3042 m) to have a temperate climate and high precipitation, these lakes get their water supply from hot springs as well as rainwater. Jebe Marra is a region that dominates the landscape with outstanding scenic beauty and pleasant climate. The diversity of forest types along the altitudinal gradient of Jebel Marra, hot springs and Niertiti waterfall of Jebel Marra attract visitors from the local, national, and international groups who are interested in a mosaic of colors of plants and trees. Deriba Crater Lake have always been given particular attention by the local communities and is considered locally and internationally as being one of the most ten beautiful Crater Lakes in the World. The site includes an outstanding diversity of habitats and plant communities that support its globally significant species and ecosystem diversity. Jebel Marra habitats are rich in floral and faunal diversity from its foot to its summit. A significant proportion of the flora occurs in the area. Plant families are represented by significant natural habitats. Jebel Marra is very rich in biodiversity: 932 taxa of flowering plants and 32 ferns, 500 plants and woody species. Abundant and diverse 295 bird species, including small and large forest birds, savannah species, grassland birds, wetland species, raptors, and ground birds as well as palaearctic migrants are found. Crater lakes, waterfalls, and wetlands that hold exceptionally high levels of species diversity to be found in the semidesert region (https://whc.unesco.org/en/tentativelists/6519/).

17.5.2 Ramsar Sites

Four Ramsar sites covering an area of 3.436.009 ha, and satisfying criteria, are designated in Sudan in 2005 (https://www.ramsar.org/). These are Dinder National Park, Dongonab Bay-Marsa Waiai, Suakin-Gulf of Agig, and Khor Abu Habil Inner Delta, which were inscribed in 2005, 2009, 2009, and 2022, respectively. In addition to species mentioned in the biosphere reserve section, each site was qualified for Ramsar criteria based on specific criteria.

17.5.2.1 Dinder National Park

Located in the center of migration routes among three continents, the park is visited by many species of migratory birds, and some of the mayas (wetlands) contain quantities of fish throughout the dry season.

17.5.2.2 Dongonab Bay-Marsa Waiai

Dongonab Bay-Marsa Waiai has an expanse of coastline including coral reefs, mangroves, offshore islands, soft-bottom mud flats, sand beaches, and hard-bottom rocky shores in addition to saltmarshes, sabkha, and khor basins. As a result, the site is rich in biological diversity, provides support to various threatened species, and provides permanent habitats, breeding grounds, and areas of refuge for various fish and shrimp. The Bottlenose Dolphin and various shark species have also been recorded within the site.

17.5.2.3 Suakin-Gulf of Agig

The site boundary follows the Suakin Archipelago, a proposed marine protected area that occupies the southeastern quarter of the site. It comprises diverse forms of wetlands such as sand flats, coral reefs, lagoons, sand shores among others, a diversity that allows for a rich range of fauna and flora to thrive at the site such as marine Hawksbill Turtle (*Eretmochelys imbricata*) and Green Turtle (*Chelonia mydas*); seabirds; commercial bony fish and shrimp species. Sightings of the Sea Cow (*Dugong dugon*) and Bottlenose Dolphin (*Turciops truncates*) have also been noted.

17.5.2.4 Khor Abu Habil Inner Delta

A typical characteristic of this site is its hundreds of temporary pools or "mayas" that appear amid the dunes during the flooding of the river (Khor Abu Habil) between July and October. It is an important site for migratory waterbirds, hosting seven threatened species including the critically endangered Rueppell's vulture (*Gyps rueppellii*) and sociable lapwing (*Vanellus gregarius*). The forest remnants of the Site are a source of products such as oil from Balanites (*Balanites aegyptiaca*) and juice from the African baobab (*Adansonia digitata*), Gum Arabic from acacias, fodder, medicinal plants, and fuel wood for the hundreds of villages in and around the site.

Sunut forest and UmShugeira Island and Taya-Basonda-Algalabat National Park are mere protected areas. Established in 1939 in semidesert ecosystem, the first protected area is a staging area for migratory birds, whereas the second protected area was proclaimed in 1994. It is within the low woodland savannah ecosystem. Flagship species are Heuglin's Gazelle (*Eudorcas tilonura*), Oribi (*Ourebia ouribi*), and the Martial Eagle (*Polemaetus bellicosus*).

17.6 Socioecological Importance

Ecosystems services and functions provided by wetland is of primary importance for human livelihood and animals alike: The fertile soil developed by trapping silt is used to grow crops and vegetables. Reeds growing in wetlands have many uses, including (i) Fodder plant for animals in summer. (ii) Construction material for roofing, thatching, and basketry. (iii) Pulp and paper production. (iv) Sewage water treatment. (v) Musical instruments, such as flutes and reed instruments. (vi) Arrows and pens (https://selfgardener.com/). Wetlands contribute to climate change mitigation; despite occupying less than 5% of global land area and less than 2% of the ocean, they store roughly 50% of all carbon buried in ocean sediments (https://www.pewtrusts.org/).

17.6.1 Fishing

Fishing is mostly practiced in wetlands. It is the source of income for many households. Nubian Flat area along the Red Sea coast is considered one of the best fisheries for Yellow-spotted Triggerfish *Pseudobalistes fuscus* in the world (https://www.wildseaexpedition.com/); other common catches include Giant trevally *Caranx ignobilis*, Two-spot Red Snapper *Lutjanus bohar*, Bluefin trevally *Caranx melampygus*, Snubnose Pompano *Trachinotus blochii*, Milkfish *Chanos chanos*, Green Humphead Parrotfish *Bolbometopon muricatum*, Shortjaw Bonefish *Albula glossodonta* (VU), and other coral species (Ali 1984).

17.6.2 Grazing

Wetlands often support various grasses and forbs that are suitable for livestock and therefore provide significant grazing opportunities for local communities; it is crucial to balance the needs of livestock with the conservation of wetland ecosystems to ensure sustainable use (https://www.farmprogress.com/weeds/strategies-for-wetland-grazing).

17.6.3 Recreational Opportunities

Many wetlands contain a diversity of plants, animals, and water features that provide beautiful places for sightseeing, hiking, fishing, hunting, boating, bird watching, and photography (https://www.nps.gov/subjects/wetlands/). Roles played by wetlands in supporting local community' use of resources sustainably are in Table 17.4.

Table 17.4 Role of wetlands in supporting local communities.

Wetland	Status	Socioeconomic importance
Dinder National Park	Biosphere Reserve/ Ramsar site	About 7200 people live within the biosphere reserve boundaries. To conserve biodiversity and use natural resources sustainably, a management plan that integrates the local community with the park is in place (Anonymous 2004). The local communities will participate in decision-making regarding the proper management of the resources to perpetuate biodiversity in the park. https://www.unesco.org/en/mab/dinder. https://rsis.ramsar.org/ris/1461
Radom National Park	Biosphere Reserve	Because people migrate and settle inside the park, the park's management objectives require the protection of the species of plants and wild animals and the integration of the local communities in the management and conservation efforts, thereby implementing the role of the Biosphere Reserve. The integration requires the necessary studies of the local population's various activities to raise their well-being. https://whc.unesco.org/en/tentativelists/6518/
The Sanganeb Marine National Park and the Dungonab Bay–Mukkawar Island Marine National Parks	Biosphere Reserve	The ecosystems of the reserve remain weakly affected by human activities. However, regional coastal development increased recently and reinforced the need to protect this landward development of marine areas. The increase in activities from residents and tourists such as anchor, boat, and diving could intensify pollution and direct damage to ecosystems. In addition, predators could affect coral communities by coral bleaching in response to climate change. Staffing, financial resources, and on-ground management should be increased to manage this large, nominated area and marine buffer zone. Commitments will be necessary to increase and sustain Government funding levels, particularly considering potential tourism use and impact. https://whc.unesco.org/en/list/262
Jebel Marra/ Deriba Caldera (crater lake)	Biosphere Reserve	As of the current date, Deriba lakes are overseen by both the local community and government authorities. As for the community, the lake represents a cultural and ritual value; and for the authorities it represents a site of natural heritage. It also has enormous potential for agricultural and forestry development in the lower plains surrounding the jebel. While the jebel springs and waterfalls are a major source of water for the downstream catchment areas, there are no acts or by-laws protecting the site. The local communities provided the necessary modest protection and management of the site. But, it is for government to propose a National Park, and process soon to be declaration as a protected area.
Dongonab Bay-Marsa Waiai	Ramsar site	The main land uses within the site are animal breeding, of camels, goats and sheep, fisheries, oyster culture, and tourist activities such as scuba diving. Potential threats arise from major land use changes because of a proposed shrimp and fish farming industry project along the southern stretch of coastline, livestock fodder production and ice plants, and overgrazing by nomads and camel herders as well as declining rainfall. https://rsis.ramsar.org/ris/1859
Suakin-Gulf of Agig	Ramsar site	Various socioeconomic activities take place within and around the site, but the most widespread practice is nomadism with camels, which are of great social and economic value. The Tokar Delta is the sole area with significant potential for cultivation in the whole coastal zone, with cotton, sorghum, millet, and vegetables. The proposed shrimp and fish-farming industry project, as well as efforts to widen the entrance and ship channels and reconstruct the port of Suakin, pose major threats to the character of the site. https://rsis.ramsar.org/ris/1860
Khor Abu Habil Inner Delta	Ramsar site	The forest remnants of the site are a source of products such as oil from *Balanites aegyptiaca*, juice from *Adansonia digitata*, and Gum Arabic from acacias, fodder, medicinal plants, and fuel wood for the hundreds of villages in and around the site. These communities practice artisanal fishing, agriculture, hunting, and nomadic livestock rearing. While the impact of communities' dependence on the site's resources is unknown, household sewage, urban wastewater, agricultural effluents, and droughts are significant threats affecting its ecological character. The site is yet to benefit from a management plan and a national legal conservation status. https://rsis.ramsar.org/ris/2485

References

Abdelkarim, A.A., Kanno, I.O., and Musa, F.I. (2024). Observation on identification of birds and reptile in Taya-Basonda–Al galabat National Park, Gedaref State-Sudan. *Journal for Nature Conservation* 79: 126614.

Ali, M.T. (1984). Fishes and fisheries of Lake Nubia, Sudan. *Hydrobiologia* 110: 305–314.

el Moghraby, A. I. 1995. Wetlands of the Nile in the Sudan. In Mahmoud, Zuheir N., Abu Gideiri, Yousif B. (Eds). Wetlands and Biodversity in Sudan. Workshop on Wetlands and Biodversity in Sudan, Khartoum. Nile Basin Initiative (Sudan Office) (accessed 22 October 2008).

Anonymous (2004). Management Plan for Dinder National Park, Suda. Higher Council for Environment and Natural Resources, Wildlife Conservation General Administration. UNDP/GEF Project SUD/98/G41 and SUD 00/014.

Ballut-Dajud, G.A., Sandoval Herazo, L.C., Fernández-Lambert, G. et al. (2022). Factors affecting wetland loss: a review. *Land* 11 (3): 434.).

BirdLife International. (2024a). Important bird area factsheet: Lake Kundi. https://datazone.birdlife.org/site/factsheet/lake-kundi-iba-sudan (accessed 31 May 2024).

BirdLife International (2024b). Important bird area factsheet: Lake Abiad. https://datazone.birdlife.org/site/factsheet/lake-abiad-iba-sudan (accessed 31 May 2024).

BirdLife International (2024c). Important bird area factsheet: Suakin archipelago. https://datazone.birdlife.org/site/factsheet/suakin-archipelago-iba-sudan (accessed 31 May 2024).

Burgis, M.J. and Symoens, J.J. (1987). *African Wetlands and Shallow Water Bodies-Zones Humides et lacs peu profonds d'Afrique*, 650. Paris: Orstom.

Claudino-Sales, V. (2019). Sanganeb Marine National Park and Dungonab Bay: Mukkawar Island Marine National Park, Sudan. *Coastal World Heritage Sites* 28: 541–547. https://doi.org/10.1007/978-94-024-1528-5_79.

Cloudsley-Thompson, J.L. (1967). Altitude and faunal distribution on Jebel Marra. *Sudan Notes and Records* 48: 133–140. http://www.jstor.org/stable/44947286.

Deniau, C.P., Mathevet, R., Gautier, D. et al. (2022). Flyways to hell? An empirical assessment of Palearctic migratory waterbird harvest practices in key wetlands of Sahel-sub-Saharan Africa. *Biological Conservation* 265: 109419. https://doi.org/10.1016/j.jafr.2023.100733.

El Khidir, S.A. and Ali, O.M. (2018). The impact of war on wildlife in radom National Park, Southern Darfur State-Sudan. *Journal of Biodiversity & Endangered Species* 6: 3. https://doi.org/10.4172/2332-2543.1000220.

Eltayeb, H., Idris, E., Adam, A. et al. (2012). A forest in a city biodiversity at Sunut forest, Khartoum, Sudan. *Egyptian Academic Journal of Biological Sciences, B. Zoology* 4 (1): 199–209.

Emerton, L. (2018). Economic valuation of wetlands: total economic value. In: *The Wetland Book* (ed. C.M. Finlayson), 1–16.

Grasel, D., Giehl, E.L.H., Wittmann, F., and Jarenkow, J.A. (2021). Patterns of plant diversity and composition in wetlands across a subtropical landscape: comparisons among ponds, streambanks and riverbanks. *Wetlands* 41 (7): 90.

Green, J., El Moghraby, A.I., and Ali, O.M.M. (1984). A faunistic reconnaissance of Lakes Kundi and Keilak, western Sudan. *Hydrobiologia* 110: 33–44.

Green, J., El-Moghraby, A.I., and Ali, O. (2009). Biological observations on the crater lakes of Jebel Marra, Sudan. *Journal of Zoology* 189 (4): 493–502. https://doi.org/10.1111/j.1469-7998.1979.tb03976.x.

Hammerton, D. (1968). Recent discoveries in the caldera of Jebel Marra. *Sudan Notes and Records* 49: 136–148. http://www.jstor.org/stable/41716898.

Hashim, I.M. and Mahgoub, K.S. (2008). Abundance, habitat preference and distribution of small mammals in Dinder National Park, Sudan. *African Journal of Ecology* 46 (3): 452–455.

Hussien, M.E.A.E. (2015). Water birds of Sinnar Dam reservoir. *Poultry, Fisheries & Wildlife Sciences* 3: 129. https://doi.org/10.4172/2375-446X.1000129.

Ibrahim, M.A. and Hashim, I.M. (2011). Biomass and habitat use of guinea fowl (*Numidia meleagris*) in Dinder Biosphere reserve, Sudan, during the dry season. *University of Khartoum Journal of Agricultural Sciences (Sudan)* 19 (3): 452–455.

Korssi, A.M.A.A. and Hamad, D.M. (2019). The assess of pools and small lakes as wetland sites in Zalingei Area, Darfur, Sudan. *Journal of Biomaterials* 3 (1): 28–32. https://doi.org/10.11648/j.jb.20190301.15.

Madani, I.K.R.A.M., Tahir, Y.F., Hamdeen, H.M. et al. (2015). Vegetation ecology and taxonomy of El-Ga'ab area, North-Western Sudan. *European Academic Research* 3 (3): 18.

Magboul, R., Khider, A., Idris, E., and Alam, T.H. (2013). Ant diversity at Sunut forest, Khartoum, Sudan. *Egyptian Academic Journal of Biological Sciences. A, Entomology* 6 (1): 43–46.

Magdi, M. and Ali, M. (2006). Macrophytes in aquatic ecosystems: from biology to management. In: *Proceedings of the 11th International Symposium on Aquatic Weeds*. European Weed Research Society. http://dx.doi.org/10.1007/978-1-4020-5390-0.

Meina, O.M. and Hashim, I.M. (2012). Storks' diversity, flock characteristic, roosting ecology and fluctuation in numbers in some mayas at Dinder National Park, Sudan, during the dry seasons 2009 and 2010. *University of Khartoum Journal of Agricultural Sciences* 20 (2): 216–228.

Mitsch, W.J., Bernal, B., and Hernandez, M.E. (2015). Ecosystem services of wetlands. *International Journal of Biodiversity Science, Ecosystem Services & Management* 11 (1): 1–4. https://doi.org/10.1080/21513732.2015.1006250.

Mohammed, A.A., Bauer, H., Coals, P. et al. (2023). Abundance of larger mammals in Dinder National Park, Sudan. *European Journal of Wildlife Research* 69: 46. https://doi.org/10.1007/s10344-023-01675-1.

Moomaw, W.R., Chmura, G.L., Davies, G.T. et al. (2018). Wetlands in a changing climate: science, policy, and management. *Wetlands* 38 (2): 183–205.

Mubarak, E.A.T., Abdalla, N.E.A., and Abdelgadir, M.M. (2018). Current status and sustainability issues of marine resources and biodiversity in Sudanese national red sea water. *Journal of Aquaculture & Marine Biology* 7 (2): 110–112. https://doi.org/10.15406/jamb.2018.07.00194.

Nikolaus, G. (1983). An important passerine ringing site near the Sudan Red Sea coast. *Scopus* 7: 15–18.

Qiu, J., Zhang, Y., and Ma, J. (2024). Wetland habitats supporting waterbird diversity: conservation perspective on biodiversity-ecosystem functioning relationship. *Journal of Environmental Management* 357: 120663. https://doi.org/10.1016/j.jenvman.2024.120663.

Raburu, P. O., Kwena, F., and Nyandiga, C. O. (2012). Challenges and Lessons Learnt on Community Based Wetland Management. http://hdl.handle.net/1834/7823

Reimold, R.J., Phillips, J.H., and Hardisky, M.A. (1980). Socio-cultural values of wetlands. In: *Estuarine Perspectives*, 79–89. Academic Press. https://doi.org/10.1016/B978-0-12-404060-1.50015-0.

Robertson, P. (2001). Sudan. In: *Important Bird Areas in Africa and Associated Islands: Priority Sites for Conservation* (ed. L.D.C. Fishpool and M.I. Evans), 877–890. Newbury and Cambridge: Pisces Publications and BirdLife International (BirdLife Conservation Series No. 11).

Salimi, S., Almuktar, S.A., and Scholz, M. (2021). Impact of climate change on wetland ecosystems: a critical review of experimental wetlands. *Journal of Environmental Management* 286: 112160.

Sawicka, B. and Egbuna, C. (2020). Chapter 1 – pests of agricultural crops and control measures. In: *Natural Remedies for Pest, Disease and Weed Control* (ed. C. Egbuna and B. Sawicka), 1–16. Academic Press. https://doi.org/10.1016/B978-0-12-819304-4.00001-4.

Sinada, F.A. and Abdelrahman, M.E. (2019). The phytoplankton and water quality of Sennar and Roseires Reservoirs, Sudan. *Athens Journal of Sciences* 6: 83–101.

Subedi, B., Anju Poudel, A., and Aryal, S. (2023). The impact of climate change on insect pest biology and ecology: implications for pest management strategies, crop production, and food security. *Journal of Agriculture and Food Research* 14: 100733. https://doi.org/10.1016/j.jafr.2023.100733.

Xu, X., Chen, M., Yang, G. et al. (2020). Wetland ecosystem services research: a critical review. *Global Ecology and Conservation* 22: e01027.

Zuheir, N. M. (2020). Contribution of Research Projects and Environmental Impact Assessment Studies to Fauna Biodiversity in Sudan. *International Journal of Environmental Sciences and Natural Resources* 24(2): 556135. DOI: https://doi.org/10.19080/IJESNR.2020.24.556135

18

Wetlands and Conservation

Exploring Biodiversity Dynamics, Economic Potentials, and Conservation Strategies in South Africa

Adeyemi Ojutalayo Adeeyo[1], Bramley Jemain Lemine[2,3], Tinyiko Rivers Nkuna[4], Titus A.M. Msagati[1], and Mercy Adewumi Alabi[5]

[1] *Institute of Nanotechnology and Water Sustainability, Florida, South Africa*
[2] *Department of Chemical Engineering, Cape Peninsula University of Technology, Bellville, South Africa*
[3] *Department of Water Science, University of Western Cape, Bellville, South Africa*
[4] *Department of Earth Science, Faculty of Science Engineering and Agriculture, University of Venda, Limpopo, South Africa*
[5] *Department of Microbiology, School of Life Sciences, University of KwaZulu-Natal, Durban, South Africa*

18.1 Introduction

The South African National Water Act (Act 36 of 1998) defines "a wetland as a land that is transitional between terrestrial and aquatic systems, having the water table at or near the surface, or where the land is occasionally covered in shallow water." Under normal circumstances, the land supports or would support vegetation that is typically adapted to life in saturated soil (Salimi et al. 2021). The term "wetland" also refers to any place that is sometimes seasonally or permanently covered in water and examples of these locations include tidal flats and areas that are inundated near rivers, rice paddies, or swamps (Collins 2006). These include human-made wetlands like waste-water treatment ponds and reservoirs as well as other kinds of ecosystems including peatlands, marshes, rivers, lakes, and coastal regions like saltmarshes, seagrass beds, and mangroves (Maua et al. 2022).

Wetlands are thought to span between 6% and 10% of the earth's surface worldwide, while just 4.7% or 1.15 million km^2 of Africa's continental area is thought to be covered by wetlands (Maua et al. 2022). They barely cover 2.4% of South Africa's geographical area but according to the report released in 2020 by the Department of Water and Sanitation in South Africa, half of South Africa's native wetland area is lost. Wetlands are the most threatened ecosystems in South Africa; out of all 791 types of wetland habitats, 48% are severely endangered, 12% endangered, 5% vulnerable, and 35% least threatened (Department of Environment, Forestry and Fisheries 2021). Just 11% of the wetland ecosystem types in South Africa are adequately protected, whereas more than 70% are not (Department of Forestry, Fisheries and the Environment 2022). The Ramsar classification of wetlands divides wetlands into three main groups: human-made, marine and coastal, and inland. These categories are further subdivided into 42 types (Gehrig et al. 2015).

Wetlands either directly or indirectly help achieve each of the 17 Sustainable Development Goals (SDGs) (Seifollahi-Aghmiuni et al. 2019). For instance, over a billion people rely on wetlands for their livelihood (in line with SDG 1 which is no poverty), 3.5 million people eat rice farmed in

Wetlands of Tropical and Subtropical Asia and Africa: Biodiversity, Livelihoods and Conservation, First Edition. Edited by Thammineni Pullaiah.

wetland paddies as their main diet (in line with SDG 2 which is no hunger), wetlands provide 266 million jobs in wetland tourism and travel (in line with SDG 8 which is decent employment and economic growth), 40% of all species on earth reside and reproduce in wetlands (in line with SDG 15 which is life on land) and in accordance with SDG 13 on climate action, while peatlands make up only 3% of the world's land area, they store twice as much carbon as all of the world's forest biomass (Gardner and Finlayson 2018). Therefore, wetlands would be a cost-effective investment if governments cater for their preservation and responsible use. However, if not judiciously used, wetlands can become stressed and limited in their natural benefits due to economic advantages (Maua et al. 2022).

Wetlands are vital to the preservation of biodiversity and the provision of ecosystem services to humankind. They provide a number of ecological services, including reducing the danger of flooding, storing carbon, regulating the temperature, and acting as water reservoirs (Du Toit et al. 2021). They provide drinking water, fish, fodder, fuel; regulate the rate of runoff in urban areas; protect shorelines from erosion, wildlife habitat; and provide recreational opportunities for society. They also recycle nutrients, purify water, lessen floods, and recharge groundwater (Xu et al. 2019). Additionally, they serve as a haven for animals, filter pollutants, and preserve rare biodiversity (McInnes and Everard 2017).

However, the function of wetlands is affected by a wide range of biotic and abiotic elements. Wetlands are under danger from climate change, which can impact a wetland environment by raising the temperature and altering the hydrological patterns, thus changing the ecosystem's biogeochemistry (Salimi et al. 2021). There has been a decline in wetland extent and quality worldwide due to the combined effects of natural and human activities (Xu et al. 2019). Urban development not only has the potential to destroy wetlands but it also frequently modifies hydrological cycles, increases pollutants, which changes wetlands, and affects the richness and composition of species (Du Toit et al. 2021). Nonetheless, wetlands conservation can benefit greatly from the prudent and traditional usage of wetlands by local populations and indigenous peoples (Chanza and Musakwa 2021).

Wetlands are referred to as hotspots of biodiversity. However, because they are compatible yet environmentally fragile systems, they are the most vulnerable ecosystems on the planet (Zhou et al. 2020). According to statistics available between 1970 and 2015, the Ramsar Convention Secretariat reported a 35% decline in worldwide wetlands (Gardner and Finlayson 2018). Disturbances to wetland soils, overgrazing, drainage for pastures and crops, improperly managed burning regimes, vegetation clearing, and industrial and urban development (including mining activities) have all contributed to the destruction of between 35% and 60% of South Africa's wetlands. According to the National Biodiversity Assessment, 88% of wetland areas are vulnerable, and fewer than 2% are properly protected, making wetlands the most threatened environment in the nation (Skowno et al. 2019).

In South Africa, wetlands are most prevalent in mesic eastern regions, particularly on the Highveld (Sieben et al. 2017). This is because rainfall is adequate to refill wetlands on a regular basis, and the plains' morphologies and the foothills of the mountains allow for the formation of small basins where water collects for extended periods of time (Job and Le Roux 2019). However, this area of the nation also has the most intensive land uses that impact wetlands such as agriculture and industrialization. As a result, many of the wetlands in this region have been degraded by eutrophication, drainage, and the input of industrial effluents, making them some of the most threatened in the nation (Sieben et al. 2017).

Owing to the various types of wetlands and the need for preservation of biodiversity, wetlands need to be protected in urban areas to lessen or eradicate the effect of anthropogenic activities and climate change on wetlands. However, management strategies as translated into action must not undermine the prescribed laws and policies that aim to promote the management of wetlands. The obligations emanating from these laws and policies may stem from local, regional, and global interventions required to advance wetland management. The soft, nonbinding, tools like the SDGs and Agenda 2063, for example, may further bolster wetland management as espoused later, and the institutions that possess the power of interpreting and applying the law play a crucial role in advancing the protection of wetlands as they further guide the action. Therefore, this chapter details the traditional knowledge of wetlands in South Africa, the biodiversity of wetlands including plants, animals, and microorganisms as well as the functions of wetlands. It further explains the mitigation strategies that can be employed against destruction of wetlands by climatic factors and management strategies to promote sustainable use and restoration of wetlands.

18.2 Wetlands of South Africa

The most widely used and accepted classification system was developed by the US Fish and Wildlife Service (USFWS). The categorization system's goal is to identify or characterize the wetland's biological units that share a set of uniform natural characteristics. The system has been used to create units for inventory and mapping, support resource management choices, and standardize vocabulary and concepts related to wetlands (Ollis et al. 2013). The "National Wetland Classification System" was the former name of the classification scheme created for the SANBI, which has been modified and is now "Classification System for Wetlands and other Aquatic Ecosystems in South Africa" (Ollis et al. 2013). For all wetlands in South Africa, SANBI has created a wetland categorization system that enables system distinction and prioritization for conservation or management objectives (Ollis et al. 2016). Wetland systems are categorized under the SANBI categorization system based on their primary biophysical drivers, or abiotic elements, which impact the wetlands' performance (Ollis et al. 2013). To distinguish between the different types of wetlands, the SANBI classification system employs a hierarchical system based on six levels (Figure 18.1). The first level divides wetlands based on their system (marine, estuarine, or inland systems) and the sixth level groups wetlands based on the characteristics of the wetlands, such as geology, natural versus artificial vegetation cover, substratum, salinity, and acidity/alkalinity (Ollis et al. 2016).

The degree of connectedness with the ocean is the main factor used to distinguish inland systems from marine and estuarine systems (Ollis et al. 2015). The area of the open ocean that lies over the continental shelf and/or the coastline that it is connected to, down to a depth of 10 m at low tide, is referred to as a marine system. Marine systems are aquatic ecosystems that are a part of the open ocean. They extend from deep water to the shore and are distinguished by their vulnerability to tidal variations and wave activity along their landward border (Ollis et al. 2016). A body of surface water is referred to as an estuary system if it is a part of a watercourse that is periodically or permanently open to the sea, if where the water level rise and fall caused by the tides is measurable during spring tides when the watercourse is open to the sea or if the salinity is measurably higher due to the influence of the sea (Ollis et al. 2016). Estuarine systems are permanent, partially enclosed bodies of water that are occasionally or constantly exposed to the sea on a decadal time frame, extending to the point where salt penetration or the upper limit of tidal action occurs (Van Niekerk and Turpie 2012). Aquatic ecosystems classified as inland systems lack any current oceanic link and are typified by the total lack of marine exchange and/or tidal impact (Ollis et al. 2013).

There is a wide range of wetlands encompassed in Ramsar definition; however, for illustrative purposes, there are five broad wetland systems that are estuaries, marine, riverine, palustrine, and lacustrine (Figure 18.2). Estuaries are wetlands where rivers meet the sea and salinity is intermediate

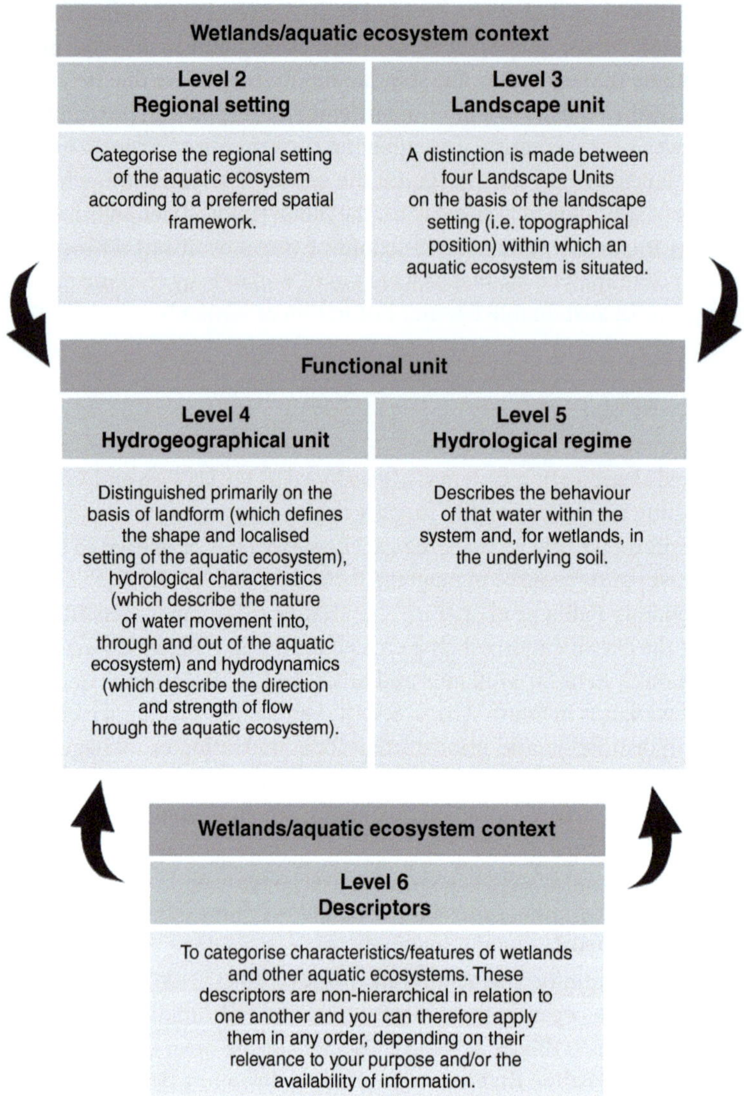

Figure 18.1 Classification of wetlands according to levels.

between salt and freshwater. Examples are deltas, mudflats, and salt marshes. Marine wetlands (e.g. shorelines and coral reefs) are not influenced by the river flows. Riverine wetlands (e.g. meadows flooded forests and oxbow lakes) are lands periodically inundated by overtopping rivers (Kanaujia and Kumar 2014). In palustrine wetland, there is permanent water (e.g. marshes, fen, and papyrus swamp) and lacustrine wetlands are areas of permanent water with little flow (such as ponds, kettle lakes, and volcanic lakes) (Kanaujia and Kumar 2014).

St. Lucia is the greatest estuary in South Africa and is the location of the country's highest population of hippos and crocodiles (Siyabona Africa n.d.). A globally significant high-altitude Ramsar site is the Blesbokspruit wetland system. It is the sole Ramsar site in Gauteng and the biggest permanent wetland in South Africa's Highveld area. De Mond Nature Reserve is an estuary wetland that serves as a crucial breeding ground for the stunning Damara tern seabird (Getaway Magazine 2024). The majority of Makuleke is located inside the Kruger National Park and forms the Limpopo River boundary between Mozambique, Zimbabwe, and South Africa. Makuleke is composed of several types of marsh and 31 dispersed pans. It was the first Ramsar site in the world – not only in South Africa – to be jointly owned and administered by a community (Getaway Magazine 2024). With five different types of wetlands, the Ndumo Game Reserve in KwaZulu-Natal makes up the biggest floodplain system in South Africa. It is most renowned for its amazing birdlife and has one of South Africa's greatest bird counts (430 species) (Getaway Magazine 2024). Kosi Bay is located within the iSimangaliso Wetland Park and is home to various coastal habitat types as well as a distinctive four-lake system that includes both freshwater and marine lakes (Peer et al. 2022).

Riverine Wetland

Palustrine Wetland

Figure 18.2 Types of wetlands based on water type.

Marine Wetland

Estuarine Wetland

Lacustrine Wetland

Figure 18.2 (Continued)

18.3 Biodiversity of Wetlands

18.3.1 Plants

Wetland plants will grow in places with abundant water. The vegetation found in wetlands is distinctly different from the terrestrial flora that surrounds it, and in order for plants to thrive in wetlands, they must have certain functional adaptations (Sosnová et al. 2010). For instance, a wetland plant's morphology necessitates a shallow root system to withstand water stress and aerenchyma tissues in the stem to deal with oxygen shortage (Mofutsanyana et al. 2020). The wetlands are home to three different kinds of plants: emergents, which make up the bulk of wetland plants, plants floating on the water's surface, and submerged plants, which grow underwater (Butt et al. 2021).

Freshwater and saltwater wetlands are home to mangrove trees and cypress trees, respectively. Saltwater marshes are home to rushes, reeds, sedges, and saltbush, whereas freshwater marshes are home to grasses, wildflowers, and shrubs. Plants that have been reported to be associated with wetlands include *Heliconia psittacorum, Hedychium coronarium, Hemerocallis flava, Canna indica, Iris pseudacorus, Zantedeschia aethiopica, Alpinia purpurata, Ludwigia inucta,* and *Polianthus tuberosa* (Sarmento et al. 2013; Toro-Vélez et al. 2016; Sandoval et al. 2019). Indicator plant species of temperate grassy wetlands include *Eleocharis dregeana, Carex acutiformis, Fuirena pubescens, Helichrysum aureonitens, Schoenoplectus brachyceras, Leucosidea sericea, Cliffortia nitidula, Themeda triandra, Cyperus fastigiatus, Kyllinga alba, Miscanthus junceus, Wahlenbergia undulata, Typha capensis,* and *Phragmites australis* (Sieben et al. 2017).

The function of plant roots and rhizomes to supply substrates for associated microorganisms, oxygenate the regions around the roots, and absorb contaminants from water is one of the most noticeable characteristics of wetlands (Shahid et al. 2020). Wetland plants primarily absorb nutrients such as nitrogen, phosphorus, and others through the vascular bundles and epidermis of their roots, which then carry the nutrients up to the stem and leaves. In addition, they supply carbon for denitrification during the breakdown of biomass, which keeps contaminants from leaching out of sediments (Shelef et al. 2013). Because of their many characteristics pertaining to the process of water treatment, plants that thrive in artificial wetlands are an integral part of the design. Radial oxygen loss (ROL), a mechanism that takes place in the root zone, makes macrophytes the primary supplier of oxygen in constructed wetlands (Wang et al. 2018).

Numerous bacteria reside in plant roots because they serve as a location for microbial attachment and emit exudates, which are carbon-based excretions that aid in denitrification, a process that improves the elimination of pollutants under anoxic settings (Shelef et al. 2013). Wetland vegetation also helps the ecosystem retain water, preventing floods in nearby rivers and streams as well as water erosion (Ramachandra et al. 2012). In South African wetlands, the plants found include all monocotyledons plant families Typhaceae (bulrushes), Cyperaceae (sedges), Juncaceae (rushes), Prioniaceae (palmiet), Eriocaulaceae (eriocaulons), Restionaceae (restios), Xiridaceae (yellow-eyed grasses), and the Poaceae (grasses) (Water Research Commission 2018).

Prionium serratum, Carex austro-africana, Carex killickii, Carex subinflata, Carpha schlechteri, Cyathocoma bachmannii, Cyperus pulcher, Cyperus textilis, Cyperus thunbergii, Epischoenus adnatus, Ficinia capillifolia, Fuirena hirsuta, Isolepis digitata, Schoenoplectus confusus, Scirpoides thunbergia, and *Eriocaulon dregei* are some wetland plants that are endemic to South Africa (van Ginkel et al. 2011; Rebelo et al. 2020). However, some plants are threatened (Figure 18.3) and may be extinct if not protected. *Ophioglossom gracillium, Pilularia bokkeveldensis, Pilularia dracomontana, Marsilea schelpeana, Marsilea fenestrata, Liparia angustifolia, Ceratophyllum submersum, Bolbitis heudelotii, Liparia laevigata, Psoralea abbottii, Cliffortia longifolia, Cliffortia nivenioides, Cliffortia*

Figure 18.3 Some threatened wetland plants in South Africa.

recurvata, Erica fontana, Erica bakeri, Grubbia rourkei, Hydrostachys polymorphia, Limonium linifolium, Dymondia margaretae, Gnaphalium declinatum, Cotula filifolia, Cotula myriophylloides, Cotula pusilla, and *Ursinia merxmuelleri* are some of threatened wetland plants in South Africa (iNaturalist 2020).

18.3.2 Animals

Wetland ecosystems are home to a wide range of creatures (Figure 18.4). Wetlands may be home to beavers, otters, bobcats, deer, minks, and muskrats, among other animals (Garrett 2022; North Carolina Division of Water Resources n.d.). Among the most prevalent marsh creatures are otters, muskrats, and minks. Muskrats are flourishing marsh creatures, in contrast to the endangered minks (Miller 2019). The brownish, semiaquatic mouse has a white patch on its chin, stiff, short hairs, scaly tail, and webbed feet. Muskrats build dome-shaped homes out of marsh vegetation and gather near cattails. They consume rush plants in addition to cattails, which are their primary food source. They are a vital component of the food chain in marshes and provide food for owls, big turtles, water snakes, and hawks (Miller 2019). Marshes are home to great blue herons, rails, and red-winged blackbirds. The heron, a bird that is frequently connected to marshes, nests above ground in trees near wetlands and wades in the water. Herons consume fish, amphibians, crabs, insects, and reptiles. In addition to birds like grouse, plover, storks, herons, and other waterfowl, wetlands are home to worms, insects (such as craneflies, damselflies, butterflies, dragonflies, mosquitoes, and deer flies), crayfish, snails, and clams because they are abundant in water and decaying plant matter. Common marsh fish include several species of bass, pike, walleye, and sunfish (Miller 2019).

Figure 18.4 Some animals found in wetlands. *Source:* f. 169169/Adobe Stock Photos, i. Sander Meertins/Adobe Stock Photos, j. slowmotiongli/Adobe Stock Photos, l. zhengzaishanchu/Adobe Stock Photos.

Wetlands are particularly important to the survival of amphibians. In wetlands, salamanders, frogs, and toads prefer to deposit their eggs in ephemeral or transient pools where there are no scavenging predators. Wetlands are home to mammals too. They serve as habitat for a broad range of animals due to their abundant plant life and water supply. There is a complete food web in wetlands due to the huge diversity of creatures there. A robust base is provided by many invertebrates, whose predators in turn create excellent prey for bigger animals (Garrett 2022). Across South African wetlands, animals found include savanna elephant, spotted hyena, brown hyena, wild dog, hippopotamus, antelope, common reedbuck, suni, red duiker, Nile crocodiles, leatherback sea turtles, loggerhead sea turtles, great white pelicans, pink-backed pelicans, woodward's batis, blue-mantled crested flycatchers, gaboon adder, rock Monitor, and Nile monitor lizards (Siyabona Africa n.d.). The White-winged Flufftail, one of the rarest and most threatened water birds in Africa, has been found at Middelpunt Nature Reserve (MNR), establishing that a breeding population exists outside Ethiopia (Abdellah 2023).

18.3.3 Microorganisms

Wetland systems' microbial populations are vital to the biogeochemical cycles that underpin wetland functioning. Microbes are essential to the food chain since they are both primary producers and decomposers (Upadhyay et al. 2017). Photoautotrophic organisms are among the producer microorganisms, which are crucial to maintaining a robust food web. Higher trophic species are broken down by bacteria when they die, which helps recycle vital energy and return it into the system as dissolved organic carbon (Rai et al. 2015). Methanotrophic bacterial dispersion and distribution, in addition to nitrogen, regulate methane cycle in wetlands (Putkinen et al. 2012).

Numerous types of microorganisms live in the water, sediments, and plants in wetland systems, and they each have a unique purpose that depends on the environment. Proteobacteria, Firmicutes, Actinobacteria, and Bacteroidetes are the four phyla that comprise most bacteria in wetlands (Cao et al. 2017). There have been reports of the isolation of *Planococcus, Psychrobacter, Bacillus, Mycobacterium*, and *Rhodobacter* species from estuaries (Li et al. 2020). Other microorganisms that have been isolated from wetlands include members of the genera *Chitinophaga, Isoptericola, Sinorhizobium, Nitrososphaera, Nitrosopumilus, Candidatus* (Wang et al., 2019), *Nitrosotalea, Nitrosoarchaeum* (Zhao et al. 2021), *Nitrosomonas, Nitrosospira, Nitrobacter, Nitrospina, Bradyrhizobium, Hyphomicrobium, Rhizobium, Rhodoplanes, Paracoccus, Methylobacterium, Gemmobacter, Brevundimonas, Roseobacter, Azospirillum, Pseudomonas, Acinetobacter*, and *Aeromonas* (Wang et al. 2022). Species of the genera *Simplicispira, Francisella, Dechloromonas, Flavobacterium, Malikia*, and *Flectobacillus* have also been reported (Ballesteros et al. 2023). *Bacillus megaterium, Rumellibacillus* sp., *Micrococcus luteus, Paenibacillus* sp, and *Bacillus* sp. were isolated from wetland settlement (Jankoski et al. 2021).

The simultaneous activity of aerobic and anaerobic microbial communities is facilitated by the near proximity of oxic–anoxic conditions, which are frequently produced by the roots of marsh plants. Because of the nutritional input and quick recycling provided by active anaerobes and aerobes, these systems are extremely productive and appealing to humans and a wide variety of other creatures. In contrast to soils and aquatic ecosystems, the richness and functionality of microbial communities in wetland systems, however, remain largely unexplored (Bodelier and Dedysh 2013). In South Africa, Proteobacteria, Bacteroidetes, Actinobacteria, and Firmicutes have been isolated in constructed wetlands (Wang et al. 2022).

18.4 Livelihoods and Economic Contributions of Wetlands

Wetlands serve both physical and economic benefits. The direct advantage of a wetland is termed wetland value, while the indirect benefit is referred to as wetland function (Adeeyo et al. 2022). Wetlands have been reported for functions such as maintenance of a high concentration of aquatic fauna, permanent resident amphibians, native and migratory bird species, reptiles, mammals, and other insect species, through which wetlands contribute to the conservation of biodiversity (Zeeshan et al. 2019). Wetland plants, particularly helophytes, offer a plentiful source of raw materials for basketry and brooms, which is one of the most noteworthy services in craftsmanship (Balfet 2019), and the variety of plants used in basketry across the world demonstrates how traditional knowledge can adapt to the environment (Fajardo et al. 2021). There are destructive and nondestructive activities on wetlands that contribute to the economy. While destructive activities are not encouraged, nondestructive activities can be maximized to improve the livelihood of people and for contribution to the economy. Wetlands are of great importance in agriculture, health, bioenergy as well as recreation and tourism.

18.4.1 Agriculture

Throughout the world, wetlands are frequently linked to agriculture, since millions of hectares of diverse wetland types sustain a broad range of activities. On the other hand, these same agricultural activities put many wetlands in danger by altering the hydrological and other natural regimes that support them as well as their biological makeup. However, the importance of wetlands in agriculture remains undoubted. In many developing nations, wetland agriculture is crucial for reducing poverty and ensuring food security (McCartney et al. 2010). Wetlands serve to achieve the SDGs of the global food security agenda by sustaining agriculture and creating jobs, serving as a supply of water for crops and cattle, and serving as a habitat for aquaculture and rice production (FAO 2019).

The way different farming systems remove biota and use water and nutrients to generate food determines how they affect wetlands. Some farming systems are naturally more efficient than others in these areas (Convention on Wetlands 2022). To fulfill the 2030 agenda for sustainable development, the United Nations' SDGs, and the 2063 agenda of the Africa Union, as well as to ensure the prudent use of wetlands in accordance with the Convention on Wetlands, it is necessary to exchange knowledge about how agriculture affects wetlands, raise awareness of the significance of healthy wetlands for agriculture, and pinpoint strategies for transforming the sector (Dixon et al. 2021).

Wetlands give people food and a supply of water for crops, cattle, and aquaculture, which directly contributes to agricultural productivity. For many people worldwide, staple foods like rice and fish are provided by human-made wetlands, such as ponds, rice paddies, and wet pastures (FAO 2019). According to an analysis of the Ramsar Site Information Service (RSIS), around 37% of Wetlands of International Importance are used as cattle feed, and more than 50% of them supply food for people. Wetlands also help agriculture by regulating the environment, which includes regulating agricultural pests, replenishing groundwater, cycling nutrients, and sequestering carbon (Verhoeven and Setter 2010).

18.4.2 Biofuel and Energy Sources

Wetlands in South Africa serve as valuable resources for income generation, aligning with global efforts to reduce fossil fuel dependency and combat climate change. By harnessing wetlands for biofuel, biofertilizers, bio-oil and other high-value goods production, communities promote economic growth, sustainability, and environmental conservation. Abundant vegetation in natural and constructed wetlands is utilized for biomass harvesting, contributing to climate change mitigation and reducing reliance on fossil fuels. Additionally, wetland-based systems for wastewater treatment offer further economic potential (Zhang et al. 2014; Avellán and Gremillion 2019; Rasool et al. 2023). Innovative technologies like floating treatment wetlands (FTWs) and microbial fuel cells (MFCs) enhance energy production and wastewater treatment efficacy, maximizing the utility of wetland resources (Colares et al. 2020, 2021).

Case studies illustrate the GDP generation potential of wetland-based livelihoods in South Africa, such as the Biogreen Project targeting invasive plant species for biofuel production (Case Study 1). Creativities like the Wetlands for Life Initiative focus on wastewater treatment and biofertilizer production, promoting sustainable agriculture (Case Study 2). Projects like the GreenTech Solutions Project showcase innovative technologies for wetland-based energy production, emphasizing economic and environmental benefits (Case Study 3). The Biogreen Project in South Africa exemplifies wetland-based biofuel production, targeting invasive plant species in wetland areas. Through partnerships with local communities and stakeholders, the project not only addresses invasive species issues but also contributes to carbon emissions reduction and SDGs. The Wetlands for Life Initiative focuses on wastewater treatment and biofertilizer production using constructed wetland systems. By selling biofertilizers to local farmers, the initiative generates revenue while promoting sustainable agriculture practices and enhancing environmental conservation efforts. The GreenTech Solutions Project integrates innovative technologies like FTWs and MFCs for wetland-based energy production. By generating bioelectricity from wetland vegetation and selling excess electricity to the grid, the project not only provides renewable energy but also creates economic opportunities for local communities. These cases demonstrate how wetland-based livelihoods contribute to GDP generation at both local and global levels. By focusing on biofuel, biofertilizers, bio-oil, and other high-value goods production, wetland initiatives drive economic growth while fostering sustainable development and environmental conservation. Through collaborative efforts and innovative technologies, wetlands emerge as valuable assets for achieving broader socioeconomic objectives.

18.4.3 Medicine

The medicinal properties of wetlands, including both plant and animal resources, have been acknowledged for centuries (Sardar et al. 2015). Traditional medicinal practices have often utilized various components of wetlands for their therapeutic benefits, such as anti-inflammatory, analgesic, and detoxifying properties (Leaman 2016). For instance, medicinal leeches found in wetlands have been historically used to address conditions like thrombosis and venous disorders (Sig et al. 2017). Expanding upon this traditional knowledge, contemporary research suggests that the therapeutic potential of wetlands extends beyond direct medicinal use. Considering the growing recognition of nature-based interventions for mental health issues, wetlands are increasingly viewed as potential contributors to psychological well-being (Maund et al. 2019).

Access to natural environments, including wetlands, has been associated with lower stress levels, higher self-esteem, and reduced anxiety and sadness (Hunter et al. 2019; Olafsdottir et al. 2020). The presence of water in these landscapes, characteristic of wetlands, has been particularly linked to enhanced well-being, surpassing the benefits of green spaces alone (Grassini et al. 2019). Research indicates that individuals living in proximity to urban wetlands often perceive these areas as therapeutic, experiencing a boost in positive emotions (Maund et al. 2019). The rich biodiversity found in wetlands may contribute to this effect, as studies have demonstrated a positive correlation between species richness and self-reported psychological wellness (Aerts et al. 2018; Mavoa et al. 2019). Therefore, preserving and promoting access to wetlands not only preserves traditional medicinal resources but also provides opportunities for nature-based interventions that support mental health.

In the context of income generation for local communities, the recognition of wetlands as potential contributors to mental well-being opens avenues for ecotourism and wellness-based initiatives. Communities living near wetlands could develop nature-based tourism programs, offering activities such as birdwatching, nature walks, and mindfulness retreats that capitalize on the therapeutic aspects of these environments. By leveraging the appeal of wetlands for mental health promotion, communities can create sustainable livelihoods while simultaneously conserving these valuable ecosystems. The income generated from wellness-based initiatives in wetland areas can have a significant impact on local economies. This impact extends beyond direct spending on activities to include revenue generated for local accommodations, restaurants, transportation services, and handicrafts. Studies have shown that nature-based tourism tends to have a higher proportion of expenditures retained within local economies compared to other forms of tourism.

18.4.4 Tourism, Recreation, and Education

The Travel and Tourism sector contributed 7.6% to global GDP; an increase of 22% from 2021 and only 23% below 2019 levels (World Travel and Tourism Council 2022). An estimated USD 925 billion is spent on tourism related to wetlands annually, with half of all foreign visitors visiting wetlands of some kind, especially those near coasts. When you include day visits for leisure and domestic travel, the economic impact of tourism to wetlands is actually significant (Ramsar 2012). Wetlands are recreational areas that significantly boost the travel and tourism sector (Musamba et al. 2012) because they draw large numbers of visitors. Nature-based tourism is an emerging business in terms of resource usage, employment, and other best practices (Karanth and DeFries 2011). In many developing nations, the tourist industry is widely acknowledged for its role in creating jobs and increasing revenue. As a result, maintaining and managing naturally occurring recreation areas, including wetlands, is an essential prerequisite to estimating the number of visitors and leisure activities that will occur there.

In every nation, wetlands such as lakes, reservoirs, and coral reefs are vital components for attracting tourists. These engage in various revenue-generating activities, including serving as tour guides, boat transporters, fishermen, and sellers (Shah and Islam 2023). The educational value of wetlands is closely related to their advantages for recreation and overall well-being. Wetland education facilities are widely available worldwide, with a growing network that can accommodate a broad range of purposes, including community participation and traditional school group trips (Wetland Ecosystem Services n.d.).

18.5 Threats and Conservation of Wetlands in South Africa

Adeeyo et al. (2022) outline various human activities that pose a threat to the proper functioning of wetland ecosystems. Additionally, climate change poses a growing threat to wetlands. Rising temperatures, altered precipitation patterns, and sea-level rise can disrupt the delicate balance of these ecosystems (James et al. 2013). Changes in water availability can lead to shifts in plant communities and affect the overall functioning of wetland ecosystems. To ensure the preservation of wetlands for future generations, it is imperative to understand the threats they face and take proactive measures to conserve and manage them effectively (Table 18.1).

18.5.1 Rising Temperatures

High temperatures present a significant concern for wetlands in South Africa as the ongoing effects of climate change drive up global average temperatures (Dallas and Rivers-Moore 2014). The implications of increasing temperatures on wetland environments are multifaceted, with changes in water flow patterns and biogeochemical processes disturbing the natural equilibrium of these crucial ecosystems (Van Niekerk et al. 2019). With rising temperatures, wetlands may undergo shifts in their roles, potentially leading to the conversion of vital services they offer into negative impacts. For example, temperature variations can influence water quantity and quality within wetlands, affecting the diverse plant and animal species dependent on these habitats for survival.

The warming climate directly impacts wetland ecosystems and compounds challenges like habitat loss. According to Hoegh-Guldberg et al. (2018), the AR5 report identifies wetlands as threatened ecosystems due to increased temperature, with impacts expected to be greater at 2 °C warming

Table 18.1 Threats against wetlands and proposed solutions.

Natural threat	Description	Proposed solution
Flooding	Periodic or prolonged inundation of wetlands by water, often due to heavy rainfall, snowmelt, or changes in river flow patterns.	Implementing natural flood management techniques such as the restoration of floodplains, construction of levees or embankments, and planting native vegetation to absorb excess water.
Drought	Prolonged periods of abnormally low precipitation leading to water scarcity and reduced water levels in wetlands.	Promoting water conservation measures such as rainwater harvesting, installation of water-efficient irrigation systems, and restoration of natural hydrological patterns through habitat management.
Salinization	Increased salt concentration in wetland soils and water bodies, often due to natural processes like sea level rise, saltwater intrusion, or evaporation in arid regions.	Implementing measures to restore freshwater inflows, such as dredging channels or installing water control structures, and planting salt-tolerant vegetation to stabilize soils and reduce salinity levels.
Invasive species	Introduction of non-native plant or animal species into wetland ecosystems, often outcompeting native species, disrupting food webs, and altering ecosystem dynamics.	Implementing invasive species management strategies such as mechanical removal, controlled burns, introduction of natural predators or competitors, and establishment of early detection and rapid response programs.
Wildfires	Uncontrolled fires that can occur naturally or due to lightning strikes, often leading to the destruction of wetland vegetation, loss of habitat, and release of carbon emissions.	Implementing prescribed burning programs to reduce fuel buildup, establishing firebreaks to contain wildfires, and conducting public education campaigns to raise awareness about fire prevention and safety measures.
Storm surges	Sudden rises in sea level during coastal storms, leading to inundation of wetlands, erosion of shorelines, and destruction of vegetation and wildlife habitat.	Implementing coastal protection measures such as constructing seawalls, restoring natural dune systems, and planting salt-tolerant vegetation to buffer wetlands from the impacts of storm surges.
Temperature extremes	Extreme cold or heat events that can stress wetland organisms, disrupt breeding cycles, and alter ecosystem structure and function.	Implementing climate adaptation strategies such as restoring wetland connectivity, creating shaded areas to reduce temperature extremes, and enhancing habitat heterogeneity to provide refuge for species during extreme events.

than at 1.5 °C. As temperatures persist in their increase, the complex logic within South Africa's wetlands faces disruption, putting the biodiversity and ecological functions they sustain at risk. Conservation initiatives must confront the obstacles presented by rising temperatures to safeguard the enduring well-being and adaptability of wetlands in South Africa.

18.5.2 Altered Precipitation Patterns

Changes in precipitation patterns present a significant peril to wetlands in South Africa, disturbing the delicate equilibrium of these ecosystems. Changes in precipitation levels and patterns can result in fluctuations in water availability, impacting the hydrology and overall well-being of wetlands (Salimi et al. 2021). The intensification of these changes by climate change heightens the vulnerability of wetlands to drying out or experiencing flooding beyond their natural adaptive capacity. Studies indicate that specific regions in South Africa, such as St. Lucia and Mkhuze, display marked rainfall gradients that influence adjacent wetlands (Gensel et al. 2022). These disparities in rainfall distribution can disrupt wetland ecosystems' complex biodiversity and ecological functions. Jury (2019) predicts that a substantial portion of South Africa, apart from the eastern coastal plains, will experience increased aridity by the year 2050.

18.5.3 Sea Level Rise

The rise in sea levels presents a considerable risk to wetlands in South Africa, leading to coastal inundation and erosion that disrupt the fragile equilibrium of these ecosystems. Most of South Africa's wetlands are situated along the eastern coastal area. Coastal wetlands serve a vital function in mitigating the effects of rising sea levels by absorbing excess water and providing a habitat for a diverse array of plant and animal species. Nevertheless, as global sea levels continue to climb, these wetlands confront the challenge of adapting to evolving circumstances.

Studies indicate that some wetlands are presently under threat from coastal flooding and erosion, which are worsened by rising sea levels, with forecasts showing a notable escalation in the number of endangered sites by 2050 across various emission scenarios (Day and Rybczyk 2019; Scholes and Engelbrecht 2021). This underscores the pressing necessity for preemptive actions to safeguard and conserve South Africa's wetlands from the adverse impacts of rising sea levels. Looking ahead, policymakers, researchers, and local communities must collaborate on implementing sustainable strategies that alleviate the implications of increasing sea levels on these essential ecosystems.

Climate change poses an increasing threat to wetlands. Escalating temperatures, altered precipitation patterns, increased frequency of extreme weather events, and rising sea levels can disrupt the delicate equilibrium of these ecosystems (Dube et al. 2023) and alter the hydrological regime of wetlands, leading to changes in water availability and quality (Kusangaya et al. 2021). Variations in water availability can trigger shifts in plant communities and disrupt the overall functionality of wetland ecosystems. These changes can harm the flora and fauna that depend on specific wetland conditions. Furthermore, climate change can exacerbate existing threats, such as habitat loss and degradation, making conserving and restoring wetland ecosystems even more challenging.

18.6 Conservation of Wetlands

Conservation efforts in South Africa are focused on protecting and restoring wetland ecosystems. These efforts consider both natural threats, such as climate change and habitat loss, and human-induced pressures, like pollution and agricultural expansion. By understanding the potential interactions between these factors, scientists and conservationists can develop strategies to mitigate the negative impacts and preserve the integrity of these valuable ecosystems. According to Sinthumule (2021), most of the communities need to be made aware of the role played by the government in managing wetlands. Conservation of wetlands is crucial for preserving these unique ecosystems and their numerous benefits.

Efforts to conserve wetlands involve various strategies. One approach is establishing protected areas, such as national parks or wildlife refuges, where wetlands are safeguarded from development and human disturbance. These protected areas provide safe havens for wildlife and offer opportunities for research, education, and recreation (Stolton et al. 2015). Additionally, restoration projects aim to recreate or enhance degraded wetland habitats by restoring hydrological conditions and reintroducing native vegetation. These initiatives help to revitalize the ecological functions of wetlands and promote biodiversity (Angelstam et al. 2017).

Indigenous knowledge systems are becoming more significant, and they are essential to addressing the contemporary world issues of ecosystem management and climate change (Chanza and Musakwa 2021). Some academics contend that, despite indigenous communities' long-standing knowledge systems for preserving the same ecosystems, their marginalization has exacerbated the accelerated degradation of ecosystems (Whitney et al. 2020). To combat ecological degradation, indigenous knowledge systems have acquired traction in policy, research, and real-world applications. Though most academics in conservation science do not view it as a solution, their ideas are in line with including it within the broader conservation agenda (Makondo and Thomas 2018).

Communities use indigenous knowledge systems to affix socioecological value systems – value systems that transcend generations and change as they try to solve modern problems – to natural resources (Whitney et al. 2020). Therefore, it is necessary to see the preservation of natural resources as ecological manifestations of particular sociocultural groups, taking into account the importance of these resources to local populations (Savo et al. 2016). Because of this, indigenous knowledge systems are crucial to achieving the objective of resource conservation, especially in the light of the changing and unpredictable climatic system that is increasing the reliance of indigenous populations on their ecosystems (Chanza and Musakwa 2021).

Collaboration between government agencies, conservation organizations, scientists, and local communities is essential for successful wetland conservation. Together, these stakeholders can develop comprehensive management plans addressing each wetland site's threats. Such plans may include measures to reduce pollution inputs, regulate water use to maintain appropriate hydrological conditions, control invasive species that threaten native biodiversity, and raise public awareness about the importance of wetland conservation. Furthermore, SANBI plays a crucial role in wetland conservation through research, monitoring, and capacity building (Driver et al. 2012). The institute works closely with local communities and stakeholders to raise awareness about the value of wetlands and promote sustainable practices that protect these ecosystems.

The government of South Africa has also taken steps to enhance wetland conservation. The National Water Act (NWA) of 1998 provides a legal framework for the protection and sustainable use of water resources, including wetlands. Additionally, the National Environmental Management: Protected Areas Act designates protected areas, including wetlands, to ensure long-term conservation. Overall, it is imperative to recognize and address these threats to ensure the preservation and conservation of our invaluable wetland environments. Efforts must be undertaken at local, national, and international levels to mitigate pollution sources, protect remaining wetland habitats from further destruction or degradation, and implement adaptive management strategies that account for the impacts of climate change on these critical ecosystems.

18.7 Management Strategies, National Policies, and Legislations Against Natural Threats

South Africa became a signatory to the Ramsar Convention in 1975, which promotes the legal force for wetland protection and conservation. The convention aims to protect wetlands now and in future, aligning with the intergenerational equity principle. Various aspects inform how international instruments are incorporated into a legal system, and South Africa follows a dualistic approach to implementing conventions (Shale 2019). In application, the Ramsar is not directly implemented but has been transformed into national laws. In the light of this, there is no direct proof that national laws have been passed solely to promote the objectives of the Ramsar Convention; this is so as legislation promoting wetland conservation and protection is located in various pieces of, mostly, environmental legislation. It is posited that a single policy or legislation focusing on wetland protection and conservation for South Africa may better streamline, demonstrate, and articulate the aim of the Ramsar Convention. However, hope was sparked by the report submitted to the Conference of the (Ramsar) Party in 2021 for formulating a National Joint Framework on Wetlands (available at http://ramsar.org accessed 21 November 2023). It is therefore deemed necessary to highlight critical aspects of the Ramsar Convention and chart the legislative landscape through the South African laws.

Many commentators (Kidd 2011; Elliot and Blackmore 2019; Glazewski 2023) have criticized the legal wetland landscape in South Africa for its uncoordinated nature and the impact on the institutional framework, which goes against the ambitions of the Ramsar Convention. Having said this, it hints at the loss of pursuit of stemming the loss and encroachment on wetlands, and Sharma et al. (2021) submit the encumbering issues relating to the lack of legislation for wetland protection. This demonstrates the power of a legislative landscape and the lack of or inherent issues relating to this, going against the grain of the environmental clause set in the Constitution of the Republic of South Africa of 1996 (hereafter Constitution). The Constitution, not a statute, is the supreme law of the land and requires that all laws and actions be in line with it.

According to the Constitution's environmental provision, everyone has the right to live in an environment that does not endanger their health or well-being. To achieve this, for the benefit of the present and coming generations, there has to be reasonable legislatives and actions that assure environmentally sustainable growth and utilization of natural resources, avoid pollution and ecological degradation, and foster conservation while seeking legitimate social and economic advancement.

Lemine (2020), with a view on wetlands, draws parallels between the benefits of wetland services and socioeconomic rights. For instance, drawing on water, food, and shelter provided by wetlands to correlative rights like access to water and food (Lemine 2020). Therefore, this advocates for protecting and conserving the environment as a legal basis to promote other rights that are consequences. Other than wetlands forming an integral part of the "environment," section 2(4)(r) of the NEMA singles out wetlands as a sensitive ecosystem that requires "specific attention in management and planning procedure, especially where they are subject to significant human resource usage and development pressure."

The NWA aims to protect, develop, conserve, manage, and control South Africa's water resources. Wetlands are water resources and are defined separately, as illustrated supra. The NWA empowers the Minister of Water and Sanitation to determine the water resources class and set resource quality objectives for them. Evidence of this is the publication of the regulations for certain catchment areas since 2016. Classifying a water resource as protected depends on whether it is, as termed in the NWA, a "significant water resource." However, there are no clear guidelines on determining this significance level, even though it is necessary to advance the protection of wetlands. The lack of clarity regarding significance poses several risks for wetlands that must be addressed.

Creating the United Nations Educational, Scientific and Cultural Organization's Man and Biosphere Program (MABS Program) may catalyze recognition of wetlands as "significant water resources." The MABS Program focuses on the relationship between man and the environment (available at https://www.unesco.org/en/mab/ accessed 1 December 2023). It is opined here that the nature of this relationship can promote intergenerational equity as envisaged in section 24(b) of the Constitution and the Ramsar Convention's objective, therefore placing it in man's hands to promote the right to an environment that is not harmful to their health and well-being.

There is no special legislative protection for the 10 biosphere regions in South Africa, like for protected areas or wetlands of international importance. However, the White Paper on the Conservation and Sustainable Use of South Africa's Biodiversity 2022 (White Paper CSB) recognizes its protection outside protected areas. This means that one need not rely on the fact that wetlands are in a buffer zone of a protected area to build an argument for its protection. The White Paper CSB further ensures that biosphere regions are valued by promoting the scaling up and resourcing of these initiatives. This is one aspect that presiding officers and policymakers can consider when guidelines or decisions regarding "significance" are required.

18.8 Future Enhancements

In the coming years, efforts to integrate wetland conservation with agricultural practices will intensify. Recognizing wetlands' crucial role in sustaining agriculture and ensuring food security, there will be a shift toward developing sustainable farming systems. Adoption of eco-friendly practices like agroecology and precision agriculture will prioritize biodiversity conservation while enhancing productivity. Advancements in water management technologies will optimize resource use, promoting long-term agricultural sustainability. With heightened global efforts to mitigate climate change, wetlands will become key contributors to bioenergy production. Increased investment in research will focus on harnessing wetland biomass for biofuel and bioelectricity generation. These initiatives will reduce reliance on fossil fuels and create economic opportunities for local communities living in wetland areas.

Recognition of wetlands' therapeutic benefits will lead to their integration into healthcare and wellness initiatives. Nature-based interventions like ecotherapy will be promoted to improve mental health and well-being. This trend will spur the development of eco-friendly healthcare facilities and wellness retreats in wetland areas, leveraging nature's healing power. Research into the medicinal properties of wetland plants and animals will lead to the discovery of novel compounds of pharmaceutical relevance. Wetlands will play a pivotal role in sustainable tourism practices, with governments prioritizing their protection and management. Ecotourism programs focused on wetlands will offer immersive experiences, highlighting their ecological and cultural significance. Wetland education programs will expand, raising awareness about conservation and fostering environmental stewardship among diverse audiences.

The avoidance of destructive activities in wetland areas will allow for the conservation of their biodiversity. Plants, animals, and microorganism in wetlands depend on controlled and maintained ecology to survive and a disturbance to their natural existing conditions can lead to migration or death, which might ultimately result in extinction. If climatic conditions can be effectively controlled and an altered wetland ecology is prevented, plants and animals in wetlands will find their natural habitats conducive and would not need alternative habitats. With conserved biodiversity, other importance of wetlands such as health uses and tourism that depends on the biodiversity of wetlands are preserved.

The promulgation of the National Wetland Policy for South Africa is welcomed; it provides an opportunity for better coordination. Where the gaps in the law exist due to fragmentation, reform must be affected before these enabling provisions are merely carried over and transplanted into the policy. The policy must be accompanied by strategies that can be used to measure the action required for proper implementation. To implement the necessary measures for wetlands protection, it is essential to identify the different individuals and groups involved, such as policymakers, administrators, users, water tribunals, and courts. The courts and tribunals play a critical role in this process as they interpret and apply legislation. Depending on the jurisdiction, their rulings may have the power to bind lower courts or serve as persuasive precedents.

Considering how South Africa will manage shared wetlands with other countries is crucial. The policies and strategies implemented must consider their obligations in the Ramsar Convention as well as other regional agreements and Agenda 2063: the Africa We Want, which outlines the vision for the future of Africa. In South Africa, this includes the Maluleke wetland and Orange Mouth Basin, for example. However, challenges may arise when one country is a Ramsar state party, and the other is not. What can be explored is how much of the Convention is incorporated into the bilateral agreement but also the extent of the crystallization of Ramsar into customary international law. As a multilateral environmental agreement, the Ramsar Convention creates an interstate wetland obligation beyond that which focuses simply on states sharing wetland resources, as explained previously. Articles 4.2 and 7.1 of the Ramsar Convention indicate the factors with which state parties must comply. These factors include sharing research, data, and knowledge, which underpin interstate cooperation and may promote intergenerational equity for wetlands through interstate relations.

18.9 Conclusion

Wetlands provide a number of ecological services, including reducing the danger of flooding, storing carbon, regulating the temperature, and acting as water reservoirs. They provide drinking water, fish, fodder, fuel, wildlife habitat, regulate the rate of runoff in urban areas, protect shorelines from erosion, and provide recreational opportunities for society. Wetlands also function to recycle nutrients, purify water, lessen floods, and recharge ground water. Additionally, they serve as a haven for animals, filter pollutants, and preserve rare biodiversity. However, pollution stands as one of the most significant challenges confronting wetlands. In addition to pollution, wetlands confront other perilous threats such as habitat destruction and degradation. Human activities, including urban expansion, agricultural encroachment, and land drainage for conversion purposes, have resulted in extensive losses of wetland habitats. Of great importance is the effect of climate change on the destruction of wetlands. Therefore, necessary measures need to be taken for the conservation of wetlands. By conserving wetlands, we can protect its biodiversity and ensure the continued existence of plants and animals for future generations.

To implement the necessary measures for wetlands protection, it is essential to identify the different individuals and groups involved, such as policymakers, administrators, users, water tribunals, and courts. Collaboration between government agencies, conservation organizations, scientists, and local communities is essential for successful wetland conservation. Recognizing the importance of wetlands, various organizations and initiatives are working toward their conservation in South Africa. The Working for Wetlands program, led by the Department of Forestry, Fisheries, and the Environment, focuses on rehabilitating degraded wetlands and implementing sustainable management practices. This program restores wetlands, controls invasive species, and involves local communities in conservation efforts. Wetland conservation is paramount in safeguarding these unique ecosystems and the array of benefits they provide.

References

Abdellah, L. (2023). Rare bird species found in South Africa's wetlands. https://www.getaway.co.za/environment/rare-bird-species-found-in-south-africas-wetlands/ (accessed 18 March 2024).

Adeeyo, A.O., Ndlovu, S.S., Ngwagwe, L.M. et al. (2022). Wetland resources in South Africa: threats and metadata study. *Resources* 11 (6): 54.

Aerts, R., Honnay, O., and Van Nieuwenhuyse, A. (2018). Biodiversity and human health: mechanisms and evidence of the positive health effects of diversity in nature and green spaces. *British Medical Bulletin* 127 (1): 5–22.

Angelstam, P., Barnes, G., and Elbakidze, M. (2017). Collaborative learning to unlock investments for functional ecological infrastructure: Bridging barriers in social-ecological systems in South Africa. *Ecosystem Services* 27: 291–304.

Avellán, T. and Gremillion, P. (2019). Constructed wetlands for resource recovery in developing countries. *Renewable and Sustainable Energy Reviews* 99: 42–57.

Balfet, H. (2019). Basketry. https://www.britannica.com/art/basketry (accessed 15 December 2023).

Ballesteros, N., Páez, L., and Luna, N. (2023). Characterization of microbial communities in seven wetlands with different anthropogenic burden using next generation sequencing in Bogotá, Colombia. *Scientific Reports* 13 (1): 16973.

Bodelier, P.L.E. and Dedysh, S.N. (2013). Microbiology of wetlands. *Frontiers in Microbiology* 4 (79): 1–4.

Butt, M.A., Zafar, M., and Ahmed, M. (2021). *Wetland Plants: A Source of Nutrition and Ethno-medicines*. Cham: Springer International Publishing.

Cao, Q., Wang, H., and Chen, X. (2017). Composition and distribution of microbial communities in natural river wetlands and corresponding constructed wetlands. *Ecological Engineering* 98: 40–48.

Chanza, N. and Musakwa, W. (2021). Indigenous practices of ecosystem management in a changing climate: prospects for ecosystem-based adaptation. *Environmental Science & Policy* 126: 142–151.

Colares, G.S., Dell'Osbel, N., and Wiesel, P.G. (2020). Floating treatment wetlands: a review and bibliometric analysis. *Science of the Total Environment* 714: 136776.

Colares, G.S., Dell'Osbel, N., and Barbosa, C.V. (2021). Floating treatment wetlands integrated with microbial fuel cell for the treatment of urban wastewaters and bioenergy generation. *Science of the Total Environment* 766: 142474.

Collins, N.B. (2006). *Wetlands: The Basics, and Some More*. Free State Department of Tourism, Environmental and Economic Affairs.

Convention on Wetlands (2022). *Wetlands and Agriculture: Impacts of Farming Practices and Pathways to Sustainability*. Gland: Convention on Wetlands.

Dallas, H.F. and Rivers-Moore, N. (2014). Ecological consequences of global climate change for freshwater ecosystems in South Africa. *South African Journal of Science* 110 (5–6): 1–11.

Day, J.W. and Rybczyk, J.M. (2019). Global change impacts on the future of coastal systems: perverse interactions among climate change, ecosystem degradation, energy scarcity, and population. In: *Coasts and Estuaries* (eds. E. Wolanski, J.W. Day, M. Elliott, and R. Ramesh), 621–639. Elsevier.

Department of Environment, Forestry and Fisheries (2021). Working for wetlands: 20 years of wetland restoration in South Africa.

Department of Forestry, Fisheries and the Environment (2022). Berg Estuary designated as South Africa's 28th wetlands of international importance.

Dixon, A., Wood, A., and Hailu, A. (2021). Wetlands in Ethiopia: lessons from 20 years of research, policy and practice. *Wetlands* 41 (2): 20.

Driver, A., Sink, K.J., and Nel, J.L. (2012). National biodiversity assessment 2011: an assessment of South Africa's biodiversity and ecosystems.

Du Toit, M.J., Du Preez, C., and Cilliers, S.S. (2021). Plant diversity and conservation value of wetlands along a rural–urban gradient. *Bothalia, African Biodiversity & Conservation* 51 (1): 1–18.

Dube, K., Nhamo, G., and Chikodzi, D. (2023). Mapping and evaluating the impact of flood hazards on tourism in South African national parks. *Journal of Outdoor Recreation and Tourism* 43: 100661.

Elliot, F. and Blackmore, A. (2019). A hitchhiker's guide to South Africa's wetland legislative framework and international obligations. *South African Journal of Environmental Law and Policy* 25: 42–72.

Fajardo, J., Verde, J., and Rivera, D. (2021). Basketry as an ecosystem service of wetlands: traditional crafts in central Spain. *Anales del Jardin Botánico de Madrid* 78: e115.

FAO (2019). *The State of the World's Biodiversity for Food and Agriculture.* Rome: FAO.

Gardner, R.C. and Finlayson, C. (2018). Global wetland outlook: state of the world's wetlands and their services to people. Ramsar Convention Secretariat. Stetson University College of Law Research Paper No. 2020-5.

Garrett (2022). 29 animals that live in wetlands (with pictures). https://faunafacts.com/animals/animals-that-live-in-wetlands/ (accessed 18 March 2024).

Gehrig, S., Harding, C., and Turner, D. (2015). *Developing Ecological Response Models and Determining Water Requirements for Wetlands in the South-East of South Australia: Task 1: Data review and methodological framework.* Goyder Institute for Water Research Technical Report Series.

Gensel, J., Humphries, M.S., and Zabel, M. (2022). Origin, transport, and retention of fluvial sedimentary organic matter in South Africa's largest freshwater wetland, Mkhuze Wetland System. *Biogeosciences* 19 (11): 2881–2902.

Getaway Magazine (2024). *8 wonderful wetlands in South Africa.* https://www.getaway.co.za/things-to-do/birding/wonderful-wetlands-south-africa/ (accessed 18 March 2024).

van Gickel, C.E., Glen, R.P., Gordon-Gray, K.D. et al. (2011). *Easy Identification of Some South African Wetland Plants*, 19–359. South Africa: Water Research Commission.

Glazewski, J. (2023). *Environmental Law in South Africa.* LexisNexis South Africa.

Grassini, S., Revonsuo, A., and Castellotti, S. (2019). Processing of natural scenery is associated with lower attentional and cognitive load compared with urban ones. *Journal of Environmental Psychology* 62: 1–11.

Hoegh-Guldberg, O., Jacob, D., and Bindi, M. (2018). Impacts of 1.5 °C global warming on natural and human systems. Global warming of 1.5 °C. *The Intergovernmental Panel on Climate Change (IPCC).*

Hunter, M.R., Gillespie, B.W., and Chen, S.Y.-P. (2019). Urban nature experiences reduce stress in the context of daily life based on salivary biomarkers. *Frontiers in Psychology* 10: 722.

iNaturalist (2020). Threatened obligate wetland species of South Africa. https://inaturalist.ca/check_lists/4304942-Threatened-Obligate-Wetland-Species-of-South-Africa (accessed 17 March 2024).

James, N., Van Niekerk, L., and Whitfield, A. (2013). Effects of climate change on South African estuaries and associated fish species. *Climate Research* 57 (3): 233–248.

Jankoski, P.R., Correa, A.P.F., and Brandelli, A. (2021). Biological activity of bacteria isolated from wetland sediments collected from a conservation unit in the southern region of Brazil. *Anais da Academia Brasileira de Ciências* 93 (3): e20191269.

Job, N.M. and Le Roux, P.A.L. (2019). Developing Wetland Distribution and Transfer Functions from Land Type Data as a Basis for the Critical Evaluation of Wetlands Delineation Guidelines by

Inclusion of Soil Water Flow Dynamics in Catchment Areas. Technical Report submitted to South African Water Research Commission.

Jury, M.R. (2019). South Africa's future climate: trends and projections. In: *The Geography of South Africa*, World Regional Geography Book Series (ed. J. Knight and C.M. Rogerson), 305–312. Cham: Springer International Publishing.

Kanaujia, A. and Kumar, A. (2014). Wetlands: significance. Threats and their conservation. *Green* 7 (3–4): 1–25.

Karanth, K.K. and DeFries, R. (2011). Nature-based tourism in Indian protected areas: New challenges for park management: tourism in Indian protected areas. *Conservation Letters* 4 (2): 137–149.

Kidd, M. (2011). *Environmental Law*, 2e. Cape Town: Juta.

Kusangaya, S., Mazvimavi, D., and Shekede, M.D. (2021). Climate change impact on hydrological regimes and extreme events in Southern Africa. In: *Climate Change and Water Resources in Africa* (ed. S. Diop, P. Scheren, and A. Niang), 87–129. Cham: Springer International Publishing.

Leaman, D. (2016). *Medicinal plants in wetlands*. In: *The Wetland Book*, 1–4. Dordrecht: Springer.

Lemine, B.J. (2020). Developing a strategy for efficient environmental authorisation of activities affecting wetlands in South Africa: towards a wise-use approach. *Obiter* 41 (1): 154–167.

Li, M., Mi, T., and Yu, Z. (2020). Planktonic bacterial and archaeal communities in an artificially irrigated estuarine wetland: diversity, distribution, and responses to environmental parameters. *Microorganisms* 8 (2): 198.

Makondo, C.C. and Thomas, D.S.G. (2018). Climate change adaptation: linking indigenous knowledge with western science for effective adaptation. *Environmental Science & Policy* 88: 83–91.

Maua, J.O., Mbuvi, M.T.E., and Matiku, P. (2022). The difficult choice – to conserve the living filters or utilizing the full potential of wetlands: insights from the Yala swamp, Kenya. *Environmental Challenges* 6: 100427.

Maund, I., Reeves, S., and Cromie, D. (2019). Wetlands for wellbeing: piloting a nature-based health intervention for the management of anxiety and depression. *International Journal of Environmental Research and Public Health* 16 (22): 4413.

Mavoa, S., Davern, M., and Breed, M. (2019). Higher levels of greenness and biodiversity associate with greater subjective wellbeing in adults living in Melbourne, Australia. *Health & Place* 57: 321–329.

McCartney, M., Rebelo, L.-M., and Sellamuttu, S.S. (2010). *Wetlands, Agriculture and Poverty Reduction*. Colombo: International Water Management Institute.

McInnes, R.J. and Everard, M. (2017). Rapid assessment of wetland ecosystem services (RAWES): an example from Colombo, Sri Lanka. *Ecosystem Services* 25: 89–105.

Miller, L. (2019). Marsh plants & animals. https://sciencing.com/marsh-plants-animals-6462874.html (accessed 18 March 2024).

Mofutsanyana, S.S., Collins, N.B., and Adelabu, S.A. (2020). Changes in plant functional composition of wetland vegetation along an aridity gradient on the Highveld plateau of South Africa G. Overbeck, ed. *Applied Vegetation Science* 23 (4): 622–634.

Musamba, E.B., Boon, E.K., and Ngaga, Y.M. (2012). The recreational value of wetlands: activities, socio-economic activities and consumers' surplus around Lake Victoria in Musoma Municipality, Tanzania. *Journal of Human Ecology* 37 (2): 85–92.

North Carolina Division of Water Resources (n.d.). Wetland animals in North Carolina.

Olafsdottir, G., Cloke, P., and Schulz, A. (2020). Health benefits of walking in nature: a randomized controlled study under conditions of real-life stress. *Environment and Behavior* 52 (3): 248–274.

Ollis, D., Snaddon, K., and Job, N. (2013). *Classification System for Wetlands cation System for Wetlands and other Aquatic Ecosystems and other Aquatic Ecosystems in South Africa*. Pretoria: South African National Biodiversity Institute.

Ollis, D., Ewart-Smith, J., and Day, J. (2015). The development of a classification system for inland aquatic ecosystems in South Africa. *Water SA* 41 (5): 727.

Ollis, D.J., Day, J.A., and Mbona, N. (2016). *South African Wetlands: Classification of Ecosystem Types.* Dordrecht: Springer.

Peer, N., Muhl, E.-K., and Janna, J. (2022). Community and marine conservation in South Africa: are we still missing the mark? *Frontiers in Marine Science* 9: 884442.

Putkinen, A., Larmola, T., and Tuomivirta, T. (2012). Water dispersal of methanotrophic bacteria maintains functional methane oxidation in *Sphagnum* mosses. *Frontiers in Microbiology* 3: 15.

Rai, U.N., Upadhyay, A.K., and Singh, N.K. (2015). Seasonal applicability of horizontal sub-surface flow constructed wetland for trace elements and nutrient removal from urban wastes to conserve Ganga River water quality at Haridwar, India. *Ecological Engineering* 81: 115–122.

Ramachandra, T.V., Aithal, B.H., and Kumar, U. (2012). Conservation of wetlands to mitigate urban floods. *Journal of Resources, Energy and Development* 9 (1): 1–22.

Ramsar (2012). Wetland tourism: a great experience. https://www.ramsar.org/sites/default/files/documents/library/ramsar-wwd2012-leaflet-en.pdf (accessed 18 March 2024).

Rasool, S., Rasool, T., and Gani, K.M. (2023). Unlocking the potential of wetland biomass: treatment approaches and sustainable resource management for enhanced utilization. *Bioresource Technology Reports* 23: 101553.

Rebelo, A.J., Jarmain, C., Esler, K.J. et al. (2020). Water-use characteristics of Palmiet (*Prionium serratum*), an endemic South African wetland plant. *Water SA* 46 (4): 558–572.

Salimi, S., Almuktar, S.A.A.A.N., and Scholz, M. (2021). Impact of climate change on wetland ecosystems: a critical review of experimental wetlands. *Journal of Environmental Management* 286: 112160.

Sandoval, L., Zamora-Castro, S., and Vidal-Álvarez, M. (2019). Role of wetland plants and use of ornamental flowering plants in constructed wetlands for wastewater treatment: a review. *Applied Sciences* 9 (4): 685.

Sardar, A., Khan, Z., and Perveen, A. (2015). Appraisal of ethnobotanical uses of the wetland plants of Punjab, Pakistan. *African Journal of Traditional, Complementary, and Alternative Medicines* 12 (4): 9.

Sarmento, A.P., Borges, A.C., and De Matos, A.T. (2013). Effect of cultivated species and retention time on the performance of constructed wetlands. *Environmental Technology* 34 (8): 961–965.

Savo, V., Lepofsky, D., and Benner, J.P. (2016). Observations of climate change among subsistence-oriented communities around the world. *Nature Climate Change* 6 (5): 462–473.

Scholes, R. and Engelbrecht, F. (2021). *Climate Impacts in Southern Africa During the 21st Century. Report for Earthjustice and the Centre for Environmental Rights.* Global Change Institute, University of Witwatersrand.

Seifollahi-Aghmiuni, S., Nockrach, M., and Kalantari, Z. (2019). The potential of wetlands in achieving the sustainable development goals of the 2030 agenda. *Water* 11 (3): 609.

Shah, S.A. and Islam, M.S. (2023). Recreational benefits of wetlands: a survey on the Dal Lake in Jammu and Kashmir of India. *International Hospitality Review* 37 (1): 94–109.

Shahid, M.J., Al-surhanee, A.A., and Kouadri, F. (2020). Role of microorganisms in the remediation of wastewater in floating treatment wetlands: a review. *Sustainability* 12 (14): 5559.

Shale, I. (2019). Historical perspective on the place of international human rights treaties in the legal system of Lesotho: moving beyond the monist-dualist dichotomy. *African Human Rights Law Journal* 19 (1): 193–218.

Sharma, S., Phartiyal, M., and Madhav, S. (2021). Global wetlands: categorization, distribution and global scenario. In: *Wetlands Conservation* (ed. S. Sharma and P. Singh), 1–16. Wiley.

Shelef, O., Gross, A., and Rachmilevitch, S. (2013). Role of plants in a constructed wetland: current and new perspectives. *Water* 5 (2): 405–419.

Sieben, E.J.J., Collins, N.B., and Kotze, D.C. (2017). Temperate grassy wetlands of South Africa: description, classification and explanatory environmental factors. *South African Journal of Botany* 113: 68–76.

Sig, A.K., Guney, M., and Uskudar Guclu, A. (2017). Medicinal leech therapy – an overall perspective. *Integrative Medicine Research* 6 (4): 337–343.

Sinthumule, N.I. (2021). An analysis of communities' attitudes towards wetlands and implications for sustainability. *Global Ecology and Conservation* 27: e01604.

Siyabona Africa (n.d.). Exploring iSimangaliso Wetland Park.

Siyabona Africa (n.d.). Wildlife of iSimangaliso Wetland Park.

Skowno, A.L., Poole, C.J., and Raimondo, D.C. (2019). National Biodiversity Assessment 2018: The Status of South Africa's Ecosystems and Biodiversity. Synthesis Report.

Sosnová, M., Van Diggelen, R., and Klimešová, J. (2010). Distribution of clonal growth forms in wetlands. *Aquatic Botany* 92 (1): 33–39.

Stolton, S., Dudley, N., and Avcioglu Cokcaliskan, B. (2015). Values and benefits of protected areas.

Toro-Vélez, A.F., Madera-Parra, C.A., and Peña-Varón, M.R. (2016). BPA and NP removal from municipal wastewater by tropical horizontal subsurface constructed wetlands. *Science of the Total Environment* 542: 93–101.

Upadhyay, A.K., Singh, N.K., and Bankoti, N.S. (2017). Designing and construction of simulated constructed wetland for treatment of sewage containing metals. *Environmental Technology* 38 (21): 2691–2699.

Van Niekerk, L. and Turpie, J.K. (2012). National Biodiversity Assessment 2011: Technical Report.

Van Niekerk, L., Taljaard, S., and Adams, J.B. (2019). An environmental flow determination method for integrating multiple-scale ecohydrological and complex ecosystem processes in estuaries. *Science of the Total Environment* 656: 482–494.

Verhoeven, J.T.A. and Setter, T.L. (2010). Agricultural use of wetlands: opportunities and limitations. *Annals of Botany* 105 (1): 155–163.

Wang, Q., Hu, Y., and Xie, H. (2018). Constructed wetlands: a review on the role of radial oxygen loss in the rhizosphere by macrophytes. *Water* 10 (6): 678.

Wang, W., Su, Y., and Wang, B. (2019). Spatiotemporal shifts of ammonia-oxidizing archaea abundance and structure during the restoration of a multiple pond and plant-bed/ditch wetland. *Science of the Total Environment* 684: 629–640.

Wang, J., Long, Y., and Yu, G. (2022). A review on microorganisms in constructed wetlands for typical pollutant removal: species, function, and diversity. *Frontiers in Microbiology* 13: 845725.

Water Research Commission (2018). Easy identification of South African wetland plants.

Wetland Ecosystem Services (n.d.). Recreation and tourism.

Whitney, C.K., Frid, A., and Edgar, B.K. (2020). Like the plains people losing the buffalo': perceptions of climate change impacts, fisheries management, and adaptation actions by Indigenous peoples in coastal British Columbia, Canada. *Ecology and Society* 25 (4): art33.

World Travel and Tourism Council (2022). Economic impact research.

Xu, T., Weng, B., and Yan, D. (2019). Wetlands of International Importance: status, threats, and future protection. *International Journal of Environmental Research and Public Health* 16 (10): 1818.

Zeeshan, A., Ashtar, H., and Amna, S. (2019). Economics Importance of Wetlands, their Benefits and Values Case of Pakistan. MRPA Paper No. 95260. https://mpra.ub.uni-muenchen.de/95260/

Zhang, D.Q., Jinadasa, K.B.S.N., and Gersberg, R.M. (2014). Application of constructed wetlands for wastewater treatment in developing countries – a review of recent developments (2000–2013). *Journal of Environmental Management* 141: 116–131.

Zhao, L., Fu, G., and Wu, J. (2021). Bioaugmented constructed wetlands for efficient saline wastewater treatment with multiple denitrification pathways. *Bioresource Technology* 335: 125236.

Zhou, J., Wu, J., and Gong, Y. (2020). Valuing wetland ecosystem services based on benefit transfer: a meta-analysis of China wetland studies. *Journal of Cleaner Production* 276: 122988.

Index

Note: *Italicized* and **bold** page numbers refer to figures and tables, respectively.

Wetlands of Tropical and Subtropical Asia and Africa: Biodiversity, Livelihoods and Conservation,
First Edition. Edited by Thammineni Pullaiah.

n

o

y

z